Paleoecology,

Concepts and Applications

Paleoecology,
Concepts and Applications

J. ROBERT DODD
Indiana University

ROBERT J. STANTON, JR.
Texas A & M University

A WILEY-INTERSCIENCE PUBLICATION

JOHN WILEY AND SONS   New York • Chichester • Brisbane • Toronto

Copyright © 1981 by John Wiley & Sons, Inc.

All rights reserved. Published simultaneously in Canada.

Reproduction or translation of any part of this work
beyond that permitted by Sections 107 or 108 of the
1976 United States Copyright Act without the permission
of the copyright owner is unlawful. Requests for
permission or further information should be addressed to
the Permissions Department, John Wiley & Sons, Inc.

Library of Congress Cataloging in Publication Data
Dodd, James Robert, 1934-
  Paleoecology, concepts and applications.

  "A Wiley-Interscience publication."
  Includes index.
  1. Paleoecology. I. Stanton, Robert J.,
joint author. II. Title.

QE720.D62   560′.45   80-19623
ISBN 0-471-04171-8

Printed in the United States of America

10 9 8 7 6 5 4 3 2 1

To our parents Edgar and Elizabeth Dodd
and Robert and Audrey Stanton

Preface

This book is intended to serve two major functions: to be a textbook for an advanced undergraduate or graduate level course in paleoecology and to be a reference work and review for the professional paleontologist. Paleoecology is both specialized and broad. It is specialized in the sense of being only one aspect in the field of paleontology, and it is broad in being concerned with many aspects of biology, sedimentology, and geochemistry, in addition to paleontology, as concepts from all these fields are applied to the study of ancient organisms and their environments. Obviously, one book cannot cover such a broad area comprehensively. We do hope that this book can serve as an introduction to major aspects of paleoecology and to the voluminous literature in the field.

We assume the reader to have at least a minimal knowledge of the various fields that contribute to paleoecology. We particularly depend on the reader's having a basic knowledge of the structure and function of the organisms with important fossil records. We also assume a basic understanding of fundamental geologic concepts and terminology as they impinge on the field of paleoecology.

Paleoecology is not a very well defined subject. It has no basis of neatly structured series of principles that make a natural organization for the subject. In fact, each person working in the field probably has his or her own ideas as to what constitutes the coverage of the subject. Consequently, the method of approach and the organization of a book on paleoecology is not well established. Two major viewpoints or approaches to the subject are perhaps most often taken: theory and application. A good deal of the early interest and research in paleoecology was on the application of fossils and their paleoecological relationships to the determination of ancient environments. Relationships observed between modern organisms and their environment were applied to paleoenvironmental interpretation without really understanding in many cases the reasons for the relationships. In recent years, much paleoecological research has been on theory, especially the application to the fossil record of modern ecological concepts (such as those that encompass population and community ecological relationships). This recent trend has been toward understanding the functioning of ancient ecosystems and their evolution rather than using the relationships to determine paleoenvironments. Clearly, both aspects are important and supplement each other. We can better determine ancient environmental conditions if we understand why the environment affects the organisms present. Also, we can better understand the reason for a certain relationship if we know the details of the ancient environment. Thus in this book we try to show a balance between theory and application, but emphasize application. We stress application because other recent books and reviews have concentrated on the theoretical aspects. No comparable recent treatment has been given to the applied approach.

In recent years much progress has been made in interpreting depositional environments on the basis of sedimentological features. Paleontologists have in some cases deferred to sedimentologists to determine the environmental conditions under which the fossils lived, but paleoecology clearly has much to contribute to environmental interpretation. This was forcefully brought home to us when we were recently in the field with a sedimentologist. After spending some time looking at a section of strata, we asked him if he could identify the depositional environment. He replied that the observed features could be produced either in deep water or in a restricted shallow water setting and that if we could tell him which of these the fossils indicated he could identify the sedimentary environment! The greatest precision of environmental interpretation comes from an integration of interpretations based on paleoecology *and* sedimentology, as well as on other approaches.

Many people have helped us in the preparation of this book. Perhaps our greatest debt of gratitude should go to the many students at Indiana University and Texas A & M University who have taken our courses in paleoecology and have helped to stimulate and then sharpen many of the ideas presented here. Special thanks are due to colleagues with whom we have discussed our ideas and some of whom have read all or parts of the manuscript. Included in this group are Alan Horowitz, Richard Alexander, Gary Lane, Don Hattin, David Kersey, Eric Powell, and Stefan Gartner. A number of people have generously allowed us to use published and unpublished figures and photographs. We especially wish to thank Ken Towe, J. D. Hudson, George Clark, Copeland MacClintock, and Allen Archer for contributing photographs.

J. ROBERT DODD
ROBERT J. STANTON

Bloomington, Indiana
College Station, Texas
January 1981

Contents

Paleoecology,

Concepts and Applications

Palaeoecology

Concepts and Applications

1
Introduction

Paleoecology has grown during the past several decades to become a major component of paleontology. In this period of time, activities in paleoecology have expanded from an initial strong focus on the reconstruction of the ancient physical environment to a wide range of topics of both biological and geological emphasis. As a consequence of this broader range of content, paleoecology is perhaps less clearly definable now than in the past.

A definition of paleoecology can be approached by first defining ecology. However, we are confronted immediately with a great many definitions of ecology, formulated on the basis of different perspectives and objectives of the ecologist. We favor the simple and concise statement that ecology is the *study of the interactions of organisms with one another and with the physical environment.* This definition encompasses the totality of processes and responses in the ecosystem and incorporates both the physical and biological aspects. By extension, it establishes the domain of paleoecology as the *study of the interactions of organisms with one another and with the physical environment in the geologic past.*

The emphasis in paleoecology, however, is distinctly different from that in ecology. This is in part because incomplete preservation of the fossil record precludes the examination of many of the standard topics of ecology. It is also, in part, because the paleoecologist is much more conscious of the role of time. Whereas the ecologist deals with processes taking place during time spans measured in years, the paleoecologist has difficulty recognizing phenomena of this short duration in the geologic record. Instead he or she generally works within a framework of thousands or millions of years. Consequently, both evolutionary processes and long-term environmental change are an integral part of the analytical perspective.

The ecologist is potentially able to examine directly the totality of the ecosystem under study; to determine the life histories and interactions of all the organisms present and to relate these biologic data to the instantaneous characteristics of the physical environment. Thus the ecologist can develop multivariate and quantitative models of the ecosystem that are both precise and realistic. In contrast, in the fossil record most of the organisms, commonly including even the most abundant taxa, are not preserved, and the succession of

1

short-term phenomena are not distinguishable. Consequently, the inadequacies of the data base in paleoecology may appear insurmountable. For questions that demand a complete and detailed record of the ancient biota, this may be true. In general, however, the capabilities of paleoecology, as compared to those of ecology are more encouraging than the data base would suggest for two major reasons. (1) Ecologic models based on the available detailed data may be precise and real, but their generality can only be tested by observing essentially contemporaneous geographic gradients in environment and biota; they can not be tested through time because of the short span of scientific observation. Time is, on the other hand, the dimension most readily available to the paleoecologist. (2) Much of ecologic study does not depend on the potentially available comprehensive data base but relies on limited components of the biota and on samples gathered only periodically or infrequently. Thus much of ecologic interpretation is based on information similar in quality and quantity to that available to the paleoecologist.

The definition we have presented highlights the two dominant subject areas in paleoecology. One is the study of organism-environment interactions and the other is the study of the more strictly biological attributes of the organisms —their individual life histories, their interactions with one another, and their integration into communities. The present state of the art and point of view in paleoecology reflect the blending of distinctive orientations of workers in different countries. On the one hand, the study from a geologic/paleontologic perspective of organisms in their modern environment (actuopaleontology) has been carried out much more strongly in Germany than elsewhere (Richter, 1929; Schäfer, 1962, 1972). This approach was little used in the United States until the years shortly after World War II. At that time, intensive study of areas of modern carbonate deposition in the islands of the western Pacific Ocean and in the Bahamas and south Florida was undertaken. Even later, until the 1960s, paleoecology in the United States was carried out largely with a stratigraphic/paleontologic orientation, with relatively little undergirding of modern ecologic data and theories. This is evident from a review of the *Treatise on Marine Ecology and Paleoecology* (Hedgpeth, 1957b; Ladd, 1957), in which the majority of the ecologic contributions focus on the environmental tolerances of organisms and the paleoecologic papers concentrate on the determination of physical aspects of the environment.

These historically different viewpoints have merged, and the more purely biologic aspects of paleoecology have become increasingly important during the past two decades. Increasing effort is being expended in trying to understand fossils as once-living organisms, in trying to understand lists of fossils as once-integrated communities and ecosystems, and in trying to understand range charts as they represent biogeographies changing through geologic time. Paleoecology has increasingly looked to ecology and biology for new approaches to be adapted and applied to fossils. Recent examples would be analysis of fossils in terms of diversity, community ecology, evolution, and trophic structure in the ecosystem, and utilizing electrophoretic techniques in characterizing population variability. Of course, information flow has not been in just one direction. Paleoecologists have made contributions to these areas as well as others such as skeletal deposition, chemistry, mineralogy, and structure. Today paleoecology is progressing on a broad front as more and more kinds of fossil evidence are being incorporated, as new analytical and computer-based techniques are being used, and as new ecologic concepts are applied.

The application of paleontologic data to reconstructing ancient environments has remained a strong force in paleoecologic work. The interpretive criteria are still largely undefined, and so their development continues to be a vital intellectual challenge. In addition, improvement in describing ancient environments is essential in order to provide a framework for describing earth history and for understanding the geologic processes that have been active. At a more economical-practical point of view, the application of paleoecology in the reconstruction of ancient depositional environments has been essential in the exploration and development of many earth resources and is the justification for most of the paleoecologic work being done and, in a very real sense, pays the bill.

OBJECTIVES AND ORGANIZATION OF THIS BOOK

This book focuses on the paleontologic techniques by which the ancient depositional environment can be reconstructed. Our objective is to provide a concise review of the current paleontologic approaches in environmental analysis. We do this by analyzing one by one the techniques available for paleoenvironmental reconstruction. We have attempted to provide a balanced and comprehensive discussion of techniques and examples relevant to all the groups of fossils. However, our own research has been largely with macroinvertebrates, and the paleoecologic literature on these fossils is larger than that on vertebrates, plants, or microinvertebrates. Consequently the preponderance of the case examples deal with macroinvertebrates. By emphasizing the bases as well as examples of specific techniques, we hope to have presented approaches in a form applicable to the full spectrum of taxonomic groups.

To describe not only present capabilities but also to evaluate their potential for future development, we have used the following format in the discussion of each approach:

1 The biological, chemical, or physical principals that form the theoretical basis for the approach are described. This provides a means for evaluating the basis of the approach for the paleoecologic analysis, and for identifying the aspects most in need of further study. The theoretical potential of the techniques is evaluated in light of the underlying assumptions and the nature of the fossil record.

2 The application of the technique is illustrated by examples representing a range of geologic ages, locations, and kinds of fossils.

3   We also illustrate the use of the techniques with examples from the Pliocene
of the Coalinga region, California. Our research in paleoecology has been
concentrated in this area of the Central California Coast Ranges. When
possible we also have used our results from this area for two reasons: (*a*) to
provide a thread of continuity between examples of paleoenvironmental
reconstruction and (*b*) to provide an integration of a number of approaches,
and thus the opportunity to evaluate the extent to which they complement
and confirm one another. The Pliocene stratigraphy and paleontology of the
Coalinga region are described in general terms at the end of this chapter.

Fossils and their characteristics at the level of the individual specimen or taxon
are discussed in Chapters 2 to 7. Thus the analysis falls in the broad category of
autecology. By contrast, the remaining chapters deal with the assemblage or
with a more inclusive grouping as the unit of analysis, and thus with topics of
synecology.

THE DATA BASE IN PALEOECOLOGY

Paleoenvironmental reconstruction depends on three ingredients: a well-
established stratigraphic framework, good taxonomy, and a comprehensive
ecologic background. The stratigraphic setting provides the spatial and temporal relationships for the comparison of fossils within geologic history. The
basic data of paleoecology are the fossils, adequately identified and correctly
positioned within the stratigraphic framework.

The necessary ecology consists of an understanding of the ways in which living organisms function within their ecosystem: how their morphology and
physiology is adaptive to their conditions of life, the ways in which they may interact with one another, and the ways in which they may modify their life history
to fit the environment. The ecologic information required by the paleoecologist
is largely at the level of natural history—a field of biology that is relatively inac-
tive in this present era of emphasis on biochemistry, cell biology, and medically
oriented topics. Consequently, the paleoecologist is commonly confronted with
the task of gathering for himself the ecologic information necessary to interpret
the fossils and their ecosystem. Ecologic data are necessary ingredients in
paleoecology because they usually provide the best basis for developing a sense
of the possible ways in which fossils could have interacted with one another and
have coped with the physical environment. The information garnered from
ecology is applied to paleoecology in many ways. At one end of the spectrum
general ecologic "laws" developed inductively from the living world are applied
deductively to the fossil record. These might be general relationships of diversity
with environmental resources or stability, for example. At the other end of the
spectrum the present day significance of a particular species or morphologic
feature is applied to the same species or biotic characteristic in the fossil record.

THE OPERATIONAL BASE IN PALEOECOLOGY

The application of ecologic data involves uniformitarianism, analogy, and simplicity. The concept of uniformitarianism has generated much discussion in geology. This has been in part because of the initial broad range of meanings assigned to it by Lyell and other early geologists as they sought to establish geology as a valid science. Uniformitarianism can be classified as either substantive or methodological (Gould, 1965). *Substantive uniformitarianism* implies that the materials, conditions, and rates of processes during earth history have remained constant. *Methodological uniformitarianism* implies that the laws of nature (such as gravity, the properties of fluid flow, and thermodynamics) have been constant in their operation through geologic time.

Rigorously defined, substantive uniformitarianism has been largely abandoned because earth materials have not remained constant in composition and proportions during geologic time, and the rates of some processes fluctuated widely beyond the ranges presently observable. Thus the present may not be a good key to the past. In addition, substantive uniformitarianism limits thought and speculation—by specifying dogmatically the limits within which geologic explanations must remain, it inhibits the generation of "outrageous hypotheses" (Davis, 1926) that turn geologic thought into novel and stimulating avenues. Methodological uniformitarianism, in contrast, is a statement of the inductive-deductive logical processes that are inherent in science in general. Consequently, Gould (1965) has argued that the term uniformitarianism is not necessary—that geology has already abandoned the substantive variety and that the methodological variety is merely the normal scientific mode of thought, for which geology does not require a special name.

In geology the fundamental laws of nature are readily applied. For example, thermodynamic principles can be applied in experimental petrology to generate phase diagrams that specify with a high degree of confidence the physical and chemical conditions of igneous and metamorphic phenomena. The basic laws of hydrodynamics can be applied to understand the fluid characteristics resulting in the deposition of a sediment with specific textural features and sedimentary structures.

In paleoecology, however, the fundamental laws by which we decipher the ancient environment generally are not easily defined. The apparently constant and inherent characteristics of the organic chemical system and of the established reproductive methods do not provide much guidance for paleoenvironmental reconstruction—they are so general they prove to be trivial. Consequently, in the absence of basic laws to guide his reasoning, the paleoecologist commonly relies on substantive rather than methodological uniformitarianism. The widespread use of the substantive approach is evident as each of the techniques is considered one by one in the chapters that follow. At the most elementary and simplistic level, the environmental interpretation of a fossil is based on the habitat characteristics of the most closely related living taxon. This procedure is described as "taxonomic uniformitarianism" by Lawrence (1971a). As an example, when a fossil oyster is found, the paleoecologist asks what the environmental tolerances are of the living equivalent taxon and then infers that the fossil also lived within this range of conditions. In that this reasoning is based on an assumed environmental constancy of the taxon, it relies on substantive uniformitarian logic and is of lower validity than interpretations based on methodological uniformitarian techniques. On the other hand, if the reasoning is based on the assumption that the name of the group of organisms serves as a transfer unit and reflects morphologic and physiologic attributes determined by external environmental conditions, then we are arguing by process-determined *analogy*. Much of paleoecologic reasoning is by analogy of this sort, intermediate in rigor and validity between the substantive and methodological approaches.

Analogy, whether for example in morphology of the individual organism, in community structure, or in population dynamics, is inferred to represent response to time-independent environmental forces. In the study of the morphology of fossils, for example, we recognize that most living animals moving in water or the air are "streamlined." By analogy we can recognize streamlining in fossils and infer from it the causative paleoenvironmental condition, as Lorenz (1974), in an article on "Analogy as a source of knowledge" has described in establishing the value of analogy in the study of culture.

Whenever we find, in two forms of life that are unrelated to each other, a similarity of form or of behavior patterns which relates to more than a few minor details, we assume it to be caused by parallel adaptation to the same life-preserving function. The improbability of coincidental similarity is proportional to the number of independent traits of similarity, and is, for n such characters, equal to 2^{n-1}. If we find, in a swift and in an airplane, or in a shark and a dolphin, and in a torpedo the striking resemblances illustrated in Fig. 1 [Fig. 1.1], we can safely assume that in the organisms as well as in the man-made machines, the need to reduce friction has led to parallel adaptations. Though the independent points of similarity are in these cases not very many, it is still a safe guess that any organism or vehicle possessing them is adapted to fast motion.

There are conformities which concern an incomparably greater number of independent details. Figure 2 [Fig. 1.2] shows cross sections through the eyes of a vertebrate and a cephalopod. In both cases there is a lens, a retina connected by nerves with the brain, a muscle moving the lens in order to focus, a contractile iris acting as a diaphragm, a diaphanous cornea in front of the camera, a layer of pigmented cells shielding it from behind as well as many other matching details. If a zoologist who knew nothing whatever of the existence of cephalopods were examining such an eye for the very first time, he would conclude without further ado that it was indeed a light-perceiving organ. He would not even need to observe a live octopus to know this much with certainty. (Lorenz, 1974, p. 230.)

The important attribute in paleoecologic logic is a reliance on the assumption that process results in analogous structures. However, to assume that

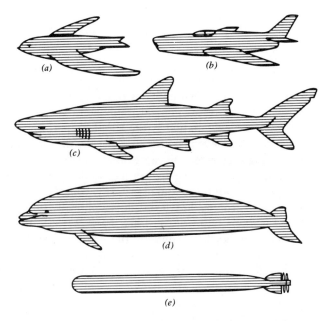

Figure 1.1 Analogy of streamlined form attributed to functional requirement for speed in each case. (*a*) Swift; (*b*) airplane; (*c*) shark; (*d*) dolphin; (*e*) torpedo. After Lorenz (1974), copyright The Nobel Foundation.

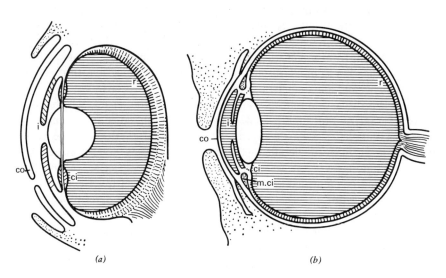

Figure 1.2 Eye in (*a*) invertebrate (octopus) and (*b*) vertebrate (man). Although independently evolved, striking similarity in detail is determined by identical function. co, cornea; ci, corpus ciliare; m.ci., musculus ciliris; i, iris; r, retina. After Lorenz (1974), copyright The Nobel Foundation.

"streamlined" shape can be equated with rapid motion on a fluid medium is simplistic, as Lorenz points out, for "streamlining" may have a variety of functional explanations. Also, different organisms may have solved a specific environmental problem in a number of ways; "streamlining" may not be a unique solution to a particular environmental situation.

The ever-present danger of oversimplifying in our search for explanations of the geologic past is inherent in the general principle of *simplicity*: everything else being equal, the best explanation is the simplest one. Simplicity in this sense is that the most probable explanation is generally the one with the fewest steps from cause through intermediate causes and effects to the final result.

The problem of understanding the logic of paleoecology relative to the traditional concepts of uniformitarianism and of critically considering the validity of types of paleoecologic reasoning has been philosophically and pragmatically analyzed by several workers (e.g., Scott, 1963; Lawrence, 1971a). Both Scott and Lawrence argue that analogy contains the error of substantive uniformitarianism. It does to a degree, but the error can be minimized by thorough and thoughtful analysis within the framework of known ecologic responses of organisms to the processes that do and may have occurred in the environment. In the end, we agree with Gilbert (1896) in his analysis of analogy and hypothesis, and believe that analogy between the fossil and living world is the primary source of paleoecologic hypothesis.

The discussion above, and paleoenvironmental analysis in general, is based on the viewpoint that the phenomena of paleontology are *deterministic*; that the fossil record is the result of specific and unique causes and that the causes can be inferred from its analysis. In recent studies Raup (1977) and others have shown that many broad phylogenetic and diversity patterns through geologic time cannot be demonstrated to be deterministic in that they are not significantly different from *stochastic* (random chance) computer-generated patterns. This work has led to much discussion in the paleontologic literature about the deterministic and/or stochastic nature of the fossil record. We believe that much of this discussion reflects confusion caused by the scale at which observations are analyzed and interpreted. When modern phenomena are analyzed at the most detailed scale, causes can be identified for effects if the scale of analysis is sufficiently detailed—the natural world is deterministic. When modern phenomena are viewed from a more distant perspective, the immediate causes cannot be identified and the proportions of alternative effects seem to fit stochastic expectations—the natural world is not clearly deterministic. Thus ecologic observations indicate that a deterministic viewpoint is valid in paleoecology, but that causes may often be indeterminate because the paleontologic data are not of sufficient detail.

The life histories and consequent abundances and distributions in space and time of living organisms can only be understood if the physical and biological parameters of their environment are studied in detail. Obviously, a great many parameters may be important, but some of the parameters are more important than others. Consequently, the explanation of observed phenomena generally involves an unstated ordering of environmental parameters by their determined or apparent importance. Thus the explanation of a complex ecologic system is simplified by presenting the effects of only the most important parameters. A basic assumption, however, is that with more detailed observations, additional, second- and higher-order parameters could be incorporated into the explanation. In paleoecology, the completeness and detail of the available data are such that only the most basic parameters can be included in the reconstruction of the environment. Thus the confidence level of the interpretation is determined by the degree to which only a very few parameters are important. This simplifying procedure should be valid in paleoecology because it is exactly that used in ecology, and in science in general. It saves us from the despair of attempting to derive from the limited paleontologic data an explanation incorporating the myriad of environmental parameters.

The characterization of humid and mesic ($\approx$ semiarid) forests of the northern hemisphere by a few temperature parameters (Wolfe, 1978) is an example of this procedure (Fig. 1.3). Many other environmental parameters could be used in addition to mean temperature and temperature range to separate vegetation types, but temperature is clearly considered dominant. Other parameters, or even other temperature characteristics would add dimensions to the figure and thus detail to the explanation of the vegetation types. Wolfe considers them either less important or dependent and correlated with those he has used. The value and justification of a simplifying method such as this in paleoecologic analysis is that its level of complexity is comparable to the general qualitative nature of available paleobotanical data. For the botanist the chart represents a starting point for further refinement. In Wolfe's analysis, and in general, the validity of the approach depends on correctly identifying the essential parameters.

THE NATURE OF THE FOSSIL RECORD

Taphonomy, the study of the post-mortem history of fossils, is an essential, critical aspect of each of the analytical methods surveyed in this book. Because of destruction by various processes after death of the organism, a potential fossil may not be preserved. Consequently, a fossil assemblage may be only a small and biased representation of the original community. Taphonomy is important in the interpretation of the fossil assemblage for two reasons. (1) It helps in understanding the relationship of the fossil assemblage to the original community and thus allows to some extent the reconstruction of the community. (2) Recognition of taphonomic processes that have formed the fossil assemblage provides insight into the depositional and postdepositional environment. Subsidiary topics within taphonomy are *necrolysis*, which deals with the decomposition of the organism upon death, *biostratinomy*, which deals with the sedimentational history of the fossil, and *fossil diagenesis*, which deals with chemical and mechanical alteration of the fossil between the time of its burial and collection.

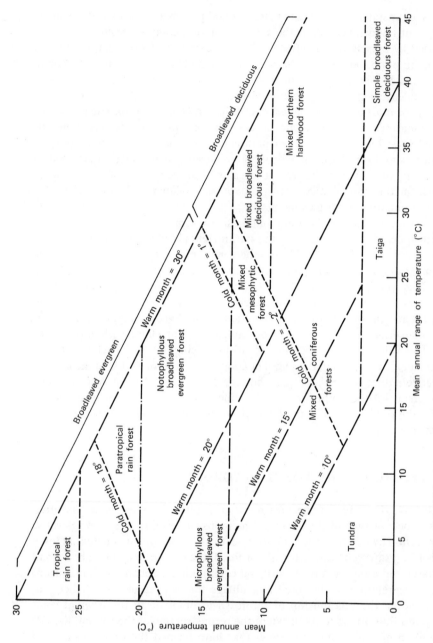

Figure 1.3 Major temperature parameters that are correlated with the different humid and mesic forests of the Northern Hemisphere. After Wolfe (1978), reprinted by permission of *American Scientist*, journal of Sigma Xi, The Scientific Research Society.

During each of these stages of the post-mortem history, mechanical, chemical, and biological processes are reshaping the original community (Fig. 1.4). Studies of these processes in modern settings have contributed significant information by which the fossil assemblage can be interpreted (Schäfer, 1962 and 1972, and Müller, 1979, for example). Comparison of specific community-assemblage pairs has provided information about the nature and magnitude of preservational bias in the fossil record (e.g. Lawrence, 1968; Stanton, 1976).

An often overlooked aspect of the study of fossil assemblages is that by recognizing the effects of taphonomic processes on the organisms, significant information about the depositional and postdepositional conditions can be determined. For example, the observation of skeletal material in modern natural settings and in flume experiments has provided information about the behavior of skeletons as clastic particles, and this information is useful not only in assessing the preservational gap between the original community and the fossil assemblage, but also in determining the energy regime of the depositional environment.

Because taphonomy has an effect on each of the methods of environmental reconstruction, we have chosen to discuss it in that individual context in each chapter rather than reviewing it exhaustively in a separate section. Discussions of taphonomy in the individual chapters are largely directed toward awareness, recognition, and interpretation of the modifying effects of taphonomic processes.

As a broader context for thinking about the processes that form from the original community the assemblage of fossils that we must interpret, two opposing views must be kept in mind. One is that a fossil assemblage accumulates slowly

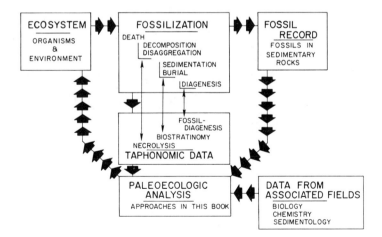

Figure 1.4 The fossil record is formed from the ecosystem by the processes of fossilization. The fossil record is interpreted to reconstruct aspects of the ecosystem by incorporating into the paleoecologic analysis concepts and data from associated fields and taphonomic data.

through the year-by-year preservation of some fraction of the community. Thus the assemblage represents a *time-averaged* sampling of a sequence of communities over a period of years and of perhaps a considerable range of environments (Fürsich, 1978). The opposing view is that preservation is in general so poor, as indicated for example by the paucity of skeletons of dead organisms accumulating on the modern sea floor, that the fossil record is much more likely the result of occasional chance preservation of an individual community. Thus an assemblage may be a fairly reasonable representation of the community existing during a short interval rather than the accumulation of meager sampling during a longer time interval. Examples of both extreme time averaging and catastrophist viewpoints can be found in the fossil record; of course the intermediate conditions of fossilization have probably been the general rule.

THE PLIOCENE STRATA OF THE COALINGA REGION

The Coalinga region is situated in west-central California, on the border between the Coast Ranges and the San Joaquin Valley (Fig. 1.5). Cenozoic sedimentary rocks are well exposed in broad folds on the flanks of the Coast Ranges and in several eroded anticlines on the adjacent valley floor. The strata are best exposed and have been most thoroughly studied in the Kettleman Hills and the Kreyenhagen Hills (Woodring, Stewart, and Richards, 1940; Stewart,

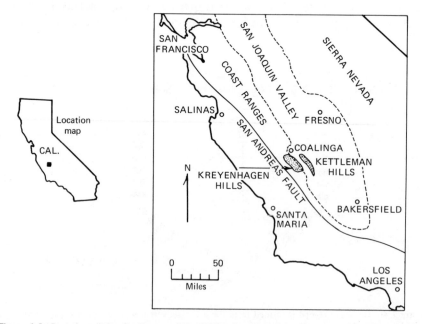

Figure 1.5 Location of the Coalinga region, California. After Stanton and Dodd (1970), *Journal of Paleontology,* Society of Economic Paleontologists and Mineralogists.

1946; Stanton and Dodd, 1970, 1972, 1976a, 1976b; Dodd and Stanton, 1975, 1976; Stanton, Dodd, and Alexander, 1979).

Western California from the San Joaquin Valley onto the continental shelf of the Pacific Ocean has been a site of rapid sedimentation throughout the Cenozoic. Tectonic activity resulting in the formation of the present Coast Ranges and in the large lateral movement along the San Andreas and related faults has also been strong during this time span. Consequently the depositional setting has been one of great local relief, of more or less isolated tectonic basins, and of multiple and diverse sediment sources in the adjacent land areas. The resulting strata are complex with numerous and abrupt facies changes occurring laterally and vertically within the area.

In very general terms, the Neogene strata were deposited in this complex paleogeographic setting during a single transgressive-regressive cycle of sedimentation. Basin formation and transgression took place in the Miocene and was relatively rapid so that the bulk of the sediments are regressive, in progressively shallowing water (Fig. 1.6). The strata exposed in the Coalinga region that will provide examples of paleoecologic analytic techniques in the chapters that follow are largely of the coarse-grained, shallow marine to nonmarine sediments. They were deposited in the late stages of the cycle, which is still continuing with the deposition of the modern fluvial and lacustrine sediments in the San Joaquin Valley.

The approximately 7000 ft thick Pliocene and Pleistocene sequence in the Coalinga region has been subdivided into the Jacalitos, Etchegoin, San Joaquin, and Tulare Formations. The total sequence consists of sandstone, mudstone, and a minor amount of conglomerate. Distinctive lithologic differences between the formations are lacking (Fig. 1.7). In a single stratigraphic section, the formations are defined and subdivided on the basis of distinctive marker beds or "zones" characterized largely by fossil content. Lateral changes in lithology and fossil content are so great, however, that correlation of formation boundaries and marker beds between sections only a few miles apart is difficult or impossible.

According to the established Pacific Coast geologic time scale based on macrofossils, the Jacalitos, Etchegoin, and San Joaquin Formations are lower,

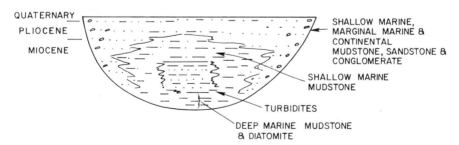

Figure 1.6 Generalized pattern of Neogene sedimentation in the Coalinga region, south-central California.

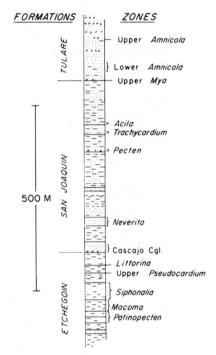

Figure 1.7 Pliocene stratigraphic section in Kettleman Hills California. Note general uniformity of lithology. After Dodd and Stanton (1975).

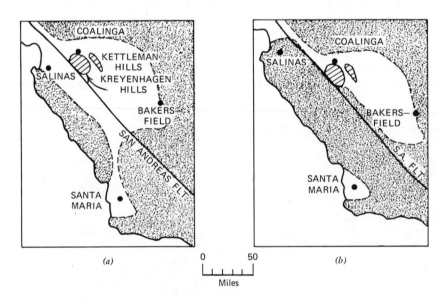

Figure 1.8 Early (a) and Late (b) Pliocene paleogeography of west-central California. After Galehouse (1967), Geological Society of America Bulletin.

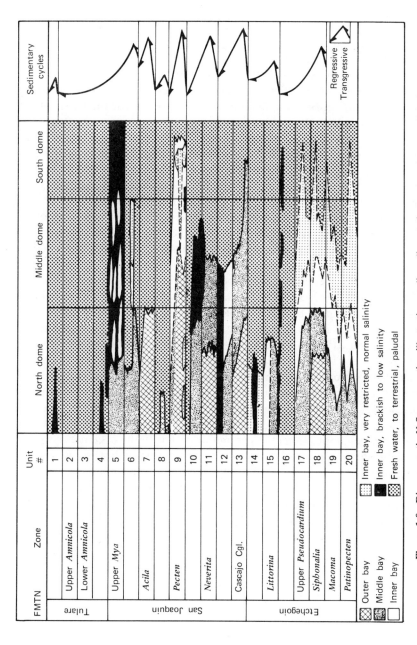

Figure 1.9 Diagrammatic N-S cross section illustrating cyclic sedimentation and environmental facies pattern. After Stanton and Dodd (1970), Journal of Paleontology, Society of Economic Paleontologists and Mineralogists.

15

middle, and upper Pliocene, respectively, and the Tulare Formation is Pleistocene. This usage will be followed here, although with improved world-wide biostratigraphy resulting from the Deep-Sea Drilling Program it is becoming clear that these age designations may not be correct. In particular, it may be that the Pleistocene-Pliocene boundary should be lowered in the section into the San Joaquin Formation, and the Pliocene-Miocene boundary should be raised into the Jacalitos Formation.

The strata were deposited within a broad embayment that was becoming progressively filled with sediment and increasingly isolated from the open ocean to the west (Fig. 1.8). At about the end of deposition of the San Joaquin Formation, the connection between the embayment and the ocean was broken, and sedimentation since then has been in nonmarine environments.

Within this generalized broad transition from a relatively open marine to a nonmarine environment, the depositional conditions fluctuated widely in short-term sedimentary cycles. These cycles are best displayed in the Kettleman Hills. The north end of the Kettleman Hills was relatively near the inlet from the open ocean. Southward, the environment became progressively more restricted, "back bay," and nonmarine. This lateral gradation can be recognized throughout the section, in both the lithology and the fossils. The environmental gradient also shifted laterally with time (Fig. 1.9). Two possible mechanisms might explain the shifting pattern of environments. (1) The extent of communication between the open ocean and the Kettleman Hills area of the embayment may have varied: when communication was good, marine conditions extended far to the south in the area; with poor communication marine conditions were restricted to the most northern part of the area. This fluctuating sequence of events forms "transgressive-regressive" cycles. Transgression was caused by apparently sudden improvement of communication with the open ocean so that marine sediments would overlie brackish to nonmarine sediments. Regressive sedimentation and progressive deterioration of the marine environment at any particular location would occur as communication was restricted. (2) Alternatively, the reintroduction of the marine environment could have been caused by subsidence within the basin and actual transgression followed by basin filling and regression.

2

Taxonomic Uniformitarianism

The most basic method of interpreting environments from fossils is by assuming that their environmental requirements were the same as those of the most closely related living representatives and transferring this environmental information from the modern to the fossil. This is based on a strict substantive application of the principle of uniformitarianism, that is, the ecology of present organisms is the key to past organisms. This method has been successfully used for decades and in fact was probably the first paleoecological method employed. The Greek scholar Xenophanes (ca. 540 B.C.) noted fossil sea shells preserved in rocks well above the level of the present Mediterranean Sea. The shells he observed were very much like (or even identical to) living molluscs that he had observed in the Mediterranean. He made the simple uniformitarian deduction that this meant that the sea had once been more extensive and had occupied the area where he observed the fossil shells (Adams, 1938).

Paleoecologic interpretation of environments such as this can be found in many places in the ancient as well as more modern literature. Leonardo da Vinci used the presence of fossil shells miles from the sea to prove the once greater extent of the Mediterranean. Edward Forbes, the so-called father of modern marine ecology and paleoecology, suggested using the uniformitarian extension of modern data on the distribution of organisms in the sea to interpret ancient environments (Forbes, 1843). He made a detailed study of the distribution of organisms in the Aegean Sea, especially in relation to depth. Forbes noted the marked difference in the organisms found in different environments and suggested that fossil organisms must have been similarly influenced by environment. Likewise, M. K. Elias (1937) used data on modern organism distribution from Forbes and others in his classical paleoecologic study of the Permian Big Blue Limestone of Kansas. Elias studied the vertical distribution of fossils through the strata in order to determine variation in water depth during deposition. His results were rather crude by modern standards, but his study was especially important as a pioneering effort in paleoecology.

Another classical and important study using taxonomic uniformitarianism is that of M. L. Natland (1933) who investigated the distribution of modern and fossil benthonic foraminifera and made paleoenvironmental interpretations of the Cenozoic sediments of the Ventura Basin in California. Natland recognized five foraminiferal assemblages off the modern Southern California coast which were differentiated largely on the basis of water depth (see Chapter 1). The distribution of fossil foraminifera in the sediments suggested marked changes in water depth which were ultimately interpreted as due to introduction of the shallow water forms by turbidity current flows. Although it was not appreciated for many years, Natland's paleoecologic study was a vital part of the evidence supporting the hypothesis that sediments can be transported by this method. His study was one of the earliest to describe examples of turbidites.

The taxonomic uniformitarian approach is certainly not limited to the invertebrates. Palynologists have made extensive use of this method in interpreting environments from the distribution of fossil pollen and spores. The technique has been especially effective in studying cores from Pleistocene and Holocene lake deposits (e.g., Davis, 1961 and 1969).

Some have criticized the uniformitarian approach on the grounds that it often does not consider why a particular taxon is restricted to certain environmental settings, that is, it uses the substantive uniformitarian approach (see Chapter 1). Evolution could occur that changes the physiology of the organism and its response to environmental parameters without affecting its morphology. Apparent examples of this have been found (e.g., Stanton and Dodd, 1970). Certainly the ideal approach would be to determine on the basis of physical laws why a given measurable feature of the fossil is related to specific environmental parameters, the methodological uniformitarian approach (see Chapter 1). (Many of the techniques discussed elsewhere in this book such as the functional morphology approach and the oxygen isotopic paleotemperature determination method are of the methodological uniformitarian type.) Such relationships are not as subject to evolutionary change; however, even the methodological approach may be complicated by the fact that through evolution organisms develop increasingly superior morphologies to cope with environmental problems. Thus the relationship noted between a given morphology and environment in the Recent may not have yet evolved in fossils of an earlier age. Another problem is that a certain general morphology may be a response to more than one environmental parameter. In spite of these complicating factors, in many cases the reason for the relationships between environmental parameter and taxon can be explained. For example, echinoderms are not found in fresh or very low salinity water because their excretory systems are not efficient enough to maintain a concentration gradient between internal fluids and the surrounding medium. This fundamental physiologic feature of the echinoderms is not likely to have evolved independently of morphology so we can be fairly confident that echinoderms have never lived in fresh or very brackish water. Although the blind application of the uniformitarian approach may involve risks, it has proved many times over to be a useful method of paleoenvironmental interpretation and will certainly continue to be widely used.

The taxonomic uniformitarian approach can be applied at many different taxonomic levels. As an example, a Pleistocene limestone may contain specimens of the coral *Acropora palmata*. This species is living today in the Caribbean area where it is restricted to very shallow (usually less than 5 m), turbulent water in the open reef environment. It is most abundant on the shallowest, most turbulent portion of the reef. Thus the depositional environment of the limestone containing this species (assuming that it is in place) can be quite specifically determined by the taxonomic uniformitarian approach. Another limestone, this one of Miocene age, may contain a now extinct species of *Acropora*. The entire genus *Acropora* of course has a much broader range of environmental requirements than any one of its species. The genus is widespread in reef and associated environments throughout the world. Two common species of *Acropora* occur in the Caribbean area, *A. palmata* and *A cervicornis*. *A. cervicornis* is most abundant in moderately turbulent water in the back reef area and on patch reefs some distance from the shelf edge in less turbulent environments than *A. palmata*. The total environmental range would be further broadened if the many Indo-Pacific species are included. Thus the taxonomic uniformitarian approach used at the generic level would yield less precise information than at the species level, but would still indicate a probable reef or reef-associated environment.

The next step up the taxonomic scale to the family level would yield still less precise information. An early Cenozoic limestone might contain *Dendracis*, an extinct genus of the family Acroporidae. The family Acroporidae today largely consists of species of the genus *Acropora* but includes a few other less common genera. All live in tropical, shallow, full salinity environments, but their total environmental range is greater than that for the genus *Acropora* and certainly for any species.

A Cretaceous limestone may contain a species of an extinct scleractinian family such as Stylinidae. The order Scleractinia includes all of the modern hexacorals from the reef building forms of the tropics to the relatively deep water, ahermatypic forms of the high latitudes. The total environmental range of the order is much broader than lower taxonomic levels. Most (but certainly not all) are tropical. Most (but not all) live in shallow water. All live in normal or near normal salinities. All require well to moderately well oxygenated water that is relatively sediment free. So the presence of a fossil scleractinian gives much less specific information although they generally suggest shallow, tropical seas.

A Paleozoic limestone may contain an extinct order, Rugosa, of the class Anthozoa. Modern anthozoans include the orders Octocorallia and Scleractinia, and several other orders without skeletons. Modern anthozoans are most abundant in the shallow tropics but are common down to considerable depths. They occur at temperatures from 0 to 40°C or above. They are most common at normal marine salinities, but some forms extend into brackish or hypersaline waters. Consequently, a very limited amount of information concerning environment can be obtained by taxonomic uniformitarianism at the class level. Even less could be said at the phylum level. Coelenterates occur in practically all aquatic environments, both marine and nonmarine.

As is clear from the example above, the substantive taxonomic uniformitarian approach increases in usefulness (or precision) with decreasing geologic age. The best results are obtained in rocks of late Cenozoic age which contain many extant species or at least genera. Only the most general applications can be made to rocks of Paleozoic age. However, the methodologic uniformitarian approach in which, for example, the morphology of the modern organism is causally related to the environment can be used with older as well as geologically younger species. To cite a single example, the relationship of the morphology of modern bryozoans to water turbulence and sedimentation rate has been applied to taxonomically very different Paleozoic forms in order to interpret environments (Boardman, 1960).

In the simplest case, the taxonomic uniformitarian approach involves transferring ecological information from a single extant species to its fossil counterpart. Greater precision might be obtained by using information for several species, in fact as many species as possible. Perhaps the ultimate application of the method is to recognize and transfer information from the modern community that is the counterpart of the fossil community. This approach is discussed in detail in Chapter 9. More commonly the uniformitarian approach has been used with several species of one or a few taxonomic groups rather than the entire community.

How can information from several species or other taxonomic units be combined into a single environmental interpretation? At one end of a spectrum of approaches the investigator may simply present the data and then make a subjective interpretation based on those data. Data for all species (such as depth or temperature distribution) may be presented in some systematic form showing overlap of ranges. Table 2.8 is an example of this format. This method is comparable to the method of concurrent ranges or assemblage zones used in biostratigraphy. Ideally all ranges will overlap in only one narrow band which allows a very precise interpretation of the environmental parameter. In practice this ideal is seldom realized. This may be due to several factors: (1) A certain amount of post-mortem mixing may have occurred. For example some shallow water species may be transported by currents into deeper water. (2) Data on modern environmental requirements may be incomplete. (3) As indicated above, some species may have evolved between the time when the fossils lived and today so that the environmental requirements have changed. In the case of nonoverlapping ranges, some attempt is usually made to explain the discrepancy and the area of greatest overlap is considered to indicate the most likely environment.

Greater precision should be attainable if quantitative information on the relative abundance of species is included in the analysis rather than simply presence or absence of the species. Environmental information from an abundant species is likely to be more indicative of depositional environment than that for a rare species which may not even be indigenous to the environment where it was buried. Not only might the abundance of the fossil species be useful in increasing the precision of the interpretation, but the relative abundance of the modern species throughout its environmental range can be taken into account. For example the rare occurrence of a modern species in water deeper than most members of the species results in an increase in the total depth range, but the *abundant* occurrence of that species probably means a shallower depth. Delorme (1971) uses this approach in his study of the depth distribution of fossil ostracodes.

Perhaps the most rigorously quantitative approach to using the uniformitarian method for several species attempted to date is that of Imbrie and Kipp (1971) on the temperature distribution of modern planktonic foraminifera. Their approach has been extensively used in connection with the CLIMAP project which has the goal of mapping the climatic pattern for various times during the Pleistocene and Holocene epochs (Cline and Hays, 1976). Imbrie and Kipp have taken data on the relative abundance of modern planktonic foraminifera in various temperature regimes and subjected them to a principal component analysis in order to establish assemblages characteristic of different climates. Then, by viewing a sample as a mixture of these assemblages in different proportions, they are able to determine a temperature for the sample. Imbrie and Kipp claim a precision within 1°C for this method when used on samples from deep sea cores. In principle this technique should be usable with other fossil groups in other geologic settings. However, the study of foraminifera in the deep sea is perhaps an ideal setting for this quantitative approach. The environment is ecologically relatively simple, the amount of mixing and diagenetic effect is minimal, and the information on modern distributions comes from the same type of samples (deep sea sediment) as those to which it is applied. A complicating factor in interpretations based on planktonic organisms is the effect of currents in transporting the organism before and after death (Weyl, 1978).

The implication throughout this discussion has been that the fossils studied are preserved in the place where they lived. As indicated in Chapter 1, this is clearly not always the case. A part of any analysis using the taxonomic uniformitarian approach (or any paleoecologic study for that matter) should be a consideration of possible mixing of fossils from several environments. On the positive side, our own experience (Dodd and Stanton, 1975) and that of others (e.g., Walker and Bambach, 1971; Peterson, 1976) have suggested that transport and mixing of fossils may not be as common as was once thought. Using a complete assemblage of fossils helps to alleviate the problem of transport by diluting the contribution of a few incidental transported specimens.

Differential preservation, perhaps due to differential chemical diagenesis, is also important in studies of this kind, particularly studies such as those of Imbrie and Kipp in which the relative abundance of an entire assemblage is used in the analysis.

Differential preservation may cause the final assemblage investigated by the paleoecologist to be quite different from the original living assemblage. The fossil assemblage preserved in this way may look much more like an assemblage from another environment. Chave (1964) gives an excellent example of this from the Upper Cretaceous Navesink Formation of New Jersey. The foraminifera in assemblages from the Navesink Formation consist entirely of planktonic species. Modern assemblages dominated by planktonic species are almost exclusively found in deep sea sediments, although planktonic species also occur in shallow water sediments along with more abundant benthonic species. The implication from the uniformitarian approach is thus that the Cretaceous assemblage was deposited in deep water. This is surprising because the formation contains sedimentary features suggesting deposition in shallow water. Its stratigraphic setting also suggests relatively shallow water. More importantly from the paleoecologic point of view, the total fossil assemblage also contains oysters which by the uniformitarian approach suggest shallow water. Chave explains the apparent dilemma as resulting from differential preservation. The assemblage contains only fossils that were originally composed of low-magnesium calcite. All evidence of the aragonitic bivalves and gastropods which were probably once in the assemblage is gone. The benthonic foraminifera that were originally high-magnesium calcite also appear to have been dissolved. The final assemblage is thus highly biased.

Perhaps the safest procedure to follow is to base interpretations using the uniformitarian approach on fossils that are present and not on the absence of fossils. In the New Jersey Cretaceous example the planktonic forams by themselves have no special depth implication. They can occur in sediments deposited at all depths. The absence of fossils normally expected in shallow water sediments leads to the erroneous interpretation. In a more general sense, the absence of fossils in sedimentary rocks anywhere should not be taken to imply that they never were there. The vast volumes of unfossiliferous sedimentary rocks in the world today could not conceivably have been deposited in environments without life. Likewise, that all skeletal remains ever produced would be preserved is inconceivable. Thus differential preservation of fossils either within a given assemblage or between volumes of rock is a fact of life which must constantly be taken into account in paleoecological studies.

The literature on the relationships between modern organisms and their environment which is potentially applicable to the geologic record is extremely large. In this chapter we only briefly summarize some of the most general relationships for large taxonomic units. The greatest use of the uniformitarian approach (especially in the Cenozoic) has been the substantive approach at the generic and specific level. A large collection of papers and monographs has developed on this approach which we cannot hope to review here.

Much of the information included in this chapter is summarized in text-books on invertebrate zoology and on paleontology and in compendia such as the *Treatise of Invertebrate Paleontology* and the *Treatise on Marine Ecology and Paleoecology*. The reader is referred to these and more basic sources for details. Taxonomic uniformitarianism is so fundamental that the basic information on the ecologic requirements of important fossil groups is included.

In this chapter we briefly discuss the mode of life and environmental relationships for each of the invertebrate phyla (plus the calcareous algae) which has an important fossil record. Our emphasis is on locomotion or attachment method and feeding methods. We presuppose a basic knowledge on the part of the reader of the morphology and biological functioning of these groups. We discuss the environmental requirements (especially emphasizing temperature, salinity, depth, substrate, and turbulence) of the group and its major subdivisions. We consider the relationship of morphology to environment as observed in modern representatives of each group. This latter approach impinges on functional morphology which is treated more fully in Chapter 5. This chapter concentrates on relationships observed in living forms whereas Chapter 5 contains deductions of function from the morphology of fossil forms. For more detail on the uniformitarian approach we have included a list of important references in Table 2.1. The taxonomic uniformitarian approach has also been extensively used for plant and vertebrate fossils. We briefly mention these groups and give references for more extensive treatments but do not consider them in the detail we give for the invertebrate groups. We do this because plant and vertebrate fossils are not as abundant as invertebrates and because we are not as familiar with the literature on vertebrate and plant fossils.

CALCAREOUS ALGAE

Organism Characteristics

The calcareous algae are of considerable interest and importance in the interpretation of depositional environments because they are commonly found in association with invertebrates, especially in the marine environment. Five algal groups are of particular importance in paleoecology: (1) the blue-green algae (especially in stromatolites), (2) the green algae (especially the Codiacians and Dasycladacians), (3) the red algae (especially the coralline algae), (4) the Chrysophyta (especially the Coccolithophoracians), and (5) the diatoms. Several other groups such as the brown algae and the charophytes are of local importance. For two recent reviews of the calcareous algae including a

Table 2.1 References to literature on ecology and taxonomic uniformitarian studies of important groups of fossil organisms

Group	General References on Ecology	Studies Using Taxonomic Uniformitarianism
Coccoliths	Haq, 1978 Paasche, 1968	MacIntyre et al., 1969 Geitzenauer, 1969
Diatoms	Burckle, 1978 Simonsen, 1972	Barron, 1973 Donahue, 1970
Stromatolites	Wray, 1977 Walter, 1976 Walter, 1977	Peryt and Piatkowski, 1977 Playford et al., 1976 Ahr, 1971
Green algae	Wray, 1977 Johnson, 1961	Conrad, 1977 Flügel, 1977a
Red algae	Wray, 1977 Adey and Macintyre, 1973	Buchbinder, 1977 Flügel, 1977b
Foraminifera	Boltovskoy and Wright, 1976 Boersma, 1978 Murray, 1973 Loeblich and Tappan, 1964	Berger and Gardner, 1975 Kennett, 1976 Imbrie and Kipp, 1971 Walton, 1964
Sponges	Bergquist, 1978 Fry, 1970 deLaubenfels, 1957	Termier and Termier, 1975
Corals	Barnes, 1974 Wells, 1957 Yonge, 1957	Frost and Langenheim, 1974 Philcox, 1971 Hubbard, 1970 Wells, 1967
Bryozoa	Woollacott and Zimmer, 1977 Ryland, 1970 Schopf, 1969	Pedley, 1976 Brood, 1972 Labracherie and Prud'homme, 1966 Lagaaij and Gautier, 1965
Brachiopods	Rudwick, 1970 Ager, 1967	Alexander, 1975 Fürsich and Hurst, 1974 Surlyk, 1972 Ager, 1965
Bivalves	Vermeij, 1978 Stanley, 1970	Strauch, 1968 Valentine, 1961

24

Table 2.1 (*Continued*)

Group	General References on Ecology	Studies Using Taxonomic Uniformitarianism
	Morton, 1967 Abbott, 1954	Baden-Powell, 1955 Durham, 1950
Gastropods	Linsley, 1978a Purchon, 1968 Fretter and Graham, 1962	Linsley et al., 1978 Linsley, 1978b Peel, 1975 Valentine, 1961
Cephalopods	Denton and Gilpin-Brown, 1973 Stenzel, 1964	Mutvei and Reyment, 1973 Cowen et al., 1973 Trueman, 1940
Ostracodes	Pokorny, 1978 Benson, 1975 Benson, 1961	Hazel, 1971 Oertli, 1971 Wagner, 1957
Crinoids	Breimer and Lane, 1978 Fell, 1966	Meyer and Lane, 1976 Breimer, 1969 Ausich, 1980
Echinoids	Kier, 1974 Durham, 1966 Nichols, 1962	Kier, 1972 Nichols, 1959 Fell, 1954

wealth of information on their ecology and paleoecology, see Wray (1977) and Flügel (1977a.)

Stromatolites

Stromatolites are produced by the interaction of blue-green algae and sedimentary processes (Walter, 1977) and are thus best termed biogenic sedimentary structures. They are sometimes considered a special type of trace fossil. Although they certainly satisfy the definition of fossils as being evidence of ancient life, they are not the sort of fossils we usually associate with that term. They are the product of the activity of many separate species, not all of which are necessarily blue-green algae. Stromatolites are formed by filamentous blue-green algae that grow on the sediment surface. The algae trap sediment on their mucilaginous sheaths and to some extent may also cause precipitation of $CaCO_3$ as a result of their photosynthetic activity (see below). Periodic alternation between sediment trapping and upward growth of the algae through the trapped sediment to form a new layer of algal filaments produces the characteristic laminated structure (Fig. 2.1). The period of the growth-trapping episodes is not always clear and is probably variable. In some cases the cycles are diurnal. During the daylight hours the algal filaments of one species grow primarily in the vertical direction, and at night growth of another species is primarily horizontal (Gebelein, 1969). Sediment grains are thus trapped between the vertical filaments during the day and bound by the horizontal filaments at night (Fig. 2.1). This process results in alternating organic-rich and organic-poor layers. Other studies suggest that the sediment layers have been produced by storms or unusually high tides so that more irregular time cycles are involved. A voluminous literature on the stromatolites has developed over the last 20 years (see Walter, 1976). Much of the interest has been due to the fact that stromatolites can be used as environmental indicators.

The role of the algae in the precipitation of the carbonate making up the stromatolites has been debated. Clearly much if not most of the carbonate is simply detrital sediment; however, some may result from precipitation due to the photosynthetic activity of the algae. Use of CO_2 by the algae in its photosynthesis results in an increase in pH and an increase in the CO_3^{2-} ion concentration with resulting precipitation of $CaCO_3$.

$$2HCO_3^- \rightleftharpoons CO_2 + H_2O + CO_3^{2-}$$

Such carbonate precipitation is in a sense biochemical but not skeletal. This biochemical precipitation process may have been more important in Paleozoic and Precambrian stromatolites than in modern forms (Monty, 1977). A few modern blue-green algal species produce skeletal carbonate surrounding their filaments. These are entirely from fresh or brackish water. In contrast, fossil calcified blue-green algae such as the upper Paleozoic *Girvinella* are common in marine rocks (Wray, 1977).

Codiacians, Dasycladacians, and Charophytians

These are benthonic plants that grow either on loose sediment with rootlike holdfasts penetrating into the sediment or are attached to hard objects. The codiacians and dasycladacians show differing amounts of calcification; many genera in fact are uncalcified. At the other end of the spectrum, the living genus *Halimeda* may be as much as 97% $CaCO_3$ by weight. They are often segmented and fragile so that usually only fragments are found in the fossil record. In many species the $CaCO_3$ occurs as individual needles rather than larger plates, and these commonly completely disintegrate on death and produce fine needles of aragonite a few μm in length that make significant contributions to the fine fraction of carbonate sediment. Because of their aragonitic composition they are often poorly preserved and difficult to identify in the fossil record. Charophytes live in fresh or brackish water. The only calcified part of the plant is the female reproductive body or oogonium. These are fairly common in lacustrine sediments and have been extensively

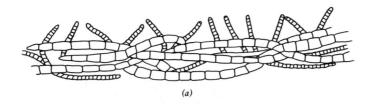

(a)

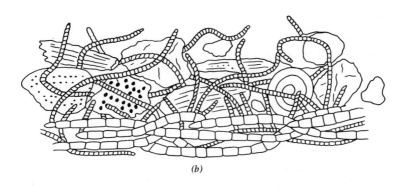

(b)

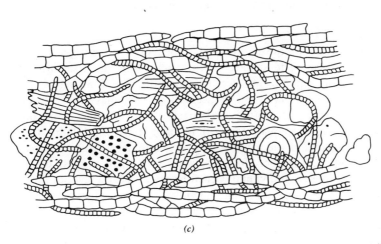

(c)

Figure 2.1 Diagrammatic sketch of day-night accretion cycle in stromatolites. (*a*) and (*b*) Daylight: upward growth and sediment trapping; (*c*) darkness: horizontal growth and sediment binding. After Gebelein (1969, Journal of Sedimentary Petrology, Society of Economic Paleontologists and Mineralogists.

used in biostratigraphy. Two additional fossil groups that may belong to the calcareous green algae are the receptaculitids and calcispheres. Receptaculitids, an extinct group found only in Paleozoic rocks, were once considered to be sponges, but Nitecki (1972) regards them as algae perhaps related to the dasycladaceans. Calcispheres are especially common in Paleozoic rocks but also occur in younger strata. The modern dasycladacian genus *Acetabularia* produces calcareous reproductive cysts that are morphologically similar to fossil calcispheres (Marszalek, 1975).

Red Algae

Most modern calcified red algae belong to the group Corallinaceae. The coralline algae have two growth forms: articulate and crustose. The articulated forms are erect plants with jointed, branched thalli and are attached to a hard substrate. Their articulated nature gives them flexibility in turbulent environments. The individual segments normally separate after death of the plant. Crustose forms are variable in morphology from branching to massive, often varying within the same species. They usually encrust on a hard substrate, but a few forms grow free. The crustose corallines are important reef builders in the modern seas and also were abundant during Cenozoic time. Sometimes the crustose types form nodules. In addition to the Corallinaceae, several other groups of red algae are of paleoecologic interest. The Solenoporaceae, which are closely related to the corallines, belong to an extinct group of red algae with an extensive geologic record (Wray, 1977).

Coccolithophorids

The coccoliths are produced by single-celled planktonic algae of the order Coccolithophoridae. The surface of the single cell, the coccolithophore, is covered by many separate platelets or coccoliths. The function of the coccoliths has been the subject of much speculation. They may serve (1) to regulate the density of the cell, acting as ballast, (2) to provide some protection against microscopic zooplankton, (3) to filter and reflect light in the very brightly lit shallows of the tropical seas, or (4) to act as lenses to concentrate the light (Haq, 1978). Coccolithophores are very small, usually on the order of 50 μm in diameter. Individual coccoliths are at most a few μm in diameter. Coccolithiphoridae are important members of the phytoplankton, being common in open marine sediments but not in shallow water or restricted marine settings. They are important sediment contributors, forming one of the main constituents of Cretaceous and younger chalks. Other calcareous platelets such as discoasters are often studied in conjunction with coccoliths, especially in biostratigraphic studies. All of these very small calcareous platelets of probably algal affinity are called the calcareous nannoplankton (Haq, 1978).

Diatoms

Diatoms are microscopic algae (mainly 50–500 μm in size) with opaline silica tests belonging to the plant division Chrysophyta. Both planktonic and benthonic species of diatoms occur in modern seas and in fresh water. Some species, meroplanktonic types, have both benthonic and planktonic stages during their life cycles (Burckle, 1978). The planktonic species are very abundant and in fact are perhaps the most abundant modern plant on earth. Thus they are the most important primary producer in most aquatic ecosystems. Where diatoms are particularly abundant, in highly productive areas such as areas of upwelling of nutrient-rich deep waters, the siliceous shells or frustrules may be a major sediment component. Diatomites or diatomaceous shales are common in some areas in rocks of Cenozoic age.

The benthonic species are less well known. Benthonic diatom species attach to the substrate by a mucilaginous material which can serve as a sediment stabilizer. Diatoms are often constituents of algal mats which may produce stromatolites.

Other groups such as the silicoflagellates, dinoflagellates, and other species occur in the phytoplankton. They are not discussed further here. For a recent discussion of these and other marine microfossil groups see Haq and Boersma (1978).

Environmental Tolerances

Temperature

Most studies of modern calcareous algae have concentrated on tropical to subtropical shallow marine species particularly as related to their association with modern carbonate sediment deposition. The calcareous algae certainly reach their peak of development in such environments but are not restricted to them. The calcareous green algae are essentially all tropical in distribution but a few species of dasycladacians have been described from temperate settings. The coralline algae occur from the tropics to the Arctic, but are best developed in the tropics. Certain genera of coralline algae, such as *Archaeolithothamnium, Lithoporella, Neogoniolithon* and *Porolithon,* are tropical and subtropical forms whereas other genera, such as *Mesophyllum* and *Lithothamnium,* are mainly cold water types (Wray, 1977). Individual species may be rather narrowly limited in their temperature tolerances. The coralline algae may be quite abundant in temperate environments to the extent of being major sediment contributors as described for one area on the west coast of Ireland (Bosence, 1976). Perhaps the coralline algae reach their peak of development on the algal or *"Lithothamnium"* ridge on the most exposed part of tropical coral reefs. Stromatolites are best developed in carbonate sediments characteristic of tropical environments but mat-forming blue-green algae may occur across a wide temperature range.

The planktonic algae are worldwide in distribution although the Coccolithophoridae reach their maximum development in the tropics. Diatoms are most abundant in temperate to arctic settings where deep, cold water is upwelling. This is probably more due to high productivity in these areas than to any direct temperature response (Burckle, 1978).

Both coccoliths (Haq, 1978) and diatoms (Burckle, 1978) have been used for paleotemperature analysis in studies of deep sea cores by the substantive taxonomic uniformitarian approach. Special care is needed for such studies with these groups because of the ease with which planktonic forms can be transported by oceanic currents. Differential preservation of these delicate fossils near or below the compensation depth in the deep sea also necessitates caution in paleoecological interpretations. Despite these problems, paleo-temperature results from coccoliths and diatoms have been shown to be consistent with results based on other fossil groups (e.g. McIntyre et al., 1972 and Burckle, 1978).

Salinity

The blue-green algae are quite salinity tolerant; thus stromatolites form across a range of fresh water to hypersaline conditions; but normal marine to hypersaline conditions are most favorable for their preservation. High-salinity conditions are apparently most favorable for their preservation because few browsing organisms that might destroy the laminations can survive there. Consequently, stromatolites are commonly associated with the supra-tidal, evaporitic, sabkah environment. One modern, fresh water, blue-green algal genus, *Plectonema,* has calcified sheaths. No calcified marine forms are known in the modern, but a number of extinct calcified genera (such as *Girvinella*) appear to have lived in normal marine conditions.

The majority of the calcareous green algae are normal marine, but a few genera, such as *Acetabularia* and *Penicillus* can tolerate brackish conditions. The charophytes are the only calcareous algae other than *Plectonema* which live entirely in fresh and brackish water and are particularly common in hard-water lakes. Practically all of the coralline algae are restricted to normal marine conditions although a few species occur in brackish water. Coccoliths are normal marine. Diatoms are abundant in waters from fresh to hyper-saline. The application of the substantive uniformitarian approach to diatom species with differing salinity tolerances offers promise as a paleosalinity determining method at least for late Cenozoic sediments (Burckle, 1978).

Depth

Ideally all algae are restricted to depths shallow enough for photosynthesis, but some species may utilize heterotrophic energy sources entirely or in part and may thus occur below the photic zone. Within the photic zone, distribution of species may be quite uneven. Stromatolites were once thought to be almost entirely intertidal, but in recent years both modern and fossil examples have been described from subtidal depths (Monty, 1977). In fact, subtidal stromatolites may have once been much more common than they are today because of the activity of grazing invertebrates which burrow through the algal mats and destroy the lamination (Garrett, 1970). Stromatolites were more common in the late Precambrian and early Paleozoic but have de-creased in abundance as grazing invertebrates evolved. Because few grazing invertebrates are able to function in the intertidal and supratidal area, stromatolites are preferentially preserved in these environments whereas they are destroyed or never form in subtidal environments. Thus stromatolites are often good evidence for intertidal and supratidal environments, particularly when they are found associated with other faunal or sedimentary evidence for these environments. However, care should of course be exercised in using stromatolites as a criterion for recognizing these environments, particularly in rocks older than middle Paleozoic (Playford and Cockbain, 1969).

The depth distribution of the calcareous algae is largely a function of their adaptation to growth at differing light levels. Some blue-green algal species are adapted to a high light intensity found in very shallow water or even the intertidal. The calcareous green algae, especially the dasyclads, are most abundant in water no more than a few meters deep; however, some species occur as deep as 100 m or more (Wray, 1977). The photosynthetic pigment of the red algae is well adapted for growth at the low light levels. They are found in water as deep as 250 m. Coralline algae are also found in very shallow water; in fact, they are most abundant there. Detailed studies of individual genera of red algae indicate that they are depth dependent (Bosence, 1976). A similar study of the green algal genus *Halimeda* showed depth dependent distribution of various species of this genus in Jamaica (Goreau and Goreau, 1973).

The coccoliths and diatoms are most common in deep water, open marine sediments although they of course live in relatively shallow water in the open ocean. Although these groups are common in deep water sediments, they are not necessarily restricted to such sediments. Many coccolith-rich chalks and diatomites apparently were relatively shallow water deposits. The phyto-plankton would probably be more common in shallow water deposits if they were not masked by nearshore terrigenous sediment.

Substrate

Benthonic diatoms are restricted to relatively shallow water as a consequence of their need to have both a solid substrate upon which to live and sufficient light for photosynthesis. Consequently the relative proportion of benthonic and planktonic diatoms might be used as a paleodepth indicator (Wornardt, 1969). Floras containing a predominance of benthonic and *meroplanktonic* (benthonic species with a planktonic stage) species are most likely to occur in shallow, nearshore conditions. The proportion of species that are *holoplanktonic* (planktonic throughout life) increases with depth and comprises nearly all of the flora in water depths greater than about 200 m. (Fig. 2.2.)

Stromatolites in a sense are the substrate, so relating their occurrence to the substrate may result in circular reasoning. However, stromatolites are usually found in fine carbonate rocks. The fineness of the substrate may in part be due to the trapping of the fine sediment by the algae. Although most commonly described from areas of carbonate deposition, stromatolites also occur in areas of terrigenous sedimentation.

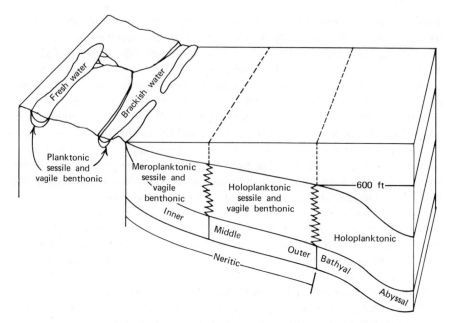

Figure 2.2 Generalized environmental distribution of planktonic and benthonic diatoms. After Wornardt (1969).

Calcareous green algae are mostly found in areas of soft sediment although a few species are adapted to attachment to hard substrates. The coralline algae are mainly adapted to attachment or encrustation on hard substrate, but a number of exceptions exist. All of the articulated varieties attach to the substrate, but several crustose species occur free.

Turbulence

Stromatolite morphology has long been known to be influenced by water turbulence (Fig. 2.3; Gebelein, 1969; Logan et al., 1974). In the Shark Bay, Australia area where they have been most extensively studied stromatolites from quiet water settings have laminated to gently undulating surfaces. Those from slightly more turbulent environments have a dome shape due to slight scour between individual "colonies" by currents. Club shaped stromatolites develop in still higher energy settings where scour prevents sedimentation between the clubs and may even cause some erosion. Oncolites form in the subtidal environment where they can be rolled around, allowing accretion on all sides. Moderately strong currents are required to roll the oncolites. Unidirectional or bidirectional currents in some cases cause the dome on stromatolites to become elongate parallel to current direction, allowing determination of ancient current directions.

Calcareous green algae are particularly common in areas with moderate to low water turbulence, especially in lagoonal and shelf settings; but some species (e.g., *Halimeda*) are fairly common in turbulent environments.

Coralline algae occur across a broad range of turbulence from the extremely high-energy setting on the windward margin of reefs to very quiet water at depths as great as 250 m. Red algae are well adapted to turbulent environments either by a flexible thallus as in the articulated genera or a massive, encrusting form as in some of the crustose genera. The morphology of some of the branching, crustose forms ranges with water turbulence from loosely branching forms in quiet water to dense, stubby branches in turbulent water (Fig. 2.4; Bosence, 1976; see Chapter 5 for a detailed discussion of this relationship). Rhodoliths, or laminated red algal nodules, also appear to be sensitive current indicators. Currents must be strong enough to occasionally turn over the rhodoliths so they can grow on all sides, but they cannot be so strong that they too frequently turn the rhodolith, inhibiting the growth of the encrusting algae. The shape and surface texture of the rhodoliths depend on current strength varying from flat and irregular in weaker currents to spherical and smooth in high currents (Bosellini and Ginsburg, 1971).

Other Parameters

Several studies, particularly those of Wray (1977), have related the distribution of the various algal groups to paleogeographic setting (Fig. 2.5). These distribution patterns are really the result of combinations of environmental parameters rather than being a single independent parameter. Wray's idealized distribution of major algal groups across a tropical shelf shows charophytes predominant in fresh water environments near the shore, stromatolites predominant in the intertidal and supratidal environments at the shoreline, green algae and some red algae predominant in the shallow shelf, lagoonal environment, and coralline algae predominant on reefs and banks at the shelf margin. The benthonic algae decrease in abundance going into deeper, open marine water, where planktonic coccoliths and diatoms are predominant. Wray (1977) cites examples from the Devonian, Pennsylvanian, and Paleocene where parts of this idealized sequence can be found.

The productivity of ancient environments is an important parameter but one that is difficult to measure. The abundance of preserved phytoplankton is of course the most direct measurement of productivity (Tappan, 1968); however, many phytoplankton species are not readily preserved. Also the abundance of preserved phytoplankton is a function of sedimentation rate as well as productivity. Areas of high productivity today, especially upwelling areas, have an abundance of diatoms in the sediments. The uniformitarian approach would thus suggest high productivity for diatom-rich sediments in the past. In a classical study of Pleistocene and Recent deep sea sediments of the equatorial Pacific, Arrhenius (1959) noted vertical variation in the carbonate content of the sediment. The carbonate consists largely of coccoliths

(a)

Sedimentary structure		Surface current velocity (cm/sec)	Sediment movement (g/hr/ft)
Rippled sand	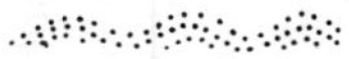	> 15–20	> 60–80
Algal mat		< 15–20	< 60–80
Algal dome		I – II	8–60 High total sediment accumulation
Algal biscuit		I – II	8–60 Low total sediment accumulation

34

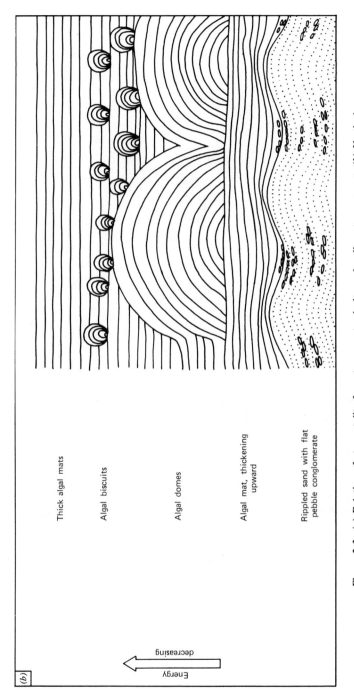

Thick algal mats

Algal biscuits

Algal domes

Algal mat, thickening
upward

Rippled sand with flat
pebble conglomerate

Energy decreasing

Figure 2.3 (*a*) Relation of stromatolite form to current velocity and sediment movement. (*b*) Vertical sequence of stromatolitic structures as current velocity decreases. After Gebelein (1969), Journal of Sedimentary Petrology, Society of Economic Paleontologists and Mineralogists.

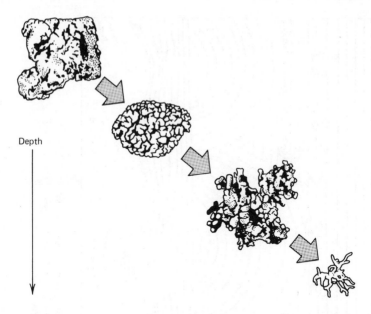

Depth

Figure 2.4 Variation with depth or turbulence of growth form of crustose red algae (a hypothetical example). After Wray (1971).

and planktonic foraminifer. He interpreted the increase in abundance of carbonate as reflecting an increase in productivity in response to more intense upwelling during glacial epochs. Many later studies of deep sea sediments have contained similar conclusions.

FORAMINIFERA

Organism Characteristics

There are two major ecological groups of foraminifera: benthonic and planktonic. The benthonic types live on all sorts of substrate from mud through sand to hard substrate and on algae or other organisms. Some encrusting forms cement themselves to the substrate. Planktonic foraminifera live suspended in the water mass at all levels, but are especially abundant in the photic zone (usually less than 50 m). Each species of planktonic foraminifera has a certain depth zone where it is particularly common. There are thousands of species of living benthonic foraminifera but only about 30 species of planktonic types. The abundance and widespread distribution of the planktonic species make this group especially useful in paleoecology, particularly the paleoecology of the open ocean. Of course they give information on surface conditions and not on the bottom depositional environment itself.

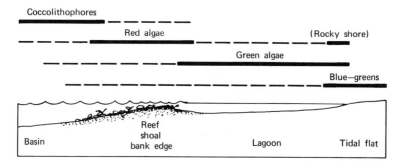

Figure 2.5 Generalized environmental distribution of various types of calcareous algae. After Wray (1971).

Foraminifera feed primarily on phytoplankton, but they also include zooplankton (e.g., copepods) in their diet. Some are scavengers and some are parasites. Some of the planktonic species and large benthonic foraminifera contain zooxanthellae, symbiotic photosynthesizing dinoflagellates, and probably in part utilize food produced by these symbionts.

Many books and review papers contain valuable information on the ecology and paleoecology of foraminifera. Particularly useful ones include Boltovskoy and Wright (1976), Hedley and Adams (1976), and Bandy (1964). To a larger extent than with most fossil groups, foraminifera have been the subject of substantive taxonomic uniformitarian studies. Examples of such studies include Natland (1933), Walton (1964), and Phleger (1960) (see Boltovskoy and Wright, 1976, for reference to many other examples).

Environmental Tolerances

Temperature

As a group the foraminifera are very tolerant of temperature variations. The group occurs all the way from tropical lagoons and tide pools where temperatures may be over 40°C to high-latitude areas where sea water is freezing, at nearly −2°C. Experimental work with foraminiferal cultures have shown similar tolerances; however, individual species are much more limited in their temperature tolerances. The distribution of planktonic species in particular seems to be temperature controlled and has been more thoroughly studied than the distribution of benthonic species. Table 2.2 (from Boltovskoy and Wright, 1976) shows the latitudinal distribution of most of the planktonic species. Factors other than temperature may be involved in this distribution, but clearly temperature is the major parameter. The distribution of benthonic species also is clearly temperature dependent. The limiting effect of temperature on the distribution of foraminifera as well as other groups may operate in three ways. (1) The most narrow limits are usually on reproduction. (2)

Table 2.2 Temperature range of modern foraminifera species[a]

Species	Arctic Antarctic	Subarctic Subantarctic	Transition	Subtropical	Tropical
Globigerina pachyderma (sinistral)	—	— - -	- -		
Globigerina pachyderma (dextral)		— - -	- -	- -	
Globigerina quinqueloba		—	—		
Globigerinita uvula	- -	- -	- -		
Globigerina bulloides	—	—	—	—	
Globigerinita glutinata	- -	- -	—	—	- -
Globorotalia scitula	- -	- -	—	- -	
Globorotalia inflata		—	—	—	
Globorotalia truncatulinoides				—	
Globorotalia hirsuta			—	—	
Orbulina universa			—	—	
Globigerinella aequilateralis			—	—	—
Globigerinoides ruber			- -	—	
Globigerina falconensis			- -	—	- -
Globigerinoides trilobus			- -	—	- -

Globorotalia menardii
Globoquadrina dutertrei
Globigerinoides conglobatus
Globigerinoides tenellus
Globigerina calida
Globorotalia crassaformis
Hastigerina pelagica
Globigerina rubescens
Globigerinoides trilobus (F. sacculifera)
Pulleniatina obliquiloculata
Globorotalia tumida
Candeina nitida
Globigerina digitata
Sphaeroidinella dehiscens
Globigerinella adamsi
Globoquadrina hexagona
Globoquadrina conglomerata

[a] After Boltovskoy and Wright (1976).

39

Growth may occur over a somewhat wider range of temperature. (3) Simple survival is likely to be possible over an even wider temperature range. A further complication is that temperature seldom operates completely independent of other variables. For example a foraminifera may be able to survive over one temperature range at a certain salinity but the survival temperatures may be different at another salinity. Not only does temperature exert great influence on the geographic distribution of the foraminifera, but it also is important in controlling vertical distribution; this is often difficult to prove decisively, though. Some foraminifera exhibit the feature of *equatorial submergence* in their distribution. A species that lives at the surface or in shallow water at higher latitudes occurs at progressively greater depths at lower latitudes. This apparently is explained by temperature control. Submergence has also been observed in many groups in addition to the foraminifera.

Many examples of morphologic features of foraminifera that vary with temperature have been described (see Kennett, 1976, for a review). Perhaps the most widely studied and used character of this type is coiling direction. Students of foraminifera have long noted that the coiled shells of certain species may coil either to the right (*dextral*) or to the left (*sinistral*) (Fig. 2.6). O. L. Bandy (1960) and D. B. Erickson (1959) independently noted that the coiling direction in modern assemblages of *Neogloboquadrina pachyderma* appeared to depend on the growth temperature. When the temperature was less than about 9°C, specimens of this species were dominantly sinistral in coiling direction. At temperatures greater than about 15°C assemblages are predominantly dextral with mixtures of dextral and sinistral occurring between 10 and 15°C (Fig. 2.7). Many studies have since been made which use coiling direction, usually in combination with other methods, to determine paleotemperature trends (Fig. 2.8; e.g., Ingle, 1967, and Kennett, 1967) and also as an aid in correlation (e.g. Bandy, 1960, and Jenkins, 1967).

Is there any functional significance to this correlation between coiling direction and temperature? This question has never been satisfactorily answered, but several suggestions have been made. Perhaps coiling direction is genetically linked with some physiologic feature of foraminifera that does have adaptive significance relative to temperature. Perhaps the dextral and sinistral forms actually represent separate species that have different temperature sensitivities. Perhaps one of the coiling directions is found in sexually produced forms and the other in asexually produced individuals. Coiling direction also varies in other foraminiferal species (e.g., *Globigerina bulloides*) but the correlation with temperature has not been so well defined as in *N. pachyderma*.

Several other features of foraminiferal morphology vary with temperature in modern species. The planktonic species in particular show several such features (Kennett, 1976). For example, *Globorotalia truncatulinoides* varies in shape from conical in tropical waters to more discoidal in cooler climates.

Sinistral   Dextral

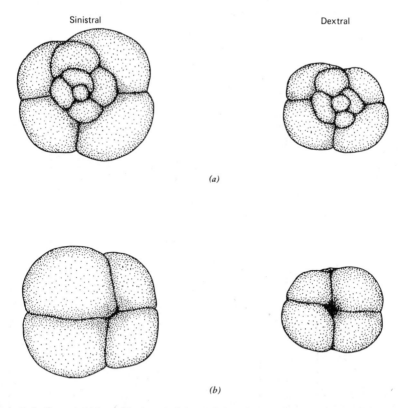

(a)

(b)

Figure 2.6 Dextral (right coiling) and sinistral (left coiling) specimens of *Neogloboquadrina pachyderma*. (*a*) Dorsal view; (*b*) ventral view. After Bandy (1960), Journal of Paleontology, Society of Economic Paleontologists and Mineralogists.

The number of chambers in the final whorl of *N. pachyderma* varies from 4 to 5 with temperature in a regular pattern. The final chamber of several species is sometimes smaller than the penultimate, in the so-called *kummerform* individuals. These forms are particularly common when the specimen has grown under marginal conditions such as at reduced temperatures. The porosity of the wall of planktonic species varies with temperature (Bé, 1968) perhaps as an adaptation to variable density and viscosity of sea water. The more porous specimens, which are thus less dense, are formed under tropical conditions where sea water is less dense and viscous than in colder climates. At least in some species, overall test size correlates inversely with temperature; although some species show the opposite relationship. In the foraminifera as in other groups of organisms, colder temperatures may slow the rate of maturation so that although the individual grows more slowly it reaches a larger final size under cold conditions.

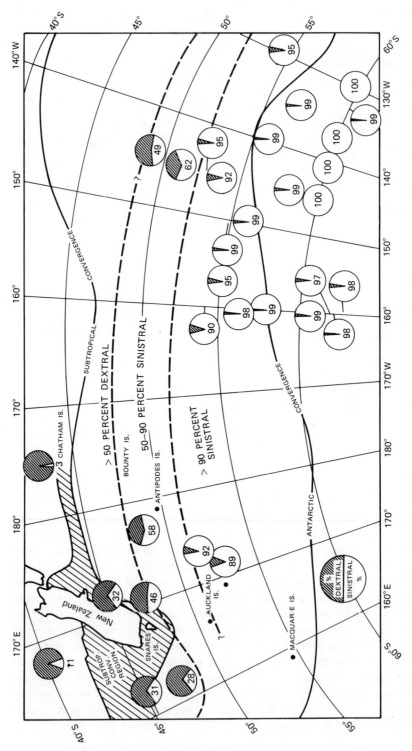

Figure 2.7 Relative proportion of sinistral and dextral specimens of *Neogloboquadrina pachyderma* from recent sediment samples from the South Pacific Ocean. After Kennett (1968).

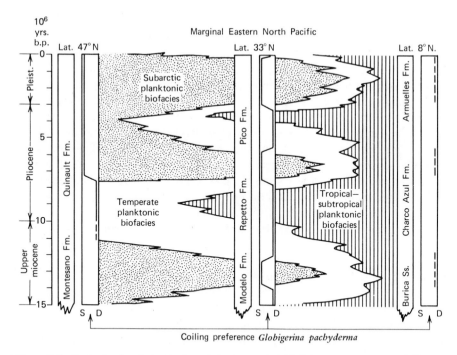

Figure 2.8 Generalized variation in the relative proportion of sinistral and dextral specimens of *Neogloboquadrina pachyderma* in three Neogene stratigraphic sequences along the marginal eastern North Pacific basin. After Ingle (1967), from Bulletins of American Paleontology, *52* (236), text-figure 41.

Salinity

The foraminifera occur over a wide range of salinity, from almost fresh to hypersaline values of 90‰ or more (Sellier de Civrieux, 1968, as reported in Boltovskoy and Wright, 1976). Indeed foraminifera have been reported from fresh water in a few cases. Some of these reported occurrences were probably actually thecamoebians rather than foraminifera. Others are perhaps instances of marine forms being isolated from their original marine habitat by some process. In any case, true foraminifera are very rare in fresh water. Several benthonic species have been studied in detail in culturing experiments and found to survive over wide salinity ranges (as much as 5 to 50‰). Forms in their natural habitat may occur in areas with salinity near zero to normal marine conditions of 35‰. Certain genera are much more tolerant of brackish conditions than others. Most of these genera can also live in waters of more normal salinity but are particularly characteristic of brackish conditions. Other genera are particularly common in hypersaline conditions. Interestingly, some of the genera common in brackish waters are also common under hypersaline conditions but not especially common under more normal salinity conditions. This is probably because these are euryhaline genera, which can compete successfully under conditions of variable salinity but are not so successful under conditions of stable salinity where stenohaline forms are better adapted. As discussed in Chapter 8, the euryhaline forms are opportunistic species and the stenohaline forms equilibrium species.

Although experimental data are not as abundant and reliable, planktonic foraminifera are clearly much more stenohaline than their benthonic counterparts. They apparently do not occur in water with salinity less than about 30‰ or greater than 40‰. A number of examples of morphologic variation within a species being correlated with salinity have been described. Ornamentation and wall thickness seem to be especially sensitive to salinity. Forms from lowered salinity habitats tend to have reduced ornamentation and thin walls (Boltovskoy and Wright, 1976). Test size also is often smaller in specimens from low-salinity settings than in forms from more normal salinity sites.

Depth

Benthonic foraminifera occur at all depths from sea level to the bottom of the deepest trenches sampled (about 10,500 m in the Tonga Trench). Depth is not a "pure" environmental variable. Many parameters change with depth (e.g., temperature, light, pressure, oxygen, $CaCO_3$ saturation). Each of these parameters can have an effect on the foraminifera, so if their distribution changes with depth, the critical parameter or parameters may not be determinable. Some species appear always to occur at a certain depth regardless of temperature or other variables, whereas many species are extremely widely distributed in terms of depth. Much of the paleoecologic research done on foraminifera dating back to the classical work of M. L. Natland has been on depth distribution; consequently a great deal is known about the bathymetry of this group of organisms. For more complete discussions of the depth distribution of foraminifera, particularly see Boltovskoy and Wright (1976), Walton (1964), and Murray (1973).

Because solubility of $CaCO_3$ increases with depth, most calcareous forms disappear at depths greater than about 3000 m. Agglutinated species increase in abundance at great depths (Fig. 2.9). In fact, the ratio of calcareous to agglutinated forms can be used as an approximate index of depth. Saidova (1961) noted that the calcareous to agglutinated ratio in the sublitoral zone of the north Pacific was usually about 100:1. The ratio drops to about 30:1 in the upper bathyal zone and to less than 1:1 in the abyssal zone.

The relative proportion of planktonic to benthonic foraminifera in the sediment is strongly influenced by depth of deposition (Fig. 2.9). Planktonic foraminifera are much less common in shallow, near-shore water than in deep, open ocean water. This may be due to the less stable environment in the near-shore area and the lesser volume of water in which the specimens can live over the shelves. Planktonic foraminifera can live to depths of several hundred meters in the open water. In shallow water only a portion of the potential space for occupancy is available, thus a smaller number of planktonic

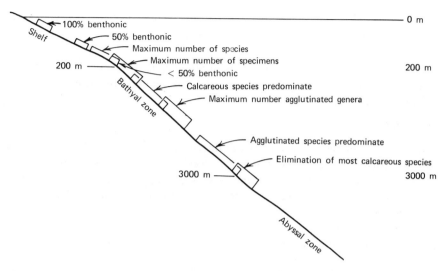

Figure 2.9 Variation with depth of proportion of benthonic and agglutinated foraminifera. After Boltovskoy and Wright (1976), Dr. W. Junk B.V., Publishers.

forams live over a given area in shallow than in deep water. On the other hand, the abundance of benthonic specimens is greater in shallow water where food on the sea floor is likely to be more abundant. The result of the operation of these factors is that the percentage of planktonic specimens in the total foraminiferal fauna increases from near zero at the shoreline to nearly 100% at depths greater than 1000 m (Grimsdale and Morkhoven, 1955). In addition to being a very effective method for determining water depth, this technique has the advantage of requiring a minimum of taxonomic work. The technique has been used by several workers including Stehli and Creath (1964) who used it to determine the paleobathymetry of the Cretaceous in the Gulf Coast area.

A number of morphological changes with depth have been described in the literature (see Bandy, 1964, and Boltovskoy and Wright, 1976, for reviews). Most of the trends are restricted to particular species or genera and often opposite trends occur in other genera. Only a few of the more broadly applicable trends are mentioned here. Agglutinated foraminifera from bathyl and deeper sites tend to be larger and have more complex interiors than those from shelf depths. Porcelaneous types are not common at depths greater than a few tens of meters, but one genus, *Pyrgo*, and related forms are found in the deep sea and are larger and more complex in morphology than shallow water forms. Among the hyaline calcareous forms, some genera increase their ornamentation in deeper water but a few genera show the opposite trend. Likewise several species become larger at depth but a few show the opposite trend. Several of the planktonic species develop a secondary calcite crust on their exteriors. This crust may be up to 50 μm thick. The crust tends to be thicker on individuals living in relatively deeper water. Some workers have indicated that the thickening is part of normal ontogenetic development, but others suggest that it is an adaptation to increase the density of specimens living in deep water. The lower paleotemperature as determined from oxygen isotopic analysis for heavily encrusted specimens supports the latter hypothesis (Hecht and Savin, 1972).

Substrate

The influence of the substrate on the distribution of foraminifera is difficult to study because other environmental factors are likely to vary with substrate (especially water turbulence). Many researchers have noted correlations between species and substrate (see Table 15 in Boltovskoy and Wright, 1976), but other workers have expressed doubt about the significance of these correlations. Many species actually live on algae and not the inorganic substrate proper. After death and decay of the algae the direct evidence of the original substrate is lost. This may lead to some of the apparent independence of some species from a particular sedimentary substrate. Substrate will influence the type of sediment grains used by agglutinated forms for the test. In general, agglutinated forms seem to use whatever sediment type is available, but cases have been described of some species selecting certain mineral types for their tests. Some examples are cited in the literature of variation of test morphology due to the influence of substrate, heavier shelled and more highly ornamented forms being found on coarser substrate.

Turbulence

Because of their small size, living foraminifera have difficulty in maintaining their position in turbulent environments. During life, the protoplasm of the individual may add some stability, but after death the empty test will behave like a sand grain. Consequently fewer foraminifera are found in sediments deposited in high-energy environments than in quieter settings. There are obvious exceptions to this, especially among some of the larger species and encrusting forms. The encrusting species *Homotrema rubrum* is a common form in the shallow reef environment of the tropical Atlantic today. Fragments of this species are responsible for the pink beaches of Bermuda. Some of the large fusulinid species of the late Paleozoic and nummulitids of the Eocene are commonly found in rocks that were apparently deposited in turbulent waters. Morphologically similar forms adhere to surfaces (such as algae) with mucilagenous material. After death and decay these specimens may be transported out of their turbulent life environment. Some examples have been cited of more robust, tightly coiled shells developing under turbulent conditions as compared to related, quiet-water forms.

Other Factors

Many other environmental parameters may also influence the occurrence and morphology of foraminifera. The intensity of light has at least an indirect influence on the distribution of planktonic foraminifera. They are most concentrated in areas of high illumination because their primary food source, diatoms, needs light for photosynthesis. Possible diurnal vertical migration of planktonic species in response to light may occur with the foraminifera sinking during the dark hours and rising during the day. Oxygen levels are probably important in some cases in controlling foraminiferal distribution although at least some species are able to survive in environments with very low dissolved oxygen concentrations. Species of the genus *Bolivina* are especially well adapted to low oxygen concentrations. Water turbidity is not usually a limiting factor on foraminiferal distribution although undoubtedly some species are better adapted to withstand turbidity than others. This factor has not been extensively studied.

SPONGES

Organism Characteristics

All sponges are sessile, epifaunal organisms that feed by filtering out of the water plankton and other suspended organic material. At least some sponges are especially adapted to utilize very small food such as bacteria. They are thus primary consumers in the trophic structure but, to the extent that they utilize nonliving organic matter, can be considered scavengers. Sponges are less well studied in the modern as well as the fossil record than most groups, although they are a fairly common faunal element and have a long geologic history. They have recently been the subject of renewed interest among paleoecologists because of their possible role in the formation of modern and fossil bioherms (Hartman, 1977 and Wiedenmayer, 1978). The sponges consist of four major classes, the Calcarea (having calcareous spicules), the Hyalospongea (with siliceous spicules), the Demospongia (with both organic and siliceous spicules), and the Sclerospongia (with siliceous spicules and a calcareous skeleton). Living specimens of the sclerosponges were only recently recognized (Hartman and Goreau, 1970). This group is of particular interest because it may include the stromatoporoids, a very important component of reef faunas at times in the Paleozoic and Mesozoic. Another extinct group, the archaeocyathids, were long considered to be sponges but are now commonly placed in a separate phylum. Each of these classes has a distinctive distribution reflecting different responses to environmental parameters (deLaubenfels, 1957).

Environmental Tolerances

Temperature

Sponges as a group are found across the spectrum of temperatures encountered in the oceans. In the modern seas the calcisponges are particularly common at relatively cold temperatures. The hyalosponges have their greatest diversity in deep, hence cold water. All modern members of the sclerosponges are tropical, and the geologic evidence suggests that the stromatoporoids were all or largely tropical in distribution. Demosponges occur in all temperature regimes (deLaubenfels, 1957).

Salinity

Sponges live in a wide range of salinities from fresh to hypersaline although they reach their maximum diversity at near-normal salinities. The greatest tolerance seems to be in the demosponges, the other groups being more restricted to normal salinities (deLaubenfels, 1957).

Depth

Sponges as a group extend across the entire depth spectrum: hyalosponges are restricted to deep water (greater than 100 m) with a few exceptions, especially in the high latitudes; calcisponges are mostly shallow water, usually less than 100 m; and the few representatives of sclerosponges are shallow water forms. The extinct stromatoporoids appear to have lived in relatively shallow water, often being associated with reefs. Some workers have speculated that the stromatoporoids had photosynthetic algal symbionts in their tissues and thus were restricted to shallow water by their need for light. Care should be taken in attempting to extend this symbiotic relationship back in time as far as the Paleozoic. Some of the sponge groups may well have changed their depth and water turbulence habitats since that time (Finks, 1970).

Substrate

Most modern sponges are adapted for attachment to a hard substrate but a few have developed a rootlike holdfast for life on a soft substrate. In the fossil record, sponge spicules are often common in fine-grained rocks (Finks, 1970) where there is no evidence for a hard substrate. The sponges could have had holdfasts for support on the sediment. They could have been attached to an organic substrate that has since decayed, or the spicules could have been transported a considerable distance from their place of origin.

Turbulence

The flexible but usually strong nature of the sponge body is well adapted to life in a turbulent environment. Turbulence is beneficial to the sponge in providing a continuously renewed supply of water with suspended food and in removing waste products. In a classical study of functional morphology in the sponges, Bidder (1923) has shown that some modern sponges have developed a morphology adapted to life in relatively quiet water. The sponge (as indeed do many suspension feeding organisms) has the problem of keeping separated the used water, from which the sponge has already filtered the suspended food and to which it has added waste products, from new water, which still contains the needed food. The sponge has evolved two features in order to deal with this problem: first, it has evolved a single *osculum* for expelling the used water at the upper end of the body (Fig. 2.10). This allows maximum separation of the excurrent water from the incurrent water, which enters at many separate *ostia* (incurrent pores) over the body. Bidder discusses this separation in terms of the *angle of supply,* that is, the angle between the excurrent and incurrent directions (Fig. 2.10). The larger this angle the greater the separation of used and new water. The second adaptation is in maximizing the distance to which the expelled water is removed from the body. This is regulated by the velocity of the excurrent water, which is in turn controlled by the pressure developed by the flagellae within the sponge and the size of the oscular opening. The effect of the size of the opening can be visualized by comparison with a garden hose. A large opening produces a large-volume but low-velocity current that does not travel far. If the hose has a nozzle with an extremely fine opening, it will produce a small, high-velocity needle of water that does not travel far and does not allow much water to escape. The optimum sized opening will produce a stream of water that will travel the maximum distance from the end of the hose. See Leigh (1971) for a detailed quantitative treatment of the sponge osculum problem.

Bidder refers to the distance to which the sponge expels its water stream as the *diameter of supply,* that is, the diameter of the smallest possible circle by which excurrent water can be returned to the ostia. Ideally then a sponge adapted for life in a quiet water environment should have a morphology to maximize the angle and diameter of supply such as a vase shape with a narrow osculum (Fig. 2.10). A sponge living in turbulent conditions would not

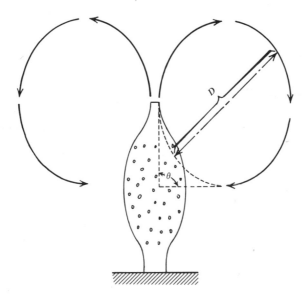

Figure 2.10 Diagrammatic quiet water sponge. *D*—diameter of supply, ⊖—angle of supply.

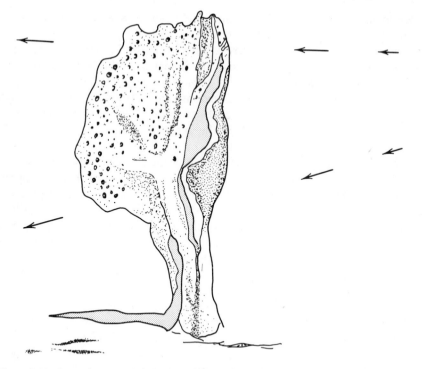

Figure 2.11 Sponge adapted for life in a unidirectional current. Arrows show the current direction. After Bidder (1923), Quarterly Journal of Microscopical Science, *67*.

COELENTERATA (CORALS)

Organism Characteristics

Three living groups of coelenterates have relatively important fossil records: scleractinian corals, octocorals, and hydrozoans. The scleractinians are by far the most important and form the major basis for our interpretations of the paleoecology of corals. Two other extinct groups, the rugose and tabulate corals are abundant in the Paleozoic. Some paleontologists would include the possibly extinct stromatoporoids among the coelenterates, but we have included them with the sclerosponges. Many of the coelenterates (especially the scyphozoans and many hydrozoans) have no mineralized skeleton and have to develop special morphologic features to separate new and used water and thus might be more varied in shape in response to other environmental requirements. A fan shape to expose a maximum area to food-bearing currents (Fig. 2.11) is common in sponges.

thus have a poor fossil record. Most of our comments in this section are confined to the scleractinian corals.

All of the corals are benthonic and many are attached to the substrate. A considerable number of modern scleractinians and most of the extinct rugose corals lived free on the sediment. All coelenterates are carnivorous, feeding largely on zooplankton. Corals are important reef formers in the modern seas and have been since late Mesozoic time. Corals were probably less important as reef formers in earlier times, but have always been common in the reef environment.

In terms of their ecology, the most important subdivision of the corals is between hermatypic and ahermatypic types. *Hermatypic* corals have symbiotic photosynthesizing dinoflagellates, zooxanthellae, in their tissue whereas *ahermatypic* forms do not (Yonge, 1957). All three of the modern coral groups have hermatypic species. In fact the symbiotic zooxanthellae are found in the tissue of several other groups of organisms including foraminifera and bivalves. The zooxanthellae have a very important effect on the functioning of the corals. Under certain experimental conditions the coral can be induced to expel its zooxanthellae. The coral may then continue to live and function at a very low rate, but can only show vigorous growth when the zooxanthellae are present (Goreau, 1959).

The zooxanthellae benefit from the symbiotic relationship by (1) using the coral's waste products, especially nitrates and phosphates, as nutrients, (2) using the coral's metabolic CO_2 for growth, and (3) having a protected location for growth. The coral benefits from the relationship by (1) removal of waste products and CO_2, (2) an internal source of oxygen as a result of photosynthesis by the zooxanthellae, (3) a source of nutrients such as glycerol and glucose, (4) an aid in the skeletal calcification process by removal of CO_2 from the calcifying fluid (Yonge, 1957).

In terms of their ecology, hermatypic corals should perhaps be considered as two organisms, a plant and an animal. Factors that affect one member of the association will indirectly affect the other so that the distribution of corals is a function of the requirements of both the coral and its zooxanthellae.

Because of this importance of the zooxanthellae to the coral, a knowledge of whether or not fossil corals were hermatypic is of considerable importance if we wish to transfer information about the distribution of modern corals to the geologic record. As the zooxanthellae are never preserved, only indirect evidence and supposition can be used to tell if ancient corals were hermatypic. Ancient scleractinian corals were in all probability hermatypic as they are so closely related to extant forms, but whether or not extinct groups such as the rugose and tabulate corals contained zooxanthellae is more debatable. The interpretation of the hermatypic nature of the extinct rugose and tabulate corals rests largely on two types of evidence: (1) the massiveness of the skeleton and (2) the presence of daily growth bands. Perhaps because of the effect of zooxanthellae on the calcification process, the skeletons of modern hermatypic corals are usually more massive than ahermatypic forms. Modern ahermatypic scleractinians are usually small in size. Encrusting forms are at most a few cm in diameter and many are solitary. Several types are branching with the individual polyps widely spaced (Wells, 1957). Another characteristic of hermatypic forms is their well developed daily growth banding which has been interpreted as resulting from the effect of the photosynthesis by zooxanthellae on the calcification process (Wells, 1963; also see Chapter 4). During the daylight hours, when the zooxanthellae are photosynthesizing, calcification is thought to occur more rapidly than at night when respiration by the algae as well as the coral produces CO_2 which inhibits calcification. The result is the day-night banding pattern that can be observed in the coral skeleton. Both the rugose and tabulate corals have well developed growth increments which have been interpreted as probably being daily. Sorauf (1974b) believes that this banding indicates a hermatypic nature at least for the tabulates. As a word of caution, some groups of organisms such as the molluscs produce daily growth bands in their skeletons without the presence of zooxanthellae (see Chapter 4). Daily growth bands in extinct corals thus may not be definitive evidence of the hermatypic nature of those corals. Although the paleoecologic and geologic evidence suggests that the environmental distribution for the rugose and tabulate corals was not identical to that for modern corals, they probably did all live in water shallow enough for photosynthesis.

Temperature

Hermatypic forms are almost entirely tropical being unable to withstand minimum temperatures of 16 to 17°C (Wells, 1957). Macintyre and Pilkey (1969) describe two species of hermatypic corals which occur off the North Carolina coast where temperatures may drop as low as 10.6°C for extended periods, but such occurrences are highly unusual. Hermatypic corals reach their maximum abundance and diversity at temperatures of 25 to 29°C and are rarely found where the temperature exceeds 40°C. Ahermatypic corals are much more tolerant in their temperature requirements, occurring in the range of about −1 to 28°C. The rugose and tabulate corals are often presumed to have had temperature requirements similar to those of modern scleractinians. Geologic evidence in general seems to support this supposition but certainly does not conclusively prove it. The geologic literature contains numerous examples of the inferring of ancient tropical environments on the basis of the presence of fossil corals.

Salinity

All corals and practically all coelenterates are marine. Hermatypic corals live only at near-normal marine salinities (Wells, 1957). They grow best in the range of 34 to 36‰, but some can tolerate at least brief exposure to salinities as low as 27‰ or as high as 48‰. Ahermatypic corals are also basically normal marine animals but they are slightly more tolerant of variable salinity than are the hermatypic forms. The role of salinity in controlling the distribution of corals is demonstrated by the coral reefs of south Florida (Fig. 2.12). The reefs are best developed where protected from intrusion of low-salinity water from the Florida peninsula by long, continuous islands. Reefs are non-existent or poorly developed in areas where gaps occur between islands allowing the outflow of low-salinity water. Suspended sediment as well as reduced salinity may inhibit coral growth in these areas (Multer, 1977).

Depth

Hermatypic corals are sensitive to light intensity because of the presence of photosynthesizing plants in their tissues. For this reason, hermatypic corals are restricted to depths less than about 150 m (Fig. 2.13) and growth is most luxuriant in water less than about 15 m deep. Individual species may be quite narrowly restricted in their depth distribution. For example *Acropora palmata,* the common moosehorn coral of the Caribbean area, is found only at depths less than 7 m (Multer, 1977). Ahermatypic corals are much more widely distributed, being found at depths as great as 6000 m (Wells, 1967). In some places they occur in such abundance as to form reeflike structures even at great depth (Feuhert, 1958). They differ from shallow water reefs, however, in their low diversity of corals and lack of calcareous algae.

The morphology of individual coral colonies also varies with depth. The modern Caribbean species *Montastrea annularis* varies from hemispherical in water less than 5 m deep to columnar in water 5 to 25 m deep to platelike in water greater than 25 m deep (Fig. 2.14; Graus and Macintyre, 1976). Several other species show similar variation (Goreau, 1963). Graus and Macintyre have developed a simulation model that demonstrates that these growth forms can be explained by differential growth rates controlled by light intensity and resulting photosynthesis by the zooxanthellae. Lecompte (1958) noted a similar variation in the shape of stromatoporoid colonies in Devonian reefs in Belgium. Massive stromatoporoids grew on the apparently shallowest portions of the reefs whereas lamellar forms grew in deeper water. Philcox (1971) described massive and platy forms of tabulate corals from the Silurian of Iowa, which he attributed to differences in sedimentation rate, but which might be better explained by depth differences.

Substrate

The majority of the scleractinians as well as the octocorals and hydrozoans require a hard substrate for attachment. However, a number of species of corals are adapted to live unattached in the substrate, often in areas of fairly rapid sedimentation. Some of these, such as *Manicina* can reorient themselves and unbury when moved about or buried by currents (Gill and Coates, 1977). Individuals of other species may develop a *circumrotary* form because of growth on all surfaces of the colony, as they are occasionally rolled over by currents. The rugose corals were apparently largely an unattached group, at least after a juvenile attachment stage. Specimens of Rugosa often have a highly variable growth form, which is an adaptation to life on a soft sub-

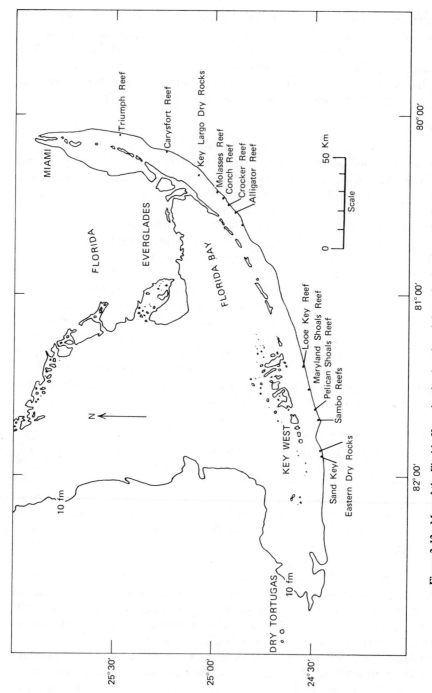

Figure 2.12 Map of the Florida Keys showing location of the major barrier reefs. Reefs are absent where turbid, low-salinity water flows from Florida Bay between gaps in the Keys.

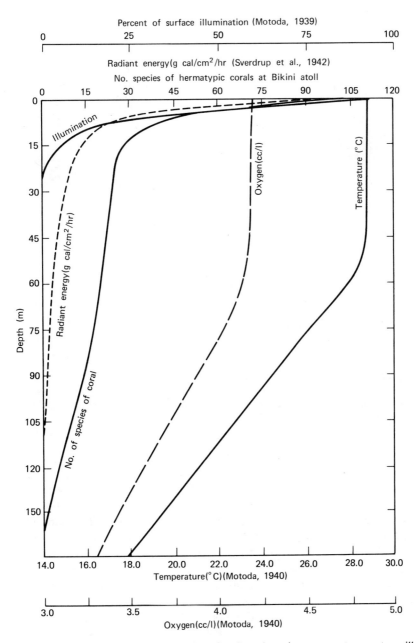

Figure 2.13 Variation with depth in number of reef coral species, oxygen, temperature, illumination, and radiant energy at Bikini Atoll. After Wells (1957), Geological Society of America, Memoir 67.

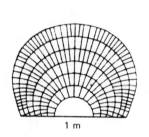

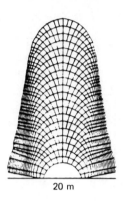

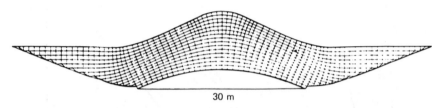

Figure 2.14 Computer simulation of the light-adapted growth form of colonies of the coral *Montastrea annularis*. Simulated forms closely approximate actual forms of specimens growing at the indicated depths. After Graus and Macintyre (1976), copyright 1976 by the American Association for the Advancement of Science, Science *193*: 895–897, text-figure 1.

strate. The coral starts growth on a small, hard object that soon loses its function as a support. When the coral topples over or is moved by some process, it resumes upward growth, often producing a sharp bend in its skeleton (Fig. 2.15). This may happen several times during the life of the individual, producing grotesque shapes (Hubbard, 1970). The shape of the base of coral colonies may be a function of substrate firmness. Based more on fossil than modern data, Hubbard (1974) concludes that colonies growing in soft substrate tend to have a more cone-shape base and forms living on firm substrate have a flatter base. The hermatypic corals are sensitive to suspended sediment in the water, prolific coral growth being restricted to clear water (Wells, 1957). This apparently did not apply to the rugose corals that are frequently found in fine terrigenous rocks which must have been deposited in turbid environments. Ahermatypic forms appear to be less sensitive to turbidity.

Turbulence

Modern hermatypic corals are usually found in relatively turbulent water. This may be because the corals require light, well oxygenated water, abundant

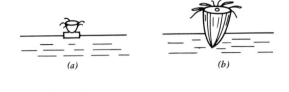

Figure 2.15 Development of rugose coral form on an unstable substrate. (*a*) Coral starts to grow on a small piece of hard substrate such as a shell fragment; (*b*) growth continues; (*c*) growing coral tips over into the sediment after it outgrows its substrate; (*d*) coral resumes growth from its new orientation.

food in suspension, and freedom from fine sediment settling on their polyps— conditions that are most likely to be found in turbulent waters. The influence of turbulence on the shape of the colony is often mentioned in the literature and was studied in detail in the early twentieth century by Vaughan (1919). Delicate branching corals are seldom found in turbulent water. Massive and heavy branching forms are characteristic of turbulent settings. Vaughan showed that massive forms of *Porites porites* from turbulent environments developed branching when transplanted in a quiet-water environment. Heavy branching forms such as *Acropora palmata* often have their branches oriented into the oncoming currents or waves (Shinn, 1963). The hydrozoan *Millepora* occurs in four morphologies in reef environments in the Caribbean area: (1) branching, (2) bladed, (3) boxwork, and (4) encrusting. Only encrusting forms occur in the most shallow turbulent environments. In progressively less turbulent settings boxwork forms occur followed by bladed, and finally branching forms. This relationship appears to be a direct reflection of the physical strength of the skeleton, encrusting types being strongest and branching weakest (Stearn and Riding, 1973). The morphology and orientation of the coral is determined by the mechanical strength of the skeleton and the stress placed on it (Graus et al., 1977; Chamberlain, 1978). In turbulent settings corals either have massive skeletons or have developed oriented branches. In settings with no current, a vertical branching pattern minimizes mechanical stress on the skeleton due to gravity. In environments with strong currents, inclined orientation minimizes stress (Fig. 2.16; see Chapter 5 for a detailed discussion of this study).

Coral morphology may also be controlled by feeding characteristic. Corals with a planar or fan shape such as the sea fans or gorgonians among the octocorals orient with the planar direction perpendicular to the current direction (Grigg, 1972). This is apparently an adaptation to feeding with the maximum surface presented to the food-bearing current. Similar orientations are found in other groups such as sponges, bryozoans, and crinoids (Warner, 1977).

BRYOZOA

Organism Characteristics

Bryozoa are rather inconspicuous but nonetheless ubiquitous members of the benthonic fauna of modern seas. They are strictly colonial. Individuals within the colony (zooids) are commonly less than 0.5 mm in size; the colony (zooarium) may reach several cm in maximum dimension. Bryozoans feed with a lophophore, the tentacles of which produce ciliary currents that bring food to the mouth. They appear to feed largely on phytoplankton and detrital organic matter. A few species contain zooxanthellae in their tissue and probably receive some of their nutrition from these algae. Most species require a hard substrate for attachment; however, the preferred substrate may be marine algae or one of the sea grasses. When the plant that provides the substrate dies and decays, the bryozoa are freed to become incorporated in the sediment. Bryozoans also attach to shells (both living and dead). Bryozoa are also important constituents of reefs (Cuffey, 1977). In many environments every available bit of hard substrate seems to contain bryozoan encrusters. This can present problems to man when the available substrate is the hull of a ship or the interior of a water pipe! In addition, a few taxa have adapted to an unattached, free-living mode of life. Not all bryozoans have calcified skeletons but the majority do. Some of the uncalcified ctenostomes bore into a calcareous substrate and thus can leave a potential fossil record as a trace fossil (Pohowsky, 1978).

Bryozoan colonies occur in many different forms. Schopf (1969) lists 18 different colonial or zooarial forms, based mainly on the cheilostomes. These zooarial forms or growth types can be grouped into four basic categories (Table 2.3): (1) erect, rigid; (2) erect, flexible; (3) encrusting; and (4) free living (unattached to a substrate). Brood (1972) described 11 different zooarial types among the cyclostome bryozoans that are similar to the cheilostome types. As is discussed below, these zooarial forms are sensitive to environmental conditions, certain forms being better adapted to a given set of environmental parameters.

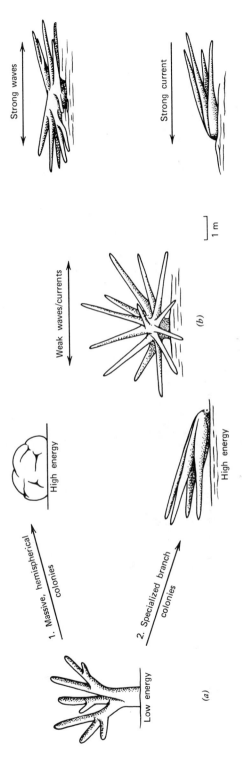

Figure 2.16 Adaptation of coral colonial form to differing conditions of current and wave energy. (*a*) Colonies that branch under low-energy conditions may either become more massive (1) or branch toward the current (2) under high-energy conditions. (*b*) In strong waves branches orient bidirectionally parallel to wave direction. They branch toward unidirectional currents. After Graus et al. (1977).

Table 2.3 Estimated association of bryozoan growth type with environmental parameters[a]

Growth Type	Substratum			Current (cm/sec)			Rate of Sedimentation (cm/10³ years)			
	Hard	Flexible	Particulate	Low	20 Moderate	100 High	Low	10 Moderate	100 High	1000 Very high
Erect, rigid										
Adeoniform	XX			X	XX		XX			
Eschariform	XX		X		XX		XX			
Reteporiform	XX			X	XX		XX			
Vinculariiform	XX			XX			XX			
Erect, flexible										
Catenicelliform	X	XX	X		X	XX		XX	X	
Cellariform	XX	X	X	X	XX	X	X	X	XX	
Flustriform	XX	XX	XX	XX	X	X	XX	X	X	

Encrusting

	1	2	3	4	5	6
Celleporiform	X	XX		XX	X	XX
Conescharelliform	XX	XX		XX	Not applicable	
Membraniporiform A	XX	X		XX	X	XX
Membraniporiform B	X	XX		XX		XX
Petraliiform	XX	X	X	XX		XX
Pseudovinculariform	XX	XX		XX		XX
Setoselliniform			XX	X		XX

Free living

	1	2	3	4	5	6
Lunulitiform			XX	XX	X	XX XX

[a]After Schopf (1969). X, occasional association; XX, frequent association.

Environmental Tolerances

Temperature

Bryozoans occur across a broad temperature range from the tropics to polar climates. Many species are stenothermal and thus can potentially be used as paleoclimatic indicators. Apparently no one has described any temperature effect on the morphology of any particular species.

Salinity

Bryozoans are found under salinity conditions ranging from fresh water to hypersaline lagoons. The freshwater forms belong to the class Phylactolaemata which is uncalcified and thus has no certain fossil record. Marine forms reach their maximum development at normal marine salinities. Their diversity drops considerably in brackish water. A very few taxa are able to withstand at least brief periods of salinity of 1‰ or less. At the other extreme, bryozoans have been described from water with a salinity of 49‰ or higher. In some species, specimens from brackish waters show a reduction of calcification and ornamentation such as spines. Some also show a reduction in size of individual within the colony (Schopf, 1969).

Depth

Bryozoans have been found at all depths in the oceans from intertidal to 8300 m (Ryland, 1970). They are most abundant and diverse on the continental shelf from 20 to 80 m. They are not common in deep water, probably because of the scarcity of hard substrate for attachment and of currents to bring suspended food to the colony. Although the depth distribution of the entire group is broad, individual species may be restricted to fairly narrow depth ranges.

In 1936 Stach first proposed that the form of the bryozoan colony varied with depth. This relationship has been confirmed and expanded by various workers since that time and has been used as a method in paleobathymetry (e.g., Labracherie and Prud'homme, 1966; Brood, 1972). Zooarial form is probably more a function of water turbulence and sedimentation than of depth per se, but as these factors, especially turbulence, commonly decrease with depth, zooarial form can be indirectly related to depth.

Schopf (1969) has summarized the relation of zooarial form to nature of the substrate, current strength, and sedimentation rate, the three factors that seem to be of greatest influence (Table 2.3). Schopf also includes a detailed description of the various zooarial types. In their classical study of bryozoans of the Rhone Delta area, Lagaaij and Gautier (1965) also include illustrations of ten of the most common types. As can be surmised from Table 2.3, the erect flexible types and certain encrusting types are most common in shallow water whereas erect rigid forms are best developed in deeper, quiet water. Cheetham (1971) and Rider and Cowan (1977) have made detailed studies of the adaptive advantages of the zooarial types.

Substrate

Most bryozoans require a hard substrate for attachment although many encrusting types actually occur on a flexible organic substrate. Several of the erect forms and some encrusting forms can also live in loose sediment by the development of rootlike rhizoids. The bryozoa are most diverse and abundant on areas of hard substrate decreasing in abundance as the sediment becomes progressively finer. Important exceptions to the need for some form of attachment are the lunulitiform types that live free on the sediment surface. They have a cap shape or broadly conical shape and live with the apex of the cone uppermost (Fig. 2.17). The larva attaches to a sand grain or some other small piece of hard substrate. The colony is equipped with many well-developed setae that are used to keep the colony free of sediment and positioned properly on the ocean floor (Lagaaij, 1963).

Most bryozoans are sensitive to suspended sediment in the water, hence are not common in areas of high sedimentation (Schopf, 1969). This is probably because their feeding mechanism is readily clogged by sediment. Their small size also makes them more easily buried then larger animals. Only a few taxa such as the free-living lunulitiforms are able to survive in areas of rapid sedimentation. Their setae apparently remove sediment as it accumulates. Erect, flexible types are also moderately well adapted to areas of high sedimentation because they can shake off sediment as it accumulates. Lagaaij and Gautier (1965) found that these types could survive near the areas of most rapid deposition in the Rhone Delta, but no bryozoans lived where deposition was greatest.

Turbulence

At least some current is beneficial to bryozoa by bringing suspended food to the colony. In fact, many bryozoans are adapted to life in very turbulent water.

Figure 2.17 Idealized lunulitiform zoarium adapted for life on soft substrate in areas of high sedimentation rates. The colony can lift itself above the substrate and shed falling sediment grains. After Rider and Cowan (1977), Lethaia, *10*.

Forms that firmly encrust a solid substrate and erect, flexible forms are best adapted to such areas of high turbulence. More delicately branched, rigid forms are found in areas of lower energy. Brood (1972) noted a number of morphological adaptations among the cyclostomes for strengthening the colony in turbulent water. He used the presence of these adaptations to help in his paleoecologic interpretation of upper Cretaceous and lower Cenozoic assemblages in Scandinavia. Forms which branch in one plane and erect fenestrate types are well adapted to areas of moderate unidirectional current (Fig. 2.18). They orient the flattened plane of the colony perpendicular to the current direction. Eddies or areas of reduced pressure produced as the current passes through the colony aid in the capture of food (Harmelin, 1975). In areas of low turbulence and low sedimentation a horizontally flattened or encrusting growth form is favored because it exposes a maximum surface area to food settling from the overlying water (Harmelin, 1975).

In the modern oceans bryozoans are a common component of the *cryptofauna* that lives in secluded, shaded recesses and caves, particularly in the reef environment. In some cases the growth form may be subtly modified in these

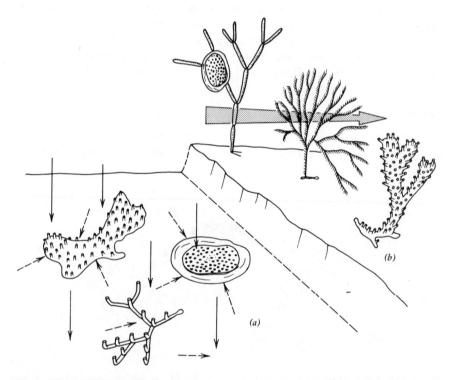

Figure 2.18 Zoarial forms of bryozoans utilizing food settling from above (*a*) and carried by currents (*b*). Large arrow shows current direction. Small arrows indicate food settling from above or being carried along the bottom by weak currents. After Harmelin (1975).

dark, low-turbulence localities. For example some species may have a fine, more delicate branching pattern in such localities than in more exposed areas (Harmelin, 1973).

BRACHIOPODS

Organism Characteristics

Although brachiopods are not uncommon in certain modern environments, they are relatively insignificant in comparison to their importance during the Paleozoic and Mesozoic. Their environmental distribution is now much more restricted than formerly. Thus, as emphasized by Ager (1967), caution must be used in extrapolating ecological interpretations from modern to the past. Or as in his delightful words, "The paleoecologist can not live by uniformitarianism alone." On the other hand, brachiopods are well suited for functional morphologic studies, and perhaps more work has been done on the functional morphology of fossil representatives of this group than any other. Also, a number of studies have related morphology of fossil brachiopods to environmental conditions as deduced from sedimentary and stratigraphic features of the rocks containing them. This section includes some of these interpretations as well as those strictly based only on modern specimens.

All modern brachiopods are sessile benthos, most being attached to the substrate by a pedicle (Rudwick, 1970). A few of the inarticulate brachiopods (e.g., *Crania*) are cemented to the substrate, and a few lie free on the substrate. A few species are adapted to life on a soft substrate by attachment with a rootlike pedicle. All brachiopods are epifaunal except for the linguloids, which are infaunal, living in permanent burrows. Ancient brachiopods were more diverse in terms of their relationships to the substrate. Many species were free living on soft substrate, and many others were cemented to a hard substrate. Some are thought to have been epipelagic, that is, attached to floating sea weed. Some species may have even had a weak swimming ability, much like the modern pectinid bivalves (Rudwick, 1970). Some of the articulates may have been semi-infaunal with only the valve margin exposed above the sediment surface. Many modern and fossil species are gregarious, being found in large concentrations at one spot and being completely lacking from environmentally similar nearby localities.

Brachiopods feed by filtering suspended plankton, organic detritus, and bacteria from the water with their lophophore. Currents are produced by the beating of cilia on the lophophore. These currents pass laterally into the mantle cavity where the food is extracted by the filaments on the lophophore and the water expelled through the center of the posterior margin. An important aspect of feeding in the articulate brachiopods is that they have no anus. Solid waste products must be expelled through the mouth, a system that is not efficient when large amounts of sediment are mixed with the food. Consequently, articulate brachiopods may supplement their diet by extracting dissolved organic compounds from the water of perhaps even living entirely on this food source (McCammon, 1969).

Environmental Tolerances

Temperature

The distribution of the brachiopods as a group is not closely controlled by temperature (Rudwick, 1970). They occur from the Arctic and Antarctic to the tropics. Modern articulates and calcareous inarticulates are most common in temperate latitudes. The linguloids are largely restricted to the tropics and subtropics (Fig. 2.19) but do occur elsewhere. These climatic distribution patterns probably have not been constant throughout brachiopod history. For example, articulate brachiopods appear to have been most common in shallow, tropical seas in the past.

Several morphologic features are correlated with temperature. The density of the punctae in articulate brachiopod shells increases with increasing temperature (Fig. 2.20), but the functional significance of this pattern is not clear (Foster, 1974). In addition, brachiopods living in cold water in the Antarctic region have thinner shells and fewer spicules in their tissue than do those from warmer areas. This may be related to the greater solubility of $CaCO_3$ and to the physiologic difficulty in secreting the shell in cold water.

Salinity

All modern brachiopods are marine; most live in near normal marine salinities of 30 to 40‰ (Rudwick, 1970). Some of the linguloids are exceptional in being able to tolerate reduced salinities (Emig et al., 1978). The deep burrowing adaptation of linguloids (Fig. 2.21) probably is an aid in escaping the effects of temporary reductions in salinity. The animal is able to retreat within its burrow to avoid contact with low-salinity water at the surface. Many examples

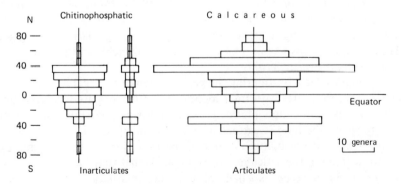

Figure 2.19 Latitudinal distribution of modern brachiopods. After Rudwick (1970).

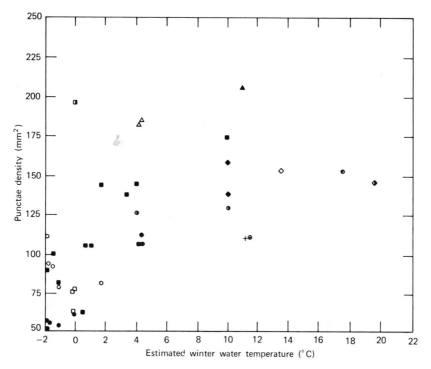

Figure 2.20 Variation of density of punctae vs. temperature in articulate brachiopods from the Antarctic region. After Foster (1974), *Antarctic Research Series, 21*, p. 20, copyright by American Geophysical Union.

from the fossil record suggest that this has always been the case (e.g., Ferguson, 1963).

Depth

Modern brachiopods occur from the lower intertidal zone to abyssal depths. They are most abundant at relatively shallow depth, on the continental shelf. A number of morphologic features have been related to depth, especially in fossils. Most of these features are basically controlled by substrate or by water turbulence, and some features that vary with depth may be controlled by the decreasing food supply with increasing depth (Fürsich and Hurst, 1974; fig. 2.22). The efficiency of the food-gathering system in the brachiopods is determined by the area of the lophophore relative to the volume of the organism. This is because the volume of the organism increases as the third power of linear dimension ($v \sim l^3$) and area only increases as the square of linear dimension ($a \sim l^2$). Increasing size decreases the surface-to-volume ratio reducing the relative proportion of lophophore surface area for feeding. Efficiency can thus be increased by decreasing the size of the organism. Deeper water taxa indeed tend to be smaller and thinner than shallow water types. Ager

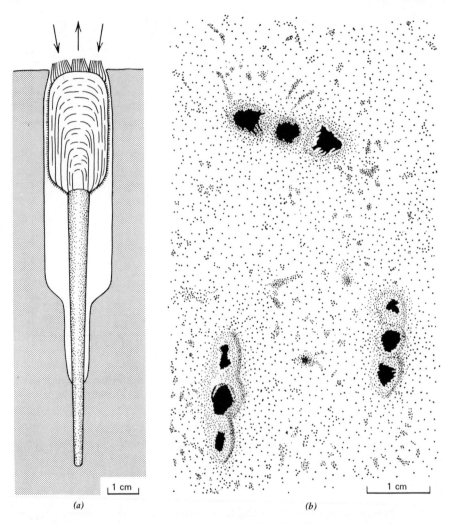

Figure 2.21 Modern *Lingula* in burrows. (*a*) Section of burrow; (*b*) burrows as seen from the surface. Arrows indicate incurrent and excurrent direction. After Rudwick (1970).

(1965) noted this same trend in deep water brachiopods from the Mesozoic. Another approach would be to increase the area of the lophophore surface in order to increase the number of current-producing cilia and food-trapping filaments. Unfortunately the lophophore is never preserved in fossils so its size can only be estimated indirectly. Two methods of estimating lophophore size are from the size and complexity of the lophophore supports (brachidia) and from the volume of the cavity between the valves, most of which is occupied by the lophophore and mantle cavity in modern forms. On this basis Fürsich and Hurst suggest the following order of increasing lophophore area for the

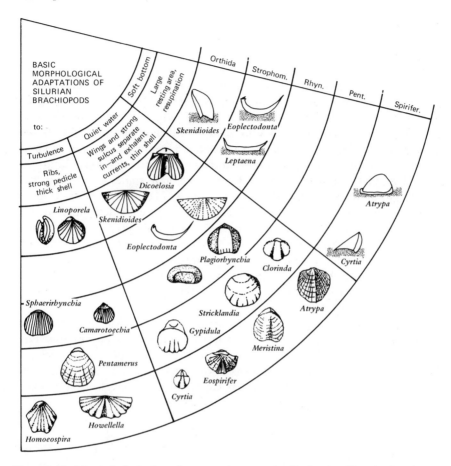

Figure 2.22 Morphological adaptations to environment in Silurian brachiopods. Strophom.—Strophomenida; Rhyn.—Rhynochonellida; Pent.—Pentameridina; Spirifer.—Spiriferida. After Fürsich and Hurst (1974).

Silurian brachiopod groups which they studied: orthids, strophomenids, rhynchonellids, pentamerids, and spiriferids. In general terms the environmental distribution of these groups supports the correlation between lophophore area and depth or water turbulence.

Substrate

Most modern and fossil brachiopods attach to a hard substrate either with the pedicle or by cementation. However, many fossil taxa apparently lived on unlithified sediment using a variety of adaptations of morphology and life habit. If the sediment is relatively firm, the animal may simply lie free on the surface (perhaps after an early stage of attachment to a small object with a pedicle) or may have a pedicle that splits into rootlets that provide attachment by penetrating into the sediment and attaching to many small grains. Brachiopods living on a soft, soupy substrate have a problem of keeping the shell from sinking into the substrate, burying the animal.

Thayer (1975a) indicated four categories of strategies that a brachiopod (or other organism) might use to keep from sinking into a soft substrate: (1) reduce the density by having a thin, smooth shell, (2) develop a broad, flat shape to distribute the body weight over a maximum area (the *snowshoe* strategy), (3) retain a small size into adulthood to reduce total mass, and (4) keep the commissure high so that feeding can continue even when the shell is largely buried (the *iceberg* strategy). Many brachiopods have developed thin, flat valves (Fig. 2.23*a*, Strategy 1). Some brachiopods have spines to help distribute their weight on the sediment (Fig. 2.23*b*, Strategy 2). Others have developed a large interarea to accomplish this (Fig. 2.23*c*). The extended hinge area or wings in brachiopods such as *Mucrospirifer* also provide support on a soft substrate (Fig. 2.23*d*). Small size (Strategy 3) is a common feature in brachiopods living on soft substrates. Another adaptation was the development of a concavo-convex shape (with the lower valve convex downward and the upper valve concave upward) to keep the commissure above the level of the sediment surface (Fig. 2.23*b*, Strategy 4).

Alexander (1975) noted many of these features in Ordovician specimens of *Rafinesquina alternata*. Specimens that lived in quiet water on soft substrates are small, broad, and flat, and have a concavo-convex shape. Some specimens also have a long hinge line. Specimens of the same species living in more turbulent water on a firmer substrate are larger, more globose with a biconvex

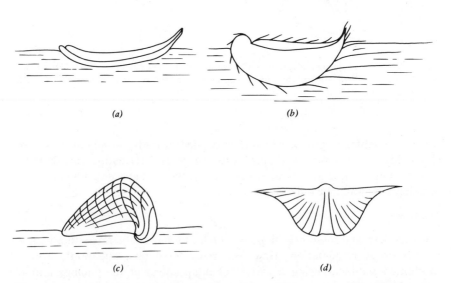

(a)   (b)

(c)   (d)

Figure 2.23 Morphological adaptations in brachiopods to prevent sinking into soft substrate. (*a*) Thin, flat, concavo-convex shape; (*b*) spines to increase surface area; also deep, cup-shaped lower valve; (*c*) large flat interarea; (*d*) extended hinge (mucronate).

shape. The shapes are so different that some earlier workers had called them different species; however, the two forms intergrade in some environments suggesting that the morphologic variation is environmentally induced (see Chapter 5 for a detailed discussion of this study).

Modern brachiopods are not adapted to life in areas with a high sedimentation rate, and are therefore most common in areas with a low to moderate sedimentation rate (Rudwick, 1970). This is perhaps because their digestive system cannot handle large amounts of ingested sediment. Sediment grains may have been a problem for fossil brachiopods also, for a number of features of brachiopod shells have been interpreted as adaptations to prevent sediment grains from entering the mantle cavity. The best known of these adaptations is the development of the ziz-zag commissure in all major articulate brachiopod groups apparently to allow increasing the open area when the valves were slightly opened without increasing the width of the gape that would allow larger sediment particles (and perhaps predators?) to enter (Fig. 2.24; Rudwick, 1964a). Other species developed spines to form a grillwork over the commissure or had a concentration of setae on the mantle margin (as indicated by numerous costellae on the shell) in order to form a baffle preventing the entrance of sediment grains and/or to act as a sensory device to detect the approach of sediment grains or predators (Fig. 2.25; Rudwick, 1968, 1970).

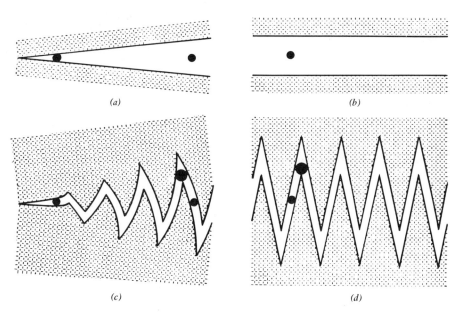

Figure 2.24 Zigzag commissure in brachiopods functioning to exclude sediment grains. (a) and (c) Commissure perpendicular to hinge axis. Open area between valves in (a) and (c) are the same. (b) and (d) Commissure parallel to hinge axis. Open area between valves in (b) and (d) are the same. After Rudwick (1964a).

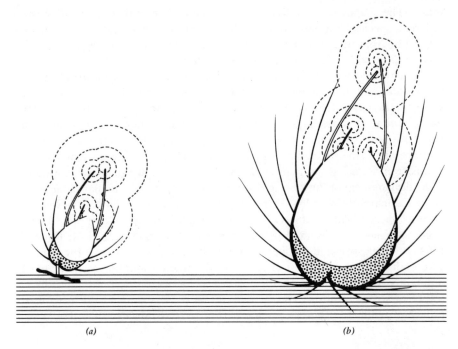

Figure 2.25 Diagrammatic sections of juvenile (*a*) and adult (*b*) specimens of *Acanthothiris* (Rhynchonellida, Jurassic) illustrating the postulated sensory function of the open anterior spines. After Rudwick (1968), Journal of Paleontology, Society of Economic Paleontologists and Mineralogists.

Turbulence

Modern brachiopods are largely confined to areas with at least moderate turbulence. This is undoubtedly in part related to the need for a hard substrate, which is more common in turbulent waters, and in part because turbulent waters generally contain more suspended food. Fossil species, however, were not as dependent on turbulent waters as are modern species, for taxa adapted to quiet-water conditions are common in the fossil record. The problem of securing food in quiet water for a brachiopod is much the same as that for a sponge and the adaptive solutions are somewhat similar. Some quiet-water brachiopod species have a well developed fold and sulcus. This functions to more efficiently separate incurrent food-bearing water from excurrent water, in effect increasing the angle of supply as in the quiet-water sponges (Fig. 2.26). A highly developed lophophore that increases current velocity would have the effect of increasing the diameter of supply.

Species adapted for turbulent water have developed methods of strong attachment to the substrate and strong heavy shells. The diameter of the pedicle

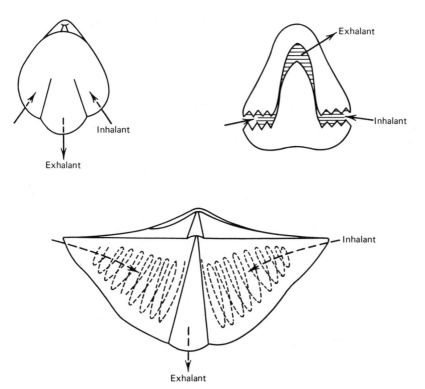

Figure 2.26 Postulated current pattern produced by brachiopods in relation to the fold and sulcus. After Ager (1963).

and thus the diameter of the pedicle opening is crudely proportional to the strength of attachment to the substrate and to turbulence. Thayer (1975b), on the basis of experimental studies with modern brachiopods, cautions that the correlation between the size of the pedicle opening and attachment strength is low. One reason for this is that some species use the pedicle not for relatively rigid attachment but as a tethering device which allows them to give with the current (Rudwick, 1970). In addition to the use of the pedicle for stability in turbulent settings other morphologic adaptations were also used. Some fossil productid species strengthened their attachment by especially modified spines cemented to the substrate (Fig. 2.27). Others increased stability by the sheer bulk of the shell, which was increased by shell thickening on the posterior portion of the ventral valve. The shells of species from turbulent environments are often thick or highly plicate to increase strength (Ager, 1965).

LaBarbera (1977) has shown that brachiopods in areas of unidirectional currents tend to align their anterior-posterior axis perpendicular to current direction because this is the most efficient feeding orientation.

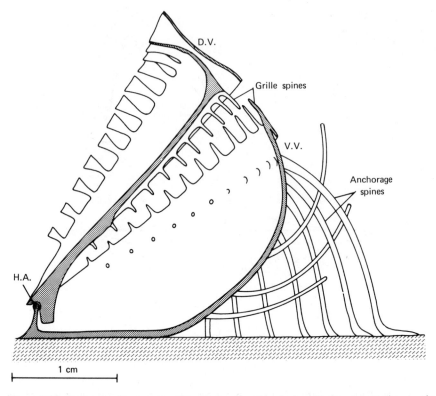

Figure 2.27 Spines in *Chonosteges* (Strophomenida, Permian) used for cementation to the substrate. D.V.—dorsal valve; V.V.—ventral valve; H.A.—hinge axis. After Rudwick (1970).

MOLLUSCS

Organism Characteristics

The molluscs are the most ecologically varied of the phyla with important fossil records. Each of the classes might well be treated separately because the ecological differences between some molluscan classes is greater than that between other phyla. The bivalves, gastropods, and cephalopods are particularly important classes of molluscs. The scaphopods, polyplacophorans, and monoplacophorans have smaller fossil records and are thus less important in paleoecology.

Bivalves

All bivalves are aquatic and practically all are benthonic; some are vagile and others, sessile. A few might be classified as nectonic (e.g., *Pecten* and *Lima*) although they are closely associated with the bottom and usually occupy a benthonic position. A few, such as *Leptopecten*, are also epiplanktonic, that is, they are attached to floating objects such as seaweed. The bivalves can be subdivided ecologically on the basis of locomotion and attachment method into seven groups (Fig. 2.28; Stanley, 1968): (1) *reclining*—lying immobile and unattached on the sea floor (*Gryphea, Exogyra*); (2) *burrowing*—actively moving through the sediment (*Mercenaria, Macoma*); (3) *boring*—boring into a hard substrate (*Pholas, Lithophaga*); (4) *bysally attached*—attached to the substrate by *byssal threads* (nylonlike organic strands) (*Mytilus, Pinna*); (5) *cemented*—attached to the substrate by secreted shell material (*Ostrea, Chama*); (6) *swimming*—nectonic animals moving through the water but usually reclining or bysally attached (*Pecten, Lima*); and (7) *nestling*—living within a preexisting cavity in a hard substrate (*Isognomon, Barbatia*). Some of these categories can be further subdivided (e.g., rapid and slow burrowers). The bivalve shell is distinctively adapted for these various modes of attachment and locomotion (Kauffman, 1969a), and as these various modes of life are best suited for differing environmental conditions, these conditions can in some cases be determined by the features of the bivalve shell (see below). Bivalves can also be subdivided into *infaunal* (living within the substrate), *semi-infaunal* (partially buried in the substrate), and *epifaunal* (on the substrate).

The bivalves are basically occupants of the second level of the food pyramid, living on suspended plankton and particulate organic material (*suspension feeders*) or fine particulate organic material deposited in the sediment (*deposit feeders*). The suspension feeders feed by passing water with suspended food through the mantle cavity, where the food is trapped by cilia on the gills and moved by the cilia to the mouth. Deposit feeders gather their food from the sediment with either specially adapted labial palps or elongate incurrent siphons (Fig. 2.28). A few are adapted for specialized food sources: *Tridacna* supplements its suspension feeding by farming masses of photosynthetic zooxanthellae in its mantle margin; *Teredo*, the ship worm, feeds on wood cellulose; and other genera have been described as parasitic and carnivorous (Barnes, 1974).

Gastropods

The gastropods are an extremely broadly adapted group (Barnes, 1974). Most are aquatic, being found in both fresh and marine water, but some are terrestrial. They can be subdivided ecologically into five groups on the basis of locomotion and attachment type (Fig. 2.29): (1) *crawling*—moving about on a soft or hard substrate including on other organisms (*Littorina, Nassarius, Acmaea*) (this group can be further subdivided on the basis of the substrate on which they occur); (2) *burrowing*—moving through the soft substrate (*Oliva, Polinices*); (3) *cemented*—attached to the substrate by secreted shell material (*Vermicularia*); (4) *foot attached*—attached for all or most of their life to one place, often on the shell of another organism (*Crepidula*); and (5) *planktonic*—suspended in the water (the pteropods, *Janthina*) (actually some are weak swimmers). As in the bivalves these groups also have adaptations for these modes of attachment-locomotion which can be used for environmental analyses (Linsley,

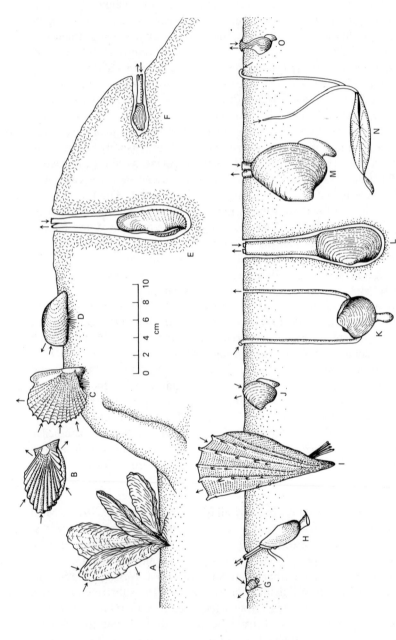

Figure 2.28 Modes of life of bivalves. A—Cemented (*Crassostrea*); B—swimmer (and recliner) (*Pecten*); C and D—attached by byssal threads (*Pinctada* and *Mytilus*); E—boring (*Pholas*); F—nestling in pre-existing cavity (*Hiatella*); G-Labial palp deposit feeder (*Nucula*); H—siphonate labial palp deposit feeder (*Yoldia*); I—nonsiphonate, byssal attached infaunal suspension feeder (*Pinna*); J—nonsiphonate suspension feeder (*Astarte*); K—mucus tube feeder (*Phacoides*); L and M—siphonate suspension feeders (*Mya* and *Mercenaria*); N—siphonate deposit feeder (*Tellina*); O—siphonate carnivore (*Cuspidaria*). After Stanley (1968), Journal of Paleontology, Society of Economic Paleontologists and Mineralogists.

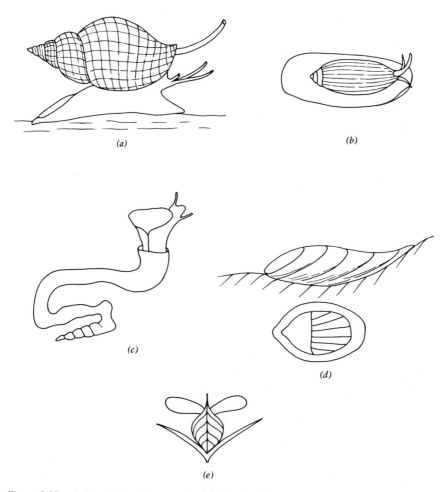

Figure 2.29 Modes of life of gastropods. (*a*) Crawling (*Buccinum*); (*b*) burrowing (*Oliva*); (*c*) cemented (*Vermicularia*); (*d*) foot attached (*Crepidula*); (*e*) planktonic (pteropod).

1978b). However the adapations in gastropods have not been as extensively studied as the bivalves. Benthonic gastropods are mostly epifaunal but a few are infaunal.

Gastropods are more varied in terms of trophic adaptations than are the bivalves. In fact Purchon (1968) recognizes 17 different feeding methods among marine, freshwater, and terrestrial gastropods (Table 2.4). Many genera are at the primary consumer level of the trophic structure, but many others are carnivores. Some gastropods feed on suspended plankton and particulate organic material by producing currents with cilia to circulate water through the mantle cavity where the food is trapped on the gills, or by producing mucus threads and nets to entrap suspended food. Several genera feed on de-

Table 2.4   Feeding methods in the gastropods[a]

Marine	Fresh Water
Herbivorous	Herbivorous
Algae	Algae
Browsers and grazers	Browsers and grazers
Raspers of rock surfaces	Raspers of rock surfaces
Suckers of cell contents	Cutters of fronds
Cutters of fronds	Particles
Particles	Collectors of organic deposits
Collectors of organic deposits	Collectors of plankton and detritus
Collectors of plankton and detritus	Carnivorous
Carnivorous	Benthonic hunters
Feeders on colonial and sedentary animals	Land
Feeders on sea anemones	Herbivorous
Feeders on fish eggs	Raspers of plant tissues
Benthonic hunters	Cutters of plant tissues
Planktonic hunters	Carnivorous
Scavengers	Predaceous carnivores
Ectoparasites	
Endoparasites	

[a] After Purchon (1968).

posited organic material on or in the sediment. Others are scavengers feeding on larger dead organisms. Many genera are herbivores feeding on algae or higher plants by rasping away the surface of the plant with their radula. Other herbivorous gastropods feed on algae and other organic material encrusted on rock or other hard substrate by scraping the surface with a hardened radula, in the process removing substrate as well as food. These gastropods may thus be important agents of erosion (Warme, 1975). Carnivorous snails feed in a number of ways. Some have a modified radula for drilling holes in bivalve or other gastropod shells. Others feed on bivalves by forcing apart the valves or chipping their edges. The most spectacular carnivores are the cone shells (*Conus*) which capture their food by harpooning it with poison darts. Cone shells may even eat small fish captured in this way. A few gastropod genera are parasites.

Cephalopods

Modern cephalopods, especially forms with an external skeleton, are greatly reduced in number and diversity relative to the geologic past. Compared to the some 10,000 species of fossil cephalopods only a single shell-bearing genus (*Nautilus*) with three species remains (Sweet, 1964). The squids and the octopi are considerably more abundant than the shell-bearing forms, but their chances of leaving a fossil record are remote. (A few genera such as *Spirula* and *Sepia* have an internal shell.) Because of this decreased abundance and diversity, the application of the uniformitarian approach is difficult. A single living genus cannot possibly represent the diversity of ecological interactions that affect the fossil forms. Relationships found in the squids and octopi may be helpful, but the question remains as to how comparable their interactions are to other groups of cephalopods. Because the uniformitarian approach is difficult, the fossil cephalopods have been extensively studied in terms of their functional morphology (see Chapter 5).

Nautilus and the squids are strong swimmers (Stenzel, 1964). The octopods are largely benthonic although they are capable of at least short bursts of swimming. The morphology of fossil cephalopods suggests that most were nectonic although many were almost certainly benthonic and a few possibly planktonic. The function of the chambered shell in *Nautilus* and the fossil nautiloids and ammonites has long been the subject of discussion and controversy. The chambers function to regulate buoyancy (Denton and Gilpin-Brown, 1973). As new chambers are formed they are osmotically pumped free of most of the trapped water and filled with gas. This counterbalances the weight of the shell and tissue so that the total animal has a density just slightly greater than sea water. The idea that the chambers are filled and emptied on a short-term basis to control buoyancy as in a submarine does not seem to be valid. *Nautilus* moves both vertically and horizontally by swimming, largely accomplished by jet propulsion through ejecting a stream of water from a constricted tube, the hyponome. The squids swim by the same mechanism.

Practically all cephalopods are carnivores preying on a wide variety of invertebrates and fish (Barnes, 1974). Indeed the cephalopods may have been one of the most important carnivorous groups in the ancient seas. The cephalopods capture prey with their tentacles and transfer it to the mouth where it is masticated by a horny beak and a modified radula. Although *Nautilus* is a capable swimmer, it depends on food (largely crustacians) that is associated with the sea floor, and this may well have been the case with many fossil cephalopods. *Nautilus* and some squids may at times act as scavengers, eating dead flesh (Stenzel, 1964). The octopods are carnivorous on benthonic forms, especially bivalves and gastropods. They may crush the shells or in some cases are adapted to drill holes in the shells. A few deep water octopods appear to be adapted to deposit feeding (Barnes, 1974).

Minor Groups

The living molluscan classes Monoplacophora, Polyplacophora, and Scaphopoda each have minor fossil records. The monoplacophorans are perhaps the ultimate in living fossils, the youngest known fossil monoplacophoran being of Devonian age. This group was thought to be extinct until the genus *Neopilina* was discovered in 1952 living in deep water (Lemche and Wingstraud, 1959). This limpetlike mollusc has many ancestral characteristics of the phylum and may belong to the most primitive molluscan group. The ecology of modern monoplacophorans is poorly known, but they appear to be deposit feeders living on the deep sea floor. The ecology of this group may have changed considerably since Devonian times so the uniformitarian approach should be used with caution.

The polyplacophorans or chitons are moderately abundant and diverse (about 600 species) in modern seas but have a sparse fossil record (Barnes, 1974). This is in part due to their adaptation to clinging to a hard substrate, often in turbulent waters. Such turbulent, hard substrate environments are not as common in the geologic record as are soft substrates. Also, because of the segmented nature of the chiton skeleton, it is almost invariably disarticulated after death. The isolated chiton plates may be more common than suggested by published occurrences. Chitons are herbivorous, feeding largely on algae growing on rocks. They use highly developed, hardened radular teeth to accomplish this scraping (Lowenstam, 1962).

The scaphopods or tusk shells burrow into the soft sediment, usually a sandy surface. They feed on foraminifera and other microorganisms in the sediment by searching and capturing it with tentacle-like *captaculae* (Gainey, 1972). Their ecology and mode of life is much like the deposit feeding bivalves (Barnes, 1974).

Environmental Tolerances

Temperature

All the molluscan classes have broad temperature ranges although smaller taxonomic groups may be much more restricted. The limited temperature ranges of molluscan genera and species have been the basis either totally or in part of many paleoclimatic studies, particularly of the Neogene (e.g., Durham, 1950; Baden-Powell, 1955; and Valentine, 1961). Nicol (1964 and 1967) and others have pointed out certain morphological trends within the molluscs, especially the bivalves, with temperature. Arctic forms are usually small (less than about 1 cm with some exceptions), simple, have little or no ornamentation (especially spines), are usually thin shelled, and have prominent growth lines. Tropical forms may be quite large, ornate, and heavy shelled. This trend may be due to gradients in the saturation or supersaturation of sea water with respect to $CaCO_3$. Because cold water is undersaturated the mollusc must concentrate $CaCO_3$ in order to form its shell (Vermeij, 1978). This involves the expenditure of more energy than the formation of a shell in tropical waters supersaturated with $CaCO_3$.

In contrast to this overall trend in size, some molluscan species as well as species from other groups have their largest individuals at the colder end of their range. This has been interpreted as due to later maturing of individuals at colder temperatures. Delayed maturing allows more time for growth despite the fact that the growth rate is probably slower at lower temperatures. Strauch (1968) noted a regular trend of increasing size with decreasing temperature in the bivalve species *Hiatella arctica*. Specimens from the tropics had an average shell length of about 6 to 7 mm whereas specimens from Greenland averaged some 45 mm in length (Fig. 2.30). Specimens of *Hiatella arctica* from western and central Europe, ranging in age from Oligocene to Recent, show a general trend toward increasing size with decreasing geologic age (Fig. 2.31). Strauch interprets this trend as indicating a gradual decrease in temperature during the Cenozoic in Europe.

The maximum size of the larval shell of many bivalve species also varies inversely with temperature (Fig. 2.32). This is apparently due to delays of metamorphosis to the adult stage in specimens growing at low temperature in a manner analogous to the delay in sexual maturation in low-temperature areas. The size of the larval shell can be measured by determining the location of the prodissoconch-dissoconch boundary in the umbonal area of the adult shell (Lutz and Jablonski, 1978).

The number of species cemented to the substrate increases in going from the Arctic to the tropics (Nicol, 1967). This may also be a result of changing degree of saturation of the water with respect to $CaCO_3$. Although color patterns are not often preserved in fossil molluscans, they may be temperature indicators. In general, tropical species are more brightly and elaborately colored than high-latitude forms. The proportion of septibranch plus protobranch bivalves to polysyringian bivalves is higher in cold than warm waters (Nicol, 1964). The proportion of infaunal bivalves relative to epifaunal types is in general higher in cold than in warm waters (Thorsen, 1957). The data do not conclusively support this, however, and factors other than temperature (e.g., substrate texture) are most likely to be important in controlling the proportion of infaunal bivalves. Many of the relationships between morphology or size and temperature noted for the bivalves also apply to gastropods, although they have not been as extensively studied. Dudley and Vermeij (1978) noted that the proportion of drilling gastropods appears to increase with increasing temperature. They noted that the percentage of shells of the gastropod *Turritella* which had been drilled increases from the temperature zone into the tropics. A number of other trends in molluscan morphology with temperature have been documented by Vermeij (1978).

Salinity

The salinity ranges for bivalves and gastropods as groups are very broad. They are found in freshwater to hypersaline environments, but they are most diverse

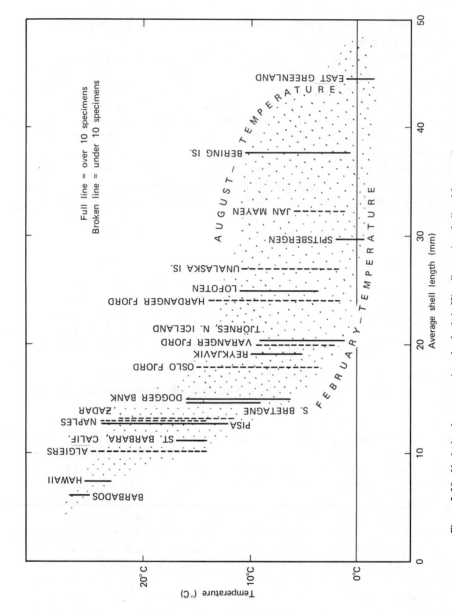

Figure 2.30 Variation in average length of adult *Hiatella arctica* shells with annual temperature range. After Strauch (1968).

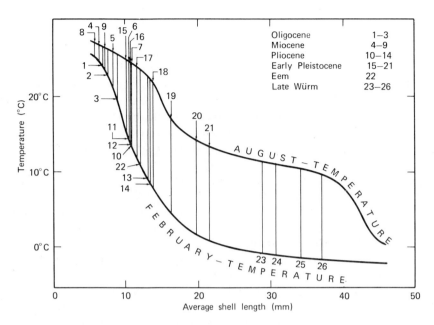

Figure 2.31 Average shell length of fossil specimens of *Hiatella arctica* plotted on temperature curves for modern specimens. After Strauch (1968).

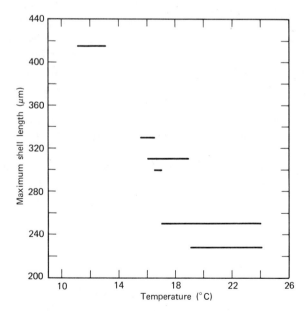

Figure 2.32 Variation in maximum larval shell length of *Mya arenaria* with environmental or culture water temperature. After Lutz and Jablonski (1978), copyright 1978 by the American Association for the Advancement of Science, Science *202*: 51-53, text-figure 1.

in full-marine environments. The cephalopods are restricted to near-normal salinity conditions (Purchon, 1968). Modern cephalopods require well circulated, oxygenated water, perhaps because they are very active as compared to other molluscs and their higher rate of metabolism requires abundant oxygen. The distribution of fossil cephalopods suggests thay they have always been restricted to full-marine environments (Sweet, 1964).

Although the bivalves and gastropods occur in the entire salinity range, individual superfamilies and genera are much more restricted in their distribution. For example among the bivalves the Unionacea and the Sphaeriacea are entirely freshwater forms. Freshwater gastropods are distributed among several taxonomic groups, but many are pulmonates (Purchon, 1968). Certain groups particularly common in brackish waters, such as the Ostreacea and Mytilacea among the bivalves and the littorinids among the gastropods, are by no means confined to brackish waters, however, but have adapted to this transitional setting from the marine environment.

Bivalves (e.g., *Mytilus, Mya, Cardium,* and *Tellina*) and gastropods (e.g., *Neritina*) living in brackish waters are often smaller than members of the same species living in more nearly full-marine conditions (Segerstråle, 1957). However, some species such as the modern *Rangia cuneata* and *Crassostrea virginica,* which are adapted to brackish conditions, may actually reach their maximum size at reduced salinities. Brackish forms may also have thinner shells which are higher in organic matrix than full-marine representatives of the species. This may be a reflection of the undersaturation of brackish waters with respect to $CaCO_3$. Eisma (1965) made a detailed study of the apparent effect of salinity on the morphology of the bivalve *Cardium edule* largely from the Dutch coast. At reduced salinities, this species grows more slowly, reaches a smaller maximum size, has a shorter ligament, and has fewer ribs than at full-marine salinities. As had been noted by previous workers, the direct correlation between number of ribs and salinity is especially strong (Fig. 2.33). The functional basis for this correlation is not certain but may result from the $CaCO_3$ undersaturation of brackish water, because more $CaCO_3$ is needed to form the additional ribs as well as to produce faster-growing, larger valves. The *Treatise on Marine Ecology and Paleoecology* contains much valuable information on the effects of salinity on the molluscan fauna. Of special interest in this regard are the chapters by Pearse and Gunter (1957), Emery et al. (1957), and Segerstråle (1957).

Depth

The molluscs as a group have a broad range of depth tolerance from intertidal to hadal. The bivalves and the gastropods are among the most abundant groups of organisms found in the oceanic trenches (Wolff, 1970). McAlester and Rhoads (1967) have noted that the depth of burrowing in infaunal bivalves is correlated with depth. Deep burrowing genera such as *Dosinia, Mya, Solen* and *Thracia* are largely restricted to shallow water environments. Shallow burrowers may also be found in shallow water, but practically all deeper water

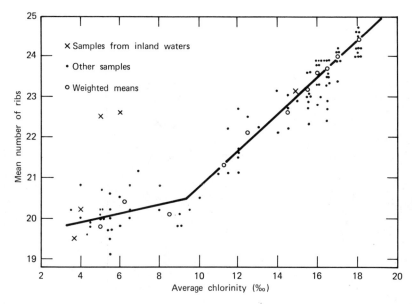

Figure 2.33 Variation in the average number of ribs in *Cardium edule* with the average chlorinity for samples from the Dutch coastal waters. After Eisma (1965).

bivalves are shallow burrowers (Fig. 2.34). Burrow depth probably is determined by environmental variability, which is clearly related to water depth. Thus shallow water bivalves burrow deeply to avoid the large fluctuations in environmental parameters such as temperature and salinity which are characteristic of shallow, and especially of intertidal, habitats. This relationship of burrowing depth to water depth is also preserved in the trace fossils (see Chapter 6). Deposit feeding bivalves are relatively more abundant than suspension feeders in deep water. Thus the deposit feeding protobranch and septibranch bivalves are most common in deep waters. Some of the same characteristics found in cold water bivalves such as small size, thin shells, and lack of ornamentation are also found in deep water forms. Because deep water is relatively cold, the cause of the characteristics is difficult to pin down. Probably the undersaturated condition of the deep water with respect to $CaCO_3$, due to high pressure and low temperature is responsible for these morphological characteristics. An abundance of pelagic gastropods (pteropods) may be an indicator of deep water just as is an abundance of planktonic foraminifera. Because of their aragonitic composition, pteropods are not as likely to be preserved as are the calcitic planktonic foraminifera, particularly in very deep water, which may be undersaturated with respect to aragonite (Herman, 1978). Some pteropod species have distinct depth ranges so that the relative proportions of these species can potentially be used to determine paleobathymetry (Herman and Rosenberg, 1969).

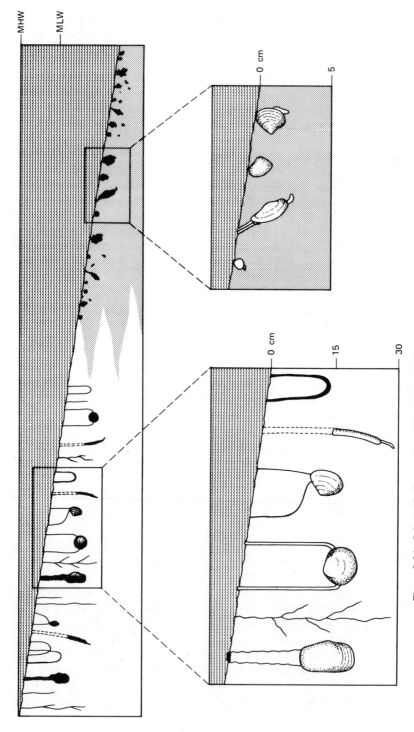

Figure 2.34 Schematic illustration of the change in bivalve burrowing pattern with water depth. Shallow water species are adapted for deep, vertical burrowing. Deeper water species are adapted for near surface burrowing. After McAlester and Rhoads (1967).

Substrate

The morphology of the bivalve shell, and perhaps to a somewhat lesser extent that of the gastropod shell, is strongly influenced by substrate, water turbulence, food supply, and, to a lesser extent, predation (Stanley, 1970). Although some extinct groups were benthonic and thus probably affected by substrate type, most cephalopods were pelagic and thus independent of substrate control. Substrate type, water turbulence, food type, and depth are interrelated and should be discussed together. For example, deposit-feeding molluscs are most common on substrates with an abundance of deposited food, which are at localities where water turbulence is low and therefore where the sediment is fine grained. Such conditions are commonly (but certainly not invariably) found in deep water.

Comprehensive discussions of the morphology of molluscs, expecially bivalves, are numerous. Papers by Stanley (1970, 1972, and others), Kauffman (1969a), and Thomas (1975 and 1978) contain extensive discussions of the relationship between bivalve morphology and substrate and other environmental parameters. Linsley (1978a and b) and Vermeij (1978) discuss the relationship between gastropod morphology and mode of life.

Cemented and byssate attached bivalves (which are invariably suspension feeders) usually require a hard substrate. They are more common in areas with turbulent water and usually a high proportion of suspended relative to deposited food, but may in some cases be found in soft substrates. In these cases they may have attached in the larval stage to sediment grains or to another shell in the soft substrate, or they may form a mat of interlacing byssal threads in the sediment (e.g., *Modiolus*). These are *endobyssate* (i.e., buried or partially buried in the substrate) and can be distinguished from *epibyssate* (i.e., attached to a hard substrate above the sediment/water interface) bivalves by a number of shell characteristics such as a broad cross sectional outline near the ventral margin (Fig. 2.35; Stanley, 1972). This shape allows a more firm, stable attachment to the substrate. Boring bivalves also obviously require a hard substrate whereas burrowers need a soft substrate.

The form of the bivalve shell can tell a great deal about the nature of the substrate and the relationship of the animal to the substrate (Stanley, 1970 and 1972). Deposit-feeding bivalves have developed certain characteristic features that adapt them to their particular mode of life (Fig. 2.28). They are usually thin shelled, streamlined, rapid-burrowing forms so that they can more easily move from place to place as the food is depleted. They are most common in low-energy, fine-grained sediment with an abundance of organic material. Such substrates tend to be soft and soupy, so the light, thin, small shells are advantageous in allowing the bivalve to maintain its position near the sediment surface (Thayer, 1975a). Tables 2.5 and 2.6, based on data from Stanley (1970), summarize the relationship of shell morphology to locomotion-attachment type, substrate, and water turbulence.

Gastropods occur on substrates of all types but are most common on hard substrates. In fact a high ratio of gastropods to bivalves is usually an indica-

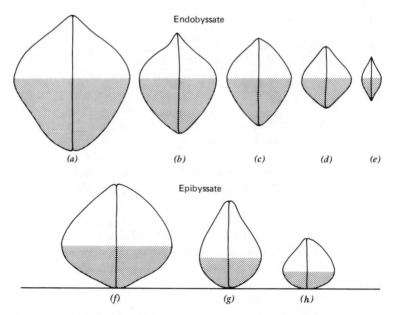

Figure 2.35 Comparison of cross-sectional shapes of endobyssate and epibyssate modern mytilid bivalves. (a) *Modiolus modiolus;* (b) *M. americanus;* (c) *M. demissus;* (d) *Brachidontes citrinus;* (e) *Amygdolum prelex;* (f) *Mytilus edulis;* (g) *Brachidontes recurvus;* (h) *B. exustus.* Shaded below point of maximum width. After Stanley (1972), Journal of Paleontology, Society of Economic Paleontologists and Mineralogists.

tion of hard substrate. The hard substrate may be algae or other plants that are not preserved, so that abundant gastropods in the absence of other evidence for hard substrate could indicate the presence of unpreserved plants. Some gastropods are adapted to burrowing in a soft substrate. Perhaps the clearest examples are the olive shells, *Oliva* and *Olivella,* which have smooth, stream-lined shells for movement through the sediment. Gastropods are not common on an unstable, shifting sand substrate and the few species which do occur there have large, heavy shells for stability. Gastropods are also uncommon on very soft, soupy substrates where they have difficulty in maintaining themselves on the surface. Table 2.7 shows the relationship of gastropod shell morphology to locomotion-attachment type and substrate.

Turbulence

Molluscs are adapted to conditions of water turbulence ranging from the extremely turbulent intertidal zone on open coasts, where they are subject to great wave action, to the essentially motionless water at depth or in bays or lagoons. The relationship between molluscan shell form and turbulence is summarized in Tables 2.5 to 2.7. Turbulent water is characterized either by hard substrate (favoring cemented, nestling, boring, or byssate filter feeders) or by an unstable shifting substrate (favoring rapidly burrowing filter feeders).

Table 2.5 Relationship of bivalve morphology to environmental conditions[a]

Locomotion— Attachment Type	Water Turbulence	Substrate	Diagnostic Shell Characteristics
Reclining	Stable bottom, often low energy	Soft	Inequivalve; bottom valve usually convex; upper valve flat; thick shells, some broad, flat shells *Placuna, Gryphaea, Exogyra, Inoceramus*
Burrowing	See subdivisions (Table 2.6)	Soft	Streamlined shell; circular to wedge shaped valve; isomyarian
Byssally attached	High to moderate energy. Relatively stable substrate	Hard or soft	Anisomyarian; maximum shell width near ventral margin; equivalved forms live with commissure oriented vertically; inequivalved with commissure horizontal. Shell elongate; byssal sinus or gape; auricle ("wing") *Mytilus, Modiolus, Anomia, Pteria, Tridacna*
Boring	Usually moderately high. But wherever hard substrate is present	Hard	Cylindrical in cross section; short, prominent spines on anterior; thin shell; large siphonal and pedal gapes *Pholas, Petricola, Zirfaea, Lithophaga*
Nestling	Moderate to high	Hard	Distorted shells; lack ornamentation *Petricola, Barbatia*
Cemented	Usually moderately high. But wherever hard substrate is present	Hard	Spines; thick, heavy shell; irregular shape; inequivalve; often anisomyarian *Ostrea, Spondylus, Chama,* rudists
Swimming	Low to moderate. Stable bottom. These forms recline or have byssal attachment when not swimming		No full-time swimmers; thin, corrugated shells; auricles; monomyarian; efficient swimmers have symmetrical auricles and a large umbonal angle; water expulsion gapes *Pecten, Lima*

[a]Mainly based on data from Stanley (1970).

Most molluscs occur in settings of intermediate turbulence. Life in high-energy settings requires special adaptation present in only a few groups. The Arcoida, for example, are particularly suited for turbulent environments because they can either resist being dislodged by their strong byssal attachment or they can rebury themselves quickly when uncovered by current scour because of their ability to burrow (Thomas, 1978). In general, relatively thick-shelled, rapid-

Table 2.6 Relationship of the morphology of the burrowing bivalves to environmental conditions[a]

Rapid burrower (shifting coarse substrate, turbulent waters, and deposit feeders on fine substrates)	Small size, slender (streamlined) cross section; cylindrical, bladelike, disklike; thin shell; little ornamentation; smooth surface; expanded shell anterior. Pointed posterior; maximum width near dorsal margin. Some thick shelled rapid burrowers in turbulent environments. Deposit feeders thin shelled; compressed in cross section. Smooth surface *Donax, Ensis, Tivela, Macoma, Tellina, Nucula*
Slow burrower (relatively stable substrates, intermediate turbulence)	Spherical outline (poor streamlining); ornamentation (radial, concentric, divaricate); rugose. Lack of features under rapid burrowers *Cardium, Anadara, Mya*
Deep, slow burrower (shallow water, stable substrate to maintain open burrow)	Thin shell; no ornamentation; large pallial sinus; may have siphonal gape; chondrophore *Mya, Panopea, Thracia*
Shallow burrower (soft, stable substrates)	Thick shell, ornamented, small or no pallial sinus *Trachycardium*

[a]Mainly based on data from Stanley (1970).

burrowing forms such as *Donax* and *Tivela* are well adapted to life in a turbulent environment with a shifting substrate.

ARTHROPODS

Organism Characteristics

Arthropods are the most diverse phylum of living organisms, but their fossil record is poor except for a few groups. The trilobites comprise one of the most abundant groups of fossil arthropods, but because they have no close living relative, little can be said about their ecology from the taxonomic uniformitarian approach. In this section we emphasize the ostracodes, the only arthropod with a fairly large fossil record and numerous modern representatives.

Table 2.7 Relationship of gastropod morphology to environmental conditions

Locomotion— Attachment Form	Substrate—Water Turbulence	Shell Characteristics
Crawling on sediment surface	Soft, stable substrate; gastropods are uncommon on unstable substrates (high turbulence) or very soft substrates (low turbulence)	"Typical" gastropod; wide range of morphologies. Usually coiled. May be highly ornamented *Murex, Strombus, Astrea*, etc.
Burrowing	Soft, stable substrate; moderate energy	Smooth, streamlined shape; no ornamentation *Oliva, Olivella, Polinices*
Cemented	Hard substrate, relatively high energy	Irregular, uncoiled or irregularly coiled except in juvenile portion *Vermicularia, Aletes*
Foot-attached	Hard substrate, often on other shells; variable turbulence conditions	Thin, flat, uncoiled shell; variable shape (conforms to substrate) *Crepidula*
Planktonic	Open water. Not controlled by substrate	Thin, small, reduced shell; some coiled; some bilaterally symmetrical Pteropods
Crawling on hard substrate	Hard substrate. Usually high energy. May be organic substrate in lower energy setting	Cap shaped (limpets) or coiled with varied morphology *Acmaea, Littorina, Nerita*

Several other groups are locally common as fossils, and other groups, especially among the crustaceans, have undoubtedly played an important role in the ecology of the past although their fossil record is sparse because of the lack of a highly mineralized skeleton. For example, in the modern seas the copepods are one of the most important organisms in the trophic structure of many communities, where they are an important primary consumer, utilizing diatoms, dinoflagellates and other phytoplankton. They are in turn the most important food source of many secondary consumers such as small fish. However, the fossil record of copepods is practically nonexistent despite the fact that they must have been abundant during the Cenozoic and perhaps much earlier as well.

The malacostracans (including crabs, shrimp, amphipods, isopods) have a modest fossil record, but by analogy with modern communities they also must have been much more common than their fossil record would indicate. When conditions are right, some malacostracans, especially crabs and lobsters, may be well preserved. The claws are most common as fossils because they are more heavily calcified than the rest of the exoskeleton. The malacostracans were probably important scavengers and carnivores as well as deposit and filter feeders in ancient communities. If we include trace fossils that were probably produced by malacostracans, their fossil record may not be so poor.

Branchiopods (especially conchostracans and cladocerans), some of which do have mineralized skeltons, have an important fossil record in fresh and brackish water sediments. Most branchiopods swim although several are benthonic and a few are planktonic (Barnes, 1974). They are largely filter feeders, sweeping plankton from the water with their appendages. Some are deposit feeders, particularly among the conchostracans, and a few are predaceous.

Trilobites and eurypterids have good fossil records but are now extinct. The trilobites have been the subject of a number of studies based largely on functional morphology (see Chapter 5). In terms of morphology and perhaps ecology, the nearest living analogue to the trilobite is the xiphosuran *Limulus*, the horse shoe crab. *Limulus* lives in shallow water on either mud or sand and feeds on burrowing invertebrates which it obtains by plowing through the sediment. Trilobites also appear to have been largely benthonic, crawling on or in the sediment, and for the most part were probably deposit feeders and scavengers. Some species were apparently swimmers, and a few may have been planktonic. These were probably suspension feeders. From their geologic occurrence and association with other fossils the trilobites appear to have been entirely marine.

The barnacles are sessile suspension feeders and parasites (Barnes, 1974). The parasites are highly modified and have no hard parts, hence no fossil record. Fossilized barnacles include three basic types: (1) forms that attach directly to the substrate (acorn barnacles such as *Balanus*), (2) forms having a fleshy stalk (goose barnacles or pedunculate barnacles such as *Lepas*), (3) forms that bore into a calcareous substrate (acrathoracian barnacles such as *Trypetea* and the boring goose barnacle *Lithotrya* discussed in Chapter 6). All of these forms are attached to a hard substrate and feed by sweeping their highly modified, featherlike appendages through the water, trapping plankton, suspended detritus, and in some cases small crustaceans such as copepods.

Eurypterids probably crawled along the bottom and some may also have been active swimmers. Their feeding mechanism is uncertain but some were probably carnivorous. Their geologic occurrence suggests that eurypterids ranged from marine into brackish and perhaps even freshwater environments.

Insects are only rarely preserved, but a few spectacular exceptions have been described (Carpenter, 1953). However, relatively little paleoecologic work has been done on these fossils.

Ostracodes are proportionately better represented in the fossil record than any other living arthropod group (Pokorny, 1978). They are largely a benthonic group although some are weak swimmers and a few fairly good swimmers. A few species have adapted to terrestrial life. Benthonic forms live on the sediment surface or on vegetation growing on the bottom. Some species belong to the *interstitial* fauna or *meiofauna*, that is, they live in the spaces between sediment grains. The ostracodes include deposit feeders filter feeders, scavengers, herbivores, and even a few carnivores. Diatoms, bacteria, protozoans, detritus, plants, and small fish are among the types of food utilized by ostracodes. They feed by creating currents and handling food directly with their variously modified appendages.

The remainder of the section on arthropods concerns primarily the ostracodes because they are the only living arthropods with a good fossil record and thus offer potential for the taxonomic uniformitarian approach.

Environmental Tolerances

Temperature

Ostracodes as well as the other major marine arthropod groups occur at all normal oceanic temperatures from the equator to high latitudes. Considerable work has been done on the temperature distribution of individual modern species (Neale, 1969) so temperature interpretations might be made on Cenozoic assemblages containing these species.

Salinity

Ostracodes are found over a broad salinity range, being common in marine, brackish, and fresh water as well as in hypersaline waters of 80‰ or higher (Benson, 1961). Specific taxa have limited tolerances, being adapted for fresh water (salinity to about 2‰), brackish waters (2–20‰), or normal marine waters (down to about 20‰). Modern freshwater ostracodes are cypridids with the exception of one cytherid genus. All have smooth, simple shells. The proportion of ornamented, ornate shells increases with salinity; however, specimens of the euryhaline genus *Cyprideis* tend to develop nodes or bulges in low-salinity environments (Sandburg, 1964). Ostracodes are the most abundant microfossil or potential microfossil in very brackish water. Benson (1961) has suggested that the ratio of ostracodes to foraminifera might be used as an index of salinity, the ratio going up as salinity goes down. In general, as in the case of most groups of organisms, ostracode diversity decreases with decreasing salinity, but abundance may be very high in low-salinity water. Ostracodes have been described from hypersaline lagoons with salinities up to 80‰.

Depth

Ostracodes live at all depths but are most common in relatively shallow water. Several features of the ostracode valve vary with depth. The size and the amount of reticulation increase with depth (Fig. 2.36) but thickness or massiveness usually decreases. Spinosity, which is probably a defense mechanism, generally increases with depth, probably because of the more intensive cropping by predators at depth (Benson, 1975). Many species of ostracodes have eye tubercules on their valves. The proportion of such species decreases and the

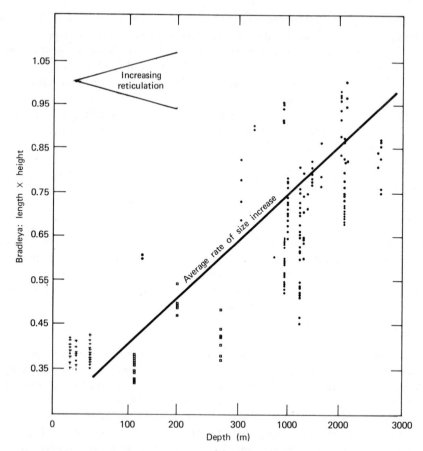

Figure 2.36 Variation in average size (length × height) with depth in species of the ostracode *Bradleya*. After Benson (1975), from Bulletins of American Paleontology, *65*, (282), text-figure 10.

size of the eye tubercules decreases as water depth increases and light intensity drops (Fig. 2.37). Essentially all ostracodes living below a depth of some 800 m are blind (Benson, 1975).

Substrate

Ostracode shell morphology is sensitive to the nature of the substrate and mode of locomotion of the animal in a somewhat analogous way to that in bivalves (Pokorny, 1978; Table 2.8). Swimming taxa have smooth, light, weakly calcified shells that are high relative to their length and may also have long spines (Benson, 1961). Benthonic ostracodes are more ornamented and those that live on a soft substrate often have a flattened venter and keel for support on the mud. Ostracodes living on a coarse substrate have reticulate, spinose shells. Interstitial species are small and robust. Many modern ostracodes live

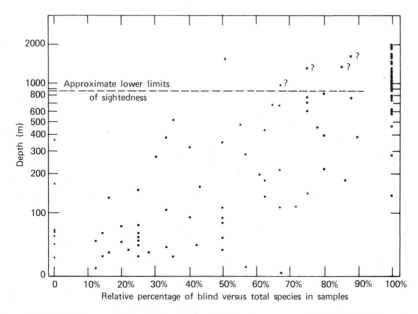

Figure 2.37 Variation with depth in relative blindness among potentially sighted species of ostracodes (those known to have eye tubercles or closely related to species with eye tubercles). After Benson (1975), from Bulletins of American Paleontology, *65* (282), text-figure 11.

Table 2.8 Morphology of ostracode shells as a function of substrate, turbulence, and locomotion method[a]

Loco- motion Type	Substrate- Water Turbulence	Shell Characteristics
Crawler	Soft substrate, low turbulence	Flat venter, frills, or keels for support
Crawler	Coarse substrate, low to moderate turbulence	Reticulate, spinose, spines encasing sensory setae. Ribbed to strengthen and add mass to valve in un- stable sand environment
Burrower	Soft substrate, low turbulence	Smooth, elongate, streamlined valve
Burrower	Coarse substrate, low to moderate turbulence	Interstitial between grains. Small, short, robust valves
Swimmers	Above substrate	Smooth, light weight valves. High relative to length. Simple hinge

[a] After Benson (1961) as modified by Tasch (1973).

on specific plants and their distribution is controlled in part by the distribution of the host plant.

Turbulence

Turbulence affects the ostracodes and barnacles by its effect on the substrate. Ostracodes are too small to be stable in very turbulent environments. Ostracodes do show adaptations for turbulence and other mechanical stresses by means of at least three methods of strengthening the valves which are similar to engineering methods for strengthening man-made structures (Benson, 1975; Fig. 2.38). Ostracodes incorporate the corrugate, arch-beam, and box-frame structure to increase the strength of the valve without the expenditure of the larger amounts of metabolic energy and $CaCO_3$ needed for more massive valves. This is especially important for ostracodes because each time they molt they must produce new valves quickly in order to regain protection and skeletal support. Other groups of organisms such as molluscs and brachiopods have not used these engineering strategies because their accreted shells can grow more slowly; hence they in general use a simple, more massive construction. These highly strengthened ostracode skeletons are not only adapted for life in more turbulent environments but also to withstand predation and to increase support for stresses produced by muscle contraction. Barnacles, on the other hand, thrive under the most turbulent conditions such as rocky, wave swept coasts. Their firm attachment to the substrate adapts them especially well to such environments. Turbulence is not essential for barnacles, however, and they are common in very quiet settings if suitable substrate is available.

ECHINODERMS

Organism Characteristics

Of the living classes of echinoderms the crinoids and the echinoids have the most important fossil record and are thus emphasized here. The asteroids, ophiuroids, and holothurians (the other living echinoderm classes) have much less impressive fossil records because their skeletons either are much reduced (holothurians) or are not readily preservable under normal conditions of fossilization (asteroids and ophiuroids). Nevertheless, they are probably more important in ancient communities than their record would indicate. The echinoderm skeleton is internal, consisting of separate plates that are surrounded by organic material. After the death of the organism the organic binding between plates decays and, except in groups with tightly interlocking plates such as the irregular echinoids, the individual plates are likely to be dispersed by currents or the activities of benthonic organisms. For this reason, complete skeletons of echinoderms are uncommon except under conditions of rapid burial.

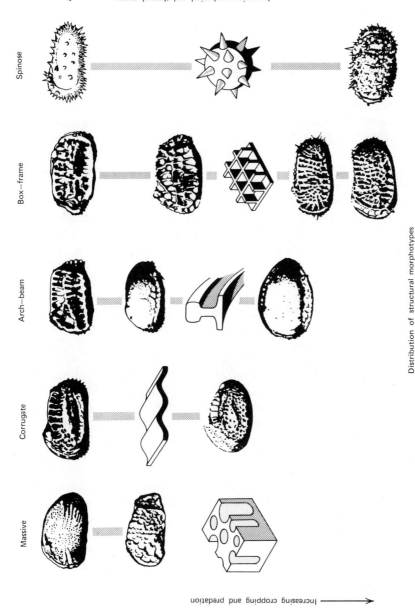

Increasing mechanical and thermal energy ←

Spinose

Box—frame

Arch—beam

Corrugate

Massive

Distribution of structural morphotypes

Increasing cropping and predation →

Figure 2.38 Hypothetical distribution of five types of structural morphology in relation to mechanical and thermal energy and cropping and predation. After Benson (1975), from Bulletins of American Paleontology, 65 (282), text-figure 23.

Although clearly not as abundant as they once were in the geologic past, the crinoids are nevertheless a relatively abundant group in some environments. Modern crinoids include both stalked and unstalked (comatulid) types (Breimer and Lane, 1978). Although the comatulids are capable of swimming, they normally rest on the bottom, firmly attached by specially adapted cirri. They use their swimming ability to find more suitable locations to attach and feed. The stalked types are either cemented to a hard substrate or have a branching root that anchors the animal in the sediment. All crinoids are suspension feeders, feeding on small metazoans, phytoplankton, zooplankton, and suspended detritus. The crinoids feed by forming a filtration fan with their arms. Three different orientations of the filter fan have been described (Meyer, 1973). (1) In the *brachial* filter fan the crinoid arms form a planar or parabolic meshwork oriented perpendicular to current direction. This orientation is found in areas of unidirectional currents. (2) In the *radial* feeding posture the arms are more randomly arranged and the *pinnules* (small side branches from the arms) are oriented radially around the arms. This orientation occurs in areas with complex current patterns. (3) In the *collecting bowl* the arms form a funnel opening upward to collect food settling vertically through the water column. This orientation is found in areas with little or no current. Food particles are captured by the tube feet which secrete mucus in which the particles are trapped. Mucus strands are then passed down the food grooves in the arms to the mouth. Crinoids are selective in the size of food they capture, with food particle size limited by the width of the food groove.

Echinoids are all members of the vagile benthos either moving on the surface or burrowing within the sediment (Durham, 1966). A few types that bore into a hard substrate may grow too large to leave their hole and thus become essentially sessile later in life. Most of the boring echinoids leave their holes to forage for food at night or during high tide. Most echinoids are either deposit feeders or herbivores although many will eat animals that cling to the rocks or sediment surface on which they browse (Barnes, 1974). A few echinoids (e.g., *Dendraster*) also feed on suspended food. Many of the regular echinoids feed by scraping algae and other encrusting organisms from the hard substrate or by scraping plants (especially kelp) with their specially adapted mouth parts (*lantern*). Other types, particularly the burrowing irregular echinoids, feed on deposited food which they select with their tube feet. The food particles are entangled in mucus strands which are passed over the surface to the mouth. A few of the deeper burrowing echinoids construct elaborate respiratory funnels for ensuring a constant supply of fresh, oxygenated water and sanitary tunnels for disposing of waste products (Nichols, 1959).

The holothurians include taxa that move on the surface of a hard or soft substrate, others that burrow within the substrate, and even a few that swim (Pawson, 1966). Most feed on deposited detritus although a number are suspension feeders. Some holothurians pass tremendous quantities of sediment through their gut in the process of feeding. One estimate is that a single holothurian may ingest up to 140 lb of sediment in a year's time. They may thus be very important in modifying or obliterating sedimentary structures as well as in removing the organic component from the sediment.

Asteroids and ophiuroids are both important constituents of modern marine communities and probably also were during much of the Phanerozoic. Both groups are active members of the vagile benthos (Barnes, 1974) and are adapted to many different feeding methods. The asteroids are particularly important as carnivores; modern starfish prey on all types of invertebrates and even small fish. Many are also scavengers, some are adapted to deposit feeding, and others are suspension feeders. The ophiuroids also feed on all of the food types.

Environmental Tolerances

Temperature

All of the echinoderm groups are adapted to a wide range of temperature. Certain echinoid genera and families (e.g., *Eucidaris* and *Phyllacanthus*) are entirely tropical (Fell, 1954). Among modern unstalked crinoids those with highly branched arms are more characteristic of the tropics whereas those with simply branched arms are more common in colder waters. Some crinoid species have longer arms in colder waters than individuals of the same species living in warmer water (Breimer and Lane, 1978).

Salinity

Practically all echinoderms are restricted to waters of near-normal salinity, although a few species of all groups except the crinoids can be found in moderately brackish water (Binyon, 1966). Abundant echinoderms generally indicate normal salinity.

Depth

All of the major modern echinoderm groups are broadly adapted for depths from sea level to hadal. The holothurians are perhaps the most abundant organism at extreme depth in the oceanic trenches. Although some textbooks have indicated that modern crinoids are basically a deep water group, they actually are most abundant in relatively shallow to moderate depths (Fell, 1966). Comatulids are particularly abundant in the shallow reef environment, and the stalked species also are most abundant on the continental slopes. Several echinoderm species respond negatively to light and many are nocturnal. The avoidance of light may be the reason some species are restricted to deep water. The crystallographic orientation of the calcite crystals in echinoid tests may be an adaptation to avoid light. Raup (1966a) has shown that a tangential orientation of the crystallographic *c*-axis admits more light through the test than does a perpendicular orientation. The tangential orientation is more common in deep water forms where lower ambient light levels make a more opaque test unnecessary.

Substrate

Crinoids occur on both hard and soft substrate. The root structure of the
stalked forms, whether cemented or branching, should indicate the nature of
the substrate. Many of the comatulids are basically restricted to hard substrate, clinging to the bottom with their cirri.

Echinoids show a number of adaptations related to substrate (Fig. 2.39).
Many of the regular (spherical, radially symmetrical) echinoids are found in
areas with hard substrate where they live in borings of their own making,
crevices, or simply on the surface. The irregular (flattened, bilaterally symmetrical) echinoids live on or in soft substrates where they usually burrow in
or at the surface of the sediment. The flattened shape of the sand dollar is
clearly a streamlining adaptation for burrowing (Kier, 1974). The sea biscuits
or heart urchins are less streamlined and are slower burrowers but many are
highly modified for life in the substrate. Another adaptation for burrowing is
the presence of perforations through the test (as in the sand dollar *Mellita*)
which allows sediment to pass through the test in burrowing rather than having to go all the way around the margins. Kier (1972) has described examples
of each of these types of echinoids from the Yorktown Formation (Miocene)
and compares them to their modern analogs (Fig. 2.39).

The spines and tube feet of echinoids show many adaptations, several of
which are related to substrate (Durham, 1966). The most obvious function of
the spines, particularly for taxa living on the surface, is for protection. Many
of these have long, sharp, but relatively sturdy spines (e.g., *Lytechinus* and
Strongylocentrotus) and some have especially long, sharp spines with poison
glands (e.g., *Diadema*). Other species have very heavy spines that are hard to
break and are especially effective in wedging the specimens in narrow cavities
or crevices in the substrate (e.g., *Eucidaris*). Other spines function to aid in
locomotion along the surface. At least in part the spines are used in boring in
some species (e.g., *Strongylocentrotus*). An unusual spine modification is
found in the genus *Plesiodiadema* which lives on soft mud and uses its long
arching spines with hooflike expansions at the end for support on the soft
substrate. The irregular echinoids such as *Dendraster* have spines that are
much reduced in size but very numerous, which aid in burrowing and in
feeding. The sea biscuits such as *Echinocardium* have perhaps the greatest
diversity of spines, which are variously adapted for protection, locomotion,
feeding, respiratory funnel building, and sanitary tube building (Nichols,
1959). In fossil echinoids the spines are not generally preserved attached to
the test, but the spine morphology can often be determined from the *tubercles*,
which are the protrusions on the test at the point of attachment of the spine
to the test (Smith, 1978).

Tube feet in echinoids are also adapted to several substrate-related functions such as locomotion, food capture, food movement, and building respiratory and sanitary tubes. Although the tube feet themselves are obviously never
preserved, the pores in the test through which the tube feet pass are in some
cases indicative of their function (Nichols, 1959; Smith, 1978).

Currents have a very strong effect on the method of feeding and the morphology of crinoids (Breimer and Lane, 1978). The morphology of a relatively heavy calyx on a slender stem presents some clear advantages to the crinoid but also some mechanical problems. The stalk allows the crinoid to live above the substrate and many potential suspension-feeding competitors. It also allows the use of a feeding fan to efficiently seine suspended food from the water. On the other hand the crinoid is mechanically unstable with a very high center of gravity on a slender supporting stalk. The crinoids have developed a number of adaptations to cope with this stability problem. Taxa living in quiet water (*rheophobic* types) have adapted the collecting bowl feeding strategy to obtain food settling vertically through the water column. A long stem is no large advantage to these forms because of low density of feeding competitors; indeed some probably lost the stem through evolution and solved the center-of-gravity problem by lowering the calyx to the sea floor! Others developed a short but heavy, rigid, vertical stem. Some have stems that are slightly flexible in the upper part only. The presence of such stems in fossils should indicate low current velocity. Crinoids living in currents (*rheophilic* types) need more flexibility in the stem to form a brachial filter fan perpendicular to the current direction. The filter fan in modern forms is often parabolic in shape with the concavity and the oral side of the calyx facing away from the current (Macurda and Meyer, 1974). The current velocity decreases on the downcurrent side of the fan and food particles can thus be captured more readily by the tube feet. The pinnules on the arms of crinoids that feed in currents in this way branch out from opposite sides of the arms, whereas crinoids in weak current areas and areas with irregular currents have pinnules branching radially from the arms. Some crinoids orient passively in the current in this manner and are dependent on a relatively rigid stem for support. Other forms can adjust the orientation of the arms in such a way as to obtain hydrodynamic lift from the current, behaving much like underwater kites. A rigid stem is then not necessary for support. Long, flexible stems among fossils should indicate currents sufficiently strong to produce lift. Finally, some species may have developed a low-density calyx by the inclusion of low-density liquids such as fats or oils in combination with thin, porous skeletal plates although no modern forms show this adaptation. Perhaps the most effective means used by the crinoids to overcome the center-of-gravity problem has been the loss of the stem and development of the ability to move to a suitable location on an elevated substrate. The substrate thus in effect substitutes for the stalk. This is the solution used by the comatulids, the most successful of the modern crinoid groups.

Turbulence

Boring of cavities by echinoids may be an adaptation to turbulent water. Many boring forms live in intertidal localities where they would easily be swept away by waves if not securely wedged in their borings. The borings also obviously serve a protective function.

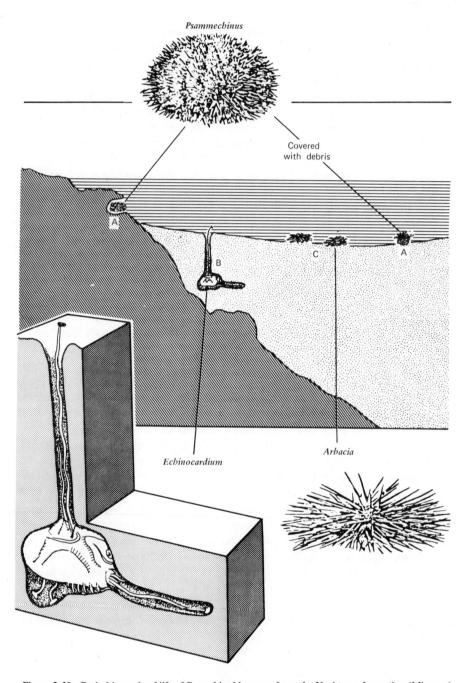

Figure 2.39 Probable mode of life of five echinoid genera from the Yorktown formation (Miocene) of Virginia. After Kier (1972).

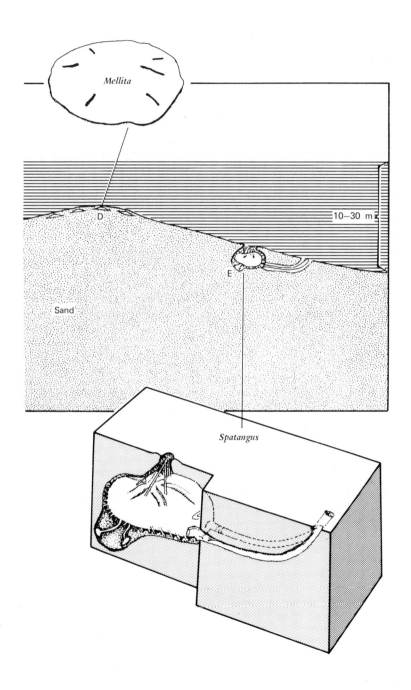

Mellita

D

10–30 m

Sand

E

Spatangus

Oldfield (1976) noted that the exterior surface of the coronal plates in echinoids is simple and unornamented in echinoids living in turbulent conditions and more highly ornamented in individuals living in less turbulent settings. The surface of those that live in burrows has an ornamentation pattern of small, blunt, conical pegs.

LAND PLANTS AND VERTEBRATES

The taxonomic uniformitarian approach has been extensively used in environmental interpretations based on land plants and to a lesser extent on vertebrates. However, these major areas of study are mentioned only briefly here. The cursory treatment does not indicate a lack of importance but rather a lack of space and of expertise on the part of the authors to adequately treat these groups.

In the case of plant fossils, we mention two areas of special importance in taxonomic uniformitarianism: palynology and leaf morphology in angiosperms. Hundreds of studies of pollen in lake sediments trace changing climates during the history of filling of the lake. Many of these lakes have formed as a result of the last Pleistocene glaciation, hence the pollen in the lake sediments records the history of climatic changes since that time. Typically the pollen composition of a mid-latitude lake changes from indicators of a harsh climate at the base of the section, with conifer, birch, and other pollen indicative of cold temperate climates, upward to pollen of deciduous trees typical of temperate climate (Fig. 2.40). Studies of numerous lakes in the eastern United States have been used to reconstruct the vegetation and climatic pattern dur-

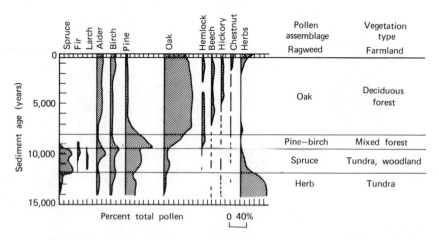

Figure 2.40 Fossil pollen assemblage in a sediment core from a lake in southern New England. Pollen percentages are somewhat generalized. After Davis (1969), reprinted by permission of American Scientist, Journal of Sigma Xi, the Scientific Research Society.

ing the Wisconsin glacial stage when the climatic zones were several hundred km south of their present location (Davis, 1966; Fig. 2.41).

The amount of rainfall also affects pollen composition, with grass pollen being more common in drier climates and tree pollen in wet. The majority of palynologic studies of this sort have been done on nonmarine sediments; however, pollen and spores are common in nearshore marine sediments and have been the subject of paleoecologic investigations (Heusser, 1978).

Angiosperm leaf characteristics are extensively affected by climatic conditions (Wolfe, 1978). Botanists have long noted that large, thick leaves with smooth, *entire* margins (i.e., lacking lobes and teeth) are more common in warm climates (Fig. 2.42; Bailey and Sinnott, 1915). Leaves from cool climates are more often small and thin and have highly incised margins. In fact, the percentage of plant species with entire margins has been shown to vary directly with mean annual temperature (Fig. 2.43). Wolfe (1978) and others have used these relationships to interpret Tertiary climates of North America

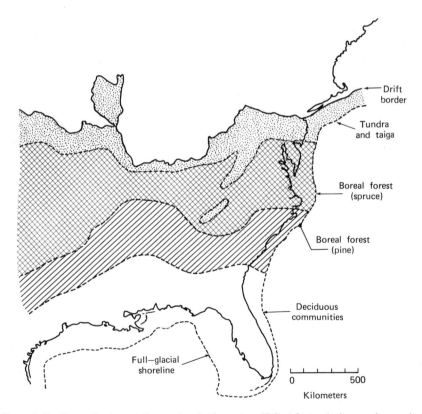

Figure 2.41 Generalized vegetation pattern in the eastern United States during maximum glacial conditions of the last (Wisconsin) glaciation as determined from pollen studies. After Whitehead (1973), reprinted by permission of the University of Washington.

Figure 2.42 Leaves with incised margin (*left*) and entire margin (*right*).

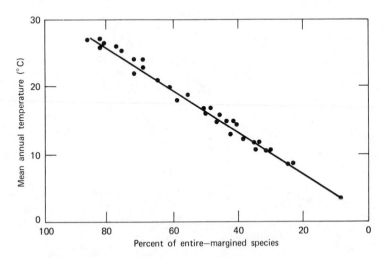

Figure 2.43 Variation with mean annual temperature in percentage of species with leaves having entire margins in forest of eastern Asia. After Wolfe (1978), reprinted by permission of American Scientist, Journal of Sigma Xi, The Scientific Research Society.

(Fig. 2.44). Dorf (1964) gives a more complete discussion of the use of the uniformitarian approach with plant fossils.

The morphology and distribution of vertebrate animals is strongly influenced by environmental conditions. The taxonomic uniformitarian approach thus potentially can be used in paleoenvironmental interpretations, especially for geologically relatively recent time. However, vertebrates have not been as extensively used as other fossil groups because of their more rapid rate of evolution (Romer, 1961). Few vertebrate species are older than Pleistocene and most genera are no older than late Cenozoic. Thus strict application of the uniformitarian principle is largely limited to Pleistocene fossils. Vertebrate fossils are also less abundant and well preserved than invertebrates.

An example of a more general application with vertebrates is the determination of paleotemperatures from the distribution of large reptiles (Colbert, 1964). The mammals and birds are of less use in temperature determination because of their warm-blooded (*endothermic*) nature. Of the terrestrial verte-

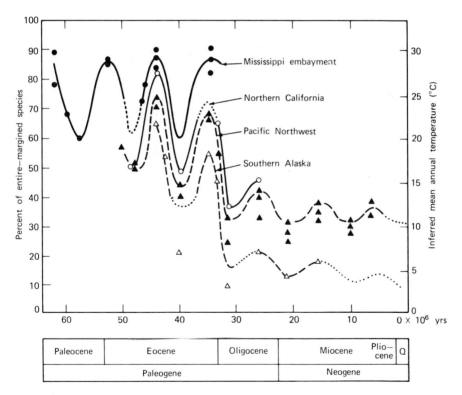

Figure 2.44 Variation in percentage of species with leaves having entire margins during the Cenozoic in four areas of North America. The curves are dotted when data are lacking or uncertain. After Wolfe (1978), reprinted by permission of American Scientist, Journal of Sigma Xi, The Scientific Research Society.

brates, reptiles and amphibians are cold blooded (*ectothermic*), and thus their body temperature is strongly influenced by the environment. Among modern amphibians and reptiles the smaller forms such as frogs, snakes, and turtles can survive in cold climates by hibernation. Large reptiles, such as the crocodilians, are restricted to the tropics and subtropics with a very few warm-temperate representatives (Fig. 2.45). Thus the presence of abundant large reptiles in the fossil record probably indicates warm temperatures. This approach leads to the conclusion that during most of the Mesozoic the tropical and subtropical zones were broader than today (Fig. 2.46). Other workers such as Bakker (1975) have explained the wide latitudinal distribution of some of the large reptiles (dinosaurs) as evidence of their being endothermic. The uniformitarian approach to paleotemperature determination is certainly more safely applied to the still extant crocodilians.

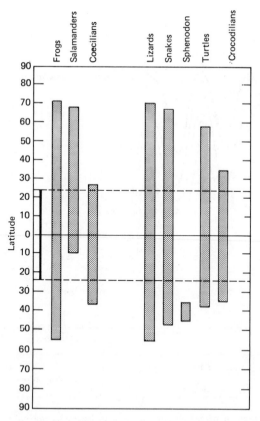

Figure 2.45 Latitudinal range of orders (and suborders of Squamata) among the modern amphibians and reptiles. The dotted horizontal lines show the tropical limits. After Colbert (1964).

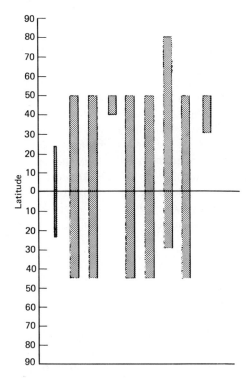

Figure 2.46 Latitudinal range of certain Cretaceous reptile groups. The narrow bar at the left shows the limits of the modern tropics. After Colbert (1964).

TAXONOMIC UNIFORMITARIAN APPROACH IN THE KETTLEMAN HILLS

The Kettleman Hills Pliocene is especially well suited for application of the uniformitarian approach. Essentially all genera and many of the species of fossils found there are still living nearby in the Pacific Ocean and the ecology of these forms has been extensively studied. Much information is readily available from sources as Keen (1971), Keen and Coan (1974), Valentine (1961), Ricketts et al. (1968), MacGinitie and MacGinitie (1949).

Temperature

We can be fairly certain about the temperature conditions during Pliocene time in the Kettleman Hills area because practically all of the Pliocene genera and species that are still extant are living along the Pacific Coast at near the same latitude of the Kettleman Hills. This suggests that the temperature could not have been much different from the present temperature at that latitude. The fauna of the *Pecten* zone (Fig. 1.9) can be used to illustrate this point. Forty-four genera have been identified from this unit. All but two of these genera include 35°N latitude (the approximate latitude of the Kettleman Hills) in their modern range (Table 2.9). One of these, the genus *Mya*, presently occurs no further south than San Francisco Bay along the east Pacific Coast. The occurrence of *Mya* in the *Pecten* zone suggests that the temperature was perhaps somewhat colder than today. However, *Anadara trilineata*, a now extinct species, is one of the most abundant bivalves in the *Pecten* zone and in other units in the Kettleman Hills Pliocene. *Anadara* does not occur north of central Baja California today. Thus *Anadara* would seem to indicate a warmer than present temperature for the Kettleman Hills Pliocene. This apparent conflict can be explained by factors other than temperature. *Mya* is probably restricted to San Francisco Bay and northward at present because it is a genus that is largely limited to quiet water, bay settings. There are few bays of any size appropriate for *Mya* for many miles south of San Francisco Bay. Thus the southern end of the range of *Mya* may be more a function of geography than of temperature. *Anadara trilineata* is an extinct species whose niche was not filled by another *Anadara* species when it became extinct. *Anadara* is not a strictly tropical species in other parts of the world. It is quite abundant along the southeastern coast of the United States, for example. Hence the occurrence of *Anadara* in the *Pecten* zone probably has no climatic significance. This points up the need for caution in using the uniformitarian approach particularly on the basis of one or a few genera.

Even if the Kettleman Hills Pliocence genera and species were not mostly still living we could deduce a generally temperate climate for the area. The lack of forms such as hermatypic corals and dasycladacian and codiacean algae suggests that the climate was not tropical. The molluscs are not in general large and ornate as are many modern tropical forms. On the other hand, they are not the mostly small, simple thin-shelled forms characteristic of cold climates.

Salinity

The uniformitarian approach suggests that salinity conditions varied during deposition of the Pliocene of the Kettleman Hills area. Some fossil assemblages contain genera such as the bivalve *Anodonta* and the gastropods *Juga* and *Amnicola* which are today restricted to fresh water. Others contain genera that are particularly common in brackish water, such as the bivalves *Mya* and *Ostrea,* the gastropod *Littorina,* and the barnacle *Balanus* and genera that are widely distributed but range into brackish water, such as the bivalve *Aequipecten.* When typically open marine forms are absent, brackish conditions most likely existed. Other assemblages include molluscan generea found only in the modern open ocean indicating more normal marine salinities.

The fauna of the *Pecten* zone reflects a wide range of salinity conditions

Table 2.9 Modern latitudinal ranges of marine *Pecten* zone molluscan genera along the west coast of North America[a]

	Degrees Latitude			
	20	40	60	80

Gastropods

- *Calliostoma*
- *Calyptraea*
- *Cancellaria*
- *Jaton*
- *Littorina*
- *Nassarius*
- *Neverita*
- *Olivella*
- *Opalia*

Bivalves

- *Anadara*
- *Chama*
- *Chaceia*
- *Chione*
- *Chlamys*
- *Florimetis*
- *Glycymeris*
- *Macoma*
- *Modiolus*
- *Mya*
- *Mytilus*
- *Ostrea*
- *Panopea*
- *Pecten*
- *Protothaca*
- *Saxidomus*
- *Semele*
- *Solen*
- *Tellina*
- *Trachycardium*
- *Tresus*

[a] Data from Keen (1937).

(Table 2.10). Fossil assemblages immediately below the *Pecten* zone contain freshwater species, indicating that the *Pecten* zone itself represents a marine transgression into the area over nonmarine strata. Assemblages in the *Pecten* zone in the northeast portion of the area contain an abundance of molluscan species indicative of full marine conditions. Also included in this area are the relatively stenohaline groups: corals, brachiopods, and echinoids. To the south, and especially on the west flank of Middle Dome, the full marine form disappear and brackish water indicators such as *Aequipecten, Ostrea,* and *Balanus* predominate (Fig. 2.47). This suggests a region of freshwater input which is producing brackish conditions in this area.

Table 2.10 Depth range, substrate, and salinity preference of molluscan genera from the Pliocene of the Kettleman Hills[a]

Genus	Depth	Substrate	Salinity
Gastropods			
Calicantharus			N
Calliostoma	0–915		N
Calyptraea	0–140	H	N
Cancellaria	25–550		N
Fluminicola			F
Jaton			N
Juga			F
Littorina	0[b]	H	B, N
Nassarius	0–365	S, M	N, B
Neverita	0–2815	S, M	N
Olivella	0–90	S	N
Opalia	0–100	H	N
Bivalves			
Aequipecten			B, N
Anadara			N
Chama	0–45	H	N
Chaceia	0–45	H	
Chione	0–45	S, M	N

Table 2.10　(*Continued*)

Genus	Depth	Substrate	Salinity
Chlamys	5–272		N
Florimetis			N
Glycymeris	0–365		N
Macoma	0–1545	G, S, M	N, B
Modiolus	0–75	H, S	N, B
Mya	0–50	S, M	B, N
Mytilus	0–40	H	B, N
Ostrea	0–35	H	B, N
Panopea	0–20	S, M	N
Pecten	5–185	S	N
Protothaca	0–45	H, G, S	N, B
Pseudocardium (E)			
Saxidomus	0–35	S, M	N
Semele	0–190	H, S	N
Solen	0–75	S	N
Tellina	0–440	S, M	N
Trachycardium	0–120	S	N
Tresus	0–35	S	N

[a] H—hard substrate, G—gravel, S—sand, M—mud, F—fresh water, B—brackish water, N—normal marine, E—extinct. Blanks indicate no data. Data from Keen and Coan (1974) and other sources. Depth ranges apply to the western North American coast only.
[b] Intertidal only

Depth

All of the extant species in the Kettleman Hills Pliocene live in shallow water today. For example, all genera from the *Pecten* zone occur in water no deeper than 25 m. In fact, practically all genera occur at essentially 0 m depth (Table 2.10). Water depth probably varied from zero to a few tens of meters at most. No morphologic features of the fauna indicate that the water was ever very deep.

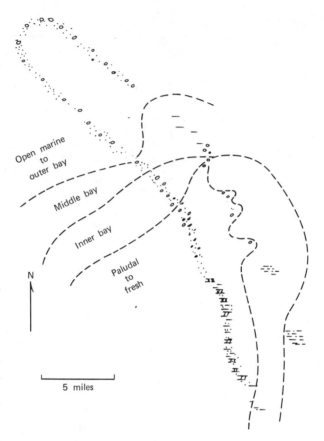

Figure 2.47 Generalized outcrop lithology and paleoenvironments of the basal *Pecten* zone. After Stanton and Dodd (1972).

Substrate

The genera and species in the Kettleman Hills Pliocene suggest a variety of substrate types (Table 2.10). Indeed direct observation of lithologies shows that the substrate varied from mud to gravel. Each fossil assemblage has to be examined on an individual basis in order to interpret substrate conditions. In most cases the faunal indications of substrate agree with the lithology in which the fauna was formed. One exception is the Cascajo Conglomerate unit which contains a varied faunal assemblage indicating a variety of substrate conditions. Clearly the Cascajo fossils have been transported and mixed.

With some notable exceptions few species requiring a hard substrate occur in the Kettleman Hills Pliocene. Groups such as the limpets, chitons, and regular echinoids that would be expected if hard substrate were present are uncommon in the fauna. The considerably greater abundance of bivalves vs.

gastropods also strongly suggests a soft substrate. Many of the gastropods such as the especially common *Neverita* live on or in soft substrate. In terms of geologic processes this suggests that submarine cementation and hard-ground formation was not occurring in the area and that there were no nearby rocky coasts or bedrock exposures. The apparent exception to the lack of species requiring a hard substrate are genera such as *Mytilus, Ostrea,* and *Balanus* which are very common in the strata. Although *Mytilus* is quite common in areas of hard substrate, it can grow in soft substrate areas (such as the Wadden Sea of northwest Europe). *Ostrea* commonly grows on other *Ostrea* shells and often forms banks in soft substrate areas (such as in the Gulf Coastal area of the United States). *Balanus* very commonly occurs in basically soft substrate areas growing on shells. Some of the gastropod species that required a hard substrate probably either lived on shells or on algae and sea grasses which have left no fossil record.

Turbulence

As in the case of substrate the turbulence indications vary considerably between fossil assemblages. Moderately turbulent, shifting sand conditions are indicated by genera such as the sand dollar, *Dendraster,* which is an efficient burrower on shifting substrates, and rapid burrowing bivalves such as *Solen* and *Siliqua.* On the other hand, genera such as *Tivela, Donax,* and *Amiantis* which are especially characteristic of the most turbulent, sandy substrate type environment are not found in the Kettleman Hills. Neither are there many genera indicative of extremely quiet conditions. A range of intermediate conditions predominated. However, note that interpretations based on the absence of fossils should be made with caution.

3

Biogeochemistry

This chapter deals with the use of chemical properties of fossils to determine paleoenvironmental conditions.

The mineral composition of fossil skeletons, their trace and minor element chemistry, and their oxygen and carbon isotopic composition are discussed. Another type of geochemical technique of considerable potential that is considered only briefly is the geochemistry of preserved organic compounds in the rock record.

The chemistry of the environment in which skeletons form has a strong influence on the chemistry of the skeleton itself. In turn, the chemistry of the skeleton and the processes by which it forms may strongly influence the environmental chemistry (Lowenstam, 1974). These interrelationships form the basis for using chemical features of fossils and shells to better understand the chemistry of the environment.

Historically the study of skeletal geochemistry has proceeded at an uneven rate. Most early references to skeletal chemistry were of an incidental nature, secondary to the main purpose of the study in which they were contained. The study of Clarke and Wheeler (1922) was a milestone and the first major American contribution to this field. Subsequently little work was done until the early 1950s. Two major developments at this time caused the field of skeletal chemistry to blossom: the oxygen isotopic paleotemperature technique developed by H. C. Urey and co-workers (1951) at the University of Chicago and the work of H. A. Lowenstam and his students (most notable K. E. Chave) on skeletal mineralogy and trace chemistry at approximately this time (e.g., Lowenstam, 1954 a and b; Chave, 1954). A concurrent milestone was the publication of a comprehensive study of the chemistry of organisms (including their skeletons) by A. P. Vinogradov (1953).

Studies of the biogeochemical properties of invertebrate skeletons, especially those based on isotopic composition, have been rather extensively and successfully used over the last 30 years (especially by geochemists and oceanographers), but they could potentially be even more widely used and refined. Many paleontologists seem to have become disenchanted with geochemical techniques (if indeed they were ever enchanted with them in the

116

first place!). This is evidenced from the brief treatment or nontreatment that these topics receive in most textbooks and reviews of paleontology and paleoecology (e.g., Raup and Stanley, 1978, and Ager, 1963). This may be in part due to the lack of understanding, training, or interest in geochemistry by many paleontologists. The complexity of geochemical relationships may also be discouraging to someone not fully conversant with the processes involved. The results may not be subject to unique interpretation. We tend to be subject to a "black box syndrome" which makes us want to be able to drop the fossil into the instrument and watch the dials point to the temperature and salinity of growth of the fossil. Finally, the negative effect of diagenesis on the chemical properties of the fossil limits the usefulness of the techniques, particularly in older fossils.

Actually these problems are not limited to geochemical techniques. Biological relationships within a community of organisms can be as complex as geochemical relationships or more so and that does not hinder our application of paleoenvironmental interpretations based on community distribution or structure. Often the functional significance of the morphology of fossils is not subject to unique interpretation, but we attempt to make such interpretations anyway. Diagenesis also has negative effects on the biological composition of fossil biotas, but we do the best we can with the imperfect record. It is hoped that this chapter will help to dispel some of the mystery behind the geochemical techniques and may encourage their wider use among paleoecologists. The paleontologist has traditionally been most comfortable describing and studying morphological features of fossils that he or she can easily see. Yet the unseen, chemical attributes of the fossil are also an important aspect of that fossil and can potentially tell the paleontologist as much about certain aspects of the environment in which the fossil lived as its morphologic features. The most complete possible description of the fossil and its environment should include chemical as well as morphological data.

The factors controlling the chemical properties of fossils can be divided into four categories: physical chemical, environmental, physiologic (genetic), and diagenetic (Dodd and Schopf, 1972). These four effects can be pictured as the four corners of a tetrahedron (Fig. 3.1). The relative importance of each of the four categories of effects in determining the chemistry of any fossil can be imagined as a point somewhere within the tetrahedron. The paleoecologist must determine the location of that point in order to completely explain the chemistry of the fossil. In practice that may be difficult or impossible, but the concept is useful in understanding skeletal chemistry. Actually these categories are not completely independent; all effects should ultimately be explainable in physical chemical terms. For example, the environmental effects on oxygen isotopic composition are the result of environmental influence on physical chemical relationships and are not truly independent. Physiologic effects that are under genetic control presumably have a biochemical basis that could be explained in physical chemical terms if they were adequately known. Diagenetic effects are

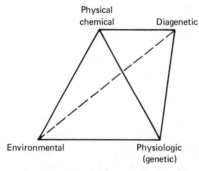

Figure 3.1 Schematic representation of the factors controlling the chemistry of fossil skeletons. Physical chemistry, physiology (genetics), environment, and diagenesis all affect chemistry. The relative contributions of these factors can be visualized as a point within the tetrahedron.

really physical chemical processes that have occurred after the formation of the skeleton. Nevertheless, this subdivision of factors is useful in facilitating discussion and is used in this chapter.

SKELETAL MINERALOGY

Strictly speaking, mineralogy is not a chemical property of the skeleton but refers to the physical arrangement of atoms in space. But this is so closely tied to the chemical nature of the atoms that for convenience we may consider mineralogy in conjunction with the chemical characteristics of fossils. The mineral composition of skeletal material has been of considerable interest to paleoecologists, both as a potential tool for paleoenvironmental analysis and because of its effect on the preservation of fossils and potential fossils. Although little success can be claimed in directly using skeletal mineralogic composition to interpret paleoenvironments, information on mineralogy has been extremely useful in interpreting geologic processes and history, especially when used in combination with other information.

Mineral Components

Many different minerals are found in invertebrate skeletons (Table 3.1). By far the most important minerals among the invertebrates are the two calcium carbonate polymorphs, calcite and aragonite; although the opaline silicates and phosphates also make major contributions. In addition, several less common minerals have also been described in biologic systems (Lowenstam, 1974) and the relatively recent discovery of several of these (e.g., Lowenstam and Abbott, 1975) indicates that probably additional minerals will be found. These minor minerals often serve a particular function for which they are better adapted than the primary mineral found in the skeleton of that group. For example, the chitons have magnetite cappings for their radular teeth (Lowenstam, 1962). This mineral is much harder than the aragonite that makes up the basic skeleton of the chitons. Some of the echinoids have dolomite in their teeth

(Schroeder et al., 1969). Dolomite is also slightly harder than the calcite that makes up the majority of the echinoid test. The same explanation might apply to the flourite gizzard plates of some gastropods (Lowenstam and McConnell, 1968). In many cases we do not know why the particular mineral is being used by the organism rather than one of the more common skeletal minerals (e.g., brucite in algae, gypsum in jelly fish statoliths). The explanation is probably related to the biochemistry of the organism. A certain mineral is simply easier to precipitate biochemically than another. The discussion in this chapter concentrates on the minerals calcite and aragonite, as they are the most common of the skeletal minerals. Lowenstam (1974) has recently published an extensive discussion of other skeletal minerals, to which the reader is referred for further details.

In addition to skeletons that consist of only pure calcite or pure aragonite, many invertebrates have skeletons consisting of a combination of these two polymorphs. For example, the skeletons of the sclerosponges are in part carbonate (aragonite) and in part opaline silica spicules (Hartman and Goreau, 1970). Fish have skeletons that are largely phosphatic, but they also have ear bones (otoliths) composed of aragonite. The minerals are always present in separate microarchitectural units rather than in intimate intermixtures. They may be separate units within the skeleton (e.g., shell layers) or they may form separate, isolated parts (e.g., fish bones and otoliths). Crystal sizes, shapes, orientations, and interrelationships within these microarchitectural units may be quite varied and complex (Bøggild, 1930), giving rise to another important area of study: the variation in these skeletal structure units and the factors responsible for that variation (Chapter 4). Skeletal calcites may contain considerable amounts of magnesium substituting for calcium (up to 30 mol % $MgCO_3$), greatly affecting the properties of the mineral. Many authors have considered high magnesium calcite (greater than about 4 mol % $MgCO_3$) as a separate mineral. Here we discuss chemical variation in the carbonate minerals separately.

The data shown in Table 3.1 are entirely for the skeletons of living organisms. The mineralogy of now-extinct groups cannot always be determined with certainty; some fossil groups that were originally aragonite may have since been converted to the more stable calcite. For example, the original mineral composition of tabulate and rugose corals and the stromatoporoids cannot be determined with certainty. All are calcite now but some workers have suggested that they were originally aragonite as are the scleractinians. However, details of their skeletal structure suggest that no major diagenetic alteration has occurred (Sorauf, 1971), and at the few Paleozoic localities where aragonite is preserved in some groups (nautiloids, bivalves, and gastropods), the tabulate and rugose corals are calcitic (Stehli, 1956).

Physical Chemical Control

That $CaCO_3$ should be by far the most common skeletal material is not surprising on a physical chemical basis for several reasons. (1) Shallow sea water is

Table 3.1 Minerals and inorganic constituents found in skeletons and hard parts of marine organisms and their phyla distribution.[a]

	Bacteria	Bacillariophyceae	Chlorophyta	Rhodophyta	Phaeophyta	Coccolithophorida	Protozoa	Porifera	Coelenterata	Bryozoa	Brachiopoda	Sipunculida	Annelida	Mollusca	Arthropoda	Echinodermata	Chordata
Carbonates																	
Calcite	?			+		+	+	+	+	+	+		+	+	+	+	+
Aragonite	+		+	+	+	?	+	+	+	+		+	+	+	+		+
Calcite and Aragonite									+	+		+	+	+	+		+
Vaterite						?											+
Monohydrocalcite														+			+
Amorphous									+				+	+	+		+
Phosphates																	
Dahllite													+	+	?		+
Francolite											+			+			+
Amorphous calcium phosphatic hydrogels							?						+	+	+		
Amorphous ferric phosphatic hydrogels													+	+		+	+

Silica: opal	+			+	+	+		
Fe-Oxides								
Magnetite						+		
Goethite						+		
Lepidocrocite			+			+		
Amorphous hydrates	+			+			+	
Sulfates								
Celestite	+							
Barite	+							
Gypsum		+						
Halides: fluorite						+	+	
Oxalates: weddellite				+		+		+

[a]After Lowenstam (1974).

saturated or supersaturated with respect to $CaCO_3$ almost everywhere and in some places $CaCO_3$ is precipitating inorganically (Milliman, 1974). (2) The precipitation of $CaCO_3$ is more readily controlled biologically than most minerals. This is because the concentration of the $CO_3{}^{2-}$ ion (and thus the degree of saturation with respect to $CaCO_3$) is strongly dependent on pH and on the concentration of CO_2, and metabolic processes readily affect these parameters. (3) Many organisms also build skeletons of opal or of phosphate minerals but the concentration of dissolved Si^{2+} and $PO_4{}^{3-}$ is much lower than Ca^{2+} (Goldberg, 1957), requiring a much larger volume of sea water to precipitate a unit mass of SiO_2 or $Ca_3(PO_4)_2$ than of $CaCO_3$. (4) Shallow sea water is usually not saturated with respect to SiO_2 or $Ca_3(PO_4)_2$, requiring strong biological intervention to precipitate opaline or phosphatic skeletons.

Calcium carbonate occurs in skeletons in one of three polymorphous forms: calcite, aragonite, or vaterite. Skeletal vaterite is rare so we concentrate our discussion on calcite and aragonite.

Calcite but not aragonite is stable under surface temperature-pressure conditions (Fyfe and Bischoff, 1965; MacDonald, 1956; Fig. 3.2). The instability of aragonite under surface conditions is indicated by its slightly larger free energy of formation than that of calcite (-269.5 vs. -269.7 at 25°C from Latimer, 1952). This is expressed in the greater solubility of aragonite (solubility product $= 4.5 \times 10^{-9}$ from Krauskopf, 1967) than calcite (solubility product $= 6.0 \times$

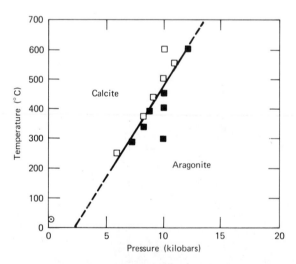

Figure 3.2 The temperature-pressure stability relationships of calcite and aragonite based on experimental studies with high-pressure-temperature bombs. The solid squares are data from experiments producing aragonite and the open squares are data from experiments producing calcite. The dashed portion of the line extrapolates the results to lower temperatures and pressures. The encircled dot shows surface temperature-pressure conditions. After MacDonald (1956), American Mineralogist, *41:* 749, copyrighted by MSA.

10^{-9} from Krauskopf, 1967). The generally decreasing abundance of aragonite in rocks of increasing geologic age is a further expression of the instability of aragonite.

On the other hand, aragonite is an extremely common mineral in modern marine sediments, probably the most common carbonate mineral in the shallow marine environment (e.g., Milliman, 1974). Much of it is of skeletal origin. This apparent paradox needs explanation. The most probable one is that because of the abundance of magnesium in sea water, aragonite is in fact not unstable in the marine environment. Precipitation of calcite in the marine environment results in the coprecipitation of a large amount of Mg in the calcite lattice (approximately 10 to 14 mol % or more of $MgCO_3$ under surface conditions). High-magnesium calcite of this approximate range of composition is a fairly common precipitate from sea water, especially as intergranular cement (Milliman, 1974)—this high-Mg calcite is more soluble and thus less stable than aragonite. Berner (1975) has shown on thermodynamic grounds that high-Mg calcite with more than 8.5 mol % $MgCO_3$ in solid solution is less stable (more soluble) at 25°C than is aragonite. Calcite containing less than 8.5 mol % $MgCO_3$ in solid solution is more stable (less soluble) than aragonite, but such low-Mg calcite cannot form in full marine sea water under tropical or subtropical temperatures because of the abundance of Mg in sea water. Consequently aragonite and high-Mg calcite form instead. Winland (1969) further presents evidence to show that high-Mg calcite is in fact converting to aragonite at shallow depths in the oceans.

An alternate explanation for the precipitation of aragonite in sea water is that the relative crystal nucleation and growth rates of calcite and aragonite are differentially affected by foreign ions. Laboratory precipitation experiments and field studies by many different workers (e.g., Berner, 1975, and Kitano, 1962) have shown that the concentration of certain ions has a strong influence on whether calcite or aragonite is formed (Fig. 3.3). The most important ion in controlling mineralogy in the marine environment seems to be Mg^{2+}. At low Mg^{2+} concentrations calcite usually forms when the solution exceeds the $CaCO_3$ solubility product. At greater Mg^{2+} concentrations, aragonite usually forms. The Mg^{2+} concentration in normal sea water (about 1330 mg/l) is considerably in excess of the amount needed to prevent formation of calcite, hence inorganic precipitation from sea water most commonly produces aragonite. Simkiss (1964) has suggested that the effect of Mg^{2+} is that of a crystal poison, that is, the Mg^{2+} ions occupy Ca^{2+} sites in the growing calcite lattice and block or greatly slow the further growth of the crystal. Due to its small size the Mg^{2+} ion is not readily accommodated in the aragonite lattice so that aragonite is able to grow much more rapidly than calcite. Other ions may have a similar crystal poisoning effect. Simkiss suggests that the PO_4^{3-} ion may also act to inhibit calcite formation by occupying the CO_3^{2-} site and that organisms may control their skeletal mineralogy by controlling the PO_4^{3-} concentration at the calcification site. Although this model has not been adequately tested by chemical determinations at the calcification site, high PO_4^{3-} concentrations

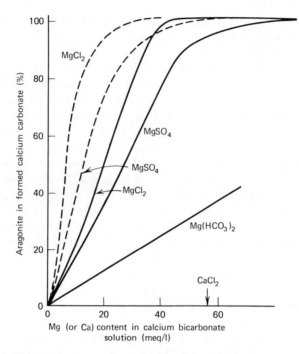

Figure 3.3 The effect of Mg compounds in solution on the mineralogy of precipitated $CaCO_3$. The solid lines are for experiments at $10 \pm 2°C$, and the dashed lines are for experiments at $28 \pm 3°C$. After Kitano (1962).

have been observed in the mantle of growing bivalves (e.g., Bevelander and Benzer, 1948).

The basic chemistry of precipitation of noncarbonate skeletal phases is in general more simple than carbonate precipitation. In most cases the water is undersaturated with respect to minerals such as opal, apatite, and magnetite. Strong physiologic control is required to increase ionic concentrations at the mineralization site within the organism so that precipitation can occur.

Genetic (Physiologic) Control

The mineralogic diversity of different groups of organisms (Table 3.1) indicates that the mineral composition of skeletal material is in part genetically controlled. The different minerals in different parts of a single organism is a further expression of the physiologic control. Mineral composition also changes during the life of the individual in some groups. For example, the larval shell of the bivalve *Crassostrea* is entirely composed of aragonite whereas the adult shell is almost entirely calcite (Stenzel, 1963). The mineralogy of the shells of the bivalve *Mytilus* changes during its life (Dodd, 1963). Possible long-term evolutionary trends in mineralogy have been described in the corals, which apparently changed from predominantly calcite in the Paleozoic to predominantly aragonite today (Lowenstam, 1974). Also, primitive gastropods seem to have been predominantly calcite whereas the more advanced forms are mainly aragonite (Lowenstam, 1964b).

One suggestion for the mechanism by which the organism controls its mineralogy is that it may modify the concentration of ions such as Mg^{2+} and PO_4^{3-} of the fluid from which calcification is occurring resulting in the ion interference effect discussed above. Kitano and Hood (1965) and Kitano et al. (1976) have shown that soluble organic constituents in the body fluids from which $CaCO_3$ is precipitating may influence the mineral produced. By adjusting both the concentration of organic compounds and Mg^{2+} and the reaction temperature, Kitano et al. (1976) were able to precipitate low-Mg calcite, high-Mg calcite, or aragonite. Perhaps these factors within organisms could also control skeletal mineral composition, although the mechanism by which this control is effected is not known.

Another possible method by which organisms control the mineralogy of their skeletons is by the influence of the organic matrix of the skeleton. Each crystal of a molluscan skeleton is surrounded by a thin, proteinaceous sheath or matrix (Wilbur, 1976; Fig. 3.4). In the growing skeleton, the organic matrix is laid down first, then the crystal is nucleated on that matrix. The crystal grows until it begins to interfere with surrounding crystals, from which it is separated by the organic matrix. The crystal stops growing when a new layer of organic matrix forms, isolating it from the *extrapallial fluid,* the liquid from which calcification is occurring. New crystals are then nucleated on that matrix surface (Wilbur, 1964). Thus the organic matrix is intimately interrelated with the calcification

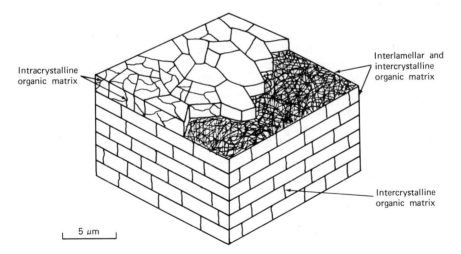

Intracrystalline organic matrix

Interlamellar and intercrystalline organic matrix

Intercrystalline organic matrix

5 μm

Figure 3.4 Diagrammatic representation of the relationship between organic matrix and $CaCO_3$ in the nacreous shell structure. After Kennedy et al. (1969), copyright Cambridge University Press.

process, especially in serving as the site upon which crystal growth is nucleated. The matrix consists largely of protein molecules, the amino acid composition of which has been studied by several workers (e.g., Hare, 1963; Degens et al., 1967; and Weiner and Hood, 1975). The amino acid composition of these proteins is different in aragonitic and calcitic skeletons and in aragonitic and calcitic portions of the same skeleton (Hare, 1963). Thus the composition of the matrix may be the controlling factor in determining mineral composition in the growing skeleton in that the amino acids may be arranged into protein molecules in such a way that side chains act as the sites upon which the Ca^{2+} and CO_3^{2-} ions are positioned when $CaCO_3$ crystals are nucleated. The organic matrix thus acts as a template on which the mineral is deposited. In some proteins the side chains are arranged to conform to the aragonite lattice and in others to the calcite lattice (Hare, 1963). This model is supported by experiments in which pieces of decalcified matrix from aragonitic bivalves were placed under the mantle of the normally calcitic genus *Crassostrea*. In 4 out of the 17 experiments, aragonite formed on the matrix despite the fact that *Crassostrea* normally precipitates only calcite except in restricted areas under its muscles (Wilbur and Watabe, 1963). In other experiments decalcified pieces of matrix were placed in calcium bicarbonate solutions from which calcium carbonate precipitated on the matrix pieces. Aragonite precipitated only on matrix from aragonitic species (Wilbur and Watabe, 1963).

The precipitation by invertebrates of noncarbonate minerals has not been studied extensively (Wilbur, 1976; Lowenstam, 1974). Limited results suggest crudely similar processes, but they include the need to concentrate the chemical constituents, which may be present in sea water in low abundances.

Environmental Control

Lowenstam (1954a and b) first noted a temperature effect on mineral composition of invertebrate skeletons. This effect is expressed in three ways (Fig. 3.5). (1) Some groups of organisms having aragonitic skeletons are far more abundant in the tropics than in the higher latitudes. The reef-building scleractinian corals are perhaps the best example of this effect. (2) Certain groups of organisms having aragonitic skeletons are found only in the tropics and semitropics. The best examples of this effect are the calcareous green algae. (3) In taxa having skeletons composed of a combination of aragonite and calcite, the proportion of aragonite increases as temperature increases. Examples of the latter effect are found in the molluscs (Fig. 3.6), annelids, coelenterates, and bryozoans. In these groups the relationship between temperature and mineral composition is specific to individual genera, species, or even subspecies (Lowenstam, 1954a, and Dodd, 1963).

The relation between temperature and the aragonite/calcite ratio is not always obvious because of the effects of other environmental parameters, particularly salinity (Lowenstam, 1954a; Dodd, 1963). For example, in the bivalve

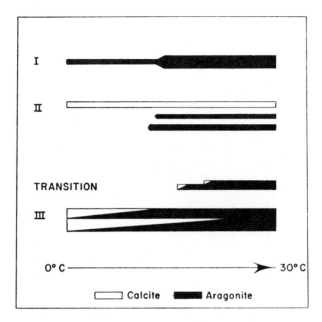

Figure 3.5 Schematic representation of three ways in which temperature affects the carbonate mineralogy of invertebrate skeletons. I—Groups of organisms having aragonitic skeletons that are now abundant in warmer waters. II—Groups with aragonitic skeletons that are found only in warmer waters, whereas calcite groups occur in both cold and warm water. Transition—groups with aragonite skeletons that secrete trace amounts of calcite in the colder part of their ranges. III—Groups containing a combination of calcite and aragonite but with increasing amounts of aragonite in the warmer part of their ranges. After Lowenstam (1954a), copyright University of Chicago.

genus *Mytilus,* reduced salinity is accompanied by an increased proportion of aragonite (Fig. 3.7) although Eisma (1965) presents apparently conflicting evidence. Other more individual effects, such as growth rate, also seem to affect mineral composition. For example, in the species *Mytilus californianus,* specimens that have grown rapidly have relatively thin shells and a relatively low percentage of aragonite (Dodd, 1963). This apparently is because the outer calcite shell layer has grown rapidly whereas the aragonitic layer that thickens the shell on the inside has grown more slowly. Other specimens that have been abraded during life by rubbing against the substrate or other specimens thicken their shell by the addition of calcite to the shell interior. These specimens thus have a low percentage of aragonite (Dodd, 1963).

Some bryozoans also thicken their skeleton by the addition of an inner aragonitic layer. Older portions of the colony have a higher proportion of aragonite because the specimens there have had more time to thicken their zooarial walls (Poluzzi and Sartori, 1974).

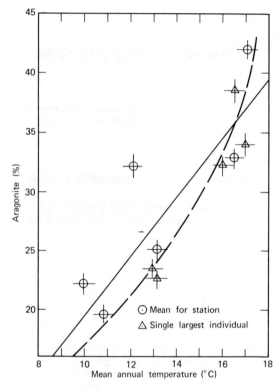

Figure 3.6 Variation in percent aragonite in *Mytilus californianus* with mean annual temperature of collecting locality. Solid line—regression line fit to data. Dashed line—trend line drawn through data points. After Dodd (1963), copyright University of Chicago.

Diagenesis

The most common diagenetic effect on the mineral composition of fossils is solution by ground water. Only a small portion of the skeletal material origin-ally buried in sediments is preserved as fossils, as indicated by the great volume of normal marine sedimentary rocks that do not contain fossils. Ground water undersaturated with respect to the mineral of which the fossil is composed will in time dissolve the fossil. Solution is especially likely to occur in a permeable rock and when the fossil is in a matrix that differs in mineralogy from the fossil. Thus a carbonate fossil buried in a quartz sand has a high probability for solu-tion. If matrix has the same mineralogy as the fossil the water will quickly become saturated with respect to that mineral by dissolving matrix as well as fossil and the fossil has a greater chance of escaping complete solution. Thus a calcite fossil buried in a limestone has a good chance of preservation.

Undersaturated water in contact with both calcite and aragonite shell material will dissolve both minerals. Eventually however, the water will become

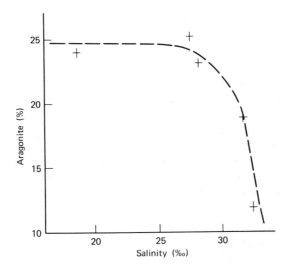

Figure 3.7 Variation of percent aragonite with salinity at collecting time in *M. edulis edulis* from Hood Canal, Washington area. After Dodd (1963), copyright University of Chicago.

saturated with respect to calcite and the calcite will stop dissolving, but the aragonite will continue to dissolve because aragonite is more soluble than calcite and so the solution has not yet reached saturation with respect to aragonite. However, as aragonite continues to dissolve, the Ca^{2+} and CO_3^{2-} concentrations of the solution rise and exceed the solubility product of calcite, and calcite should then precipitate. Eventually all the aragonite should dissolve and be reprecipitated as calcite. This process has been demonstrated in the laboratory (Kitano et al., 1976). Mixtures of calcite and aragonite fossils are thus chemically unstable and should become increasingly rare with time as indeed they do. The most ancient examples of preserved aragonitic fossils are in situations where they are effectively removed from contact with ground water such as buried in asphalt or fine shale (Stehli, 1956).

If the solution of the aragonite fossil occurs before lithification, the resulting void will collapse and all trace of the fossil may be lost with the possible exception of an imprint. If lithification had already occurred when the fossil dissolved, a mold would be preserved and if the mold is later filled in with another mineral, a cast of the original fossil will be preserved. Fossil assemblages commonly consist of a mixture of originally calcite fossils with original preservation and calcite casts of originally aragonite fossils (Horowitz and Potter, 1971). In some cases molluscs show original preservation of some shell layers and casts of others indicating an original mixed mineralogy. Also common are assemblages consisting entirely of originally calcite forms, the aragonite forms apparently having been removed by solution in an unlithified matrix (Lawrence, 1968; Chave, 1964). Obviously care must be taken in making paleoecologic interpretations based on such greatly modified assemblages.

Diagenesis may result in replacement of one mineral by another. The most common replacement minerals are fine-grained quartz, dolomite, and pyrite. Calcite is a common replacement mineral of originally siliceous fossils. Differential preservation of the assemblage through diagenesis is a matter of concern to the paleoecologist, but unfortunately the processes are poorly understood, although some pioneering work has been done in this field (e.g., Siever and Scott, 1963; Murray, 1964).

The diagenetic processes described above take place after burial. Solution of fossils also occurs before burial or soon after burial in the surficial sediments. Except in the polar regions, the shallow waters of the oceans are saturated or supersaturated with respect to $CaCO_3$ (Broecker, 1974) so carbonate skeletons are not dissolved there. With increasing depth, however, the oceans become undersaturated with respect to $CaCO_3$ so that solution occurs below the *lysocline* (the zone of rapid increase in $CaCO_3$ solution). The lysocline for calcite depends on oceanic condition and ranges from 3500 m depth in the Pacific to 5000 m in the Atlantic (Broecker, 1974; Fig. 3.8). Because it is slightly more soluble than calcite the lysocline for aragonite is shallower, ranging from about 500 m in the Pacific to 2500 m in the Atlantic. Berger (1970 and others) has pointed out that this solution cannot only be selective between calcite and aragonite shells but at least in the case of planktonic foraminifera can also selectively remove the more delicate shells while leaving the more robust forms. This differential preservation has obvious implications for paleoecologic interpretations. In the case of the planktonic foraminifera, the more delicate shells are usually those living in the shallowest, warmest water so if they are selectively dissolved from an assemblage, the remaining heavier shelled forms give the impression of colder surface-water conditions than actually existed (Berger, 1970).

The oceans are essentially everywhere undersaturated with respect to opaline silica (Broecker, 1974). Siliceous skeletons in the oceans thus begin to dissolve as soon as they are exposed to sea water. As in the case of carbonate skeletons, the heaviest, most robust forms are most likely to be preserved. Solution of opal continues within the sediment so that unless sedimentation is rapid complete loss of opaline fossils is likely (Johnson, 1976).

Less is known about the diagenesis of phosphates in the oceans. The oceans appear to be in general saturated at least with respect to the carbonate apatite mineral dahlite (Lowenstam, 1974). Phosphate minerals are common constituents of oceanic sediments, but phosphate diagenesis has been little studied. Even less is known about the diagenesis of other biologically formed minerals.

Geologic Application of Skeletal Mineralogic Data

Paleotemperatures have never been successfully determined on the basis of skeletal mineralogy for three reasons. (1) The mineralogy-temperature effect differs from species to species and perhaps even within genetic variants of a single species. Therefore, a given proportion of aragonite may correspond to a certain temperature in one species and to another temperature in another

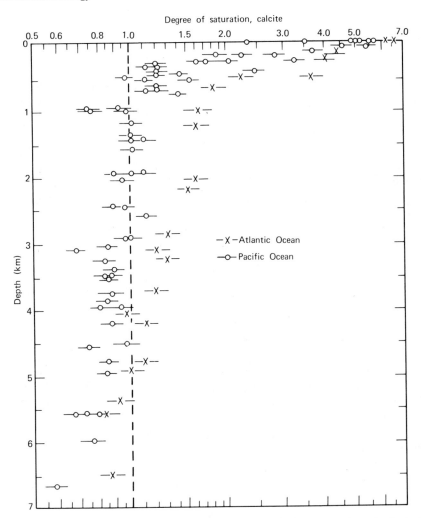

Figure 3.8 Degree of saturation with calcite of sea water samples from various depths in the Atlantic and Pacific Oceans. After Li et al. (1969), Journal of Geophysical Research, *74:*5521, copyright the American Geophysical Union.

(Dodd, 1963). This limits the technique to species that still have living representatives for which the precise temperature-mineralogy relationships can be determined. Extinct species could only be used if calibrated against some independent method of paleotemperature determination. (2) Several factors in addition to temperature influence skeletal mineralogy, and the interaction of these factors may be complex. (3) Because aragonite is not stable in diagenetic environments, finding well preserved specimens is difficult.

Although use of skeletal mineralogy may not be a practical method for detailed paleotemperature determination, it can be of value in identifying temperature trends. The presence of aragonitic taxa not found outside of the tropics, or the abundance of aragonitic taxa not normally abundant outside of the tropics would suggest a tropical climate and could be used in establishing an end point for such general trends. Gradients in the aragonite/calcite ratio in skeletons with mixed mineralogies could also be used. This might be done even with fossils in which the aragonite had converted to calcite if the originally aragonitic nature of the converted unit could be determined from its method of preservation (Bathurst, 1964).

Probably the most useful paleoecologic application of skeletal mineralogy data is in evaluating the degree of preservation of fossil assemblages. The previously mentioned examples from the New Jersey Cretaceous (Chave, 1964) and the South Carolina Oligocene (Lawrence, 1968) are just two examples that show how differential removal of aragonitic fossils can markedly change the composition of the original skeletal assemblage.

The most extensive use of skeletal mineralogy data has been in the study of the origin and diagenesis of carbonate sediments and rocks (Milliman, 1974; Bathurst, 1971). The mineral composition of a carbonate sediment is largely determined by the organisms that contribute their skeletons to the sediment because the major portion of carbonate sediments is of biological origin (Lowenstam, 1974). The relative proportion of calcite and aragonite as well as the amount of Mg in the calcite has a large bearing on the way in which the sediment behaves during lithification. The solution of aragonitic skeletal grains can result in the development of secondary porosity in carbonate rocks, a process of great practical import in the development of oil reservoirs. Bathurst (1971) has extensively reviewed the relationship of skeletal mineralogy to carbonate sediment formations and diagenesis.

Lowenstam (1974) has emphasized the importance of organisms in adding a variety of minerals to sediments. Calculations show that on a global basis enormous quantities of opaline silica, phosphate minerals, fluorite, and magnetite are contributed by organisms to the sediments. Skeletal siliceous deposits (diatomites) and phosphates (phosphate rock) are sometimes of economic importance.

Skeletal Mineralogy of Fossils from the Kettleman Hills

We have not attempted to determine paleotemperatures on the basis of skeletal mineralogy of specimens from the Kettleman Hills. The reasons why we have not attempted to do so will help to point out the limitations of the method. The most thoroughly studied modern genus in terms of the relationship of shell mineralogy to temperature is *Mytilus*. Two species of this genus are found in the Pliocene and Pleistocene sediments of the Kettleman Hills, but the specimens are not suitable for determining the paleotemperature. *M. coalingensis* is widespread in the section; however, it is an extinct species and may not exhibit the same temperature-mineralogy relationships as do the modern species. In addition, complete specimens that are needed to make a mineralogic determination are difficult to collect. The upper *Mya* zone, in the upper part of the section, contains specimens that probably belong to the modern species, *M. edulis*. Fragments of this species are often common, but complete, well preserved specimens are rare. A further complication in using these specimens is that the water in which they grew was probably brackish, giving the possibility that variation in salinity as well as temperature may have affected the mineralogy, making a unique interpretation of the results impossible.

Although the preservation of fossils in the Kettleman Hills Pliocene is usually good, the effect of mineralogy on differential preservation can occasionally be observed. Many of the fossil assemblages appear to have all carbonate fossils, both calcitic and aragonitic forms, well preserved. On the other hand, many stratigraphic intervals contain no fossils although they apparently were deposited under normal marine conditions in which potential fossils should have lived. Sometimes these intervals contain molds of fossils indicating that the carbonate shells, both calcite and aragonite, have been dissolved. In a few instances fossil assemblages occur containing only fossils that were originally calcite. For instance, some lenses and beds in the *Pecten* zone contain abundant specimens of *Ostrea* and plates of barnacles and occasional fragments of *Pecten* and *Mytilus*. All of these are originally calcite forms; no aragonitic shells are preserved. Also found are occasional assemblages composed entirely of large barnacles and *Mytilus* fragments. Some assemblages consist entirely of echinoids, another calcite group. The latter assemblages may not always be the result of differential preservation, however, because modern assemblages consisting entirely of echinoids do occur.

TRACE CHEMISTRY

The minerals precipitated by organisms are never absolutely pure but contain trace amounts of foreign ions. They may contain a considerable concentration of the foreign ion, several mol percent in some cases. In these examples they may be more appropriately called minor or accessory elements rather than trace elements. The ions may be present in solid solution in the crystal lattice, absorbed to the crystal surface, incorporated in the organic matrix, or incorporated as separate mineral phases (perhaps trapped as impurities from the surrounding environment) (Dodd, 1967). The concentration of trace elements in a fossil is a function of the physical chemistry of the skeletal formation process, of environmental variables, of the physiology of the organism, and of the diagenetic processes. Although these factors apply to all skeletal minerals, the study of skeletal trace chemistry has dealt largely with carbonates. Most of this research has been on the concentration of the two most abundant trace metals, Mg and Sr (Fig. 3.9) and we also concentrate our discussion on these elements. The reason for the greater abundance of these elements are threefold. (1) The ionic radii of Mg^{2+} and Sr^{2+} are close enough to that of Ca^{2+} to allow ready substitution in either the calcite or the aragonite crystal lattice. (2) The ionic charge is the same as that of Ca^{2+}, again facilitating substitution. (3) Mg^{2+}, and to a somewhat lesser extent Sr^{2+}, are abundant in natural waters, especially sea water.

Physical Chemical Factors

Water Chemistry

Other factors being equal, the Mg/Ca and Sr/Ca ratios in a carbonate skeleton are proportional to those ratios in the water in which the carbonates formed. In

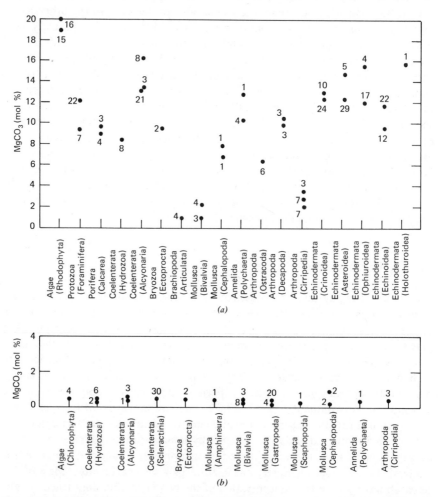

Figure 3.9 (a) Mean value of $MgCO_3$ in calcite skeletons; (b) mean value of $MgCO_3$ in aragonite skeletons; (c) mean value of Sr/Ca atom ratio in calcite skeletons; (d) mean value of Sr/Ca atom ratio in aragonite skeletons. The numbers by the points are the numbers of analyses used to determine the mean. After Dodd (1967), Journal of Paleontology, Society of Economic Paleontologists and Mineralogists.

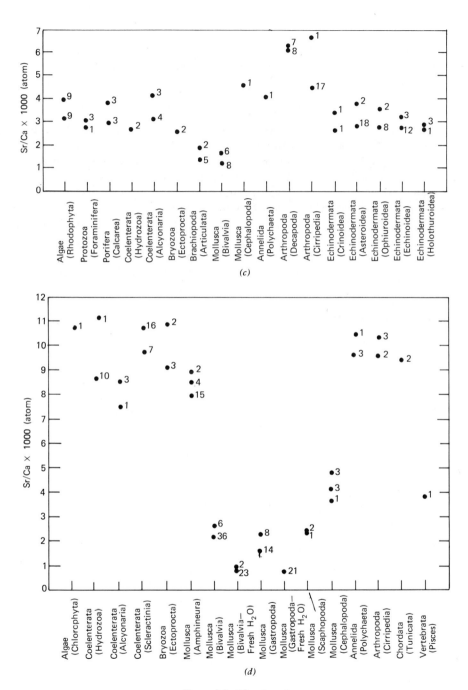

Figure 3.9 (*Continued*)

other words, the higher the concentration of a trace element relative to Ca in the water in which a shell grows, the higher will be the concentration of that element in the shell. This probably applies for many other trace constituents as well and can be expressed mathematically by

$$(M/Ca)_{skeleton} = K(M/Ca)_{water} \tag{3.1}$$

in which M is the molar concentration of either Mg or Sr (or some other cation) and Ca is the molar concentration of calcium. K is a proportionality constant commonly called the distribution or partition coefficient. The concept of distribution coefficients is especially useful in studying the trace chemistry of inorganic precipitates (e.g., Kinsman and Holland, 1969). This relationship is demonstrated for Sr in skeletal carbonates by experimental work done by Odum (1951) who varied the Sr/Ca ratio of the water in which the freshwater snail, *Physa* was growing. The Sr/Ca ratio in the water and in the snail shell were correlated, and strontianite ($SrCO_3$) formed in the shell if the Sr/Ca ratio of the water was sufficiently elevated. Buchardt and Fritz (1978) have performed similar experiments to determine the distribution coefficient for freshwater gastropods. (Fig. 3.10).

The Mg/Ca ratio in experimentally precipitated calcites is more difficult to

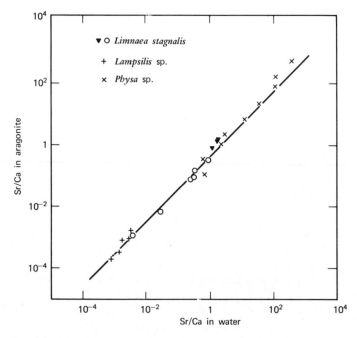

Figure 3.10 The Sr/Ca ratio in aragonite snail shells and the water in which the shells grew. After Buchardt and Fritz (1978), copyright American Association for the Advancement of Science, Science *199*:292, text-figure 2.

study, but Füchtbauer and Hardie (1976) determined the distribution coefficient for Mg/Ca in calcite under simulated marine conditions. The results are in general agreement with the Mg/Ca ratios found in naturally occurring high-Mg calcite cements that are forming in the modern oceans (Schlager and James, 1978).

The distribution coefficient has also been determined for other trace elements relative to Ca in calcite. Of special interest because of its abundance in skeletal carbonates and potential paleoenvironmental application is the Mn/Ca ratio. Raiswell and Brimblecombe (1977) have determined the distribution coefficient for this ratio in aragonite.

The distribution coefficient concept is useful in paleoecology for two reasons. (1) If the trace element to Ca ratio (M/Ca) in a fossil can be determined and if the distribution coefficient for that M/Ca ratio is known, the M/Ca ratio of the water in which the fossil grew can be determined. M/Ca ratios in natural waters depend on environmental conditions and can yield useful information about depositional environments. (2) The distribution coefficient is not constant but is temperature dependent for some ions. Thus if the M/Ca ratio is known for both the fossil and the water in which it grew, the distribution coefficient can be determined, and if the relationship of the coefficient to temperature is known, the temperature of formation can be determined. This is directly analogous to the approach extensively used in igneous and metamorphic petrology to determine temperatures of mineral formation. To complicate the problem however, the Mg/Ca ratio of the water may have a physiologic as well as physical chemical effect. For example, Lorens and Bender (1976) showed that the Mg/Ca ratio in *Mytilus edulis* shells grown in artificial sea water increased almost exponentially with increasing Mg/Ca ratios in the water in which they grew. The increase would have been linear if only simple physical chemical partitioning had been involved.

Mineralogy

Other factors being equal, Mg concentration is greater in calcite than in aragonite skeletons and Sr concentration is greater in aragonite than in calcite skeletons (Fig. 3.9; Clarke and Wheeler, 1922; Chave, 1954; and Thompson and Chow, 1955).

The reason for these differences is that the small Mg^{2+} ion (radius 0.66 Å) substitutes for Ca^{2+} (radius 0.99 Å) more readily in the calcite lattice, which has sixfold coordination and is isostructural with magnesite ($MgCO_3$), than it does in the aragonite lattice. Likewise the larger Sr^{2+} ion (radius 1.12 Å) substitutes more readily in the aragonite lattice, which has eightfold coordination and is isostructural with strontianite ($SrCO_3$).

Occasionally the physiologic effect in the molluscs apparently overpowers the mineralogic effect so that Sr concentration may actually be higher in calcite than aragonite, but this is clearly an exception to the more general rule (Lowenstam, 1964a; Dodd, 1965).

Physiologic (Genetic) Factors

Biologically formed calcite and aragonite often do not have the same trace element concentrations as inorganically precipitated minerals. This means that the distribution coefficients for biologically formed minerals (*biological distribution coefficients*) may be different from those for inorganic precipitates. The reason for this is not known but must be related to the physiology of the skeletal formation process. The physiologic effect is shown by (1) differences between different groups of organisms (phylogenetic effect), (2) differences that develop during the life of the individual organism (ontogenetic effect), and (3) differences between skeletal units within a single organism (microarchitectural effect).

Phylogenetic Effect

The most obvious evidence for physiologic control is the fact that different groups of organisms growing under the same environmental conditions have skeletons of different trace chemical composition (Fig. 3.9). On the basis of their Mg concentration, skeletal calcites can be separated into two groups: a low Mg-calcite group with less than 5 to 6 mol % $MgCO_3$ and a high Mg-calcite group with higher concentrations of $MgCO_3$. These two groups are not as distinct and well defined (Fig. 3.9) as is sometimes implied, although they are often considered to consist of two distinct mineralogies. The low-Mg groups are the molluscs, low temperature brachiopods, coccoliths, and planktonic foraminifera. With minor exceptions other groups have high-Mg skeletons.

Skeletal aragonites can similarly be divided into high- and low-Sr groups. The high-Sr group, typified by the corals and green algae, is characterized by Sr/Ca atomic ratios of about $8-11 \times 10^{-3}$, which is similar to but on the average slightly above the Sr/Ca ratio in sea water. This ratio is also similar to that found for aragonite precipitated inorganically from sea water. The low-Sr group, consisting of the molluscs (exclusive of the amphineura) and the fish (which have aragonitic otoliths), has Sr/Ca ratios mostly in the range of $1-3 \times 10^{-3}$. The cephalopods and fish are slightly higher than the rest of the low-Sr group but distinctly below the high-Sr group.

The biochemical precipitation process must occasionally deviate from the inorganic process as indicated by the differences between these Sr/Ca ratios observed in skeletal carbonates and the value predicted from the experimentally determined distribution coefficients. The Sr/Ca ratio for aragonite inorganically precipitated from normal sea water at 25°C should be about 10^{-2}, close to the value actually found for the high-Sr skeletal aragonites (Fig. 3.9) but about five times higher than the low-Sr group. The Sr/Ca ratio in calcites precipitated inorganically from normal sea water at 25°C should be about 1.3×10^{-3}. Molluscan calcites have about this value, but most other skeletal calcites have a Sr/Ca ratio of about twice this amount.

Although the Sr and Mg concentrations in the skeletal carbonates of the various groups of organisms show no well defined trends with phylogenetic position, Chave (1954) recognized a generalized pattern toward lower-Mg concentrations in skeletal calcite of physiologically more highly organized groups. A similar trend is also found for Mg in aragonite (Fig. 3.11; Lowenstam, 1963). Nevertheless, several groups of organisms are clearly exceptions to these general trends for example, the low-Mg planktonic forams and coccoliths and the high-Mg echinoderms. Also, with some notable exceptions (the annelids, cirripeds,

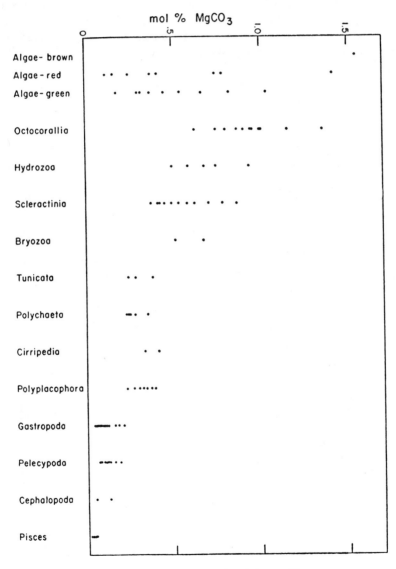

Figure 3.11 Variation in mol% $MgCO_3$ in aragonite skeletons. The groups are arranged in order of increasing phyletic complexity. After Lowenstam (1963), copyright Rice University.

and tunicates), the strontium content of biogenic aragonite is usually higher in the more primitive groups.

The observed trends for both Sr and Mg suggest that the physiology of calcification has evolved toward increasingly pure $CaCO_3$. Lowenstam (1964b) analyzed aragonitic gastropods of Recent, Cretaceous, and Pennsylvanian age (Fig. 3.12) and found a progressive increase in the Sr/Ca ratio with age which he interpreted to indicate that gastropods have been evolving toward lower Sr/Ca ratios through time. A similar trend toward decreasing Sr/Ca ratios with time has been noted in the nautiloids (Hallam and Price 1966, 1968a), but this may be because aragonite tends to add Sr during diagenesis (Ragland et al., 1969). On the other hand, some well preserved brachiopods at least as old as Mississippian seem to have the same Sr/Ca ratios as Recent forms (Lowenstam, 1961). Older fossils were not included in these studies because of the lack of adequate preservation.

Ontogenetic Effect

In addition to the phylogenetic effect on trace element composition, at least some bivalves show an ontogenetic effect. The Sr/Ca ratio in the calcitic outer shell layer of the bivalve genus *Mytilus* decreases during the life span of the individual (Fig. 3.13; Dodd, 1965). Nelson (1964) has noted that specimens of

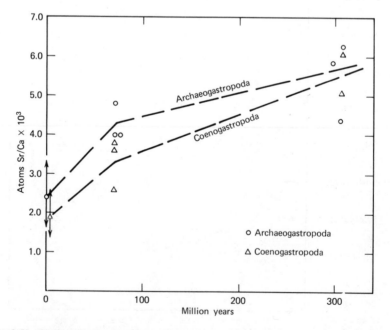

Figure 3.12 Sr/Ca ratios in aragonite shells of modern, Cretaceous, and Pennsylvanian gastropods. After Lowenstam (1964b).

some genera of freshwater bivalves at a certain age (generally 5-10 years) abruptly increase the Sr/Ca ratio of their shells.

Microarchitectural Effect

As is discussed more extensively in Chapter 4, the skeletons of many types of invertebrates consist of two or more different types of structure which are arranged in separate units within the shell. These units may have different mineralogic as well as trace element compositions (Dodd, 1966); this is another expression of physiologic control over these properties.

Method of Physiologic Control

Odum (1957b) has proposed that the Sr/Ca ratio in carbonate skeletons is a function of the degree of isolation of the calcification site from the external environment. By his model, the greater the isolation of the calcification site, the more Sr will build up in the fluids from which calcification takes place, hence the higher will be the Sr/Ca ratio. Odum indicates that the two factors primarily involved in the isolation of the calcification site are the amount of tissue between the site and the external medium and the efficiency of the circulatory system. In support of this model is the higher Sr/Ca ratio in the physiologically less advanced organisms and in certain large organisms that have a relatively large amount of tissue between the calcification site and the surrounding water. Although this model has some attractive features it also has serious drawbacks, perhaps the greatest of which is that for calcification to continue, isolation must be broken down by the introduction of fluids to renew the Ca supply.

Others have attempted to explain differences in trace element concentration as resulting from differential growth rates. An increase in the Sr/Ca ratio with increased growth rate has been noted both in some freshwater and marine bivalves (Nelson, 1963 and Zolotarev, 1974). A positive correlation between

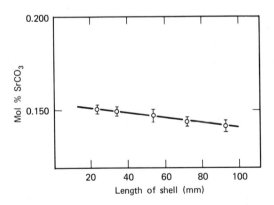

Figure 3.13 Variation with length of the specimen of mol% SrCO$_3$ in the last formed portion of the outer prismatic layer of *Mytilus californianus* from Santa Monica, California. After Dodd (1965).

growth rate and Mg content in invertebrate skeletons has also been noted (Moberly, 1968, and Zolotarev, 1974). Rapid growth might be expected to allow less time for differentiation between Ca and Mg or Sr during precipitation and thus result in a Sr/Ca and Mg/Ca ratio closer to that in sea water for more rapidly growing forms. Moberly (1968) uses essentially this argument to explain the positive correlation observed between Mg content and growth rate in various skeletal calcites. He suggested that the rate of diffusion of ions through membranes to the calcification site may also be involved in the apparent correlation between growth rate and Sr and Mg concentration. The smaller Mg^{2+} ion might diffuse more rapidly than the larger Ca^{2+} ion thus increasing the Mg/Ca ratio of the calcifying fluid during rapid growth causing a positive correlation between Mg and growth rate. This model is not applicable to the Sr^{2+} ion which is larger and should diffuse more slowly than the Ca^{2+} ion resulting in a negative correlation between Sr/Ca and growth rate.

Growth rate models seem to have some merit, but at present their state of development can account at best for only part of the observed physiologic effect. A better model to explain the physiologic effect will have to await a fuller understanding of the calcification process in general.

Environmental Factors

Temperature

The distribution coefficients for Mg and Sr relative to Ca in inorganically precipitated calcite and aragonite are temperature dependent. The Sr concentration decreases with increasing temperature in both calcite (Holland et al., 1964) and aragonite (Fig. 3.14; Kinsman and Holland, 1969). The Mg concentration in inorganically precipitated calcite increases with increasing temperature (Füchtbauer and Hardie, 1976). Similar temperature effects on biological distribution coefficients should also be reflected in the trace chemistry of skeletal carbonates, and such effects can indeed be demonstrated. However, no universal relationship between trace chemistry and temperature can be established because the values of the biological distribution coefficients differ from those for inorganic precipitates and even differ between taxonomic groups of organisms. Therefore, the temperature trace-element relationship must be determined for each taxonomic group. A given temperature-trace element relationship may hold for an entire class or phylum (Lowenstam, 1961) but others may only apply to a genus or species (Dodd, 1965).

The effect of temperature on Mg concentration in calcite is particularly well documented (e.g., Clarke and Wheeler, 1922; Chave, 1954; and Zolotarev, 1974). Mg concentration correlates positively with temperature in many groups of organisms and is especially apparent in forms characterized by a generally high Mg content (Fig. 3.15). The effect of temperature on the Sr concentration was discovered considerably later and has been demonstrated only for a few taxa (Lowenstam, 1974). Temperature and Sr content are negatively correlated in calcitic echinoids (Pilkey and Hower, 1960). They are positively correlated in

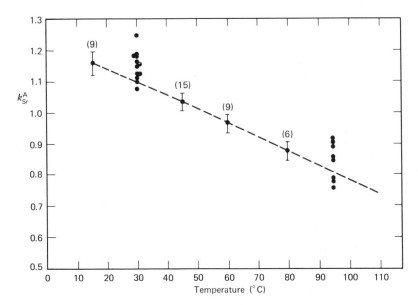

Figure 3.14 Variation with temperature of the distribution coefficient between Sr and Ca in aragonite precipitated from sea water. Numbers in parentheses indicate the number of experimental runs included in the average value. After Kinsman and Holland (1969).

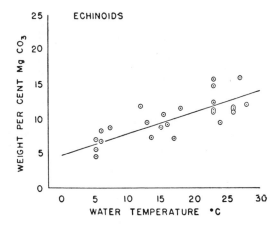

Figure 3.15 Variation with temperature of Mg concentration in echinoid skeletons. After Chave (1954), copyright University of Chicago.

articulate brachiopods (Lowenstam, 1961). Temperature and Sr are positively correlated in the calcitic outer layer of *Mytilus* but negatively correlated in the aragonitic layer (Fig. 3.16; Dodd, 1965). Corals show a negative correlation between Sr and temperature in their aragonitic skeletons which is very close to that found in inorganically precipitated aragonite (Smith et al., 1979). The variability in the correlation between temperature and Sr in different groups of

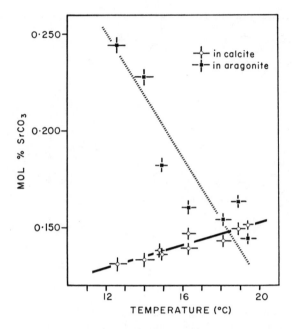

Figure 3.16 Variation with temperature of Sr in the last-formed portion of the calcite outer prismatic layer (open circles) and aragonite nacreous layer (solid squares) of *Mytilus*. After Dodd (1965).

organisms points up the difficulty in using these trace elements as paleoecologic tools. Biological distribution coefficients differ greatly from inorganic distribution coefficients in absolute value and in the direction of their variation with temperature.

The temperature-trace element relationship must be largely under physiologic control, but no very convincing models have been proposed to explain the nature of this physiologic control. Moberly (1968) and others (e.g., Kolesar, 1978) have used the growth rate model to describe the temperature effect on Mg and Sr content of skeletal carbonates because increase in temperature usually results in increase in growth rate and consequently an increase in Mg content by this model. However, this model does not explain the negative correlations between Sr and temperature found in several aragonitic groups, which of course most closely approximates that for inorganic precipitates.

Water Chemistry

Because of the partitioning effect of trace elements vs. Ca between the solution and the carbonate skeleton, the trace element to Ca ratio in the water in which the organism grew is critical in determining the trace chemistry of skeletons. Any factor affecting the trace element to Ca ratio in the water will have an influence on skeletal trace chemistry. This effect is very apparent in comparing the Sr/Ca ratio in mollusc shells from marine and fresh water. Fresh waters usually have low Sr/Ca ratios relative to sea water (Odum, 1957a) which is reflected in the generally lower Sr/Ca ratio in freshwater clams and snails as compared to marine forms (Odum, 1957b). In marine skeletons, differences in trace chemistry due to the water chemistry effect should be minor because both Sr/Ca and Mg/Ca ratios are nearly constant in sea water (Riley and Tongudai, 1967). The water chemistry is likely to be very critical in the Sr concentration in freshwater shells as Sr/Ca ratios may be quite variable in freshwater (Skougstad and Horr, 1963 and Odum, 1957a). As in the case of the Sr/Ca ratio, the Mg/Ca ratio is much lower in freshwater than in sea water (Culkin and Cox, 1966), and is clearly reflected in the generally lower Mg/Ca ratio in freshwater aragonitic bivalve shells as compared to marine aragonitic bivalve shells. The Mn/Ca ratio in shells also reflects the fact that the Mn/Ca ratio is usually higher in freshwater than in sea water (Turekian, 1969). Consequently, the Mn/Ca ratio is higher in freshwater bivalve shells than in marine forms (Vinogradov, 1953).

Eisma et al. (1976) have recently reviewed studies of the relationship of the trace chemistry of mollusc shells to salinity. Several of these studies demonstrate a correlation between trace chemistry and salinity (Fig. 3.17; Leutwein and Waskowiak, 1962; Dodd, 1965; Davies, 1972; and Neri et al., 1979) whereas others show no relationship (e.g., Thompson and Chow, 1955; and Hallam and Price, 1968b). Furthermore, the correlation between a given trace element and salinity is positive in some species but negative in others.

Shell chemistry is also correlated with salinity in groups other than molluscs. The brachiopods (Lowenstam, 1961), barnacles (Gordon et al., 1970), and echinoids (Harris and Pilkey, 1966) all show correlations between various trace elements and salinity. This variability between taxa indicates a strong physiologic control on the salinity-trace element relationship and limits its usefulness as a paleoenvironmental indicator.

As indicated above, on the basis of the distribution coefficient concept a

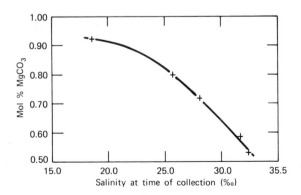

Figure 3.17 Variation with salinity of Mg in the last-formed portion of the outer prismatic layer of *M. edulis* from Hood Canal, Washington area. After Dodd (1965).

freshwater shell should have a lower Sr/Ca and Mg/Ca ratio than a marine shell. By extension brackish shells might be expected to have intermediate values for the Sr/Ca and Mg/Ca ratio, and this is probably the case for shells from water of very low salinity. However the Sr/Ca and Mg/Ca ratio of brackish waters is not a linear function of salinity since most of the Ca, Sr, and Mg in brackish water comes from the sea water. Thus the sea water dominates the composition (and trace element ratios) in brackish water and the Sr/Ca and Mg/Ca ratios are nearly constant above a salinity of about 10‰ (Figs. 3.18 and 3.19. Although the ratio of trace elements to Ca in shells may be of limited use in determining paleosalinities, these ratios in freshwater shells could be quite useful in studying the trace chemical composition of freshwater bodies.

Diagenetic Factors

A large literature has developed on the topic of carbonate diagenetic effects on trace chemistry. (See Bathurst, 1971, and Veizer, 1977 for reviews.) We can only briefly cover the topic here. Diagenetic alteration becomes an increasingly serious problem the older the fossil. Mg appears to be more subject to diagenesis than Sr and is so mobile that probably a very small proportion of fossils have been preserved with their original Mg content. Clarke and Wheeler (1922) found that all the fossils of originally high-Mg calcite groups that they examined were low in Mg. Lowenstam (1961) found no fossil brachiopods that he could be certain still had their original Mg composition; only those from cold water faunas that were probably originally low in Mg may have been preserved. For this reason, probably the best possibility of using Mg for paleotemperature determinations is in originally low-Mg calcite groups such as the molluscs.

The original Sr concentration in carbonates appears to have a much better chance of preservation. Detailed studies indicate that some calcite fossils (brachiopods and bivalves) ranging from Mississippian to Pleistocene in age have retained their original Sr composition (Lowenstam, 1961; Dodd, 1966; Stanton and Dodd, 1970). The evidence for preservation of the original Sr composition in aragonitic skeletons is not as good although Cretaceous and Pennsylvanian aragonitic gastropods seem to have their Sr composition preserved (Lowenstam, 1964b). Diagenetic alteration of elements other than Mg and Sr has been little studied although Cook (1977) demonstrated apparent leaching of B from fossil oysters.

Diagenesis will change the chemical composition of the fossil until it is in physical chemical equilibrium with the diagenetic environment. The controlling factor is probably the ionic ratios in the fluids around the fossil. The Mg/Ca ratio of ground water is almost invariably much lower than in sea water (Livingstone, 1963). Consequently the Mg/Ca ratio in most marine carbonates is much higher than the equilibrium values in the diagenetic environment, and during diagenesis Mg is lost from these carbonates. The Sr/Ca ratio of ground water is variable but usually is considerably less than that of sea water (Odum, 1957a). Consequently, skeletal carbonates that are in approximate physical

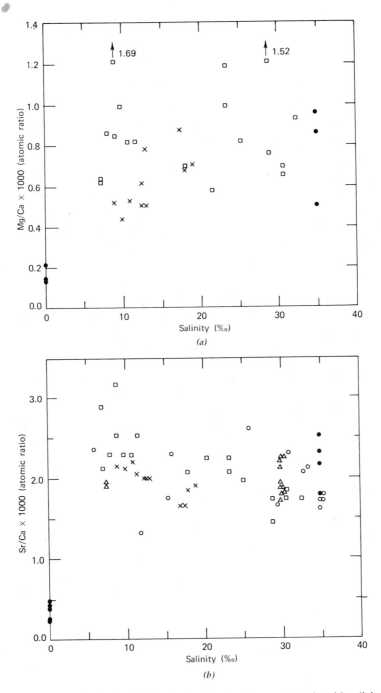

Figure 3.18 (a) Variation in the Mg/Ca atomic ratio of molluscan aragonite with salinity of the water in which they formed. (b) Variation in the Sr/Ca atomic ratio of molluscan aragonite with salinity of the water in which they formed. Data from several published sources.

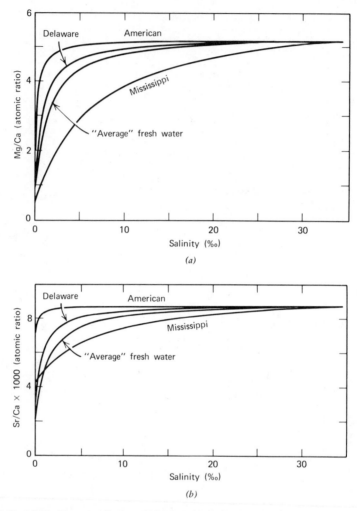

Figure 3.19 (*a*) Mg/Ca ratios that would be found in mixtures of fresh water from three rivers and "average" fresh water with normal sea water. (*b*) Sr/Ca ratios that would be found in mixtures of fresh water as in (*a*) with normal sea water.

chemical equilibrium with sea water, such as low-Sr calcites and high-Sr aragonites (Kinsman, 1969), tend to lose Sr during diagenesis. On the other hand, the Sr/Ca ratio of low-Sr aragonites is much lower than that in aragonite in physical chemical equilibrium with sea water and in fact lower than equilibrium values for most ground water. Thus the Sr content of these aragonite skeletons will tend to increase during diagenesis (Ragland et al., 1969). Veizer (1977) considered the effect of diagenesis on Sr concentrations and concluded that originally low-Sr calcite is likely to be the least altered.

Because diagenesis by ground water appears to be the dominant cause of alteration of skeletal trace chemistry, unaltered fossils are most likely to be found in a location or matrix where they have been shielded from contact with ground water (Lowenstam, 1961). Actually, deep sea fossils that have always been in contact with sea water should have the best chance for preservation. A fine-grained impermeable matrix around the fossil or, even better, a coating of asphalt (Stehli, 1956; Hallam and O'Hara, 1962) provides the best matrix for preservation.

Several criteria can be used to attempt to answer the difficult question of whether the trace chemical composition of a shell has been diagenetically altered. The question is a difficult one because the fossil may be chemically altered but may appear to be physically unaltered even under microscopic observation.

1   If the fossil is physically recrystallized, it is clearly chemically altered.

2   If the chemical composition of the fossil is outside of the range of variation of modern representatives of its group, it is probably altered, especially if the composition of the fossil is approximately in physical chemical equilibrium with its diagenetic environment. The possibility must be considered, however, that the changed chemical composition is due to evolutionary change in the characteristic chemical composition of that group through geologic time.

3   Variation in the chemical composition between different portions of a single skeleton (the ontogenetic effect) also suggests preservation of the original composition. This is especially good evidence if the pattern of variation is similar to that found in the modern organism. For example, modern specimens of the bivalve *Mytilus* show a cyclical variation in the Sr content of the outer prismatic layer probably resulting from seasonal temperature changes (Dodd, 1965). They also show a slight overall decrease in the amount of Sr progressing from the beak toward the posterior margin. These same features have been found in Pleistocene (Dodd, 1966) and Pliocene (Stanton and Dodd, 1970) *Mytilus* suggesting preservation of the original Sr composition. Curtis and Krinsley (1965) have pointed out that diagenesis should tend to make the chemistry of an altered fossil more uniform than that of the original skeleton. This is because the fossil is reequilibrating to an environment that is uniform around the entire fossil whereas the environment under which the skeleton formed was variable, both in terms of different calcification sites and seasonal variation.

4   Probably the best criterion for preservation is the demonstration that a paleoenvironmental interpretation based on skeletal chemistry is in agreement with an independent method of arriving at the same conclusion. For example, if a paleotemperature determined from the Sr content of a brachiopod is approximately the same as that based on an analysis of the biota associated with the brachiopod, the chances are good that the Sr com-

position is original. One might argue that if an independent method of paleotemperature determination is available, there is no need to bother to use trace chemistry to determine a paleotemperature. Probably no paleo-temperature determination method is worthy of complete confidence, so if two or more methods can be found to give the same answer, confidence in each is bolstered. Furthermore, the geochemical method may yield details such as the temperature range and a single, average temperature not possible with other methods.

Geologic Application of Trace Chemical Relationships

Application Types

Data on the trace chemistry of skeletal carbonates have been used in the study of a number of different types of geologic problems. (1) Interpretation of the trace chemistry of carbonate sediments depends on a knowledge of the trace chemistry of the organically derived carbonates that usually comprise the principal portion of the sediments. The major features of the trace element composition of the sediments can be explained in terms of the relative proportions and taxonomic position of the major biotic contributors to the sediment (see Milliman, 1974, for examples). (2) Studies of the diagenesis of carbonate sediments and rocks are also greatly aided by a knowledge of the trace chemistry of the skeletal materials that usually are their major constituent. The approximate starting chemistry of an altered carbonate can usually be predicted from the biotic composition of the rock. Knowing what was the original composition allows one to propose models to explain the observed chemical composition. Such an approach has been used in many studies of carbonate diagenesis (see Bathurst, 1971, for examples). (3) Skeletal carbonates form an important part of the geochemical cycle of several elements (Lowenstam, 1974), especially C, Ca, Mg, and Sr; hence an understanding of the chemistry of the carbonates is an aid in the understanding of the geochemical cycles of these elements. Studies by Revelle and Fairbridge (1957), Odum (1957a), and Turekian (1964) are examples of how data on skeletal trace chemistry have been used in considering geochemical cycles. (4) Of particular interest to the paleoecologist is the use of skeletal chemical data to make environmental interpretations. Trace element concentrations might be used to distinguish between freshwater and marine fossils or perhaps to determine salinity. As noted above, Mg and Sr concentrations are correlated with temperature in many groups of organisms. The Mg and Sr concentration in fossil representatives of these groups should then be usable for making paleotemperature interpretations.

Problems

Three major problems must be considered and/or overcome before trace chemically determined paleotemperatures can be used.

1 Diagenesis may be an extremely serious problem as discussed above, particularly in dealing with older fossils and with Mg concentrations. The effect of diagenesis should be evaluated by using as many of the criteria for recognition of diagenesis discussed above as possible.

2 Evolutionary change in the temperature-chemistry relationship is impossible to recognize except by using extant genera or species or by comparing paleoenvironmental conditions determined from trace element composition with those determined by an independent technique. At our present state of knowledge the evolutionary factor must be determined independently for each taxon before paleotemperatures based on trace element chemistry can be used with confidence. Brachiopods would appear to be very stable in their trace chemistry, probably not having changed since Mississippian time (Lowenstam, 1961). The aragonitic gastropods studied by Lowenstam (1964b) and nautiloids studied by Hallam and Price (1966 and 1968a) have gradually evolved toward decreasing Sr/Ca ratios with time. The fact that the trace element concentration in mollusc shells differs from genus to genus if not species to species (Turekian and Armstrong, 1960) indicates that the trace chemistry of the molluscs has been evolving. This problem can be avoided by doing lateral studies within rocks of the same age to recognize gradients.

3 The use of Mg/Ca or Sr/Ca ratios to determine paleotemperatures depends on the assumption that these ratios in sea water were the same when the fossils lived as they are today. Based on his study of fossil brachiopods, Lowenstam (1961) states that at least the Sr/Ca ratio has been constant since Mississippian time. Odum (1957a) had earlier concluded on the basis of his study of the Sr geochemical cycle that the Sr/Ca ratio has been constant since early in the earth's history. Turekian (1964) proposes that the Sr/Ca ratio may have been decreasing slightly since late Mesozoic time because of the effect of planktonic foraminifera on the geochemical balance. Chilingar (1956) considered the question of possible changes in the Mg/Ca ratio in sea water by examining the bulk chemistry of carbonate rocks of different ages, but the evidence is too inconclusive to make any positive statements about the Mg/Ca ratio in sea water through time. To date we have no compelling evidence that these ratios have changed, at least since Paleozoic time.

Examples

Paleotemperature determinations based on Sr/Ca ratios have been made on fossils of Pleistocene, Pliocene, Cretaceous, Permian, and Mississippian age. The Pleistocene determinations were made on specimens of the bivalve, *Mytilus californianus* from terrace deposits at five localities along the California and northern Baja California coast (Dodd, 1966). These temperatures were compared with those determined on the basis of the associated faunas, oxygen isotopic composition, shell structural relationships, and shell mineralogy. At four of the five localities the Sr paleotemperatures appeared to be reliable. At one locality the temperature was apparently too low, suggesting diagenetic loss of Sr. The paleotemperatures based on shell mineralogy were always low, suggesting loss of aragonite. The temperatures recorded by the fossils in this study were the same as or slightly warmer than present temperatures at those localities, suggesting an interglacial time of origin for the terraces.

Lowenstam (1961) determined paleotemperatures from Sr/Ca ratios in fossil brachiopods of Pliocene, Cretaceous, and Permian age. Mississippian samples were included but some were apparently slightly altered. Lowenstam (1964c) presented a more detailed discussion of the Permian paleotemperatures in a later paper. One sample from Australia was taken from an interbedded marine-tillite sequence (Lyons Group). The paleotemperature from this sample was relatively low ($7.7°C$). Samples from the NoonKenbah Formation, stratigraphically above the tillites, gave paleotemperatures ranging from 17.4 to $26°C$. This suggests a gradual warming at that site during a portion of Permian time.

Berlin and Khabokov (1966 and 1974) determined paleotemperatures for the Cretaceous of Russia on the basis of the Mg concentration in belemnites and other fossils. Their results are questionable as they did not consider physiologically controlled differences between groups of organisms or diagenesis.

Wider use of Sr/Ca paleotemperatures should be possible, especially in the upper part of the Cenozoic where diagenesis and evolutionary changes in the temperature–Sr/Ca relationship is less likely to be a problem. Well preserved brachiopods of any age would apparently offer good potential although adequate preservation in pre-Cenozoic rocks is rare. More study of both modern and fossil skeletal material would obviously greatly enhance the usefulness of the technique. At this stage Sr/Ca paleotemperatures should probably only be used in conjunction with an independent paleotemperature determination method. Finally, low-Sr calcites would seem to be more useful for paleotemperature determinations than aragonites, which appear to either pick up or lose Sr more readily.

Because of its lack of diagenetic stability, Mg appears to be a less promising paleotemperature tool than Sr. However, low-Mg calcites have not been adequately tested and eventually may prove useful in paleotemperature work.

Trace Chemistry of Fossils from the Kettleman Hills

Paleotemperatures were determined on the basis of the Sr concentration in the outer calcite layer of specimens of the bivalve *Mytilus* from five of the zones in the Pliocene section in the Kettleman Hills (Stanton and Dodd, 1970). This genus was used because of the detailed information available concerning the temperature-Sr relationship for modern specimens of the genus (Dodd, 1965). By so doing we hoped to overcome any problem of genetically related differences in trace chemical relationships between different bivalve genera. The necessary assumption that the Sr/Ca ratio in the sea that occupied the Pliocene embayment was the same as in the modern open ocean is probably valid because the Sr/Ca ratio in modern oceans is nearly uniform and we have found no evidence of an unusual chemical composition of the Kettleman Hills sea. A more difficult problem is to determine that the fossils have not been diagenetically altered but several lines of evidence suggest that they were not. (1) The specimens all appeared physically well preserved, without recrystallization. (2) The Sr concentrations are within the range for modern specimens of *Mytilus*. (3) The Sr concentration varies within individual specimens in a manner similar to that in modern specimens of *Mytilus* reflecting seasonal temperature variation (Fig. 3.20). (4) Finally, the paleo-temperatures measured are in good agreement with those suggested by other nonchemical techniques.

In the analysis, serial samples from individual shells were used to determine the temperature range during the growth of the specimen, to estimate seasonal temperature variation. The mean temperatures measured for the five zones range from 12.8 to 14.6°C (Fig. 3.21). These temperatures are about the same as those found today off the central California coast. This agrees with the results obtained from the taxonomic uniformitarian approach (Chapter 2) as well as other methods. The temperature range measured in four of the zones is somewhat greater than that found on the coast at present, but a larger range in temperature would be expected in a bay where oceanic circulation is not so effective in suppressing seasonal temperature differences.

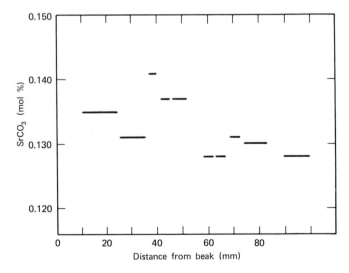

Figure 3.20 Variation in Sr concentration with distance from the beak in the outer calcite layer of a single large specimen of *Mytilus coalingensis* from the upper *Pseudocardium* zone. After Stanton and Dodd (1970), Journal of Paleontology, Society of Economic Paleontologists and Mineralogists.

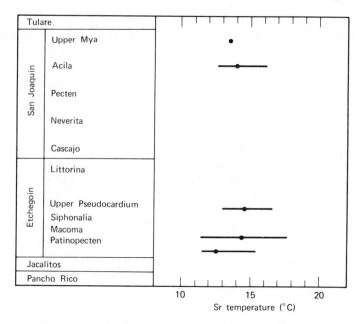

Figure 3.21 Temperature means and ranges as determined from the Sr content of the outer calcite layer of specimens of *Mytilus coalingensis* and *M.* cf. *M. edulis* (upper *Mya* zone) from zones in the Kettleman Hills section. After Stanton and Dodd (1970), Journal of Paleontology, Society of Economic Paleontologists and Mineralogists.

ISOTOPIC TECHNIQUES

The oxygen isotopic determination of paleotemperatures is the best known and most widely applied of the various geochemically based techniques of paleoenvironmental interpretation. This is because the theory behind its operation is largely explainable in physical chemical terms and the effects of physiologic factors are relatively minor.

The idea of using the oxygen isotopic composition of fossils to determine paleoenvironmental conditions was first developed by H. C. Urey (1947). The original application that he had in mind was to differentiate between marine and freshwater fossils, but from a consideration of the theory of isotopic behavior the possibility of determining palcotemperatures also soon became apparent. Although theory indicated that the oxygen isotopic composition of fossils should vary with the growth temperature of the skeleton, the then-existing mass spectrometers were not capable of measuring the small isotopic differences that would result. A mass spectrometer capable of measuring differences as small as one part in 10,000 was eventually developed by Urey and his colleagues at the University of Chicago and the first paleotemperature determinations were published (Urey et al., 1951).

Very precise determinations of carbon isotopic composition became possible as a byproduct of the oxygen isotopic project. Although the factors controlling carbon isotopic composition of fossils are more complex and less predictable than those controlling the oxygen isotopic composition, use can be made of the carbon isotopic results in paleoenvironmental interpretation, particularly in differentiating marine and nonmarine conditions.

Oxygen isotopic composition of fossils has been more extensively studied than any of the other geochemical properties because physiologic and diagenetic effects on isotopic composition are minimal. Thus interpretation is largely a matter of considering the interplay of the physical chemistry of the system as mediated by the environmental effects. The problem then becomes a matter of interpreting which environmental characteristics produced the physical chemical effect.

Isotopes of other elements could potentially be used in paleoecologic studies. Some work has been done on the measurement of Ca and Mg isotopes in skeletal materials and many studies have been made of sulphur isotopes, which also show the effects of biological activity. As these studies do not involve skeletal materials, we do not consider them here.

Physical Chemistry

The most familiar applications of the study of isotopes in geology are based on the decay of radioactive isotopes such as U^{238}, K^{40}, and C^{14} in order to determine geologic ages. The applications discussed in this chapter are concerned with the stable, nonradioactive isotopes of oxygen and carbon. Approximately 99.76% of atmospheric oxygen consists of the isotope O^{16}, 0.04% is O^{17}, and 0.20% is O^{18} (Nier, 1950). Two stable isotopes of carbon, C^{12} and C^{13}, also occur in natural materials in the approximate percentages of 98.9 and 1.1. The radioactive isotope C^{14} also occurs in nature in very small amounts, but is not considered in this chapter.

Because isotopes of an element differ only in the number of neutrons in the atomic nucleus, they behave very similarly, but not identically, in chemical reactions. The slight difference in weight also causes them to behave differently in physical reactions. Both chemical and physical differences result in differences in the proportion of oxygen isotopes in skeletal material. The greater the difference in atomic weight, the greater is the fractionation by physical and chemical processes. Thus in the case of oxygen, differences in relative abundance between O^{18} and O^{16} will be greater than those between O^{17} and O^{16} or O^{17} and O^{18}. In the subsequent discussion we will only consider differences in the relative abundance of O^{18} and O^{16} because these are easiest to measure, but the principles involved also apply to differences in the relative concentration of O^{17}.

Both oxygen and carbon atoms are present in carbonate skeletons in the $CO_3{}^{2-}$ ion. During the skeleton formation process these ions were presumably in chemical equilibrium with the oxygen and carbon in the water in which the skeleton formed through a series of chemical reactions:

$$CO_2 + H_2O \rightleftharpoons H_2CO_3 \rightleftharpoons H + HCO_3^- \rightleftharpoons 2H^+ + CO_3^{2-} \qquad (3.2)$$

Due to their slight chemical differences, the different isotopes of oxygen and carbon do not behave identically in this series of reactions and thus in order to minimize the free energy of the system the isotopic ratios in the various components of the reactions differ. As we are only interested in the end members of this series of reactions we can simplify the equation to show the isotopic exchange between water and the carbonate ion:

$$H_2O^{18} + \tfrac{1}{3} CO_3^{16=} \rightleftharpoons H_2O^{16} + \tfrac{1}{3} CO_3^{18=} \qquad (3.3)$$

The equilibrium constant, K, for this reaction can then be written as

$$K = \frac{[H_2O^{16}] [CO_3^{18=}]^{1/3}}{[H_2O^{18}] [CO_3^{16=}]^{1/3}} \qquad (3.4)$$

where the bracketed quantities indicate molecular concentrations. For simplicity, and by convention, the *fractionation factor* α is usually used in discussion of isotopic separation or fractionation. The fractionation factor for the reaction above is

$$\alpha = \frac{(O^{18}/O^{16})_{CO_3^{2-}}}{(O^{18}/O^{16})_{H_2O}}$$

or

$$(O^{18}/O^{16})_{CO_3^{2-}} = \alpha \, (O^{18}/O^{16})_{H_2O} \qquad (3.5)$$

Where $(O^{18}/O^{16})_{CO_3^{2-}}$ means the ratio of O^{18} to O^{16} in the carbonate ions and $(O^{18}/O^{16})_{H_2O}$ means the ratio of those isotopes in the water. If the oxygen isotopes behave identically chemically, both α and K would equal 1. In fact at 25°C, $\alpha = 1.021$ for this reaction (McCrea, 1950), indicating that O^{18} concentrates slightly in the carbonate relative to the water. As discussed below, the temperature dependence of the fractionation factor makes possible the determination of paleotemperatures. Note the similarity between the fractionation factor in isotopic studies and distribution coefficients in trace chemistry studies. The fractionation factor for oxygen isotopes varies with temperature just as does the distribution coefficient between a trace element and Ca in calcite and aragonite. The same types of relationships also apply to carbon isotopes. The carbon isotopes will fractionate between CO_2 and CO_3^{2-} in much the same way as do the oxygen isotopes.

This relationship in (3.5) shows that at a given temperature the O^{18}/O^{16} ratio of the carbonate should vary directly with the O^{18}/O^{16} ratio of the water with which the carbonate has equilibrated. In other words, in (3.5), if α has a fixed value, the $(O^{18}/O^{16})_{CO_3^{2-}}$ value must vary directly with $(O^{18}/O^{16})_{H_2O}$. Thus any factor causing $(O^{18}/O^{16})_{H_2O}$ to change will be reflected in the $(O^{18}/O^{16})_{CO_3^{2-}}$ value. Epstein and Mayeda (1953) and many subsequent workers have shown that the $(O^{18}/O^{16})_{H_2O}$ ratio indeed does vary and in a predictable way that can be correlated with environmental factors. The basic cause of the variation is that in the physical process of evaporation, water molecules fractionate on the basis of their oxygen isotopic content. This fractionation can be schematically represented as

$$(H_2O^{16})_L + (H_2O^{18})_V \rightleftharpoons (H_2O^{18})_L + (H_2O^{16})_V \qquad (3.6)$$

where $(H_2O^{16})_L$ is the abundance of this molecular species in the liquid phase, $(H_2O^{18})_V$ is the abundance of that molecular species in the vapor phase, and so on. The fractionation of the isotopes can then be represented as

$$\alpha = \frac{(O^{18}/O^{16})_L}{(O^{18}/O^{16})_V} \qquad (3.7)$$

At 25°C, the α value for this physical reaction is 1.008. That is, the ratio of H_2O^{18} to H_2O^{16} in the liquid is 0.8‰ higher than the H_2O^{18}/H_2O^{16} ratio in the vapor phase. This equilibrium relationship for the two types of water molecules holds both for the evaporation process and the precipitation process. This physical fractionation is due to the difference in vapor pressure of the H_2O^{16} molecule which has a slightly higher vapor pressure, hence evaporates into the vapor phase more readily than does the heavier H_2O^{18} molecule. When water evaporates the vapor phase will have an H_2O^{18}/H_2O^{16} ratio which is 8‰ lower than that in the liquid water. Likewise, when liquid water condenses from vapor as rain, the liquid will have an H_2O^{18}/H_2O^{16} ratio which is 8‰ higher than the ratio in the liquid from which it formed. This is analogous to a mixture of water and ethyl alcohol. The alcohol has a lower vapor pressure than does water, hence evaporates more readily. If the vapor above a cocktail were analyzed, it would have a slightly higher alcohol-to-water ratio than would the cocktail itself.

Measurement of Isotopic Ratios

Both oxygen and carbon isotopic ratios are measured by mass spectrometry using an instrument that is basically the design of A. O. Nier. This system requires that the sample be in the form of a gas. Carbon dioxide is used for the gas because of the ease with which it can be prepared and handled. The CO_2 is prepared from the sample by reaction with orthophosphoric acid. The method of preparation and purification is critical because chemical fractionation occurs during the acid reaction.

$$6H^+ + 2PO_4^{3-} + 3CaCO_3 \rightleftharpoons 3CO_2 + 3H_2O + 3Ca^{2+} + 2PO_4^{3-} \quad (3.8)$$

As can be seen from this reaction, only two thirds of the oxygen in the sample goes into the CO_2. The rest reacts with the hydrogen ions to form water. Fractionation of the oxygen isotopes between the water and CO_2 occurs in the reaction, but the CO_2 can be used as a reliable sample of the original carbonate if the reaction is always conducted under the same conditions (fractionation adds a constant error to all samples and standards). Detailed discussions of sample preparation techniques, are given by Bowen (1966). More recently Shackleton (1973) has introduced a modified, somewhat simpler method of sample preparation.

The CO_2 is introduced into the mass spectrometer through a small gas leak (Fig. 3.22). It enters the source where it is ionized by electron bombardment to CO_2^+ ions. The ions are accelerated in an electrostatic field and collimated and emerge as an ion beam that passes through a magnetic field. The ions are deflected by an amount that depends on their mass, the lighter ions being deflected more than the heavier. Collectors for each mass to be measured are placed at the appropriate spot beside the magnetic field. The ions discharge on the collectors and the amount of charge, which can be measured by the electronic circuitry, will be proportional to the number of ions of each mass. Precision of the isotopic measurements is improved by comparison of the sample with a standard by alternately running sample and standard. This comparison is facilitated by a special magnetically operated valve allowing rapid switching between sample and standard. The usual maxi-

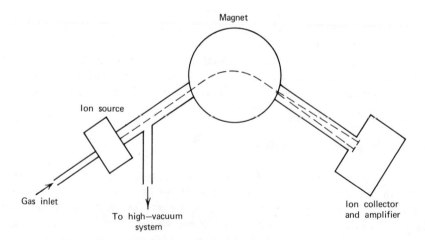

Figure 3.22 Schematic representation of mass spectrometer.

mum precision claimed for analyses using this technique is $\pm$ 0.1‰ (standard deviation).

The ionic masses of special interest in oxygen and carbon isotopic studies are masses 44, 45, and 46. Mass 44 is the most common and consists entirely of C^{12}, O^{16}, O^{16}. Mass 45 consists largely of C^{13}, O^{16}, O^{16} and Mass 46 is largely C^{12}, O^{16}, O^{18}. The ratio of Mass 46/Mass 44 will be a close approximation of the O^{18}/O^{16} ratio and the Mass 45/Mass 44 ratio will approximate the C^{13}/C^{12} ratio. These mass ratios are obviously not perfect measures of the isotopic ratios because all of the isotope of interest is not in the ions of the mass measured. For example a small portion of the O^{18} will be in the following molecular species: C^{16}, O^{18}, O^{18}; C^{13}, O^{16}, O^{18}; and C^{13}, O^{18}, O^{18}. None of these ions will be measured, but they will contain a very small portion of the total O^{18}. Likewise, the O^{16} is distributed among several molecular species, all much less abundant than C^{12}, O^{16}, O^{16}. Similar distribution problems exist in connection with the C^{13}/C^{12} determination. Finally, the presence of small amounts of O^{17} further complicates the problem. Some of the ions of Mass 45 will come from C^{12}, O^{16}, O^{17} and not from C^{13}, O^{16}, O^{16}. The error produced by these distribution problems is small and Craig (1957) has derived correction formulae for them.

For most purposes the difference in the O^{18}/O^{16} or C^{13}/C^{12} ratio between samples is of more interest than is the absolute value of the ratio. Thus the values reported in the literature are usually δ (delta) values or parts per thousand or per mil (‰) deviation from a standard. The δO^{18} value is mathematically defined as

$$\delta O^{18} = \left(\frac{(O^{18}/O^{16})\ sample - (O^{18}/O^{16})\ standard}{(O^{18}/O^{16})\ standard} \right) 1000 \qquad (3.9)$$

The δC^{13} value is similarly defined. The δ value must be referred to a certain standard. Many different standards have been used, but for paleotemperature work, δ values are usually referred to the PDB standard. This standard gets its name from the fact that it was prepared from specimens of the belemnite *Belemnitella americana* collected from the Peedee formation of South Carolina. Actually the supply of the PDB standard has long been exhausted and samples are now calibrated against secondary or tertiary standards whose relationship to PDB has been determined. Commonly in the more recent literature, samples (especially water samples) are referred to the SMOW standard (Craig, 1961). SMOW is the acronym for Standard Mean Ocean Water. As the name implies, this standard is average open marine sea water. The δC^{13} values are almost always referred to PDB.

Physiologic (Genetic) Factors

Fortunately for the use of oxygen isotopic analysis for paleoenvironmental interpretation, physiologic effects on isotopic composition are minimal and usually can be readily detected, at least in modern skeletal material. The oxygen isotopic composition of inorganic carbonates can be theoretically (Urey, 1947) or experimentally (McCrea, 1950) determined. Any departure of the oxygen isotopic composition of a modern shell from inorganic carbonate formed under the same environmental conditions should be due to physiologic factors. These have been called *vital effects* (Urey et al., 1951) and nonequilibrium precipitation in the literature. Fortunately several important groups of organisms in the fossil record such as the molluscs, brachiopods, and foraminifera seem to have little physiologic control. To avoid the problem of vital effects, paleotemperature studies have been largely confined to these three phyla. Some studies (e.g., Tourtelot and Rye, 1969, and Duplessy et al., 1970) have shown that physiologic effects do sometimes exist even in these groups. Large physiologic effects are noted in some groups such as the·calcareous algae, corals and echinoderms. Several explanations for nonequilibrium precipitation are possible. The oxygen in the carbonate skeletons in some of these groups may in part come from metabolic CO_2 (e.g. Weber and Raup, 1966) and thus may reflect the isotopic compositions of the food and fractionation during metabolic or photosynthetic processes. The presence of symbiotic zooxanthellae in the tissues of the scleractinian corals and other groups clearly affects the calcification process and probably also the isotopic composition of the carbonate (Weber and Woodhead, 1970). Another possibility is that in some groups calcification may be so rapid that equilibrium with the surrounding water is not attained. In recent years several studies of the oxygen and carbon isotopic composition of such groups has been undertaken as a technique for better understanding the calcification process (e.g. Weber and Woodhead, 1970 and Land et al., 1975).

Carbon isotopic composition may also be affected by physiologic processes. The same factors that affect oxygen isotopic composition, such as the production of metabolic CO_2 and photosynthesis by zooxanthellae, will also affect carbon isotopic composition.

Environmental Factors

Temperature

The extent of fractionation of different isotopes of an element in a chemical reaction is temperature dependent. The lower the temperature the more differently isotopes of an element behave. This can be seen from the thermodynamic relationship for the equilibrium constant of the reaction

$$\ln K = -F°/RT \qquad (3.10)$$

where K is the equilibrium constant, $F°$ is the standard Gibbs free-energy change for the reaction, R is the gas constant, and T is the absolute temperature. (See any textbook in physical chemistry for a more detailed explanation of this relationship.) The smaller the value of T the greater will be the value of K and thus the greater the fractionation. K goes to infinity as T approaches absolute zero. Theoretically the isotopic species should separate completely at absolute zero. As T increases, $\ln K$ approaches zero and thus K will approach unity, meaning no fractionation. As indicated above, the fractionation factor α for the reaction involving H_2O and $CO_3{}^{2-}$ at 25°C is 1.021 whereas α at 0°C is 1.025 (McCrea, 1950). The fractionation is such that the O^{18} tends to concentrate in the carbonate relative to the water in which it formed. Figure 3.23 shows the pattern of variation with temperature of the O^{18}/O^{16} ratio in the carbonate if the O^{18}/O^{16} ratio of the water is constant. The relationship of temperature and fractionation indicates that one should be able to determine the paleotemperature by determining the amount of fractionation, for example, the difference in O^{18}/O^{16} ratio between a fossil and the water in which it formed. But how can we determine the O^{18}/O^{16} ratio in the water of an ancient sea? Obviously we cannot, and so we must assume some value for this ratio in the water. This is clearly a serious weakness of the method, but various relationships allow us to make reasonable estimates under many circumstances. The most important factor allowing us to make reasonable estimates of the isotopic composition of water is the high volume of water in the oceans. The open, well-mixed part of the modern ocean varies little in its oxygen isotopic composition; localized, small scale processes have little effect on the oceanic ratio. Thus if paleotemperature determinations are restricted to specimens that have grown in the open ocean the assumption is usually made that the composition of the water was the same as the modern open ocean. A correction should be made, however, for the water held on the continents in glacial ice, as will be discussed below.

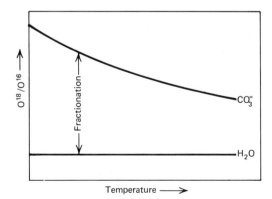

Figure 3.23 Schematic representation of the difference in the O^{18}/O^{16} ratio in water and carbonate precipitated in equilibrium with that water at different temperatures. The separation of the lines is a function of temperature-sensitive fractionation.

Epstein et al. (1951) empirically determined the relationship between the O^{18}/O^{16} ratio in mollusc shells and growth temperature by analyzing a series of samples that had either been grown in temperature-controlled aquaria or collected at natural sites where the temperature had been carefully monitored. The O^{18}/O^{16} ratio of the water in which the shells grew was also determined to take into account any variation in this ratio. Over the temperature range of about 5 to 30°C, the relationship between the O^{18}/O^{16} ratio and temperature is practically linear (Fig. 3.24). Craig (1965) has slightly modified the original temperature equation to the form

$$T(°C) = 16.9 - 4.2\,(\delta O^{18}{}_S - \delta O^{18}{}_W) + 0.13\,(\delta O^{18}{}_S - \delta O^{18}{}_W)^2 \qquad (3.11)$$

where $\delta O^{18}{}_S$ is the deviation of the O^{18}/O^{16} ratio in the sample from the PDB standard in parts per thousand and $\delta O^{18}{}_W$ is the deviation of the O^{18}/O^{16} ratio in the water in which the sample grew from the PDB standard. This form of the temperature equation is in most common use today although other modifications have been suggested (e.g., Shackleton, 1973).

Urey and his co-workers (1951) suggested a method for avoiding the necessity of assuming an O^{18}/O^{16} ratio for the water. They suggested the possibility of determining the temperature relationship for fractionation of oxygen isotopes between water and some compound other than carbonate (such as phosphate or silica). Perhaps one could determine the relationship

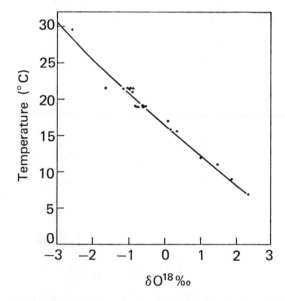

Figure 3.24 Variation in the O^{18}/O^{16} ratio of skeletal carbonate (‰ deviation from the PDB standard) with growth temperature. After Epstein et al. (1953), Geological Society of America Bulletin.

between fractionation and temperature for phosphatic fossils. Of course, this relationship would also require a knowledge of the O^{18}/O^{16} ratio of the water in which the fossil grew. In effect, this would give us two equations with the same two unknowns, the temperature and the O^{18}/O^{16} ratio of the water. Simultaneous solution of the equations should thus allow us to determine both unknowns. This relationship is shown graphically in Fig. 3.25. In effect the difference in the O^{18}/O^{16} ratio between the carbonate and phosphate that have formed in the same water is temperature dependent. Longinelli (1966) has in fact developed this method, but unfortunately the fractionation of oxygen isotopes in the phosphate-water system is quite similar to that in the carbonate-water system. Some work has also been done on establishing a paleotemperature scale based on siliceous skeletons (Labeyrie, 1974). Especially with the experimental errors in making the O^{18}/O^{16} measurements, only very low-precision paleotemperature measurements can be made by the technique based on differences in the O^{18}/O^{16} ratio between phosphate or silica and carbonate. In Fig. 3.25 this is shown by the fact that the slopes of the phosphate, silicate, and carbonate curves are nearly the same. Similar determination of the temperature of formation of igneous and metamorphic minerals by analyzing the fractionation of oxygen between different coexisting mineral phases has become a common practice (Faure, 1977).

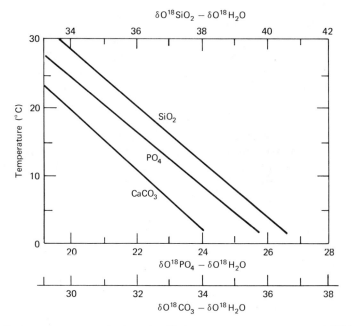

Figure 3.25 Empirically determined relationship between water temperature and O^{18}/O^{16} ratios in the shells of carbonate, phosphatic, and siliceous organisms. After Hecht (1976), with permission from Foraminifera, Vol. 2; copyright Academic Press Inc, (London) Ltd.

Carbon isotopes are obviously subject to the same physical chemical principles as oxygen and other isotopes and fractionate in chemical reactions such as those shown in (3.2). Thus, in theory, paleotemperatures could be determined from the C^{13}/C^{12} ratio in carbonate fossils. Two factors make such a paleotemperature method less practical than one based on oxygen isotopes. (1) The amount of fractionation between the isotopes of an element is dependent on the relative mass difference between the isotopes, and the mass difference between C^{13} and C^{12} is obviously less than that between O^{18} and O^{16}. (2) The C^{13}/C^{12} ratio in the dissolved bicarbonate is more variable than is the O^{18}/O^{16} ratio and is thus more difficult to reliably estimate in order to determine the amount of fractionation in the precipitation of $CaCO_3$. The lower variability of the O^{18}/O^{16} ratio is due to the enormous mass of oxygen in the reservoir of the oceans with which the oxygen in the carbonate is in equilibrium. The mass of dissolved carbon in the oceans is orders of magnitude less and thus the carbon reservoir is much more likely to be affected by local chemical processes than is the oxygen. Nevertheless, temperature variation in fractionation of carbon can be detected in nature (Emrich et al., (1970). The C^{13}/C^{12} ratio in a carbonate skeleton is more likely to reflect differences in CO_2 or HCO_3^- with which the CO_3^{2-} is equilibrated. Many factors affect the carbon isotopic composition of the CO_2 and HCO_3^- making precise interpretation of the carbon isotopic composition difficult.

Salinity

The oxygen isotopic composition of the hydrosphere varies considerably because of the different vapor pressure of H_2O^{16} and H_2O^{18}. This variation is largely due to fractionation that occurs during evaporation and as water vapor condenses back to the liquid phase as discussed in the section on physical chemical factors. Imagine a mass of water vapor that has originated over the ocean. That vapor should initially have an O^{18}/O^{16} ratio which is 8‰ lower than the ratio in the sea water from which it formed. When the vapor first starts to condense, the rain should be preferentially enriched in H_2O^{18} because its lower vapor pressure favors its condensation into the liquid phase. Indeed the rain should have an O^{18}/O^{16} ratio which is 8‰ higher than that in the vapor and thus the same as the ratio in the sea water from which the vapor formed. The removal of H_2O^{18}-enriched water as rain from the vapor will result in the remaining vapor being somewhat more depleted in H_2O^{18}. The next rain to fall will still have an O^{18}/O^{16} ratio which is 8‰ higher than the vapor, but it will have a lower ratio than the first rain because it formed from vapor that had been somewhat depleted in H_2O^{18}. As this process continues the vapor and thus the rain falling from it become increasingly depleted in H_2O^{18}. The last rain (or snow) to come from our imaginary vapor mass should be very light indeed. This process of progressive lowering of the O^{18}/O^{16} ratio during condensation can be described theoretically by the Raleigh equation (Epstein, 1959) which is shown graphically in Fig. 3.26. The x-axis shows the percentage of the original vapor mass that has con-

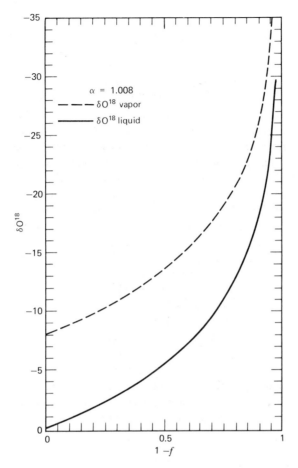

Figure 3.26 Variation in the O^{18}/O^{16} ratio of water and vapor with degree of condensation of the water-vapor system. The condensed phase is continuously removed from the system. f—fraction remaining in vapor phase. See text for detailed explanation. After Epstein (1959).

densed to rain or other precipitation, and the y-axis shows the O^{18}/O^{16} ratio (expressed as δO^{18}). The solid line shows the variation in the composition of the liquid water as the precipitation process proceeds and the dashed line shows the composition of the vapor phase. The two lines are always vertically separated by 8‰, the fractionation in the evaporation-condensation process.

This precipitation model has many implications for the pattern of O^{18}/O^{16} ratios in precipitation and freshwater bodies resulting from it (Dansgaard, 1964). The condensation of water vapor is of course largely dependent on cooling, the atmosphere being able to retain less water vapor as it cools. Secondary factors are the time required for condensation to occur and the presence of nucleation sites. Rain condensing from oceanic vapor should be heavier (have a higher O^{18}/O^{16} ratio) near the shore and become lighter inland as the vapor progressively condenses. As the vapor rises and cools in going over mountains, the rain should become lighter. As the vapor gradually cools in going from low to higher latitudes the condensation process gradually goes further to completion and the rain (or snow) becomes progressively lighter. The lowest O^{18}/O^{16} ratios described are those from polar snow (Epstein, 1959). Thus to a first approximation the O^{18}/O^{16} ratio of fresh water varies directly with the temperature at which condensation occurs. This relationship has been used in studies of glacial snow and ice. Winter snow can be readily differentiated from summer snow on the basis of its O^{18}/O^{16} ratio and, the approximate minimum and maximum temperature of the year can be determined from the O^{18}/O^{16} ratio in the ice, allowing determination of the relative temperature history on a glacier for the last several thousand years (Dansgaard et al., 1969).

The very low O^{18}/O^{16} ratio of glacier ice leads to some complications in paleotemperature determinations. As is shown schematically in Fig. 3.27, water is withdrawn from the oceans to produce the glaciers. The glaciers in effect store an excess of O^{16} on the continents causing a rise in the O^{18}/O^{16} ratio in the oceans. Craig (1965) estimates that if all glacial ice were to melt the O^{18}/O^{16} ratio of the oceans would be lowered by 0.5‰. If the volume of glacial ice were to increase, the O^{18}/O^{16} ratio in the oceans would rise due to preferential removal of H_2O^{16}. Craig (1965) estimates the maximum difference in the oceans in O^{18}/O^{16} ratio between glacial and interglacial ages to be about 1.5‰. This is equivalent to the fractionation difference produced by a 6°C change in temperature. Clearly a correction is necessary before a comparison can be made between paleotemperatures for different times in the Pleistocene and for any time when glaciers of significant size have existed. Variation in the O^{18}/O^{16} ratios in fossil foraminifera from deep sea cores may

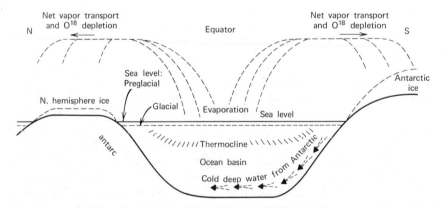

Figure 3.27 Schematic diagram illustrating global changes in the O^{18}/O^{16} ratio of sea water between glacial and nonglacial times. During glacial times ice, which is depleted in O^{18}, is retained on land. This causes the O^{18}/O^{16} ratio of the water remaining in the oceans to rise. After Hudson (1977) Scottish Journal of Geology, Scottish Academic Press Limited.

be more a reflection of variation in isotopic composition of the oceans than of temperature variation (Shackleton, 1967).

The evaporation of sea water continually preferentially removes H_2O^{16}. The O^{18}/O^{16} ratio might thus be expected to constantly increase in the seas. But of course isotopically lighter water is continually being returned to the seas by rivers so that a steady state has been reached. Circulation is vigorous enough in the open ocean that the O^{18}/O^{16} ratio varies relatively little. In restricted, nearshore areas the composition may vary markedly. The O^{18}/O^{16} ratio may be high in restricted lagoons in which evaporation exceeds the inflow of fresh water and will be low in brackish water bays and estuaries. The O^{18}/O^{16} ratio in sea water thus correlates crudely with salinity; in an individual estuary the ratio may correlate rather precisely with salinity (Mook, 1971). If sea water of a constant oxygen isotopic composition mixes with fresh water which likewise has a constant oxygen isotopic composition, the O^{18}/O^{16} ratio in the resultant mixture should vary linearly with salinity (Fig. 3.28). As discussed above, the O^{18}/O^{16} ratio in a carbonate skeleton forming at a given temperature has a constant relationship to the ratio in the water from which it formed. Thus if the oxygen isotopic composition of the water varies, this should be reflected directly in the composition of the skeleton (Fig. 3.29). If the temperature is constant across the salinity gradient, paleosalinities should be determinable from oxygen isotopic analysis of fossils. The use of this technique for paleosalinity determination requires knowing the O^{18}/O^{16} ratio of the sea water and fresh water that are being mixed to form the brackish intermediates. The sea water composition might be assumed to be that of normal, open marine water, but estimation of the freshwater composition will be much more difficult. One solution to this problem is to determine the O^{18}/O^{16} ratio in a fossil that is known to have grown in contact with the fresh water that was responsible for the dilution.

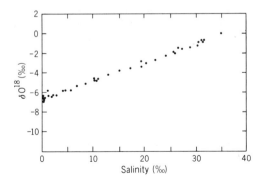

Figure 3.28 Variation of δO^{18} values with salinity for water samples from the western Scheldt estuary, the Netherlands (data from Mook, 1970). After Dodd and Stanton (1975), Geological Society of America Bulletin.

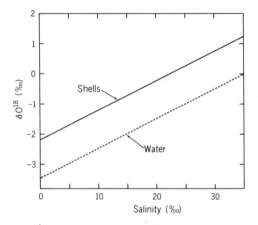

Figure 3.29 Relation between δO^{18} and salinity for shells and water at 12°C assuming simple mixing of full marine and fresh water. After Dodd and Stanton (1975), Geological Society of America Bulletin.

An ingenious but more complex system of determining paleosalinities which utilizes both oxygen and carbon isotopic composition has been developed by Mook (1971). He found that if he made a plot of the δO^{18} vs. δC^{13} values for modern shells collected at various salinities, they lay along a straight line (Fig. 3.30). This is because both the δO^{18} and δC^{13} values for the brackish water are linear functions of the δO^{18} and δC^{13} values of the sea water and fresh water mixed to yield the brackish intermediates. Shells from another bay or estuary in the same region (the coastal area of the Netherlands in the case of the work of Mook) lay along a different straight line on the δO^{18} vs. δC^{13} plot. The second line intersected the first at a δO^{18} and δC^{13} value which corresponds to the value for an open marine shell. This is because the sea water involved in both mixtures has the same oxygen and carbon isotopic composition whereas the fresh water for the two estuaries had different compositions. From the δO^{18} value at the intersection of the lines one can calculate a paleotemperature for the area. This value for the sea water can be used along with the value from a freshwater shell from each of the estuaries to calculate paleosalinities using the method described above.

Carbon isotopic composition also varies with salinity (Mook, 1971; Eisma et al., 1976). The C^{13}/C^{12} ratio is usually higher in marine shells than in freshwater forms, and skeletons of brackish water forms are likely to have intermediate values. This is because marine bicarbonate is derived from the decay of marine organic matter and equilibration with atmospheric CO_2, both of which are characterized by relatively high C^{13}/C^{12} ratios. Bicarbonate in fresh water on the other hand is largely derived from the decay of terrestrial and freshwater organic material, which is characterized by low C^{13}/C^{12} ratios. Also freshwater bodies (except perhaps for large lakes) have not had sufficient time to reach equilibrium with atmospheric CO_2. The carbon isotopic compo-

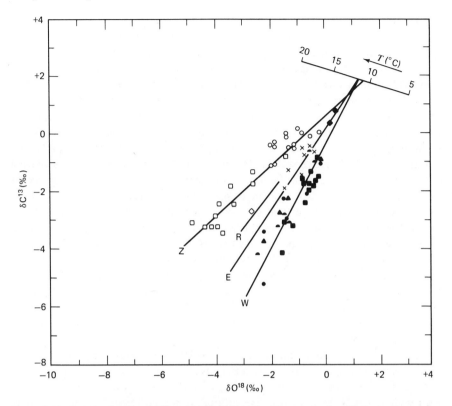

Figure 3.30 Relation between δC^{13} and δO^{18} of skeletal carbonate for four Netherland estuaries: Western Scheldt (W), Eastern Scheldt (E), Mouth of the Rhine (R), Zuiderzee-Waddenzee (Z), The line with the temperature scale indicates the isotopic composition of full marine carbonates formed at those temperatures. After Mook (1971).

sition is also altered by solution of previously deposited carbonate sediment or ancient carbonate rock with which the water body is in contact. This usually has the effect of raising the C^{13}/C^{12} ratio in the bicarbonates.

Carbon isotopes fractionate markedly in various organic processes, especially during photosynthesis. These processes differ between marine and nonmarine and aquatic and terrestrial plants (Degens, 1969). A number of studies of the carbon isotopic composition of the organic compounds in sediments and sedimentary rocks have indicated that this can be a useful method for distinguishing marine from nonmarine rocks and even distance from shore (Sackett and Thompson, 1963).

Diagenesis

Unfortunately for the paleoecologist the oxygen and carbon isotopic composition of fossils has often been altered diagenetically. In a very general sense, the older the fossil, the more likely it is that it has been altered. Unaltered fossils as old as Paleozoic are uncommon; however, preservation in Mesozoic fossils is more common, and is quite common in Cenozoic fossils. The lithology of the matrix in which the fossil is buried is as important in determining the isotopic preservation of a fossil as it is in determining trace-element preservation. In general the less permeable the sediment, the greater is the probability that the original isotopic composition of the fossil will be preserved. Some exceptional cases of preservation such as in the Pennsylvanian Buckhorn Formation of Oklahoma result from the early impregnation of the matrix with asphalt, forming an extremely effective seal against water movement through the formation (Stehli, 1956). Alteration of the oxygen and carbon isotopic composition occurs as a result of reequilibration of the oxygen and carbon in the carbonate of the fossil with that in the water and dissolved carbonate in the diagenetic environment. The same physical chemical rules apply during the alteration process as during the original formation of the fossil. Both the isotopic composition of the water and the temperature of alteration are likely to be very different from those that prevailed during formation of the fossil, hence the reequilibrated composition will usually be different from the original. Because fresh water usually has a lower O^{18}/O^{16} ratio than sea water, marine fossils that are altered by contact with fresh water will have their O^{18}/O^{16} ratio lowered and the apparent paleotemperature raised.

Oxygen isotopic composition is likely to be more extensively altered than carbon isotopic composition. This is because the amount of oxygen in the carbonate of a fossil is likely to be much less than that in the diagenetic pore water with which it is in contact because ground water movement causes the pore water to be continuously renewed. The composition of the carbonate-water system will thus be dominated by the composition of the water. The amount of carbon in the bicarbonate of the altering water will be much less than the oxygen, and the carbon isotopic contribution of the fossil to the carbonate-water system will be greater. More extensive alteration will be necessary before the original carbon isotopic composition is completely obliterated.

Alteration of the isotopic composition does not necessarily affect the physical appearance of the fossil, so if the extent of alteration is slight it may be difficult to detect. A number of criteria (similar to those discussed in connection with trace chemical alteration) have been used or suggested for detecting alteration (Dodd and Stanton, 1976).

1 Is the skeleton morphologically preserved? Does it still show its original microstructural relationships or is it replaced by secondary calcite? This is a necessary but not sufficient test; that is, a fossil that does show a recrystallization texture will not have its original oxygen and carbon isotope composition preserved, but a fossil that is apparently well preserved morphologically may in some cases be altered.

2 Is the original mineral composition of the fossil preserved? This test is

especially used for aragonitic fossils. An originally aragonitic fossil that has been altered to calcite will likely be isotopically altered.

3   Does the isotopic composition of the fossil contrast with that of the surrounding matrix? Alteration in contact with diagenetic water should affect carbonate in the matrix to the same extent as carbonate in the fossil if the process has gone to completion. This test is of course limited to fossils preserved in a carbonate-containing matrix.

4   Is the paleotemperature based on the O^{18}/O^{16} ratio "reasonable," that is, between 0 and 30°C for normal marine organisms? This test is useful in detecting gross alteration but would not detect minor alteration nor would it be adequate for fossils of organisms that lived in water that differed from normal marine in its isotopic composition.

5   Does the isotopic composition vary within the fossil, especially in such a way that can be related to seasonal variation in environmental conditions? This is an excellent way to test for alteration and was first used by Urey et al. (1951) to test for isotopic preservation in a belemnite from Jurassic rocks of Scotland. Variation in the oxygen isotopic composition within the fossil was interpreted as reflecting temperature variation during the life of the organism. Diagenesis should remove the original variations and make the composition of the fossil more uniform.

6   Is the interpretation based on the isotopic results geologically reasonable in comparison with interpretations based on independent evidence? Many of the studies of oxygen and carbon isotopic composition of fossils have used this criterion. The work by Emiliani (1955, and other papers) and several other workers on the oxygen isotopic paleotemperatures from foraminifera in deep sea cores makes extensive use of this test. The temperature variation pattern shown by isotopic analysis is in agreement with the pattern of variation in foraminiferal assemblages, coiling directions, and sedimentological properties.

Geologic Applications of the Isotopic Technique

Paleotemperature

The earliest and best known applications of isotopic data have been in the determination of paleotemperatures. Tens if not hundreds of studies have been made reporting paleotemperatures since the first such study was published by Urey et al. (1951). A sampling of these studies are referenced in Table 3.2. We cannot give a comprehensive review of all of these studies here but discuss only a few as examples.

The first fossil group to be studied with the isotopic technique was the belemnites. This group was selected because of its heavy, chemically stable skeleton composed of large, low-Mg calcite crystals. These features maximize the chance of preservation of the original isotopic composition of this group.

Another advantage of the belemnites is that they were apparently normal marine, relatively open water forms. This minimizes problems associated with the isotopic composition of carbonate forming in brakish and/or restricted environments. The first results published were from a belemnite from the Jurassic of the Isle of Skye (Urey et al., 1951). A sequence of samples was analyzed from a single specimen which appeared to show seasonal variation in the growth temperature, confirming the preservation of the isotopic composition in the specimen.

Later work by Lowenstam and Epstein (1954) gave results from Cretaceous belemnites from many locations in North America and Europe. These speci-

Table 3.2   Examples of oxygen and carbon isotope paleoenvironmental studies

Age	Fossil Group	Reference
Pleistocene	Molluscs	Valentine and Meade, 1961
	Molluscs	Masuda and Taira, 1974
	Foraminifera	Emiliani, 1955, 1966, etc.
	Foraminifera	Kennett and Shackleton, 1975
Tertiary	Coccoliths	Margolis et al., 1975
	Foraminifera	Devereux et al., 1970
	Foraminifera	Saito and Van Donk, 1974
	Foraminifera	Savin et al., 1975
	Foraminifera	Shackleton and Kennett, 1975a & b
	Brachiopods	Lowenstam, 1961
	Molluscs	Dorman and Gill, 1959
	Molluscs	Dorman, 1966
	Molluscs	Devereux, 1967
	Molluscs	Lloyd, 1969
	Molluscs	Buchardt, 1978
Cretaceous	Foraminifera	Saito and Van Donk, 1974
	Molluscs	Tourtelot and Rye, 1969
	Molluscs	Forester et al., 1977
	Molluscs, belemnites	Lowenstam and Epstein, 1954
	Molluscs	Bowen, 1961a
Jurassic	Molluscs	Tan and Hudson, 1974
	Molluscs, belemnites	Bowen, 1961b
	Molluscs, belemnites	Naydin et al., 1966
	Molluscs, belemnites	Stevens and Clayton, 1971
Permian	Brachiopods	Lowenstam, 1964c

mens showed both a stratigraphic and a geographic variation in temperature. The general stratigraphic variation pattern for the upper Cretaceous was for a gradual increase in temperature from Albian through Coniacian-Santonian time followed by decreasing temperatures through the end of the Cretaceous (Fig. 3.31). In general the Cretaceous temperatures were about the same as or warmer than temperatures at the same latitude today. A comparison of paleo-temperatures for Maastrichtian time from many North American and European localities showed a poorly defined temperature gradient with a general decline in a northward direction. Lowenstam and Epstein (1954) also determined paleo-temperatures from various other fossil groups, especially bivalves and brachio-pods, as well as from bulk chalk samples. These temperatures were usually higher than belemnite temperatures and were considered possibly to be partially altered diagenetically. Another interpretation is that the mobile belemnites spent part of their lives in deeper, colder water than the other fossil groups.

Bowen (1966) published results of analyses of belemnites, especially Jurassic specimens, from many localities around the world. One of the main objectives of this study was to document the effect of continental drift on the planetary temperature gradient. Bowen noted some such effects, but in general the data for any given geologic time are too sparse to adequately delineate the planetary temperature gradient. More recently Spaeth et al. (1971) have questioned the assumption that the original isotopic composition is adequately preserved in belemnites for paleotemperature work. They suggest that a considerable

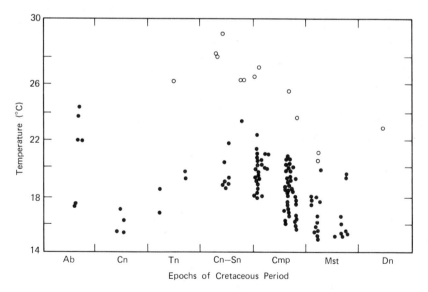

Figure 3.31 Paleotemperatures determined from the O^{18}/O^{16} ratio of brachiopods (open circles) and belemnites (closed circles) from the Cretaceous of western Europe. After Lowenstam and Epstein (1954), copyright University of Chicago.

amount of secondary calcite may occur in fossil belemnites. However, various lines of geologic and paleontologic evidence confirm the general paleotemperature trends suggested by analysis of the belemnites.

Closely following the belemnite studies have been a long series of studies of foraminifera, especially of specimens from deep sea cores. The pioneer in these studies is Cesare Emiliani. Foraminifera from deep sea cores are nearly an ideal subject for study by the isotopic method. The planktonic types, which have received the bulk of study, are composed of chemically stable low-Mg calcite. The foraminifera are also preserved in a very stable environment in contact with sea water rather than highly $CaCO_3$ undersaturated fresh water. The deep sea specimens grew in the open ocean in contact with well-mixed, average sea water. When conditions are right for their preservation, specimens are likely to be abundant through a considerable section of core. Most of the earlier studies of foraminifera have concentrated on Pleistocene and Holocene specimens in an effort to clarify the temperature history of the oceans during that time span. One of the big surprises of the early work in this field was the discovery by Emiliani (1955) of many temperature cycles during the Pleistocene rather than the then traditionally accepted four or five based on continental Pleistocene stratigraphy. Emiliani has continued to refine his work on the Pleistocene temperature history and has developed a standard temperature curve for Pleistocene and Holocene time (Fig. 3.32; Emiliani, 1966). Although some have argued with the details of this curve, the usefulness of the oxygen isotopic technique for studying Pleistocene climatic history is widely recognized.

Determination of paleotemperatures for the Pleistocene is difficult because of the uncertainty of the O^{18}/O^{16} ratio of the oceans. Variation in the mass and average O^{18}/O^{16} ratio of the glacier ice can have a significant effect on the O^{18}/O^{16} ratio in the oceans. In order to calculate a paleotemperature for a Pleistocene fossil a correction for the composition of the water must be made which requires an estimation of the volume and isotopic composition of the ice. No general agreement has been reached on what these correction factors should be; however, most of the present workers (e.g., Dansgaard and Tauber, 1969, and Shackleton and Kennett, 1975b) believe that most of the

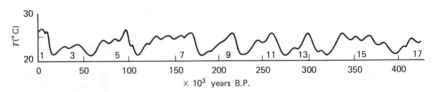

Figure 3.32 Generalized variation pattern in temperature and O^{18}/O^{16} ratio of surface water in the central Caribbean during the last 425×10^3 years. The numbers below the curve refer to time divisions recognized on the basis of the oxygen isotopic temperature data. After Emiliani (1966), copyright University of Chicago.

variation in the O^{18}/O^{16} ratio in Pleistocene foraminifera is due to variation in ice volume and not temperature. The foraminifera may thus be largely monitoring ice volume and not temperature. In terms of studying the geologic history of glaciation, this is equally as valuable if not more so (Broecker and van Donk, 1970).

Another important contribution by Emiliani in his early work on the foraminifera has been in measuring the oceanic bottom water temperatures through the Cenozoic from deep sea benthonic forams. Because the bottom water is the most dense, cold water in the oceans, it necessarily originates in the polar regions. Thus the bottom water temperatures give us a measure of the minimum sea water temperatures present on earth at a given time. Emiliani (1954) showed that bottom water temperatures declined from a high of about 10°C in late Cretaceous time to near zero by Pliocene time. Others (e.g., Savin et al., 1975, and Shackleton and Kennett, 1975b) in more recent years have refined our knowledge of this trend (Fig. 3.33).

With the advent of the Deep Sea Drilling Project (DSDP) the time scope of isotopic paleotemperature studies of the deep sea expanded greatly. Cores can be obtained from the deep sea that extend back into the Mesozoic. Oxy-

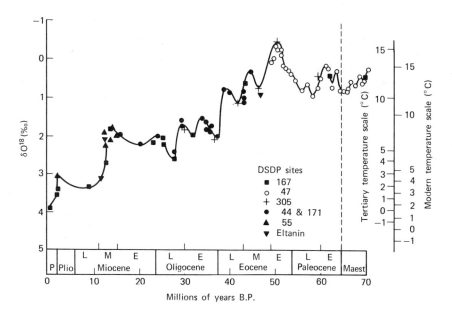

Figure 3.33 Oxygen isotopic composition of multispecies assemblages of benthonic foraminifera. The Tertiary temperature scale applies to samples older than middle Miocene. This scale is based on the assumption that there were no extensive glaciers at this time. The modern temperature scale applies to the present oceans. This scale assumes the glaciers were comparable in volume to those existing today. Temperatures for Middle and Late Miocene and Pliocene lie between these two scales. Abbreviated time intervals are P-Pleistocene, Plio-Pliocene, Maest-Maastrichtian. Illustration courtesy of S. Savin and R. G. Douglas.

gen isotope paleotemperatures have been obtained from many of these cores giving the general pattern of temperature change for approximately the last 60 million years (Fig. 3.35; Savin et al., 1975; Shackleton and Kennett, 1975b). As was also indicated by the belemnite data, the temperature dropped in late Cretaceous time before rebounding slightly in Paleocene-Eocene times. The temperature again dropped sharply in Oligocene time before rising once again in the Miocene and then dropping rather abruptly into the Pleistocene.

Pleistocene deep sea paleotemperatures have also been the subject of detailed study as the basis of oxygen isotopic analysis of foraminifera. Rather detailed maps of temperatures at various times in the Pleistocene have been prepared as part of the CLIMAP program (Cline and Hays, 1976). These temperatures are in part determined on the basis of isotopic analysis.

A number of studies have been conducted on fossil bivalves, gastropods, and brachiopods, but these groups have not received the work that the foraminifera have. These groups are often shallow water, inshore forms occurring in relatively restricted habitats. The estimated O^{18}/O^{16} ratio for the water is thus often less certain. On the other hand, many species of these groups do have diagenetically stable skeletons of low-Mg calcite and have proved useful for oxygen isotopic studies, particularly in the Cenozoic. An especially noteworthy example of a study in part involving older fossils is that of Lowenstam (1961) in which he demonstrated that the oxygen isotopic composition of the oceans has apparently not changed markedly since Pennsylvanian time. Dorman and Gill (1959) studied the variation in temperatures through the Cenozoic in Australia based on analysis of fossil bivalves. The pattern that they observed is very similar to that found later in forams from deep sea cores.

Buchardt (1977, 1978) has recently reported paleotemperatures for the upper Cretaceous and Cenozoic for the shelf area of the North Sea on the basis of analysis of mollusc shells. His results show the same general climatic trends as do the planktonic foraminifera in deep sea cores. A sharp cooling in early Oligocene time is especially prominent (Fig. 3.34).

Another approach to measuring paleotemperatures besides analysis of fossils would be to analyze ancient water. At least in a general way the oxygen isotopic composition of fresh water is a function of temperature. In general the colder the temperature of condensation of water the lower will be the O^{18}/O^{16} ratio. This is because the colder the air mass, the less water vapor it can hold. As the air cools it gradually loses its water and the more water is lost, the lighter the remaining vapor will be (Fig. 3.27). Thus, as indicated above, the isotopically lightest water is found at the poles. Dansgaard et al. (1969) have analyzed ice from cores in glaciers in Greenland to determine the temperature variation pattern for the last 100,000 years or so. The glacier ice faithfully records the lower temperatures during the last glacial age and the rising temperature into the present interglacial age.

A related approach is to determine the oxygen isotopic composition of freshwater carbonates that formed in equilibrium with the temperature-dependent rain and snow. The best source of such carbonates is deposits in

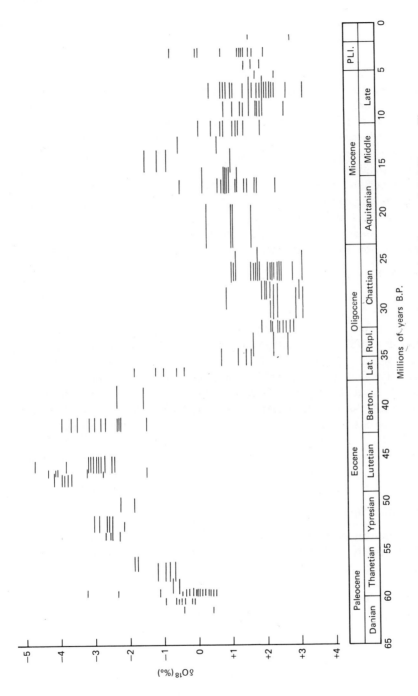

Figure 3.34 Variation in the O^{18}/O^{16} ratio of Cenozoic molluscan shells from northwest Europe. The length of the horizontal bars indicate the uncertainty in age determinations. After Buchardt (1978).

177

caves or *speliothems*. A number of studies of speliothem carbonates have traced temperature variation patterns (e.g., Harmon et al., 1978).

One of the principal uses of oxygen and carbon isotopic data in recent years has been in the correlation of deep sea cores and even shallow water deposits (Shackleton and Matthews, 1977; Mangerud et al., 1979; and Heusser and Shackleton, 1979). As data on the isotopic composition of foraminifera in deep sea sediments have accumulated over the years, certain distinctive patterns of variation of composition with time have developed (Emiliani, 1972). This allows correlation between cores on the basis of well logs of isotopic composition (Fig. 3.35). Perhaps 70% of the variation in oxygen isotopic composition of planktonic foraminifera during the Pleistocene and Holocene is due to variation in the oxygen isotopic composition of the ocean water as glaciers of the world waxed and waned. These changes in isotopic composition should be worldwide and thus should make an excellent basis for correlation (Shackleton and Opdyke, 1973 and 1976).

Paleosalinity

Isotopic studies of paleosalinity have not been as extensive as those of paleotemperature. In general the studies that have reported paleosalinity results have done so only in general terms and not in absolute values. Kennett and Shackleton (1975) have documented an abrupt decrease in salinity in the Gulf of Mexico between about 17,000 and 13,500 years BP. They interpret this as being due to a large volume of glacial melt water coming down the Mississippi River. Kammer (1979) also postulates low-salinity water to explain oxygen isotopic results from planktonic forams from the northern California area. Similar studies elsewhere have identified low-salinity conditions on the basis of oxygen and carbon isotopic analysis of forams. Tan and Hudson (1974) have worked extensively with fossil bivalves from the Jurassic Great Estuarine Series of Scotland. They have used the oxygen and carbon isotopic composition for paleosalinity as well as paleotemperature interpretations.

Other Applications

Several applications have been made of oxygen isotopic data in addition to paleotemperature and paleosalinity determination and correlation. For example, Eichler and Ristedt (1966) have used both oxygen and carbon isotopic data to study the life history of modern *Nautilus* (Fig. 3.36). Oxygen isotopic temperatures from the earlier formed part of the shell suggests that young *Nautilus* lives in water of about 100 m depth. At maturity they apparently migrate to deeper water; the oxygen isotopic temperatures of the mature shell suggest 200 to 300 m. Carbon isotopic results suggest dietary changes when the animal hatches from the egg and when it migrates to deeper water. The preferred depth habitat of planktonic foraminifera has been studied by many workers (e.g. Shackleton and Vincent, 1978). Different species of foraminifera give differing oxygen isotopic temperatures, species living in the shallower

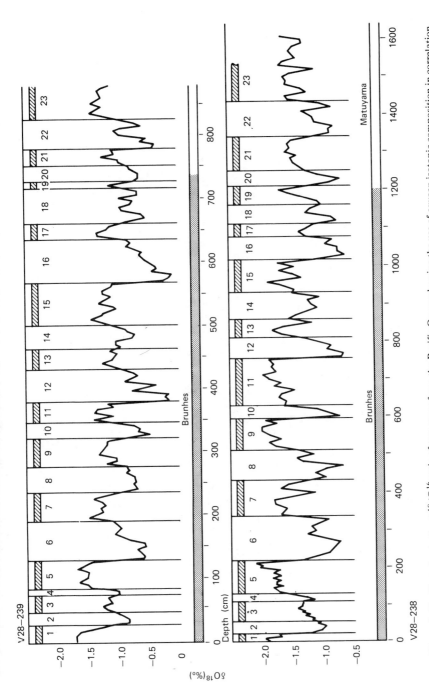

Figure 3.35 Variation in the O^{18}/O^{16} ratio of two cores from the Pacific Ocean showing the use of oxygen isotopic composition in correlation. The dark and light bars below the curves show the paleomagnetic record. Dark—normal magnetization; light—reversed magnetization. The numbers above the curves refer to the oxygen isotopic temperature time divisions of Emiliani (see Fig. 3.32). After Shackleton and Opdyke (1976). Geological Society of America Bulletin.

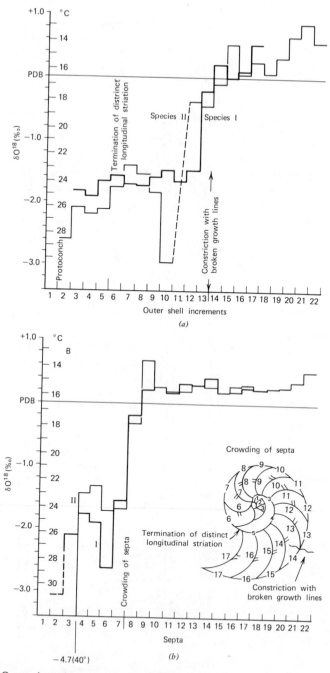

Figure 3.36 Oxygen isotopic composition and temperature equivalents in outer-shell increments (*a*) and septa (*b*) from two *Nautilus* specimens. The graphs show variation in oxygen isotopic composition during growth of the shell. After Eichler and Ristedt (1966), copyright American Association for the Advancement of Science, Science *153:*734–736, text-figure 1.

surface waters recording higher temperatures than those living in deeper, colder water.

Oxygen isotopic analyses have been extensively used in the study of diagenesis of carbonate rocks. For example, the oxygen isotopic composition of carbonate cements may indicate equilibrium with sea water, proving submarine lithification (e.g., Schlager and James, 1978). In other cases the isotopically light oxygen in cements and even the bulk rock may indicate diagenesis in contact with fresh water (e.g., Gross, 1964). Scholle (1977) has used the oxygen isotopic composition of chalks as an indicator of temperature and depth of burial during diagenesis (Fig. 3.37).

Carbon isotopic values also vary with depth in open ocean samples. Planktonic forams living in the near-surface water have higher C^{13}/C^{12} ratios than do benthonic forams living at great depth. This is related to two factors. (1) The carbon isotopic composition of the dissolved carbon (mostly in the form of HCO_3^-) in surface waters usually appears to be close to equilibrium with atmospheric CO_2 which results in relatively high C^{13}/C^{12} values. The dissolved carbon in deep waters is in part derived from the decay of organic material settling from the surface. This organic material has a relatively low C^{13}/C^{12} ratio resulting in a lower ratio in carbonate formed in equilibrium with it (Craig, 1970). (2) The carbon isotopic composition of dissolved bicarbonate in surface waters may also be elevated because of the preferential photosynthetic removal of C^{12} by the phytoplankton (Weiner, 1975). In any event the variation in the C^{13}/C^{12} ratio with depth might be used as a paleoecological tool for at least approximate depth determination under appropriate

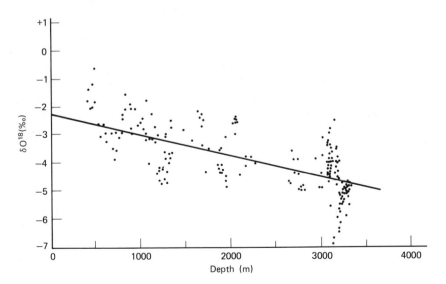

Figure 3.37 Variation with depth of burial of oxygen isotopic composition of North Sea chalk samples. After Scholle (1977), American Association of Petroleum Geologists, *61*:1000.

conditions. Berger and Killingley (1977) have used C^{13}/C^{12} ratios in foraminiferal tests to determine what they call paleofertility of the oceans. High C^{13}/C^{12} ratios are characteristic of high fertility, that is, high rates of photosynthesis. This is because increased photosynthesis removes C^{12} causing the C^{13}/C^{12} ratio of the remaining bicarbonate to rise.

This review of the literature on the use of oxygen and carbon isotopic analysis of fossils for paleoecologic interpretations is by no means complete but should serve to indicate the usefulness of this technique.

Isotopic Geochemistry of Fossils from the Kettleman Hills

Fossils from the Pliocene of the Kettleman Hills are clearly not ideal for determining paleotemperatures. These fossils lived in a semirestricted embayment that was at times brackish and even fresh. Thus the O^{18}/O^{16} ratio of the water at a given time and place is uncertain. On the other hand, conditions are suitable for using oxygen (and carbon) isotopic composition to determine paleosalinities. As a first approach to this problem we have analyzed fossils from several stratigraphic levels through the Kettleman Hills section (Stanton and Dodd, 1970; Fig. 3.38). In general the O^{18}/O^{16} ratio decreases upward, probably reflecting the general pattern of decreasing salinity with time during deposition of the section. Normally, in order to determine a paleotemperature, we must assume an O^{18}/O^{16} ratio for the water in which the fossil lived. However, if by some independent means we could estimate the paleotemperature we could solve the empirical paleotemperature equation (equation 3.9) for the O^{18}/O^{16} ratio of the water. These values could then be used to interpret paleosalinities. For temperature estimates we used the values determined from the Sr content of fossil *Mytilus*. As previously indicated these temperature values are in fairly close agreement with temperatures determined from the taxonomic uniformitarian analysis of the fauna, adding to our confidence in them. When these temperatures are substituted into the empirical temperature equation δO^{18} values for the water are determined that show a general decline stratigraphically upward (Fig. 3.38). The values for the lower part of the section are near zero relative to the PDB standard. This is about the same as modern open ocean water. The upper values approach those determined from a freshwater bivalve shell which is presumed to have grown in freshwater characteristic for this area. The generalized salinity can be determined from this relationship.

In a more detailed study we analyzed fossil *Pecten* and *Ostrea* from several localities within two single stratigraphic units, the *Pecten* zone and the upper *Mya* zone. The purpose of these analyses was to document and quantify lateral variation in salinity within a particular time interval. If we assume that all of the fossils analyzed lived at the same temperature, the variation in the δO^{18} values for the fossils should be entirely due to the variation in the δO^{18} value of the water. If the temperature is constant then the fractionation factor, α, in (3.7) will also be constant. Thus changes in the O^{18}/O^{16} ratio

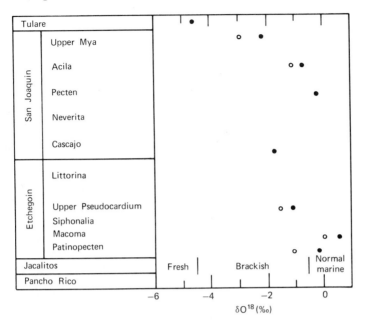

Figure 3.38 Oxygen isotopic composition of molluscan fossil samples (solid circles) and calculated values for the water in which the shells formed (open circles) from zones in the Kettleman Hills section. After Stanton and Dodd (1970), Journal of Paleontology, Society of Economic Paleontologists and Mineralogists.

of the water will directly affect the O^{18}/O^{16} ratio of the shells. To establish end members for the salinity gradient for the *Pecten* zone, we used a fossil from an assemblage that appeared to be open marine and a freshwater bivalve from immediately below the *Pecten* zone. We do not positively know that the sample we selected to represent open marine indeed did grow under exactly normal marine conditions; however, the nature of the associated fauna strongly suggests that this is the case. Also the geographic location at the north end of North Dome suggests that it was nearest to the open ocean. Ideally the freshwater end member should also have come from the *Pecten* zone, but we found no specimens there. The oxygen isotopic composition of the water was not likely to have changed during the deposition of the small amount of sediment separating the freshwater bivalves from the *Pecten* zone. The end members were then plotted on a graph of δO^{18} vs. salinity (Fig. 3.29). The salinity at which a *Pecten* zone fossil grew can then be determined by finding the salinity corresponding to the δO^{18} value on the graph. The δO^{18} data and salinity values determined from them (Fig. 3.39) show a pattern of highest salinity at the north end of the hills, lowest values on the west flank of the hills, and intermediate values to the south. These results suggest that the opening to the bay was to the northwest and that fresh water from streams was flowing in from the west or southwest. The

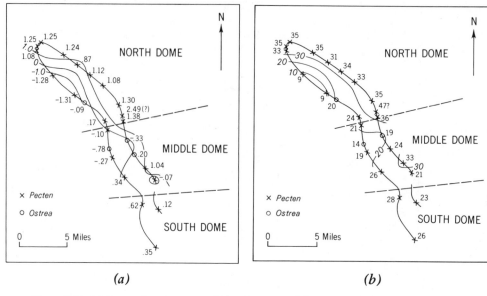

Figure 3.39 (a) Oxygen isotopic composition of samples of *Pecten* and *Ostrea* from the *Pecten* zone of the Kettleman Hills; (b) paleosalinity values calculated from the δO^{18} composition of samples from the *Pecten* zone. After Dodd and Stanton (1975), Geological Society of America Bulletin.

faunal data of various sorts support this interpretation. Carbon isotopic data (Fig. 3.40) also suggest a pattern of more marine conditions to the northwest with less marine conditions to the southeast. The lowest δC^{13} values occur here and the highest values are on the east flank of North Dome.

We used a similar approach to interpret the salinity pattern in the upper *Mya* zone. This zone contains no normal marine faunas in the Kettleman Hills area, however; and the stratigraphically closest freshwater bivalves are some 250 m above the upper *Mya* zone. For these reasons the salinity measurements are probably less reliable than those for the *Pecten* zone. No clear salinity pattern appears for the upper *Mya* zone on the scale of our study (Fig. 3.41). The bay seems to have been irregularly brackish throughout the Kettleman Hills area during upper *Mya* zone time.

As previously indicated, Mook (1971) developed a system for determining a paleotemperature for an area even when the water is in part brackish. He made plots of δO^{18} vs. δC^{13} data for two or more bays or estuaries in an area. The intersection of the lines for these data corresponds to the δO^{18} and δC^{13} values for a full-marine fossil and thus the δO^{18} value of the intersection could be used to determine a paleotemperature for the area. In the Kettleman Hills there is only one bay so this technique cannot be used directly. However we can make δO^{18} vs. δC^{13} plots for the one bay at slightly different geologic times when the fresh water diluting the bay had different isotopic composi-

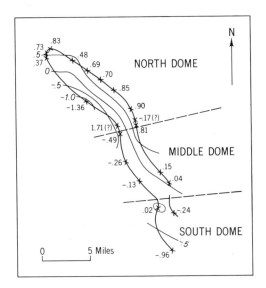

Figure 3.40 Carbon isotopic composition of samples from the *Pecten* zone. After Dodd and Stanton (1975), Geological Society of America Bulletin.

tions. The intersection of these two lines should then correspond to a fossil growing under normal marine conditions which can then be used to determine a paleotemperature. When we use this approach with the *Pecten* zone and upper *Mya* zone data (Fig. 3.42) we obtain a temperature of about 12°C. This value is quite close to that suggested by the various other approaches discussed in this book.

ORGANIC BIOGEOCHEMISTRY

In addition to inorganic or mineral components, skeletal materials contain organic compounds in amounts from a few hundreths of a percent to several percent (Hare and Abelson, 1965), Proteins are the most common type of skeletal organic compound in most invertebrates although small amounts of other classes of compounds, particularly polysaccharides and lipids, are also found (Wyckoff, 1972). The amino acid composition and their proportions in skeletal proteins are primarily controlled by phylogenetic position and skeletal mineral composition and structure (Hare, 1963; Hare and Abelson, 1965; and Degens et al., 1967). Environmental factors, especially temperature and salinity, also appear to be involved but this has yet to be studied in detail (Degens et al., 1967).

Most work on the amino acid composition of organic material in fossil skeletons has been concerned with the degradation of the molecules with time. Potentially the extent of this degradation can be used to determine the

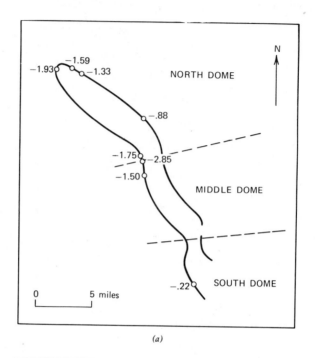

(a)

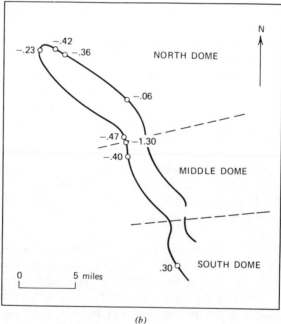

(b)

Figure 3.41 (a) Oxygen isotopic composition of samples from the upper *Mya* zone; (b) carbon isotopic composition of samples from the upper *Mya* zone. After Dodd and Stanton (1975), Geological Society of America Bulletin.

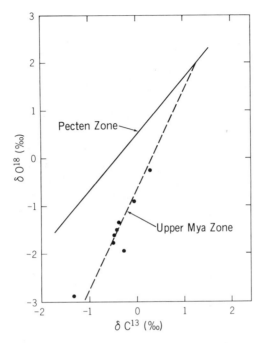

Figure 3.42 Relationship between δO^{18} and δC^{13} for samples from the upper *Mya* zone and the *Pecten* zone. After Dodd and Stanton (1975), Geological Society of America Bulletin.

age of the fossil or the temperature of the diagenetic environment because amino acids have different stabilities relative to temperature such that the ratio of the more stable to less stable amino acids should increase with time (or with diagenetic temperature for fossils of equal age). One of the most used relationships for the study of amino acid degradation has been the alteration (epimerization) of the amino acid isoleucine to alloisoleucine. The molecular structure of these two molecules are mirror images of one another, isoleucine being the *L* or left-handed form and alloisoleucine being the *D* or right-handed form. Organisms only produce the *L* or isoleucine molecule. However nonbiological synthesis or alteration produces equilibrium proportions of isoleucine and alloisoleucine:

$$L\text{-isoleucine} \rightleftharpoons D\text{-alloisoleucine}$$

At equilibrium the ratio alloisoleucine/isoleucine equals approximately 1.3. The ratio of these two molecules in a fossil is a function of the extent of alteration or epimerization. The extent of alteration is in turn a function of time, temperature, and to some extent the position of the amino acid in the protein molecule (Kriausakul and Mitterer, 1978). Thus, if the temperature history during alteration is known and the other factors compensated for, the age of the fossil can be determined. If the age of the fossil is known, the temperature of alteration can be determined, provided it has been constant. This technique is limited to Pleistocene or at best Neogene fossils because older fossils should have been completely altered with resulting equilibrium amounts of isoleucine and alloisoleucine. This and similar techniques have been used in attempts to date vertebrate (Bada, 1972) and invertebrate fossils (Mitterer, 1975) as well as sediment samples (Bada et al., 1970).

The study of the organic chemistry of fossils is in its infancy but is potentially a valuable paleoecologic and paleontologic tool. The organic chemicals contained in sediment and sedimentary rocks also are in part controlled by environmental conditions. These *chemical fossils* offer considerable potential in paleoenvironmental analysis (e.g., Huang and Meinschein, 1979).

4

Skeletal Structure

Paleontologists and paleoecologists have traditionally been primarily concerned with the external morphology of fossilized skeletons. However, study of the internal properties of the skeletons can also be useful in better understanding the organism that formed the skeleton and the conditions of its environment. By internal properties we mean the chemical, mineralogical, and physical characteristics of the skeleton. The chemical and mineralogical properties of skeletal materials were discussed in Chapter 3. In this chapter we propose to discuss the physical arrangement of mineral crystals within skeletal materials and the influence of the growth environment on this arrangement.

This is a relatively new field of study and has not yet received the attention that it merits. Promising indications have been found of correlation between the growth environment and skeletal structure but extensive applications of these to the fossil record have not been made. Many references to the skeletal structure of fossils can be found scattered through the paleontologic literature, but most are more or less incidental to the study of external morphology. An early exception is the pioneering work of Carpenter (1844). Shell structure studies were really placed on a systematic, modern footing by Schmidt (1924) and Bøggild (1930). Skeletal structure studies received considerable impetus with the increasing interest in the petrology of carbonate sediments and rocks in the late 1950s. Because fossil fragments are a major constituent of carbonate rocks, the identification of these fragments on the basis of their skeletal structure becomes critical. Several important publications that survey the skeletal structure of major taxonomic groups appeared at least in part as a result of this interest (e.g., MacClintock, 1967; Majewske, 1969; Horowitz and Potter, 1971). However, these studies were largely concerned with describing the structure and to a certain extent the origin, but did not deal extensively with environmental implications of shell structure. The greatest interest in shell structure vs. environment has been the result of studies of banding in molluscan shells, work that was pioneered by Barker (1964) and Pannella and MacClintock (1968). The extent of interest and the potential for interpretation of skeletal banding is reflected in the papers on this subject published in a recent symposium volume (Rosenberg and Runcorn, 1975).

The application of skeletal structure to paleoecology is still in its infancy and should be an area of considerable progress in the near future.

Skeletal structure can be examined on at least two levels of detail. At the finer level the structure is studied in terms of the detailed shape and arrangement of the individual crystallites making up the skeleton. This has sometimes been called skeletal ultrastructure. At a coarser level the interrelationships of larger units of relatively uniform structures within the skeleton are studied. This has sometimes been called the microarchitecture of the shell. As an example of this difference in scale, one might study the details of the size, shape, and orientation of aragonite crystals within the mother-of-pearl or nacreous layer of a bivalve shell. On the other hand, one might also study the shape and distribution of the nacreous layer itself within the shell. Both levels of study may potentially provide useful information about the growth habitat of the bivalve. Both levels of study are considered in this chapter.

The detailed structure of skeletal materials differs considerably between different taxonomic groups. However, structures may be quite uniform within some groups, such as the echinoderms, so that even small fragments of skeletal material can be identified as belonging to that phylum. The factors determining the skeletal structure are very poorly known, but clearly they are largely genetically controlled physiologic and biochemical factors. To a certain extent the structure is influenced by the physical conditions of the growth site such as through direct physical interference between growing crystals. Environmental factors are obviously also involved, as they affect the physiology of the organism.

GROWTH MECHANISM

Skeletal structure is a function of the mechanism of growth of the skeleton. Skeletal growth can be viewed at two scales just as can the skeletal microstructure that results from that growth. On the finer scale, skeletal growth can be studied in terms of how an individual crystallite grows. Considerable progress has been made in better understanding this growth or mineralization process, but much remains to be learned (Wilbur, 1976). Although agreement is not universal, the mineralization model shown in Fig. 4.1 is widely supported.

Three factors in the mineralization process are involved in determining skeletal structure. (1) The chemical composition of the solution from which the minerals precipitate (and the factors controlling that chemistry) is obviously important. The mineralization site may be within cells in some of the simpler organisms or between the living tissue and previously formed shell in others. It is always to some extent isolated from the surrounding water (or air for terrestrial organisms) allowing the chemistry to differ from that of the surroundings. (2) The organic matrix is important in determining the structure of the skeleton (Towe, 1972). The model shown in Fig. 4.1 proposes that crystal growth starts on the matrix, which acts as a template controlling both

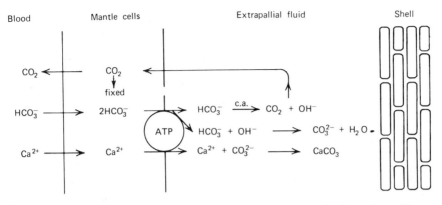

Blood   Mantle cells   Extrapallial fluid   Shell

Figure 4.1 Model showing possible method of $CaCO_3$ mineralization in the molluscs. The carbonate is secreted on an organic matrix that forms from the extrapallial fluid. The composition of this fluid is controlled by the biochemical reactions occurring in the mantle. The Ca^{2+} and CO_3^{2-} are brought to the mantle by the body fluids and/or they come from the external medium (sea water in the case of marine molluscs). Figure supplied by K. M. Wilbur.

the mineral composition (see Chapter 3) and at least in part the physical configuration of the crystals. The ions from which the crystals grow are attracted to specific sites on the matrix. Thus the arrangement of these sites determines both growth location and crystal structure. (3) To some extent the physical constraints of the growth environment determine the size and shape of crystallites. The crystallites can only grow a certain distance until they begin to interfere with the growth of other crystallites. The polygonal shapes of the crystals in some of the structures probably results from this interference and grow to fill the available space. Raup (1968) has likened the growth of crystals in some invertebrate skeletons to the growth of soap bubbles that interfere with one another, producing polygonal shapes (Fig. 4.2).

On a larger scale the microarchitecture of a skeleton is a function of the overall growth pattern. Organisms show four basic methods of skeletal growth (Raup and Stanley, 1978). The most widespread method is *accretion*, in which growth is by addition of mineral material to the preexisting skeleton. All of the previously formed skeleton is retained throughout life and is constantly being enlarged. Growth by accretion allows the organism to be continuously protected throughout life. It also permits continuous use of the entire skeleton; none of it ever has to be discarded. The accretion method of growth is ideal for the paleoecologist because the skeleton contains a continuous record of growth throughout the life of the organism.

A skeleton can grow by accretion only if it is in contact with living tissue. This and the necessity of building from the foundation of the old skeleton place definite limits on the shape and structure of the skeleton. Organisms having this mode of growth all live in a tube or some modified form of a tube (Fig. 4.3). The tube may be coiled (gastropods, cephalopods, etc.), partially

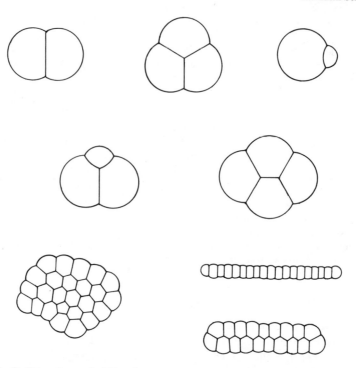

Figure 4.2 Outlines of soap bubbles that are similar in shape to crystals and plates formed during skeletal mineralization. The polygonal shapes result from interaction between the bubbles (or crystals) as they grow to fill the limited space. After Raup (1968), Journal of Paleontology, Society of Economic Paleontologists and Mineralogists.

filled with structures such as septa and tabulae allowing the organism to live only near the distal end of the tube (bryozoans and corals), partitioned into chambers with the animal occupying only the last chamber (cephalopods), doubled into two rapidly expanding tubes placed open end to open end and hinged to shut out the outside world (bivalves and brachiopods), grown together in colonies (corals and bryozoans), or equipped with a trapdoor (gastropods and cephalopods). The tube can grow in length and expand at its distal margins and it can increase the thickness of its walls by addition to the tube interior (Fig. 4.3). No other directions of growth are possible. All shells growing by accretion have a basic two-layered structure, although some are modified to one layer or to three or more. As the skeleton accretes, one layer forms at the growing margin of the tube and the other forms on the inner surface of the tube. The inner layer may grow by addition directly to the crystals of the outer layer so that the two are indistinguishable. On the other hand, the inner layer may develop different structures in different places within the tube, giving one or more extra layers or units. The geometry of the skeletal microarchitecture can be analyzed in terms of skeletal growth patterns.

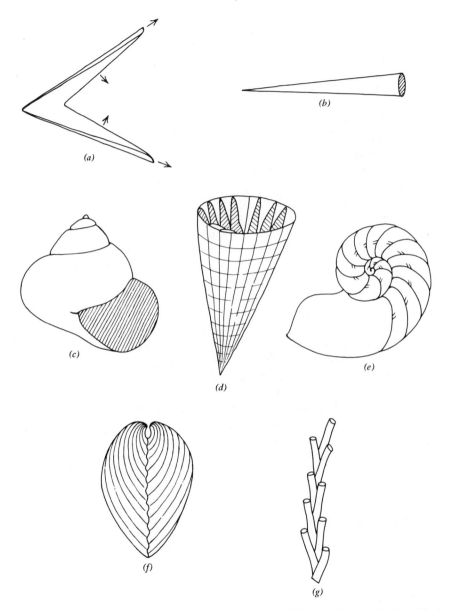

Figure 4.3 Shells shaped as cones and modified cones formed by accretion. (*a*) Section of basic cone showing growth at margins which lengthens the cone and growth on the inner surface which thickens the cone. This produces a basic two-layered microarchitecture. (*b*) A gently tapering cone (annelid). (*c*) A coiled cone (gastropod). (*d*) A cone filled with structures (coral). (*e*) A cone divided into chambers (cephalopod). (*f*) A double, rapidly expanding cone (bivalve). (*g*) Multiple colonial cones (bryozoan).

The second method of skeletal growth is by *molting*. This growth form is restricted to the arthropods and probably is a major reason for their evolutionary success. In molting, the skeleton is periodically discarded and a new, larger, and perhaps differently shaped skeleton develops. This method allows much more variety of shape because the skeleton can be essentially molded to the body. It has a disadvantage, however, in that the animal must periodically rebuild an entire new skeleton. While doing so it may have to abandon normal functioning and is without skeletal protection. The basic framework of the arthropod skeleton is organic (chitinous). In groups with a mineralized skeleton this framework is strengthened by the precipitation of calcite or apatite.

The third method of skeletal growth is by *addition of skeletal elements*. The simplest example of this is the addition of spicules to a spicular skeleton (sponges, alcyonarians, holothurians). A more complex example is the echinoderm skeleton, which is enlarged by the addition of new plates. The new plates are formed in the apical region and are in effect pushed down the side of the skeleton as newer plates are formed above them. Echinoderm plates are surrounded by and permeated with living tissues. This allows each plate to also grow by accretion throughout the life of the specimen. Thus in effect the echinoderms grow by a combination of addition of plates and accretion.

The fourth type of growth has been called *modification*. This method is used in the vertebrates. It involves growth on all sides and even within the skeleton and resorption of material so that both the shape and size of the skeleton can change as needed. In terms of variability of shape this is the ultimate skeleton. It does, however, require an internal position in order to develop. Thus it provides little protection to the organism. As a result, the vertebrates have developed skin tissue and its various modifications such as scales, feathers, and horny plates for protection.

Little is known about the relationship between shell structure and environment in most invertebrate groups. The three groups that have received the most attention in this regard are the foraminifera, corals, and molluscs. Thus we concentrate our discussion on these three groups in this chapter. Table 4.1 includes a listing of some recent and/or important studies or reviews of skeletal microstructure in the important fossil groups.

FORAMINIFERA

Three fundamental wall structures have been described in the foraminifera: *agglutinated* (arenaceous), *hyaline,* and *porcellaneous* (Fig. 4.4). The agglutinated structure consists of sediment grains acquired from the surrounding environment and embedded in a predominantly organic matrix. The porcellaneous structure consists of randomly oriented, elongate calcite crystals with a thin surface veneer of tangentially oriented crystals (Fig. 4.4A; Towe and Cifelli, 1967). It is nonporous and dense as compared to the hyaline structure.

Table 4.1   Selected examples of studies of skeletal structure

Fossil Group	Reference
Algae	Johnson, 1961 Wray, 1977
Foraminifera	Hay, Towe, and Wright, 1963 Towe and Cifelli, 1967 Hansen, 1979
Anthozoa	Kato, 1963 Sorauf, 1971 Sorauf, 1972
Bryozoa	Taverner-Smith and Williams, 1972 Ross, 1976 Sandberg, 1977
Brachiopods	Williams, 1968 Williams and Wright, 1970
Molluscs	Bøggild, 1930 MacClintock, 1967 Kennedy et al., 1969 Majewski, 1969
Arthropods	Travis, 1960 Levinson, 1961 Bourget, 1977
Echinoderms	Raup, 1966a Donnay and Pawson, 1969

The hyaline structure consists of somewhat larger elongate calcite crystals which are oriented perpendicular to the test surface (Fig. 4.4A; Towe and Cifelli, 1967). Both a granular structure and a radial structure have been described in the hyaline group, but Towe and Cifelli (1967) minimize the importance of this distinction. The hyaline wall is penetrated by numerous pores that parallel the crystals. Detailed discussions of the microstructure of foraminiferal walls are included in the work of Hay, Towe, and Wright (1963) and Towe and Cifelli (1967).

These three basic wall types are crudely correlated with growth environment (Greiner, 1969 and 1974). In the Gulf of Mexico, foraminifera with agglutinated structure are especially common in low-salinity and relatively low-temperature bays and estuaries such as Mobile Bay and Mississippi

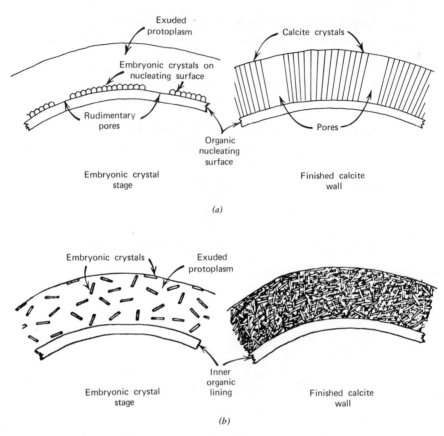

Figure 4.4 Schematic drawing of hyaline (*a*) and porcelaneous (*b*) wall type in foraminifera. After Greiner (1974), copyright President and Fellows, Harvard College.

Sound (Fig. 4.5). They are also relatively more abundant in deep water than shallow and in high latitudes than low. Foraminifera with porcellaneous walls are especially abundant in shallow, warm environments with normal to elevated salinity such as Laguna Madre on the south Texas coast and parts of Florida Bay in the Florida Keys area. Foraminifera with hyaline structure are most common in areas with temperature and salinity conditions more intermediate between these extremes such as San Antonio and Matagorda Bays. Greiner explained this distribution as being the result of differing solubility of $CaCO_3$. Because the solubility of $CaCO_3$ varies with both temperature and salinity, cold and low-salinity waters are generally unsaturated with respect to $CaCO_3$ and thus, presumably, $CaCO_3$ tests would be more difficult for the foraminifera to secrete, even with physiologic intervention. Sea water in areas with high temperature and salinity is usually considerably supersaturated with respect to $CaCO_3$. The relatively massive, randomly nucleated porcellaneous structure

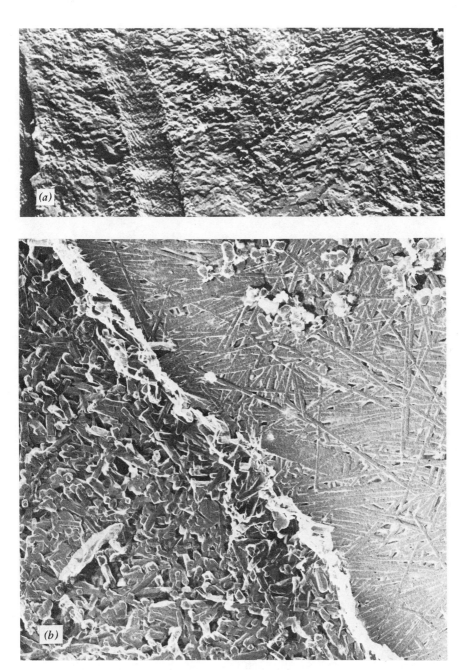

Figure 4.4A Electron photomicrographs of basic foraminifera wall types (*a*) Hyaline wall (*Cibicides refulgens*). Note pore in left-hand side of photo. × 5000. (*b*) Porcelaneous wall (*Quinqueloculina seminulum*). Upper right is shell surface, lower left is interior of wall. × 14,000 (*c*) Surface of agglutinated test (*Karreriella bradji*). Note the circular coccoliths incorporated into the test. × 5600. Photos supplied by K. M. Towe.

Figure 4.4A   (*Continued*)

can perhaps be more readily secreted in such supersaturated waters. In sea water which is saturated or only slightly supersaturated with respect to $CaCO_3$ the porous but more highly organized hyaline structure is secreted. This is a reasonable but simplistic explanation of the observed distribution pattern, for numerous exceptions can be found with porcellaneous foraminifera occurring in unsaturated waters and agglutinated taxa occurring in saturated

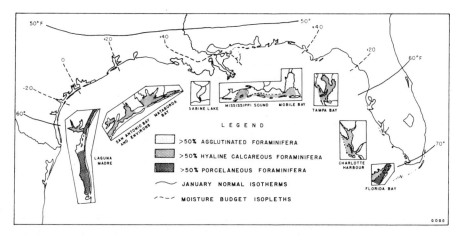

Figure 4.5 Variation in proportion of foraminifera with agglutinated, hyaline, and porcelaneous walls in selected bays and lagoons in the northern Gulf of Mexico. After Greiner (1974), copyright President and Fellows, Harvard College.

and highly supersaturated conditions. Nevertheless, the generalized distribution can be useful in paleoecological interpretations.

ANTHOZOA

The skeletons of modern scleractinian corals are composed of spherulitic clusters of aragonite needles (Fig. 4.6; Sorauf, 1972). The clusters in turn are arranged in vertical series forming elongate, fan-shaped trabeculae. Among the extinct subclasses of corals, the Rugosa have a skeletal structure very similar to the scleractinians (Sorauf, 1971). Octocorals, with a spicular skeleton, have a structure of elongate calcite crystals subparallel to the long axis of the spicule (Sorauf, 1974a). To date, most work on skeletal structure in the corals has dealt with detailed description, relation to the calcification process, and variation between taxonomic groups. This provides the necessary background for detailed studies of relationships between structure and environment.

Temperature and light appear to be the major environmental controls on the structure of coral skeletons. The skeletal characteristics that have been most studied in relation to the environment are the external features of the epitheca, which are probably related to skeletal structure, and to the spacing of internal features in colonial coral skeletons.

Growth banding is very prominent in corals and can be closely related to environmental parameters. Banding occurs on two scales in the epitheca of the Rugosa and some genera of the Scleractinia (Fig. 4.7; Wells, 1963). On the larger scale, constrictions and expansions with a spacing of about 1 cm appear to be the result of annual variation in growth. The expansions presumably represent summer growth when conditions are especially favorable, allowing expansion of the coral polyp. Constrictions probably represent winter growth or growth during times of less favorable conditions. On a much smaller scale, fine ridges with a spacing of a few micrometers are the result of daily growth cycles, probably related to variation in light intensity. Calcification in corals is much more rapid during the daylight hours than at night, presumably because of the effect of the symbiotic algae or zooxanthellae within the coral tissue. During the daylight hours these algae take up CO_2 in their photosynthesis and thus raise the pH, aiding precipitation of $CaCO_3$

Figure 4.6 Skeletal structure in modern scleractinian corals. (*a*) Aragonite needles in spherulitic arrangement (*Cladocora caespitosa*). Note banding. × 3000. (*b*) Spherulitic aragonite needles and growth banding (*Balanophyllia malounensis*). × 750. Photographs courtesy of J. E. Sorauf.

Figure 4.6 (*Continued*)

(Goreau, 1959). The daily bands can be grouped into fortnightly, tidal, and lunar cycles (Scrutton, 1964).

Each of these cycles has an environmental cause and thus is potentially useful in environmental reconstruction. One major application has been to determine the length of the day in the geologic past. Because of tidal friction, the speed of rotation of the earth should be gradually diminishing (Wells, 1963). The effect of this has been to gradually decrease the number of days per year. Astrophysicists (Runcorn, 1975) have in fact estimated the rate of this deceleration and have estimated the length of the day and thus the number of days per year for the past. Banding in coral skeletons offers an independent means of making this calculation (Fig. 4.8). Conversely, once the number of days per year during geologic time has been determined the age of the fossil corals should be determinable. This can be done by counting the number of daily growth lines between the annual expansions or constrictions. Wells (1963) determined that there were about 400 days in the year in the Devonian from counts of bands on the epithecae of specimens of the rugose coral, *Helliophyllum*. He checked this method by counting growth bands on modern specimens of several scleractinian genera; the modern specimens did indeed have about 365 bands per year. An alternative approach is to determine the number of days in the synodic (lunar) month. This number also has

Figure 4.7 Growth banding on the surface of the rugose coral *Zaphrentoides pellaensis* from the Lower Chester (Mississippian) Pella Formation, Mahoska County, Iowa. × 5. Specimen supplied by A. S. Horowitz.

decreased with time as predicted from the decrease in the earth's rotation (Scrutton, 1964).

Determining the number of days in the year from fossil corals will probably not be a very practical method of age determination for several reasons. (1) Counts of growth bands can be made only on the epitheca of very well preserved specimens. (2) The counting procedure is somewhat subjective because some bands may be indistinct and the precise location of the maxima of the annual expansion or the minima of the constrictions is hard to determine objectively. (3) The precision of this method is not as good as more conventional biostratigraphic methods. Nevertheless, the technique has confirmed the predicted slowing of the earth's rotation and has been an especially notable example of cooperation between geophysical and paleontological researchers.

Annual cycles have also been described by T. Y. H. Ma (1958 and others) in the skeletons of tabulate corals. These cycles consist of a regular pattern of closely spaced tabulae alternating with more widely spaced tabulae (Fig. 4.9). The wide spacing may be the result of rapid growth under optimal conditions, probably in the summer, and the more closely spaced tabulae develop under

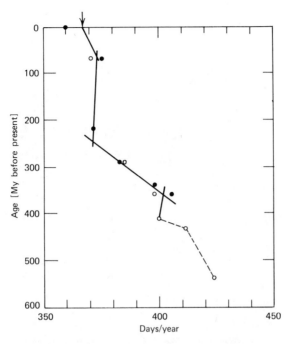

Figure 4.8 Number of days per year as estimated from growth rings in fossils. Data from various published sources. After Creer (1975).

more marginal conditions, probably in the winter. Ma used this model in an attempt to locate the position of the equator as related to the drifting continents during various times in the Paleozoic. He reasoned that specimens living near the equator should have rather uniform growth conditions and thus should not develop banding in their skeleton, but that specimens living far from the equator, where seasonality is pronounced, should have a well developed banding pattern. Thus the banding pattern for a given time in the geologic past should be correlated with latitude and should help to locate the position of the equator. This approach has not been widely used because of the scarcity of reliable data on banding in the corals, but Ma's data have been reinterpreted in light of plate tectonic theory by Fischer (1964). The model appears to be basically sound and potentially useful but needs to be more extensively tested. One problem that has developed from more recent work (Weber et al., 1975a) is that banding occurs in modern corals even in the equatorial zone.

Growth rate of corals can be determined from the growth banding. Growth rate is in part under environmental control and can thus be used to make paleoenvironmental interpretations. These data can also be used for analysis of population dynamics. Wells (1963) and Johnson and Nudds (1975) determined growth rates for Devonian and Carboniferous corals. The Car-

Figure 4.9 Banding in a tabulate coral caused by cyclic spacing of tabulae. × 1. Photo courtesy of Allen Archer.

boniferous corals from England grew more rapidly than the Devonian ones from New York. This difference has been explained as due to England being nearer the paleoequator during the Carboniferous than was New York during the Devonian. Other factors must also be considered, however, in the interpretation of growth rates. For example, Johnson and Nudds also note that monthly banding is less well developed in specimens from shaly than limey beds. Reduced light intensity in muddier and/or deeper water may be the cause for this difference.

Because banding in modern corals is often difficult to recognize, the use of x-radiographs has proved valuable. Specimens of scleractinian colonies have been studied by making thin slices of the coral colonies and transmitting x-rays through them onto x-ray sensitive film (Fig. 4.10). The photographs thus produced reveal alternating light and dark bands that appear to be related to variation in the density or compactness of the skeleton (Knutson et al., 1972). The work of Weber et al., Deines, et al. (1975) and Weber, White, et

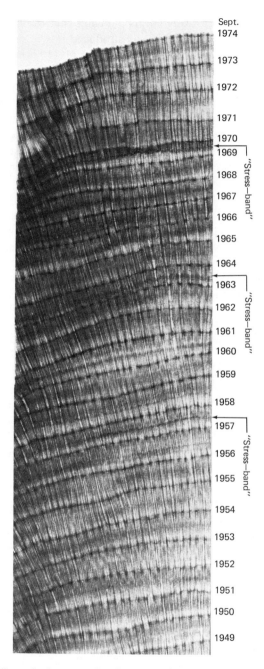

Figure 4.10 X-radiograph of a core taken from a specimen of *Montastrea annularis* living on the Hen and Chickens Reef, Florida Keys. The dates indicate the year of formation of the bands as determined by counting downward from the growing surface. "Stress Bands" are formed during temperature extremes. × 1. Photo courtesy of J. H. Hudson.

al. (1975) and others with living corals indicates that the dense bands, which probably result from slower growth, are produced in late summer or early autumn. In addition to annual bands, Hudson et al. (1976), in studying corals from the Florida Keys, described disturbance bands that apparently usually resulted from extremely low winter temperatures.

Seasonal temperature changes are not the only cause of banding, for Knutson et al. (1972) and Buddemeier et al. (1974) have noted annual growth bands in x-radiographs of corals from the tropical Pacific where the temperature varies little seasonally. These bands are probably caused by variation in light intensity between the rainy (and cloudy) summer season and the remainder of the year. Spacing between growth bands, and thus growth rate, has also been ascribed to water depth, perhaps as it is correlated with light intensity. Weber and White (1977) used the seasonal pattern of density variation in coral skeletons to determine growth rates in two species of the genus *Montastrea*. They found that the growth rate in specimens of *M. annularis* from shallow water correlates directly with the average temperature for the growth locality (Fig. 4.11). Growth rate determinations from fossil corals can thus potentially be used for determining paleotemperatures. In

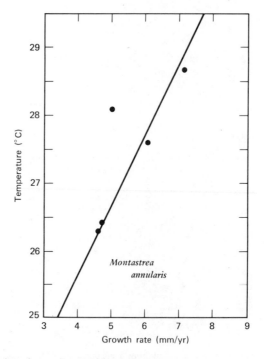

Figure 4.11 Variation of growth rate with mean annual temperature for shallow water colonies of *Montastrea annularis*. The line is a least-squares regression line fit to the data. Growth rates were determined from X-radiographs showing annual banding. After Weber and White (1977).

fact, by counting annual bands, Hudson and Shinn (1977) determined the growth rate for a large Pleistocene fossil specimen of coral and found it to be considerably slower than conspecific modern specimens. This suggests that the fossil coral grew in either deeper and/or colder water than the modern specimens.

Study of density bands in corals is in many ways similar to the study of tree rings in the science of dendrochronology. Hudson et al. (1976) have in fact suggested the term sclerochronology for the study of coral banding. This approach holds considerable potential as a paleoecologic tool for determining water depth, temperature, and perhaps other factors. An added advantage to studying banding in corals is that events occurring during the life of the coral can be put into a time framework that is longer than for other invertebrates. Large coral heads contain the record of hundreds of years of growth. By correlating the banding pattern between coral specimens (as is routinely done between trees in dendrochronology) the record can potentially be extended to cover even longer time periods.

MOLLUSCA

Shell Structural Types

Skeletal structure is more varied in the molluscs than in any other group of organisms. Perhaps in part as a consequence, skeletal structure has been more extensively studied in the molluscs than in any other group. Several aspects of skeletal structure of the molluscs, especially the bivalves, seem to be influenced by environmental conditions; the influence of environment on banding patterns has especially been the subject of extensive research in recent years.

Seven common basic shell structural types (Fig. 4.12) have been described in the molluscs (Taylor et al., 1969) and several other minor types and variants have also been recognized. The basic types are described here in only very general terms. More complete descriptions can be found in Bøggild (1930), MacClintock (1967), Majewske (1969), and Taylor et al. (1969).

Nacreous structure (mother-of-pearl) is composed of polygonal to rounded platelike crystals of aragonite that either are arranged in sheets separated by layers of organic matrix and are parallel to the growing surface or are arranged in columns of crystals perpendicular to the growing surface (Fig. 4.12*a*). This structure resembles bricks and mortar in a wall, with the bricks being the aragonite crystals and the mortar the organic matrix between crystals.

Prismatic structure consists of polygonal columns of calcite or aragonite oriented perpendicular to the growth surface (Fig. 4.11*b*). The prisms are separated by organic matrix.

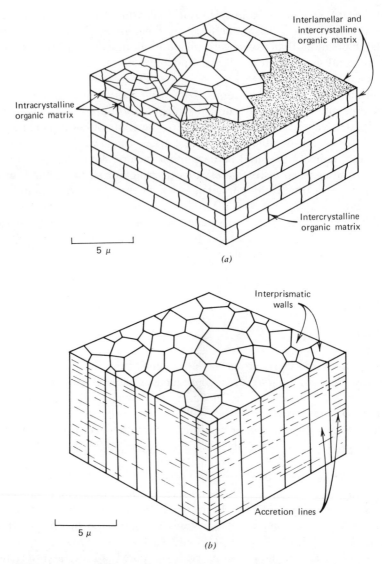

Figure 4.12 Molluscan shell structure types. (*a*) Nacreous structure; (*b*) prismatic structure; (*c*) foliated structure; (*d*) crossed-lamellar structure; (*e*) myostracal structure. After Kennedy et al. (1969), copyright Cambridge University Press.

Foliated structure consists of elongate crystals or laths of calcite arranged in sheets that are subparallel to the growing surface (Fig. 4.12*c*). This structure is similar to the nacreous but of different mineralogy and less regularity.

Crossed-lamellar structure is composed of aragonite laths that are mutually parallel in lamellae or blocks that interpenetrate in such a way that laths in neighboring lamellae are inclined at a high angle (Fig. 4.12*d*).

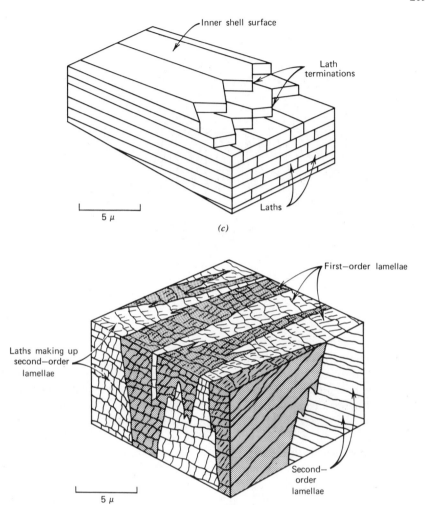

Figure 4.12 (*Continued*)

Complex crossed-lamellar structure is an aragonite structure similar to the one above but with the lamellae arranged in many different orientations.

Homogeneous structure consists of very small (1–3 μm) equidimensional granules of aragonite with parallel or nearly parallel crystallographic orientation.

Myostracal structure is composed of irregular prisms or blocks of aragonite (Fig. 4.12e). It forms under the muscle attachment area of the mollusc.

These shell structural types are widely but not randomly distributed among the molluscs. For example, the nacreous structure is particularly common among taxonomic groups that are considered to be phylogenetically

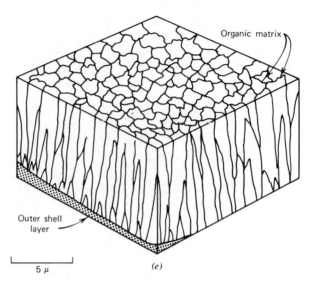

Figure 4.12 (*Continued*)

primitive and the crossed-lamellar and complex crossed-lamellar structures are most common among the more advanced groups. Thus the distribution of structural types among the bivalves is probably determined in part by phylogeny. It is also related to the mode of life, hence is sensitive to environmental influence (Table 4.2). Taylor and Layman (1972) determined various mechanical properties of the bivalve shell structures and attempted to relate these to function in the organism. They found that the nacreous structure was the most resistant to breakage in compression, impact, and bending tests, whereas the composite prismatic, crossed-lamellar, and complex crossed-lamellar structures are the hardest. These properties may in part explain the observed distribution of structural types among the different modes of life, but other factors such as the rate and efficiency of biochemical formation of the shell must also be involved. The nacreous structure, perhaps because of its strength, is common in bysally attached forms (Mytilacea) that are often exposed to strong current and wave action as well as mechanical crushing by predators. Nacreous structure is also common in shallow burrowing freshwater bivalves (Unionacea) where the structure may be advantageous in being less soluble in water that is likely to be unsaturated with respect to $CaCO_3$. The reduced solubility of the nacreous shell results from the high organic content which shields the $CaCO_3$ from contact with the water. The nacreous structure is also found in some very thin shelled bivalves where its high strength would be an advantage.

The foliated structure is common in cemented bivalves (Ostreacea) and some bysally attached forms (Anomiacea). The advantage of the foliated structure may be that it can be deposited quickly to form thick shells which

Table 4.2 Mode of life and shell structure combinations for bivalve families

	Aragonite Prisms and Nacre	Calcite Prisms and Nacre	Foliated	Composite Prisms c.l. and c.c.l.	Crossed-Lamellar and Complex c.l.	Homogeneous
Free-living epifaunal			••			
Byssate		•••••	•••		•••••	
Cemented	•••		•••		•	
Boring		•			•••••	•
Shallow burrowing	•••••			•	•••••	•••••
	•••••				•••••	•
					•••••	
					•	
Deeper burrowing	••••			••••	•••	••••
				••••		
				•		

[a] In cases in which a family has two modes of life, the family is entered twice. After Taylor and Layman (1972). Each dot represents one family with a combination of life habit and shell structure.
[b] c.l.: crossed lamellar; c.c.l. complex crossed lamellar.

are resistant to predators and mechanical destruction. The foliated structure in the Pectinacea is strong enough and flexible enough for these thin-shelled, free-swimming types. The crossed-lamellar and complex crossed-lamellar structure is found in many different bivalves but is especially common in burrowing taxa where its hardness may be an advantage in resisting abrasion by the sediment. The crossed-lamellar and complex crossed-lamellar structures are also found in boring bivalves in which shell hardness should really be valuable. The homogeneous structure, also relatively hard, is found largely in burrowing bivalves.

Little is known about variations within these basic shell structure types in response to environment. Extra myostracal layers in the nacreous structure of the bivalve *Brachidontes recurvus* seem to result from life under stress situations (usually periods of reduced salinity) (Davies, 1972). A similar variation in the nacreous shell structure of *Geukensia demissa,* a modern shallow water bivalve, may result from alternations of well developed, typical nacreous structure in the summer months, smaller pitted crystals in the autumn, and cessation of growth and solution of the nacreous surface in winter. The solution, caused by acids formed by anaerobic respiration in the animal, results in a fine-grained structure in the shell that is really a degenerated nacreous structure. When growth resumes in the spring a thin layer of aragonitic prisms (much like the myostracum) results from the vertical stacking of nacreous aragonite crystals. Finally, normal nacreous structure begins to form later in the spring or summer. Thus the nature of the structure is a function of temperature as it affects the behavior and the respiration pattern of the animal (Lutz and Rhoads, 1977). Factors other than temperature (such as reduced salinity) may also produce extra myostracal layers similar to those observed by Davies (1972). Zones of slightly finer crystal size that occur in the prismatic structure of *Mytilus* appear to result from summer growth (Dodd, 1964).

The shell structural types are arranged in discrete microarchitectural units within the shell. The shape of these units is a result of the method of growth of the shell. This can be illustrated by describing the growth of the bivalve shell. Shell growth in the other molluscan classes is a modification of this basic pattern. The bivalve shell grows by accretion in two general areas (Fig. 4.13): the mantle margin and the general mantle surface, the two areas being separated by the pallial line. Growth at the mantle margin results in growth at the margin of the shell and is thus largely responsible for increasing the length and width of the shell. Growth on the mantle surface results in thickening of the shell. Mantle margin growth in a sense produces the framework, and growth at the mantle surface plasters and thickens the basic frame. Commonly, the shell structural types in these two areas are different and result in distinctive microarchitectural units. For example, growth at the mantle margin may produce the prismatic structure which thus forms an outer layer of the shell, and growth at the mantle surface may produce the nacreous structure which lines and thickens the shell. The basic two-layered calcareous structure may be complicated by subdivision of the inner layer into two or more subdivisions that may have different structural types.

In addition to the outer calcified layer produced at the mantle margin, there is a thin organic layer, the *periostracum,* forming the outermost portion

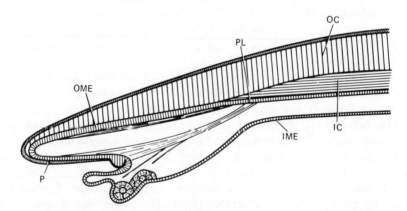

Figure 4.13 Diagrammatic cross section of bivalve shell and mantle. P—periostracum; OC—outer calcified shell layer; IC—inner calcified shell layer; OME—outer mantle epithelium; IME—inner mantle epithelium; PL—pallial line. After Wilbur (1964).

of the shell and the surface on which the outer calcareous layer is deposited. The periostracum is produced by the innermost of three folds of the mantle margin (Fig. 4.13). It wraps around the mantle margin and onto the exterior surface of the calcareous shell itself. As growth proceeds, the periostracum in effect unrolls from the mantle margin as the shell grows outward underneath it. The outer calcareous layer grows by nucleation on the inner surface of the periostracum and as it thickens it pushes the adjacent mantle margin away from the periostracum. Thus the area that was originally at the shell margin is left behind and eventually the pallial attachment migrates past it and inner shell layer is secreted. The distinctive myostracal shell structure forms under the attachment of the muscle to the shell so that the location of the muscle attachment throughout the life of the animal can be traced by this structure.

This method of growth produces an inner layer that thickens from a feather edge near the shell margin to a maximum in the beak area where the layer has been forming for a longer period of time. The outer layer is thickest at the pallial line and thins both toward the shell margin and toward the beak area.

The relative proportion of these shell units may change with environmental conditions. In species with shells composed of a combination of calcite and aragonite, the two minerals are present in different shell layers or different subdivisions within a shell layer. Changing proportions of these minerals as influenced by environment are thus reflected in changing proportions in the shell microarchitectural units. In modern specimens of *Mytilus californianus* (Fig. 4.14), for example, the aragonite/calcite ratio, as determined by x-ray diffraction, and the relative proportions of the aragonitic nacreous structure and the calcitic prismatic structure, as determined from polished

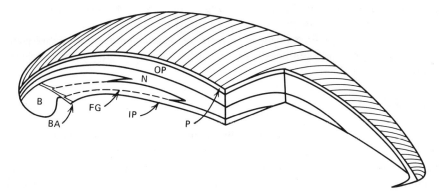

Figure 4.14 Three-dimensional cutaway view of a valve of *Mytilus californianus* showing shell structural features. P—periostracum; OP—outer prismatic layer; N—nacreous layer; IP—inner prismatic layer; BA—blocky aragonite layer; FG—fine-grained zone; B—beak area. After Dodd (1964), Journal of Paleontology, Society of Economic Paleontologists and Mineralogists.

sections and nitrocellulose peels of cross sections of the shells (Dodd, 1964), are strongly correlated. Both parameters are in part controlled by growth temperature (see Chapter 3). Specimens of *M. californianus* from relatively warmer water have a shell consisting of an outer calcite prismatic layer and an inner aragonite nacreous layer. In colder waters this species develops an inner calcite prismatic layer nearest to the beak area which is really a subdivision of the inner layer of the shell (Fig. 4.14). The extent of the inner prismatic layer is greater in colder regions, with specimens from cooler waters having an inner prismatic layer that is much more extensive than the nacreous layer. The growth temperature can be estimated from the type of structure found in this species (Fig. 4.15). Care must be taken in estimating growth temperatures from the shell structure of *M. californianus* because shells that have been abraded or bored during the life of the specimen are often patched or strengthened by the deposition of prismatic shell material on the inner surface of the shell. This occurs even in specimens growing in warm waters.

Another unusual feature in the structure of the *Mytilus californianus* shell is the interdigitation of the inner prismatic and nacreous units (Fig. 4.14). This must result from migration of the boundary between those parts of the mantle surface that secrete the two structures. This migration is apparently a response to temperature. When the temperature drops in the fall and winter the area of secretion of the calcite inner prismatic layer expands, that is, the boundary between calcite and aragonite formation extends further toward the ventral margin. When the temperature rises in the spring and summer, the area of aragonite deposition expands; that is, the boundary between calcite and aragonite formation moves closer to the beak of the shell. Each expansion of the one structure into the other records a season of growth, and by counting the number of interdigitations the age of the shell can be determined. Growth rate can be determined from the age and size of the shell. Growth rate in *M. californianus* is in part a function of the average temperature of the growth locality and thus the intertonguing feature of the structure of the shell can be used to determine growth temperature (Fig. 4.16).

The structures of Pleistocene specimens of *Mytilus californianus* has been used to interpret paleotemperatures for various locations along the southern California coast (Dodd, 1966).

Skeletal Banding

The feature of the structure of mollusc shells that has attracted the most attention in recent years is the banding or growth increment pattern observed in the shells of several species of bivalves. This banding is similar to the annually and daily produced growth features in coral skeletons discussed above. Prominent growth lines or shelving on the surface of bivalve shells have been recognized for many years as probably annual features caused by the slowing or stoppage of growth during the winter months, and have in fact been used in studies of age and population dynamics. Weymouth (1923),

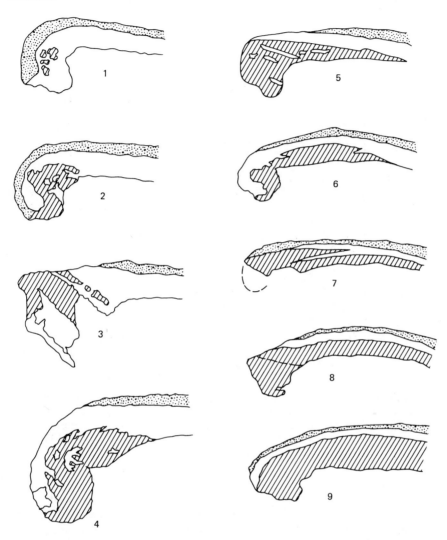

Figure 4.15 Tracings of photographs showing structural types in *Mytilus californianus.* The higher-number structural types are characteristic of colder growth temperatures. Stippled areas are calcite outer prismatic layer; clear areas are aragonite nacreous and myostracum layers; and lined areas are calcite inner prismatic layer. × 2. After Dodd (1964), Journal of Paleontology, Society of Economic Paleontologists and Mineralogists.

for example, recognized not only growth ridges but also variation in the internal structure of the shell of the pismo clam, *Tivela stultorum,* which was reflected in variation in its opacity to transmitted light. Barker (1964) published the first detailed description of internal banding within bivalve shells. He described banding cycles of five orders or scales ranging from a few μm to several mm. He interpreted these banding cycles as probably due to daily tidal

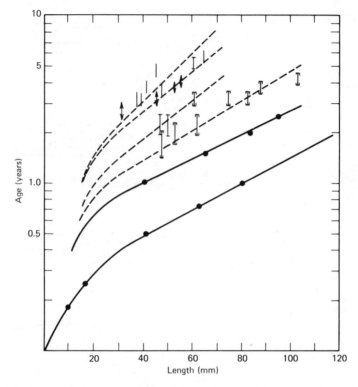

Figure 4.16 Representation of growth rates of *Mytilus californianus* at various localities, as determined from skeletal microstructure. The upper lines (slower growth rates) are based on specimens from colder localities. The upper four lines were determined from shell structural features. After Dodd (1964), Journal of Paleontology, Society of Economic Paleontologists and Mineralogists.

cycles, day to night variations, lunar monthly tidal cycles, seasonal storm patterns, and annual temperature variation. The larger scale banding patterns are produced by regular variation in the smaller scale banding, the basic, smallest-scale banding being produced by variation in the amount of organic material or conchiolin in the shell. Pannella and MacClintock (1968) recognized a similar pattern of banding in specimens of *Mercenaria mercenaria* as well as other bivalve species (Fig. 4.17). They conducted a series of growth experiments in which they notched shells of living specimens, returned them to their natural environment, and recollected them at a later date. The count of the daily cycles in the shell banding beyond the notch was the same as the number of days between notching and killing of the specimens, confirming their interpretation of the origin of these bands. They were also able to recognize the effects of storms known to have occurred during that time period as well as effects of the seasons, spawning cycles, and lunar monthly cycles in the banding pattern. The physiologic mechanism by which the bivalve produces the bands is not known but apparently is related to variation in the

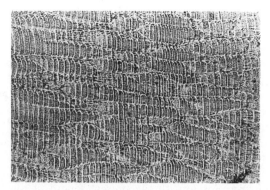

Figure 4.17 Longitudinal section of the outer shell layer of *Mercenaria mercenaria* showing daily growth bands. The growing edge of the shell is to the right. The outer surface of the shell is toward the top of the photo. × 200. Photograph courtesy of Copeland MacClintock.

relative rate of formation of organic matrix or conchiolin and calcium carbonate. The organic matrix may be forming at a relatively more constant rate than the $CaCO_3$ so that, at times when little or no $CaCO_3$ forms, an organic-rich band occurs in the shell. On the other hand, the banding may result from periodic solution at the shell surface. When the bivalve shell is closed, the animal respires anaerobically and produces acids that are neutralized by solution at the shell surface. The organic matrix that was associated with the dissolved carbonate is left as a residue on the shell surface. When the valves are reopened and the animal resumes aerobic respiration, calcification is also renewed, and the organic matrix residue is incorporated within the new shell material, giving it a high organic content and a different color. The alternating pattern of shell opening and closing (and of respiration) varies regularly with environmental conditions (especially the tidal cycle and day and night) so that the banding that results from this behavioral pattern correlates with environmental conditions (Lutz and Rhoads, 1977).

The growth lines on the outer surface of bivalve shells have also been recognized and studied in detail. Davenport (1938) first suggested that the fine growth lines on the surface of *Pecten* shells recorded daily growth intervals, and noted annual lines on *Pecten* shells resulting from winter cessation of growth. He proposed that the number of daily lines between the annual lines should increase with increasing temperature, reflecting a longer growing season. This potential method for determining paleotemperature trends has never been adequately tested on fossil *Pecten*. Clark (1968) later confirmed the daily origin of the fine growth lines by observation of *Pecten* specimens growing in aquaria as well as in nature. The same cycles and events that are recorded in the internal shell structure are preserved in these growth lines. One difference is that occasionally the *Pecten* specimen does not produce a growth line during the day so that the number of observed growth lines for any time interval is commonly a few less than the number of days.

One obvious use for the banding pattern in fossil bivalve shells is to determine the number of days in the year in much the same manner as has been done with the corals (see above). This has been done and the results are in general agreement with those from the corals (Pannella et al., 1968), although they show a somewhat more complex pattern of variation through time in the length of the day. Kahn and Pompea (1978) studied growth lines in nautiloids in order to determine variation in the length of the lunar month. They found that in modern *Nautilus* one growth line is formed each day. They further observed that once during each lunar month *Nautilus* forms a new chamber by secreting a septum in its shell. By counting the number of growth lines between septa they could determine the number of days in a lunar month. For modern *Nautilus* there are about 30 growth lines between the septa. This compares closely with the present length of the lunar month of 29.53 days. Sanders and Ward (1979) have questioned these results, suggesting the need for additional work.

The shell of a bivalve or any organism having a shell that grows by accretion can be considered as a recorder of the life history of the organism, because shell material that was produced throughout the life of the individual is preserved in the shell. The banding pattern contains the record of the growth of the animal and to the extent that environmental and biological factors control growth, variations in these factors will be recorded in the banding pattern.

Rhoads and Pannella (1970) discuss the factors controlling the banding pattern and indicate some of the paleoecological and paleobiological applications that can potentially be made from a study of the banding pattern. They also demonstrated a number of environmental effects on the banding pattern of modern specimens of bivalves. They showed that growth rate in *Mercenaria mercenaria* is dependent on substrate type, being higher on a sandy substrate than in mud. The time of death of *Gemma gemma* was studied and mortality was found to be highest in autumn and early winter. In fossils such information could tell us something about the biology of the species and would also imply seasonal climates. Species of the genus *Nucula* at 3 to 4970 m depth were compared. The banding in the shallow water forms was found to be much more pronounced and irregular than the deep ocean specimens. This suggests that shell banding in fossils can potentially provide information on paleobathymetry.

Variation in growth rate as determined from banding might also be used to determine relative paleotemperatures as growth rate usually varies directly with temperature. Paleolatitude might also be estimated by variation in the development of seasonality in the banding pattern in much the same way as attempted by Ma in his work with fossil corals. The occurrence of storms or other high-stress situations in the geologic past might be suggested by the presence of irregularly spaced interruptions or slowing of growth.

Periods of exceptionally high temperature may cause slowing of growth rate or disturbance zones in the shell banding pattern. Kennish and Olsson (1975) noted such effects in the shells of *Mercenaria mercenaria* that were growing in a bay near an area of cooling-water discharge from a power station. They suggested that the shell structure of this species could be used as a natural recorder of thermal conditions near such installations.

Hudson (1968) made paleoecological conclusions based on the shell structure and mineralogy of shells of a Jurassic mytiloid from Scotland. He determined that both the inner and outer shell layers were composed of nacreous structure. Modern species of *Mytilus* having entirely nacreous shells are all tropical or subtropical in distribution. The inner layer of the fossil specimens shows sublayering which Hudson interpreted as probably forming by seasonal growth, suggesting that environmental conditions varied seasonally. This combination of warm temperature and seasonality may indicate that the bivalves grew under subtropical conditions.

The number of daily growth increments in the annual fast-growing band and slow-growing band of some species varies with latitude (Hall et al., 1974). In *Tivela stultorum* the number of daily growth increments in the slow-growing band increases with decreasing latitude and increasing temperature (Fig. 4.18). The number of daily increments in the fast-growing band decreases with decreasing latitude. Hall (1975) also studied modern and fossil (Pliocene)

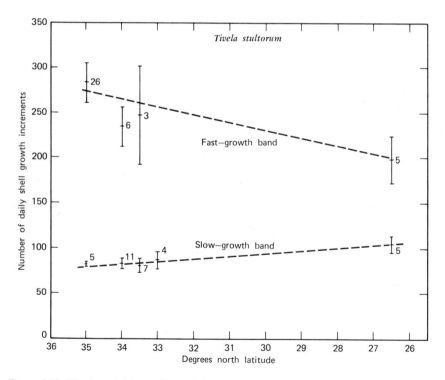

Figure 4.18 Number of daily shell growth increments per slow-growth and fast-growth band in *Tivela stultorum* from the Pacific Coast of North America. The numbers refer to the number of individual bands from which growth increment counts were made. After Hall (1975).

specimens of *Callista chione* from northern Italy. The slow-growing band in the fossil specimens of this species have more daily growth increments than do the modern specimens. They may thus indicate a warmer temperature for this interval of Pliocene time at this locality.

Coutts (1970 and 1975) has made an archeologic application of banding in *Chione stutchburyi* shells collected from archeological sites in New Zealand. By studying the banding pattern in the shells he was able to determine the time of year when the bivalves were collected for food and thus the season of occupation of the midden sites. Many of the sites were occupied only during the winter.

To date, most work on shell banding has been confined to the bivalves because banding there is better developed than in other groups. However, the methods can potentially be used on any organism that forms its skeleton by accretion. Perhaps different techniques of sample preparation will aid in the study of these groups. Some effort has been made in studying banding patterns in stromatolites (Pannella, 1972), although the reliability of data from this group has been questioned (Park, 1976).

Very prominent annual growth banding has been described in echinoderm plates (Fig. 4.19; Pearse and Pearse, 1975). This group might also offer promise in yielding paleoecologically useful information. Otoliths (fish ear bones) also show an annual banding pattern.

SKELETAL STRUCTURE IN FOSSILS FROM THE KETTLEMAN HILLS AREA

Alexander (1974) used banding in the bivalve *Anadara* to interpret water depth in the Pliocene sea in the Kettleman Hills. He noted that some *Anadara* shells clearly showed a fine banding pattern much like that described by Barker (1964) and others as being daily in origin. Distinct bands of about the proper spacing to represent fortnightly cycles could also be recognized. The regularity and distinctness of these probable monthly bands in *Anadara* was not uniform in the Kettleman Hills. By comparison with the banding on modern bivalves, the more regular banding pattern probably reflects relatively deeper water where growth was less strongly influenced by tidal cycles. Those specimens with very irregular banding patterns were similar to modern specimens of *Mercenaria mercenaria* from the intertidal zone. Within the *Pecten* zone of the San Joaquin Formation the *Anadara* specimens with the most regular banding pattern came from the northwestern end of North Dome, the area considered to be nearest the open ocean. The distinctiveness of banding also differed stratigraphically. Banding was generally less distinct in specimens lower in the section, indicating deeper water. This may be due to a gradual change in the depth preference of *Anadara* with time toward a shallower environment.

Shell structure may yield much paleoecologically useful information, but

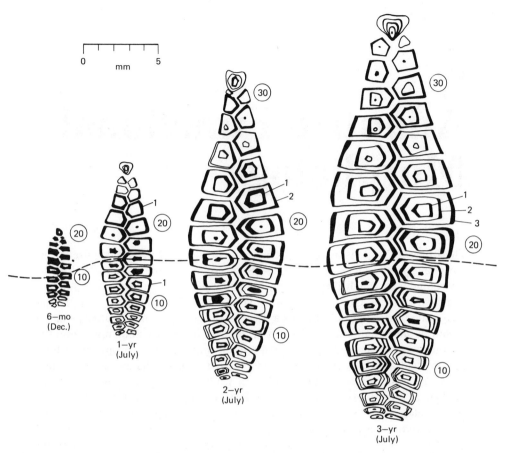

Figure 4.19 Diagrammatic representation of interambulacral growth zones of *Strongylocentrotus purpuratus* at six months, one year, two years, and three years of age. Dark areas represent translucent zones. The end of the three years of life is marked on representative plates. Plate position numbers, indicated in the circles, show how plates appear to shift towards oral side of test as new plates are added aborally. The dashed line shows the position of the ambitus (greatest test diameter). After Pearse and Pearse (1975).

its potential has just begun to be realized. The shell of an organism that grows by accretion is analogous to a tape recorder. It preserves the record of the events during the entire life of the organism as the shell material forms at the growing edge much as the tape rolls past the recording head. The shell is also somewhat analogous to a crustal plate which is formed at the mid-oceanic ridge or rise and then is displaced away from the rise. In the case of the accreting shell the growing edge moves or progressively recedes from the shell margin once it forms. The growth bands in the shell are somewhat like the magnetic bands in the oceanic crust formed by magnetic polarity changes. The task of the paleoecologist is to learn how to play back the record that is recorded in the shell structure.

5
Adaptive Functional Morphology

The fitness of organisms to their environment has long been recognized. The morphology, physiology, and life history of organisms are so well tuned to the environment in which they live that for many people it has been a significant demonstration of God's providence. With the development of evolutionary theory the fitness has become less a demonstration of original creativity than of a continuing dynamic and progressive process of change that maintains the organism in a stable and well adapted relationship with its environment as both the physical and biological aspects of the environment change through time.

Three basic methods can be used to study the adaptive functional morphology of a fossil organism. (1) If the fossil belongs to an extant species or is closely related to a living species, the relationship of morphology to function can be observed directly. This is the method of *homology*. Often the function of a given morphological feature is obvious (e.g., the wedge shape of the clam for ease of burrowing, the claws of crabs for crushing prey, or the stem of a crinoid to elevate the calyx above the sea floor); but in other cases, even in living organisms, the function is not obvious (e.g., why are coral branches inclined into the current? why does *Nautilus* have chambers? why do the arms of crinoids form parabola-shaped nets?). To some extent these questions can be answered from careful observation, but even with living organisms simple observation may not explain why a particular morphology is better adapted than another. (2) The function of the morphologic feature in the fossil may be explained by *analogy* to similar morphologies in extant organisms that are not closely related to the fossil. For example, the streamlined shape of the ichthyosaurs was probably an adaptation for rapid swimming as it is in living porpoises. The enrolled condition often found in trilobites is a protective mechanism as it is in living isopods (pill bugs). The spines in some extinct brachiopods functioned to prevent sinking into soft substrate just as do the spines in some modern bivalves. (3) The function of the morphologic feature may be determined more indirectly through use of the *engineering* or *paradigm* approach. This method, which is described in detail below, basically involves a

222

logical, engineering-type analysis of the function of particular structures. The method may simply involve logical reasoning about possible functions, or it may involve more rigorous quantitative comparison with engineering structures. In some cases models of the morphologic feature are tested using various engineering techniques.

This chapter is mostly concerned with the engineering or paradigm approach, although many of the examples given also use homology and analogy as supplemental evidence.

All aspects of an organism, including physiology and behavior as well as morphology, are adapted to the environment. The paleoecologist is of course largely limited to the study of the morphology of fossils and the adaptation of this morphology to the environment, that is, the *adaptive functional morphology*. The study of adaptive functional morphology has great potential applicability in the determination of ancient environments because it essentially depends on an engineering analysis of an organism as a machine adapted to function under specified conditions. In concept, this type of analysis does not depend on comparison of the fossil to closely related living species and thus is independent of the assumptions of taxonomic uniformitarianism (Chapter 2). This approach has been used with apparent success with such extinct groups of fossils as the richthofenid brachiopods (Rudwick, 1961, and Grant, 1972), the rudistid bivalves (Skelton, 1976), and dinosaurs (Farlow et al., 1976). In practice some knowledge of the functioning and adaptation of modern counterparts is useful. For example, the classical study by Nichols (1959) of adaptation in Cretaceous echinoids benefited from comparison with modern echinoids. Studies of streamlining and wall strength in ammonites (Chamberlain, 1976, and Westermann, 1973) have benefited from comparison with modern *Nautilus*. The adaptive functional morphologic approach merges with taxonomic uniformitarianism in the case of analysis of the adaptations of living groups of organisms. Several cases discussed in Chapter 2 as examples of taxonomic uniformitarianism could have been discussed in this chapter as adaptive functional morphology. For example, the colonial form in bryozoans (Table 2.3) or morphological adaptation to prevent brachiopods from sinking into soft sediment (Fig. 2.23) could be discussed in this chapter just as well as in Chapter 2. Because of its conceptual independence from taxonomic uniformitarianism, adaptive functional morphology is especially useful for paleoecologic interpretations of Paleozoic biotas with only distantly related living representatives.

THEORY

The observation of the fitness of living organisms to their environment provides the initial basis for interpreting the morphology of fossils in order to determine ancient environments. A firmer theoretical basis, including causative mechanisms, is provided by the modern theory of evolution by natural selection. We know that organisms are adapted to their environment; therefore, in analysis of a fossil we can ask ourselves "For what environment is this morphology adapted?" If we can correctly answer that question we can predict the environmental conditions under which the fossil lived.

In any population, the full genetic content of the population will potentially be transmitted in unchanged proportion to the next generation unless outside forces operate. This is a statement of the *Hardy-Weinberg law*. However, all the individuals in the population are not equally likely to survive because of genetic and corresponding physiologic and morphologic differences which will cause them to be differentially adapted to the environment. Through the selection process only the best adapted individuals form the breeding population that gives rise to the next generation and provides the genetic material that will determine the genetic and, therefore, morphologic characteristics of that generation. Selection is the "outside force" operating to change genetic composition which, according to the Hardy-Weinberg law, should otherwise remain constant.

The *phenotype* of the organism, that is, its morphology, physiology, behavior, or any other expression of its genetic composition or *genotype*, will vary due to natural selection. Organisms with the same genetic composition may also have variable morphology, physiology, or behavior because of direct influence of the environment. This is *ecophenotypic* variation. Often these two types of variation are difficult to distinguish. For our purposes in paleoecology, the genetic cause of the variation may not be important. We only assume that the organism is adapted for its environment and attempt to explain the way in which the morphology is related to the environment.

Under conditions of long-term stability, the genetic and thus morphologic composition will remain constant although intraspecific variability may be large or small. Under changing environmental conditions, however, the genetic composition of the population will also change so that the optimal adaptive functional morphology exists at any time. Thus an evolutionary dynamic equilibrium is maintained by maximizing efficiency at any time for the species within the existing environment.

This model of morphologic adaptiveness, universal among organisms of all kinds and based on evolutionary theory, has proved to be very useful in explaining observations in the fossil record. Its potential in paleoecology is particularly great because it is relatively time independent. Given a specific anatomical-physiological system, changes in the external environment should evoke a limited range of morphologic adaptations. For example, plants of many different lineages are adapted to life in dry habitats, that is, they are *xerophytes*. All have similar adaptations for obtaining and conserving moisture, such as shallow but extensive root systems; thick, fleshy stems to store water; reduced leaves to keep their surface to volume ratio low; and chlorophyll in the stems to make up for loss of leaf surfaces. Presumably, xerophytic plants were similarly adapted in the past.

Morphologic adaptations are explained environmentally only within the framework of the organism's anatomical-physiological system. Seilacher (1970) has pointed out that function is only one of three factors controlling morphology (Fig. 5.1). The historical-phylogenetic factor limits the range of possible morphology by specifying both the general anatomical-physiological system and the materials used to construct the skeleton. In building a bridge, the engineer is limited by the materials available and, historically, by the level of technology. Likewise, for example, the potential morphology of an echinoderm is limited by certain characteristics: an internal skeleton of plates composed of single crystals of calcite with an open meshwork construction, a water vascular system, a stenohaline physiology, and a pentaradial symmetry. Consequently, the morphologic result of the adaptive response of an echinoderm is likely to be very different than that of an organism in another phylum with different phylogenetic constraints on the body plan. The analysis of morphology as a functional entity must take place within the framework of the possibilities defined by these constraints. Seilacher (1970) also points out that certain morphologic features have no particular adaptive function but result from the method of growth of the organism. For example, the polygonal shape of coralites in a coral colony may simply be the result of physical interference of the growing coralites as they impinge on one another. The polygonal shape may thus have no functional significance. Seilacher calls this the architectural factor. See Raup (1972) for a more extensive discussion of factors affecting morphology.

ANALYSIS

The logical steps in the analysis of morphology as an adaptive functional entity were formalized by Rudwick (1964b) in the paradigmatic method. Although

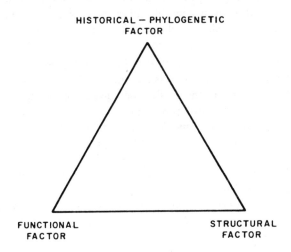

Figure 5.1 The three major factors controlling morphology. The position of the morphologic feature in the triangle represents the relative influence of the factors. After Seilacher (1970), Lethaia, *3*.

most studies of functional morphology do not specifically follow the five steps of the formal paradigmatic procedure, all or some of the steps are implied in the analysis.

1   The structure to be explained must be identified and described. It may consist of the total form or, more commonly, of a limited part of an organism. Recognition, in general, has depended on the structure being abnormal. The aberrant morphology of rudist bivalves as compared to other bivalves or of the giraffe with its extreme neck length as compared to related mammals are examples. Rigorous specification of the structure is important in the analysis and has resulted in numerous efforts to describe form in simplified and mathematical terms (Raup, 1966b; Berger, 1969; and Kaesler and Waters, 1972).

2   All possible functions for the structure ideally should be recognized. This is perhaps the most difficult step in the analysis, for how can we hope to imagine the functional aspects of an extinct organism whose physiology is unknown and whose life history can only be guessed? This step may be aided by comparison of the fossil with living relatives (homology) or unrelated forms with similar morphology (analogy). The more remote in age or relationship the modern analog, the more speculative will be the imagined function.

The paradigmatic method views the organisms in a very mechanical way, as an integrated assemblage of well-adapted individual parts. The structure of interest is assumed to have a function. Vestigial organs are not recognizable and are perhaps uncommon in fossils, as they appear to be among living organisms, so this may not be a serious problem. Organs or structures that may be of neutral or even negative value, however, may develop by being genetically linked to highly adapted structures. Their interpretation as adaptive features would be erroneous.

3   The ideal structure or paradigm to fulfill each of the possible functions conceptualized in the preceding step is formulated. The historical-phylogenetic aspect of the organism, the basic body plan, physiology, and building materials must be taken into account at this step and constrain the range of imagined structural paradigms. An additional consideration is that a functional need may be solved by a number of quite different and independent means. The conventional paradigm for speed in a quadruped is quite different for the placental cheetah or gazelle (long forelegs and hindlegs) than it is for the marsupial kangaroo (powerful hind legs with a large tail for balance). Also, a functional need may be solved by the combination of several structures and by physiologic and behavioral adaptations not evident in the fossil record. For example, one solution to the problem of lack of water in the desert by plants is for otherwise rather ordinary appearing plants to grow rapidly after one of the infrequent rainfalls, complete the life cycle, and produce seeds that do not germinate until the next rain. Such a

plant would appear much like a plant from a humid region, but it would be well-adapted to life in the desert.

The initial formulation of the paradigm is a mental construction based on analogous structures in other organisms or on engineering considerations. Experimental analysis of structures, using physical models or actual specimens, has become an important source of information at this step of the method. Calculations of strength and performance in engineering terms may also be useful.

4   The actual structure is compared with the paradigm. The function of the most similar paradigm is then assumed to be the function of the actual structure. To avoid being simplistic, this interpretation must consider again the complicating factors noted in step 3. In addition, the analysis should not consider an organism simply as a collection of parts, each adapted optimally for specific functional needs. Rather, evolution leads toward maximum efficiency in the total development of the organism internally and within its total environment. As a result, individual functions may act at less than their independent optimal levels, and corresponding individual structures are less than the paradigm, so that the total organism will be a well-integrated and harmoniously organized entity. Steps 2, 3, and 4 are very much interdependent and in actual practice would be considered together although their separation is useful in explaining the method.

5   From the inferred function, the causative environmental demand is determined.

The paradigmatic method is obviously highly speculative and involves interpretations that cannot be proven. Yet the framework of the method is inherent in all adaptive functional morphologic studies in paleoecology. The results are most tenuous when the organism being analyzed is extinct, with no closely related living relatives, and morphologically unusual so that even morphologically analogous but unrelated organisms are also lacking. The resulting conclusions of such a study are at best reasonable and possible but cannot be demonstrated to be the only possible conclusions. They are sufficient but not necessary. If the study is of a particular species and structure, the solution is an ad hoc one, suggesting a function of a particular structure but having no general application.

EXAMPLES OF FUNCTIONAL MORPHOLOGY STUDIES

The best way to demonstrate the use of the adaptive functional morphology approach is by discussing examples. The paleoecological literature contains an abundance of examples, a few of which are listed in Table 5.1. We discuss a few of them in detail in this section.

Table 5.1 Selected references to studies of adaptive functional morphology

Fossil Group	Reference
Calcareous algae	Bosence, 1976
Foraminifera	Bé, 1968
	Greiner, 1974
Sponges	Bidder, 1923
Coelenterata	Philcox, 1971
(Corals)	Stearns and Riding, 1973
	Hubbard, 1974
	Graus and Macintyre, 1976
Bryozoa	Condra and Elias, 1944
	McKinney, 1977a
	McKinney, 1977b
	Rider and Cowen, 1977
Brachiopods	Rudwick, 1961
	Rudwick, 1968
	Grant, 1972
	Fürsich and Hurst, 1974
Molluscs	Trueman, 1964
	Jefferies and Minton, 1965
	Stanley, 1970, 1977
	Carter, 1972
	Westermann, 1973
	Chamberlain and Westermann, 1976
	Linsley, 1978b
	Peel, 1978
	Skelton, 1979
Arthropods	Clarkson, 1969
	Benson, 1975
	Campbell, 1975
Echinoderms	Lane and Breimer, 1974
	Smith, 1978
	Welch, 1978
	Seilacher, 1979
Vertebrates	Hopson, 1975
	Stein, 1975
	Thomson, 1976

Calcareous Algae

Functional morphology in the calcareous algae has been little studied; however, a notable exception is the study of variation in the shape of the thallus of two coralline algal species conducted by Bosence (1976) on specimens from a bay in western Ireland. Bosence noted considerable variation in the shape of these two species, *Lithothamnium corallioides* and *Phymatolithon calcareum*, which could be correlated with environmental conditions. He describes three basic shapes of algal thalli: spherical, ellipsoidal, and discoidal (Fig. 5.2.). The algae

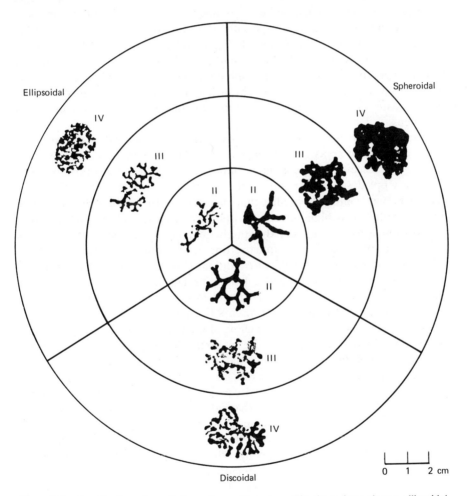

Figure 5.2 Classification of growth forms in coralline algae. The three shape classes, ellipsoidal, spheroidal, and discoidal, are based on the overall shape of the thallus. Within each class the branching density varies from open branching (II) through frequent branching (III) to dense branching (IV). Category I is unbranched. After Bosence (1976).

can also be subdivided into four categories in terms of their frequency of branching: unbranched, few branches, frequent branches, and dense branches. A quantitative method can be used to differentiate these shape and branching categories.

Bosence used two techniques to demonstrate the functional significance of the various algal morphologies. He noted the environmental distribution of living algal specimens (the homology or taxonomic uniformitarian approach), and he conducted experiments in a wave tank with actual specimens and models (the paradigm or engineering approach). In the wave tank experiments, the minimum velocity of water movement that would cause movement of the algal specimen or model was determined. Specimens and models with the ellipsoidal shape are most easily moved, especially when the long axis of the ellipsoid is parallel to the wave crests. A slightly higher velocity is required to move the spheroidal specimens, and the flat-lying discoidal specimens are most resistant to movement by the waves. The minimum velocity for transport of model specimens decreases with increasing branch density. This is apparently because of the greater cross-sectional area exposed to currents in the highly branched specimens.

These experiments suggest that algae with the discoidal shape are best adapted for life in areas with strong currents. They would be least likely to be moved. The discoidal shape is most common in specimens from turbulent settings on rippled sands, supporting the supposition. On the basis of the experiments one would also predict that specimens with few branches would be better adapted to a high-current environment than densely branched types. Observation does not support this prediction, however. Specimens with few branches are more common in quiet settings and densely branched individuals are most common under turbulent conditions (Fig. 2.4). This relationship appears to be nonadaptive. Actually, densely branched specimens may be more resistant to breakage than sparsely branched types and thus the relationship may be adaptive. Bosence did not discuss this possibility.

Rather than the shape and branching of the algae becoming adapted to the environment by natural selection, the environment seems to directly induce the variation in morphology. The discoidal shape develops because of frequent turning of the specimen by waves. Branches cannot grow on the underside of the algae. When the specimen is turned over by current or wave action, the top becomes the bottom and growth on that side stops. Frequent turning results in the discoidal shape because continuous growth is only possible around the edges of the disc. Frequent turning also results in abrasion and breakage of the tips of the branches. This induces branching. Thus the most highly branched forms are those that have been moved frequently by wave action.

This is an example of morphology adjusted to the environment but perhaps not truly adaptive morphology. Nevertheless, observation of coralline algal shapes in the fossil record may be a useful tool for environmental reconstruction.

Foraminifera

A number of examples of variation in morphology with environmental parameters have been described in modern planktonic foraminifera, and several of these examples are discussed in Chapter 2. The functional significance of this variation is not always clear, but a few examples can be readily explained in functional terms. The regular variation with latitude in porosity of the tests of planktonic foraminifera would seem to be such an example (Bé, 1968). The most porous tests are found in foraminifera from low-latitude, warm-water localities. Warm water has a lower density and viscosity than cold water, thus a porous, light test appears to be an adaptation to keep from sinking in the warm water.

Sponges

The sponges were the topic of one of the earliest functional morphological studies, but little work has been done in more recent years. Bidder's (1923) study of the relationship of sponge morphology to currents is a classical example of the application of the engineering approach. This work is discussed in detail in Chapter 2 (Figs. 2.10 and 2.11).

Corals

The study of the structure and orientation of coral colonies in response to waves and currents by Graus et al. (1977) is an excellent example of a study combining the engineering and homology approaches. The authors have made calculations on the basis of hydrodynamic relationships in order to show how morphology and orientation of a coral colony minimize the stresses produced by waves, currents, and by the weight of the colony itself. They in effect developed a morphological paradigm. Then they checked the predictions of their model against actual morphologies and orientations of speciments of a coral species, *Acropora palmata* from the island of St. Croix in the Caribbean.

Corals often live in environments with strong currents and waves and thus might be expected to show some morphologic adaptations to minimize the potentially destructive effects of the high hydrodynamic stresses found in such environments. In fact, the literature contains many references to preferred orientations and variable morphologies in corals in response to waves and currents (see Chapter 2). However, the qualitative ideas about the cause of the environmental control on coral morphology had not previously been rigorously tested.

Graus et al. (1977) approached the problem by considering the effect of a current on a single branch of a coral which they approximated by a "rough" cylinder with its center of mass and dynamic pressure at its midpoint. The stress on that cylinder can then be divided into three components (Fig. 5.3): (1) the ef-

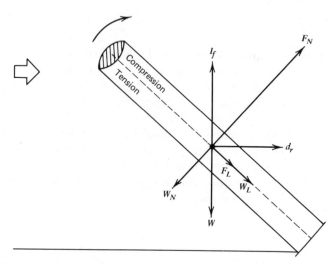

Figure 5.3 Forces acting on a branch in a current moving from left to right. The heavy arrow shows the current direction: d_r—drag produced by the current; l_f—lift produced by the current; w—weight of the branch. These forces can be resolved into components normal to and longitudinal to the long axis of the branch. W_N—normal weight component; F_N—vector sum of normal lift and drag components; F_L—vector sum of longitudinal lift and drag components; W_L—longitudinal weight component. The branch is bent upward by these forces (thin arrow), causing compression on the upper side and tension on the lower side. The neutral axis is shown by the dotted line. After Graus et al. (1977), published with permission of the American Association of Petroleum Geologists.

fect of gravity pulling down on the cylinder (W), (2) the drag produced by the current flowing past the cylinder (d_r), and (3) the lift produced by the current (l_f). These three components can be added vectorially to give the components of stress normal to and perpendicular to the axis of the cylinder (W_N, F_N, W_L, and F_L in Fig. 5.3). Stress in the coral will be minimal when the forces acting perpendicular to the long axis of the cylinder (i.e., W_N and F_N) are equal. If these forces are not equal the cylinder (or the coral branch that it models) will tend to break off at the base.

If there is no current, stress is minimized when the long axis of the cylinder is vertical. In this orientation there are no lateral forces; the only stress is compression of the cylinder due to its own weight. In a current the cylinder will tend to topple over if it is in the vertical orientation because of the drag produced by the current pushing against it. If the cylinder leans into the current the weight of the cylinder can counterbalance the hydrodynamic force so that W_N is equal to F_N. The stronger the current the greater will be the inclination from the vertical. We have all experienced this effect when we have tried to walk into a strong wind. We find progress easier when we lean forward into the wind. Graus et al. have considered these relationships quantitatively and derived equations that allow them to calculate the stress under various orientations and current strengths. They do not consider the stress on lateral branches that do not incline in the plane of the current. As they point out, lateral branches will only add to the stress produced by hydrodynamic drag. They cannot counterbalance this effect. When we walk into the wind, holding our arms out to the side would only increase the resistance. Corals adapted for life in strong currents should have few laterally oriented branches.

The morphology of the cylinder as well as its orientation has an effect on stress. The stress on a short, thick cylinder is less than on a long, slender cylinder of comparable mass. The stress is also less on a streamlined, elliptical "cylinder" than on a true cylinder with circular cross section or a branch with a square or rectangular cross section. As mentioned in Chapter 2, a coral may adopt one of two strategies in response to life in a current: it can become massive (or even better, encrusting) like the short, thick cylinder or it can branch but orient into the current (Fig. 2.16).

Graus et al. made a similar study of the effect of wave stress on the coral colony. A nonbreaking wave produces currents near the bottom both in the direction of wave movement and a backward movement as the wave passes. After considering this two-direction pattern, the authors conclude that stress is minimized when the branches are steeply inclined either in the direction of wave motion or away from the direction of motion (Fig. 2.16). Breaking waves can also place great stress on the coral in the direction of breaking. The effect of this process on cylinders has apparently not been studied in detail, but clearly a vertical cylinder would receive more instantaneous stress from a breaking wave than an inclined cylinder, which does not receive the force of the breaker along its entire length instantaneously. By this model coral branches in the breaker zone should incline toward the breakers.

In order to see if the paradigm is actually realized in nature, the authors measured shape and orientation of colonies along a traverse through a fringing reef in Isaac Bay, St. Croix in the Virgin Islands. They divided the reef into three zones on the basis of hydrodynamic conditions (Fig. 5.4): (1) an outer strong wave zone (SWZ) on the ocean side of the reef crest, (2) a current zone (CZ) in the breaker zone on the reef crest, where breaking waves result in a shoreward, unidirectional current, and (3) a weak wave zone (WWZ) shoreward from the reef crest. In the SWZ the banches of *Acropora palmata* are strongly inclined both into and away from the waves. The colonies contain few lateral branches and are thus highly elliptical in shape. In the CZ the branches incline toward the ocean at high angles and the colonies are elliptical. In the WWZ the branches are moderately inclined into and away from the wave direction. The colonies are less elliptical than in the SWZ. The observed pattern is thus in agreement with that predicted from the engineering analysis.

In part, the orientation of branches in *Acropora palmata* (and other branching corals) may simply be the result of breaking off of branches that are not oriented to withstand the stresses. Branches that gradually curve into the current suggest that the current in some way directly induces oriented growth. This is a subject that needs more investigation, perhaps by experimentally reorienting coral specimens.

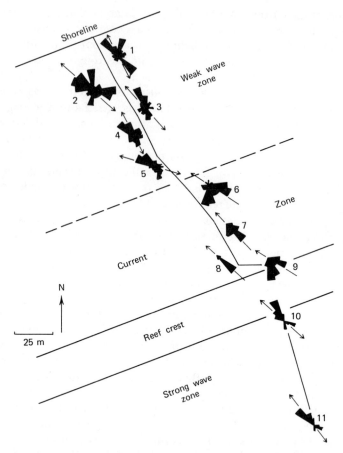

Figure 5.4 Horizontal branch distribution in *Acropora palmata* along a transect across a fring-
ing reef in Isaac Bay, St. Croix, Virgin Islands. Each segment of the rose diagrams contains 20°.
The maximum branch frequency group for each colony is standardized to a unit radius. The
arrows show mean current (single-headed arrows) or wave (doubled-headed arrows) direction.
After Graus et al. (1977), published with permission of the American Association of Petroleum
Geologists.

   The hydrodynamic study indicates that perhaps the best solution to the
problem of stress produced by currents and waves is to develop an encrusting
morphology. Many corals and other organisms have adopted this strategy. But
a branching morphology, even with branches inclined, would hardly seem as
good a strategy for minimizing stress. Why then do corals branch? This is a
clear example of a given morphology being a compromise between functional
adaptations. Branching gives a definite advantage in exposing more surface
area for feeding and allowing the colony to grow into unoccupied areas above
the substrate. Thus inclined branching is almost certainly a compromise between adaptation for feeding efficiency and stress reduction.

Bryozoans

A number of functional morphologic studies have been made of the bryozoa. Most of these studies have involved speculation about the function of particular structures without the actual testing of physical models. An exception is the study of Stratton and Horowitz (1975) in which the authors observed the flow of water through and around bryozoan models in order to investigate the feeding mechanism of certain Paleozoic fenestrate forms. The relationship of the fenestrate colonial shape to currents has long been the subject of discussion in the literature with some workers suggesting that the bryozoans feed by active current production by the animals (Cowen and Rider, 1972) and others favoring a more passive feeding method from ambient currents (Elias and Condra, 1957). Stratton and Horowitz (1975) use hydrodynamic theory of the flow of fluids through screens and observed the flow of water through and around models to study the passive feeding mechanism.

Hydrodynamic theory of flow around screens indicates that the velocity of flow will increase as the currents flow around the sides of the screen. Zones of low velocity occur immediately in front of and behind the screen. The spacing of the lines normal to flow direction in Fig. 5.5 indicates relative velocities. The more closely spaced lines indicate reduced velocities and widely spaced lines show increased velocity. An area of reduced velocity is indicated behind the screen (or bryozoan) that may have aided in feeding. Flow through the screen also results in a zone of reduced pressure behind the screen. The screen acts as a barrier to laminar flow in the water and creates a zone of turbulence with eddying of water behind the screen.

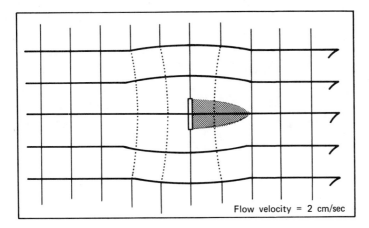

Flow velocity = 2 cm/sec

Figure 5.5 Two-dimensional flow net of water through and around a fenestrate bryozoan frond. The shaded portion of the diagram is a reduced energy zone. The spacing of the lines normal to flow direction indicates the relative velocity of the current. Close spacing indicates slowed velocity and wide spacing an increased velocity. After Stratton (1975).

Stratton and Horowitz prepared four plexiglass models of bryozoan fronds for testing in an open flume. The models were all similar but differed in the longitudinal sectional shape of the openings or fenestrules through the models. Figure 5.6 shows the four shapes used. Model number one is most similar to actual bryozoan specimens. The other models were tested in order to see if model one gave some advantage to the bryozoans as compared to other possible shapes. The models were oriented perpendicular to the current in a flume with a 2 cm/sec current. The pattern of water flow was studied by injecting small amounts of $KMnO_4$ dye into the water 1 to 3 cm in front of the models. About one third of the dye passed through model one, the remainder going around the sides. After passing through the model, the dyed water eddied back, forming a "curtain" of dye immediately behind the model. This curtain persisted for 5 to 8 sec before dispersing. Water passing through and around the other models behaved in a similar manner except a smaller portion of the dye passed through models three and four and the curtain was not as effectively held to models two and four. Thus the model most similar to an actual bryozoan was most effective in forming a zone of low-velocity water behind the frond.

The authors conclude that fenestrate bryozoa of the type they modeled could have effectively fed by having the polypide-bearing side (obverse side) of their fronds facing downstream in a unidirectional current. Water carrying sus-

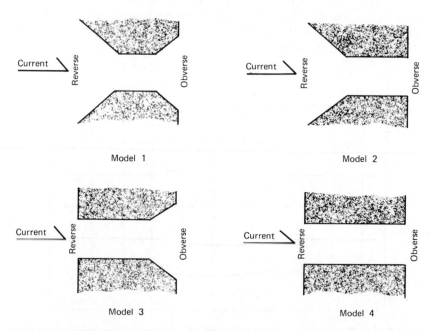

Figure 5.6 Longitudinal sections of four models of bryozoans. The minimum diameter of all openings is 1.1 mm. The obverse side contains the openings from which the animals protrude in living specimens. The reverse side has no protruding bryozoan animals. After Stratton (1975).

pended food particles would pass through and around the frond and be slowed in a low-velocity zone on the down-current side of the frond. The individual animals growing on the down-current side of the frond could then easily extract the food from this slow-moving water with their tentacles. This feeding mechanism would only be effective in low-velocity currents. Currents greater than about 5 cm/sec broke the low-velocity curtain. The presence of this type of fenestrate frond might thus indicate low-velocity, unidirectional currents.

Not all fenestrate bryozoans would necessarily require these environmental conditions. Some fenestrates were probably encrusting forms. Others such as the spiral *Archimedes* (Condra and Elias, 1944) and lyre-shaped *Lyropora* (McKinney, 1977a) grew in different colony shapes and thus were presumably adapted to differing habitats. Modern bryozoans with a colonial form most like the fenestrates have been observed in the Mediterranean to produce currents by action of the animals themselves (McKinney, 1977a). This could also have occurred in the Paleozoic fenestrates. As the Paleozoic fenestrates are only distantly related to the modern cheilostome bryozoans, the feeding mechanism may have been quite different. At any rate, the Paleozoic fenestrates appear to have been well adapted to the passive feeding method that would have been more energy efficient than producing a self-generated current.

Brachiopods

The brachiopods have been the subject of many functional morphologic studies. One of the foremost students of functional morphology and the originator of the paradigm method, M. J. S. Rudwick, has pioneered in studies of brachiopod morphology. Perhaps the type example of the paradigm method is Rudwick's study of the function of the conical ventral and thin, caplike dorsal valves of richthofenid brachiopods (Rudwick, 1961). This classical study has been cited in many subsequently published texts and reviews (e.g., Ager, 1963, and Valentine, 1973a). Rudwick's interpretation that the dorsal valve was used as a flaplike pump to bring food-bearing water to the brachiopod has been challenged on the basis of more recent work by Grant (1972). Although Rudwick's interpretation in this particular case may be incorrect, the usefulness of his approach first used in this study has been demonstrated many times over.

Alexander (1975) used a combination of three approaches in analyzing the functional significance of the shape and size of specimens of the Ordovician brachiopod *Rafinesquina*. First he considered, on a theoretical basis, the adaptive reason for a particular morphology. Next he used the sedimentary features of the strata in which the fossils were found and their stratigraphic setting as a method for determining the environmental conditions under which the fossils lived. In this way morphology could be indirectly related to the environment. Finally he conducted studies of models in a flume to determine the stability of shells of differing morphologies in currents.

The maximum size of *Rafinesquina* varies considerably between different populations. Several environmental factors theoretically might affect specimen size in brachiopods or other organisms. This has been a frequent subject of study in connection with so-called dwarf or micromorph faunas (e.g., Mancini, 1978a). Small size may be the result of a limited food supply. Less food is needed to maintain a small body than a large one. A high concentration of suspended sediment may also limit size by clogging the food-gathering system of the brachiopods and thus food intake. The concentration of dissolved oxygen in the water is a further possible factor. Small specimens with a high surface-to-volume ratio would have a relatively larger amount of exposed tissue with which to gather oxygen and would thus have an advantage in oxygen-poor environments. Small specimens would also have an advantage on soft, fluid substrates in being less likely to sink than larger specimens. Other authors have suggested additional explanations for small size (see Mancini, 1978a). Large size would be favored in a more turbulent environment where these factors are not likely to be limiting and where large size would increase resistance to over-turning by currents.

Three morphologic parameters in *Rafinesquina* were measured by Alexander: perimeter-to-volume (*P/V*) ratio, length-to-height (*L/H*) ratio, and alation. The *P/V* ratio is the ratio of the distance along the commissure to the volume of the specimen. It is a measure of the thinness of the brachiopod. A high *P/V* ratio indicates a thin specimen with large surface area but little body volume. The *L/H* ratio is the ratio of the distance between the anterior and posterior margins, perpendicular to the hinge line, to the maximum distance from the commissure to the pedicle valve exterior. It is a measure of the con-vexity of the pedicle valve, low *L/H* ratios corresponding to convex valves. Very convex valves that drastically change curvature near the anterior margin are termed *geniculate. Alation* is measured by the ratio of the length of the hinge to the width of the valve at mid-length. It is a measure of the elongation of the hinge or "wingedness."

Alexander reasoned that a high *P/V* ratio would be an adaptation for low oxygen and/or low food concentrations. A thin, broad body plan gives a larger surface area for respiration and food gathering, a useful adaptation in these environments. A high *L/H* ratio corresponds to a flat shell that presents a max-imum area in contact with the sediment surface. This should be adapted for support in a soft substrate (the snowshoe effect). A low *L/H* ratio corresponds to a more convex shell that maintains the commissure above the sediment sur-face and should be well adapted to an area with a high sedimentation rate. The high commissure would stay above the depositing sediment. (Others have sug-gested that this bowl-shaped shell would be adapted to a fluid substrate by keeping the commissure from sinking below the sediment: the iceberg effect.) High alation could be adaptive either to soft substrate by increasing the surface area of the shell for support on the substrate (snowshoe effect) or it could in-crease stability relative to overturning in turbulent environments.

Alexander measured these three parameters, as well as length as a measure of size, for specimens of *Rafinesquina* from seven stratigraphically separated localities (Fig. 5.7). He interpreted the depositional environment of each of

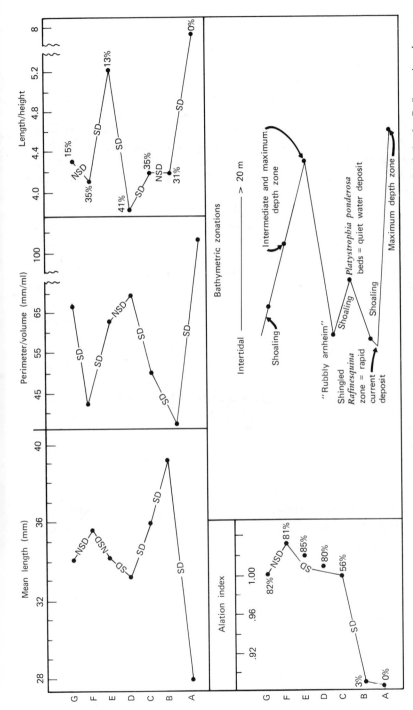

Figure 5.7 Variation in mean length, mean perimeter–volume ratio, mean length–height ratio, and alation index for *Rafinesquina alternata* among the stratigraphically successive samples A through G. The percentage values for length–height ratio and alation index indicate the frequency of geniculate specimens and alate (index >1.00) specimens respectively. NSD and SD indicate no significant difference and significant difference between successive means based on the student *t*-test at the 0.05 probability level. After Alexander (1975), Journal of Paleontology. Society of Economic Paleontologists and Mineralogists.

239

these localities largely in terms of variation in water depth and then attempted
to explain the size and shape of specimens as related to these depositional environments. Small size, thin (high P/V ratio) and compressed (high L/H ratio)
shells characterize specimens from deeper water, soft substrate environments.
Large, inflated (low P/V ratio), moderately convex (moderate L/V ratio) shells
are found in high-energy environments. Specimens with high convexity (low
L/H ratio) come from an environment that is interpreted to have a rapid
sedimentation rate. Alation does not seem to vary regularly with environment
but increases through time. Alexander interprets this to indicate that alation
evolved as an adaptation that aided both in support on a fluid substrate and in
stability in a turbulent environment. The field data thus support the theoretical
interpretation.

Finally, Alexander investigated the effect of size and shape of specimens on
stability in currents by testing brachiopod models in a flume. Fourteen models
were prepared that included all of the common shapes found in the natural
populations and a range of sizes (Fig. 5.8). The models were first placed on
sand in the flume with their hinge line perpendicular to the current direction.
The current velocity was gradually increased, and the velocity at which each
model was overturned by the current was noted (Fig. 5.8). In this experiment all
models overturned at velocities between 1 and 2 m/sec with no particular pattern in terms of size or shape. Another experiment was performed with the
hinge parallel to the current direction. Many of the specimens were more stable
in this orientation, with currents up to 3 m/sec being required to overturn them.
The alate and convex (or geniculate) alate forms were the most stable. This suggests that these morphologies could be adapted to life in strong currents provided that the orientation with the hinge parallel to current is the life orientation. More recent research by LaBarbera (1977) suggests that many
brachiopods did (and do) orient their shells in this position.

Gastropods

A study of the mode of life of some Paleozoic frilled gastropods by Linsley et al.
(1978) is a good example of the use of analogy between an extinct group or
groups of gastropods and a distantly related modern gastropod group.

The modern gastropod genus (*Xenophora*, the carrier shell) (Fig. 5.9), is
both distinctive in appearance and in mode of life. It covers its shell with shells
and shell fragments of other types of organisms, particularly bivalves and
gastropods. Undoubtedly, one function of this constructional method is for
camouflage. The shell is difficult to see against a substrate containing many
dead shells and shell fragments. *Xenophora* also uses elongate shells (often
high-spired gastropods) or shell fragments to form a regular series of spines or
props around the margin of the shell. These props function to hold the shell
slightly off the bottom. The animal hangs down from the shell and feeds on
deposited organic material in the sediment. This is apparently an effective way
of feeding, but the snail must sacrifice the ability to clamp its aperture against

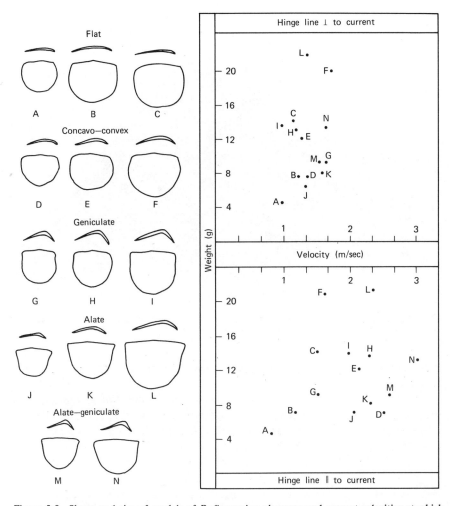

Figure 5.8 Shape and size of models of *Rafinesquina alternata* and current velocities at which they overturned. In the hinge line perpendicular orientation, the anterior margin faced upstream. After Alexander (1975), Journal of Paleontology, Society of Economic Paleontologists and Mineralogists.

the substrate for protection. The ring of props around the aperture does provide some protection and in addition the snail can draw well back within its shell, which it closes with an operculum. *Xenophora* moves about slowly and spends long periods of time in one spot. When it moves it does so by affixing its foot firmly to the substrate, raising the shell momentarily above the substrate before allowing it to topple forward with a push of the foot. The *Xenophora* mode of life requires a firm substrate in order to keep the shell suspended above the substrate, and an adequate supply of deposited food. They are mainly found liv-

Figure 5.9   Modern specimen of *Xenophora*. Photograph provided by R. M. Linsley.

ing in shallow, tropical, carbonate substrates, frequently in reef-associated environments.

Another modern genus, *Tugurium*, has a similar morphology and mode of life except that it does not cover its shell with foreign objects. It secretes its own spines which function in the same manner as the shell props in *Xenophora*. Certain other gastropod genera, such as *Astraea* and *Guildfordia*, also have spines; but they apparently function as protective devices (*Astraea*) or snowshoe-like supports in soft substrate (*Guildfordia*). They clearly cannot act as props because they extend laterally, more nearly perpendicular to the coiling axis. The spines in *Xenophora* extend at an angle downward and project below the bottom of the shell.

The Middle and Upper Silurian snail genus *Euomphalopterus* is crudely similar in overall shape to *Xenophora*, but it does not have a covering of foreign shells and has a continuous frill instead of spines or props (Fig. 5.10). The downward deflection of the frill would have supported the rest of the shell above the sediment surface much like the spines in *Xenophora*, preventing the shell from clamping to the substrate. By analogy to *Xenophora*, *Euomphalopterus* must have lived on a firm substrate, feeding on deposited food. Several other Paleozoic gastropod genera, ranging from Ordovician to Permian in age, also had frills or spines that apparently had the same function. In some genera, however, the spines extend outward, perpendicular to the coiling axis, indicating protective or supportive functions as in *Astraea* and *Guildfordia*.

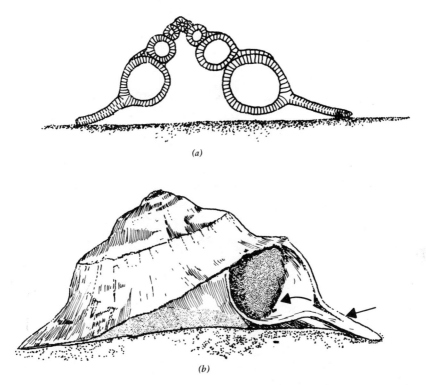

(a)

(b)

Figure 5.10 (*a*) Reconstruction of a cross-section of *Euomphalopterus* showing the propping effect of the frill; (*b*) reconstruction of *Euomphalopterus* in its presumed life position and the position of the inhalent current. After Linsley et al. (1978), *Lethaia*, *11*.

Bivalves

Many studies of functional morphology do not utilize rigorous experimental modeling or engineering theory approaches. Perhaps the most common approach is simply to use logical reasoning to derive a most likely explanation of the function of a particular structure. An analysis of the functional morphology of the rudistid bivalves of the family Hippuritidae by Skelton (1976) is an example of this approach. It is also an example of a study of an extinct group of organisms of unusual morphology that have no close living relatives. Such studies are perhaps the most challenging because the unusual morphology suggests unusual functioning for which there is no modern analog.

The Hippuritidae and the other rudist families (such as the Radiolitidae, Fig. 5.11) comprise a group of inequivalve, sessile, epifaunal heterodont bivalves that flourished in the Tethyan seas during Cretaceous times. They became extinct at the end of the Cretaceous and left no close relatives either in phylogenetic or ecological terms. The hippuritid shell consists of a conical to

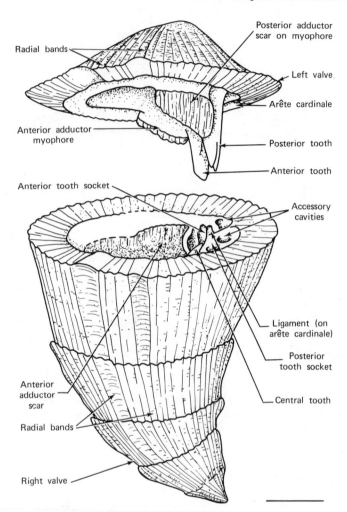

Figure 5.11 General morphology of the radiolitid rudist bivalves. The left (free) valve is shown above and the right (attached) valve below. Scale bar in lower left 1 cm. After Skelton (1979).

cylindrical right valve, often of massive proportions, with a much smaller caplike left valve. They often grew cemented together in huge aggregations forming reefs or reeflike bodies. Their geographic distributions in a broad band parallel to the present equator suggests that they required warm, tropical conditions. Their sedimentary and stratigraphic setting indicates that they lived in shallow, usually clear water of normal salinity. The reason for their unusual shape and their method of feeding has not been clear.

The body cavity in the conical to cylindrical right valve is very small in comparison to the total volume of the shell (Fig. 5.11). The earlier-formed portion of the shell is filled with tabulae which apparently served as successive platforms supporting the visceral mass that was suspended from the upper or left valve. The size of the body cavity was further reduced by the thick shell, the massive tooth and socket structure, and two or three infoldings or pillars of the valve wall.

The left valve formed a lid capping the opening of the cylinder of the right valve. It was equipped with massive teeth and myophores for muscle attachment that extended into the right valve. The left valve of the hippuritid rudists contains a complex system of ramifying canals that lead from the upper surface of the valve to the valve margin giving the valve a "honeycombed" appearance.

Skelton (1976) noted that growth lines in the shell structure indicate that the canals in the left valve must have been lined with living mantle tissue. This tissue was probably ciliated tubes that drew in water and suspended food from the outside and pumped it through the canals to the shell (and mantle) margin (Fig. 5.12). Here the food particles were trapped on mucus strands and passed laterally along a troughlike structure on the right valve margin to the ctenidia and mouth. The bivalve probably had no foot so that the mouth could be located in the mid-ventral region. The filtered water could then be ejected through the slight gape between the valves. The canal system served as a sieve device to exclude predators and sediment grains while permitting entry of small suspended detritus and microorganisms used as food. The current direction may have occasionally been reversed to clean out any sediment that would tend to clog the system. The very short but apparently strong adductor muscles would not have allowed the valves to open more than a fraction of a millimeter, but this would be adequate for egress of the filtered water and waste products. The gape probably was not wide enough for more conventional filter feeding. Skelton (1979) speculates that the radiolitid fed with small tentacles on the mantle margin that extended outside of the shell.

The hippuritids were well adapted to their niche in the reef and high-turbulence, shallow water environment. Their massive shells, especially when cemented to the substrate, gave stability in the turbulent environment. They apparently grew upward rapidly in competition with their neighbors for the food-laden water above; hence they developed their tall, cylindrical shape. The upward growth was facilitated by the development of tabulae in the right valve. Finally, their feeding system was well-adapted to a habitat with an abundance of suspended food but with predators capable of attacking more conventional clams. The rudist was securely encased in its thick right valve with a tightly fastened lid and could filter food from the water without having to gape its valves even as much as a millimeter. The rudists seem to have been so well adapted that one can hardly imagine what might have caused their extinction.

Cephalopods

A good example of the use of the engineering approach to functional morphologic studies is provided by the pioneering study of Kummel and Lloyd (1955) and by the later study of Chamberlain (1976) of streamlining in cephalopod

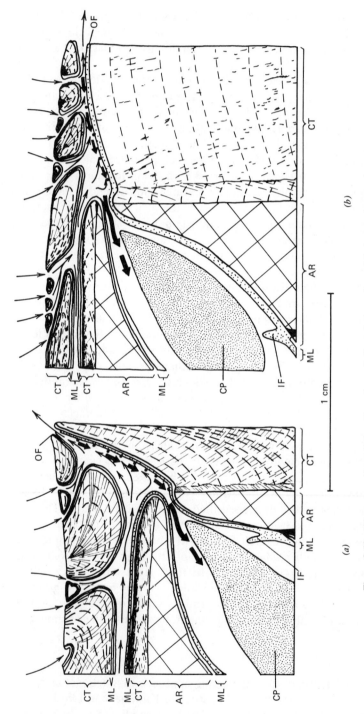

Figure 5.12 Diagrammatic radial sections through the valve margin and their associated soft parts in the ventral region of (*a*) *Hippuritella toucasi* and (*b*) *Vaccinites dentatus*. The left valve is on the top and the right valve beneath. The thin arrows indicate the water currents generated by the cilia of the mantle linings to the radial canals, and the thick arrows show the direction of ciliary transport of the particles trapped from these currents on the right mantle lobe margin. AR—aragonitic shell layer; CP—ctenidial and/or palp mass; CT—calcitic shell layer; IF—inner mantle-marginal fold; ML—mantle lobe; OF—outer mantle-marginal fold. After Skelton (1976), Lethaia, *9*.

shells. Ammonites and nautiloids are presumed to have been swimmers, as is the one living genus of nautiloids, *Nautilus*. Variation in the morphology of the shell suggests that they may have been varied in their swimming ability. Some forms perhaps spent more time swimming, whereas others rested on the bottom some of the time. The question then arises as to what features of the shell might have been adaptations for more efficient swimming. Chamberlain attempted to answer this question by building models of ammonites and nautiloid shells which he tested in a hydrodynamic trough that was designed for testing streamlining in ship hull design. He constructed plexiglass models which he attached to a carriage that moved the specimen through the water at predetermined speeds. He also attached dye particles to the leading edge of the specimens. He could then photograph the pattern of water motion around the specimen from streamers of dye.

The measurement of streamlining and thus of swimming efficiency is made by determining the drag force, D_F, on the specimen or a model of the specimen:

$$D_F = \frac{1}{2}\rho\, V^2 A C_D$$

where ρ is the density of the fluid (water), V is the velocity of the specimen, A is the area of the specimen (body size), and C_D is the drag coefficient. The drag coefficient is dependent on the body shape or streamlining and thus is a measure of streamlining. The lower the value of C_D for a specimen, the lower will be the drag force and thus the more efficient the swimming. A species that spends a great deal of time swimming or must swim rapidly would benefit from the lowest possible C_D value.

The C_D value can be determined experimentally from the cephalopod models by measuring the drag force in the hydrodynamic trough. Velocity and density of the fluid can also be measured. The area of the specimen can be calculated allowing the C_D value to be calculated with the equation above.

The flow pattern study showed that the water flowing past the leading edge of the model shells first flowed smoothly over the surface (the flow was "attached" to the shell in hydrodynamic terminology). The flow broke up into turbulent eddies as the current passed over the umbilicus. Further turbulence occurred at the trailing edge of the shell, especially as the water passed the aperture (Fig. 5.13).

In order to show the quantitative relationship between the drag coefficient and shell shape, the shell shape must be described in quantitative terms. Four parameters (S, W, D, and F) were used to describe the shape of the coiled shells, three of them (S, W, and D) being parameters used by Raup (1966b) in his quantitative description of coiled shells. S (as defined in Fig. 5.14) is a measure of cross-sectional shape of the whorl. An S value of 1 indicates a circular or equidimensional cross section and lower S values indicate a more compressed cross section. W is a measure of the whorl expansion rate or the ratio of the diameter of successive volutions. Shells with large W values are more globose than those with small values. D (Fig. 5.14) is a measure of the relative

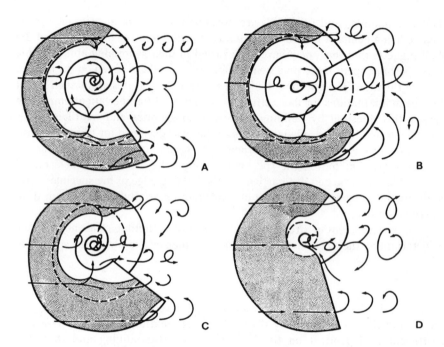

Figure 5.13 Sketch of flow lines around shells and shell models. A—lytoceratid shell model; B—serpenticonic shell model; C—widely umbilicate oxyconic shell model; D—*Nautilus pompilius* shell. The area of boundary layer attachment is shown by stippling. The dashed lines mark the umbilical shoulder (widest point in cross-section) on the outer whorl. After Chamberlain (1976), copyright The Paleontological Association.

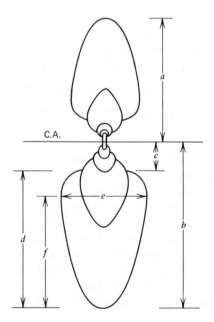

Figure 5.14 Definition of geometric parameters of shell form. $W = (b/a)^2$; $D = (c/b)$; $S = (e/d)$; $F = f/(b + a)$; C.A. is coiling axis. After Chamberlain (1976), copyright The Paleontological Association.

distance of the whorl from the coiling axis. The larger the D value, the larger will be the umbilicus (depression around the coiling axis). The fourth parameter in the Raup system, the translation rate T, is a constant in this study. Almost all ammonites and nautiloids are planispirally coiled ($T = 0$), that is, they do not translate down the coiling axis producing a spire as in most snail shells. The F value (Fig. 5.14) as defined in Chamberlain's study is a measure of the position of the umbilical shoulder, or the position of the greatest width of the shell relative to the diameter of the shell.

In general, the C_D value increases with increasing values for W. Shells with a large W value present a larger cross-sectional area in the direction of motion and thus increase the drag force. The C_D value also increases with increasing D value. A large D value corresponds to a large umbilicus, and, as indicated by the flow patterns, turbulence occurs when the current "detaches" from the surface on passing over the umbilicus. Larger values for S produce higher C_D values. Small S values correspond to more compressed shells that present a smaller cross-sectional area in the direction of movement. Finally, C_D values decrease with increasing F value. Large F values allow the current to stay attached to the shell for a longer relative distance as it passes over the shell. Thus the most streamlined shell would be compressed with a reduced umbilicus, a low whorl expansion rate, and high F value. The lowest measured C_D value was approximately 0.1 and the highest about 1.0.

One might suspect that the soft tissues of the organism extending from the aperture would affect the streamlining. Chamberlain considers the possibility but concludes that the effect is likely to be small unless the tissues extending beyond the aperture are quite large. This is unlikely based on several lines of evidence, particularly a comparison with modern *Nautilus*.

Chamberlain points out that the lowest C_D value measured was still about 10 times larger than that for a well-streamlined fish or squid. The most streamlined swimming shape is fusiform (cigar-shaped). This points out the limitation placed on the swimming efficiency of shelled cephalopods by the historical-phylogenetic factor. The nature of growth of the ammonite and nautiloid shells limits the possible shape and streamlining. By maximizing the values of S, W, D, and F the shell can approach a maximum efficiency within the constrains of those parameters. But the relatively large cross-sectional area, umbilicus, and large aperture all conspire to increase the drag relative to the more efficient fusiform shape. This relatively inefficient shape for swimming perhaps was in part responsible for the extinction of the ammonites and decline of the nautiloids.

Another factor limiting the swimming ability of the shelled cephalopods is the positioning of the swimming thrust and drag force (Fig. 5.15). Due to the location of the hyponome at the bottom of the aperture, the swimming thrust acts on the lower part of the shell. The drag force acts over the entire shell, but the summation of the force is centered at the center of the shell. The drag force and thrust act in opposite directions so that they produce a rotational moment on the shell. This results in the rocking motion that can be observed in the

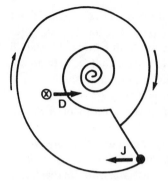

Figure 5.15 Effect of drag and thrust on shell orientation during movement. The shell is shown moving from right to left. Drag (*D*) acts to the right from the center of dynamic pressure. Thrust (*J*) acts to the left in the direction of motion from the point of application by the hyponome at the ventral margin of the aperture. These forces cause a clockwise rotation shown by the curved arrows. After Chamberlain (1976), copyright The Paleontological Association.

swimming motion of modern *Nautilus*. An increase in force needed to increase the swimming velocity would eventually result in rotation of the shell out of the swimming orientation and would thus act as an upper limit to the swimming velocity.

Trilobites

Trilobites are an especially challenging subject for functional morphology studies because they have no closely related living representatives. The analogy approach can be used to some extent by comparison with various relatively distantly related arthropod groups, especially the xiphosaurans with their modern representative *Limulus*, the horseshoe crab. The trilobites also have a complex skeleton that was apparently closely related to the functioning of the body so that a great deal of information about function should be obtainable from the paradigm method. A study by Stitt (1976) of the functional morphology of the Late Cambrian trilobite *Stenopilus pronus* is a good illustration of the type of approach that can be used in trilobite studies. Stitt primarily used the paradigm method to interpret the most likely function of various morphologic features, but he supplemented this with analogy to modern crustaceans, millipedes, and even bivalves.

    Stenopilus pronus has several unusual morphologic features. It has a large, smooth, highly convex cephalon with a prominent anterior arch (Fig. 5.16). The eyes are relatively small and are directed laterally with little or no vision in other directions. The articulation of the almost bulbous cephalon with the thorax suggests that the anterior-posterior axis of the cephalon must have been perpendicular to that of the thorax (Fig. 5.17). Thus, if the cephalon was horizontal, the thorax must have been vertical.

    The thorax has a broad axial lobe with axial rings increasing in convexity toward the posterior. The pleural spines on the thorax decrease in size toward the pygidium and are absent in the most posterior segments. The thoracic segments are unusual in having the ability to overlap and telescope. The posterior thoracic segments and especially the pygidium are covered with ter-

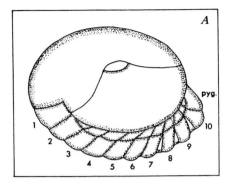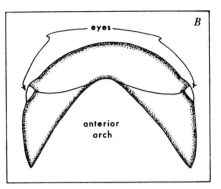

Figure 5.16 Trilobite *Stenopilus pronus*. A—lateral view in the enrolled position (X 14.5). B—anterior view of cephalon (X 18). After Stitt (1976), Journal of Paleontology, Society of Economic Paleontologists and Mineralogists.

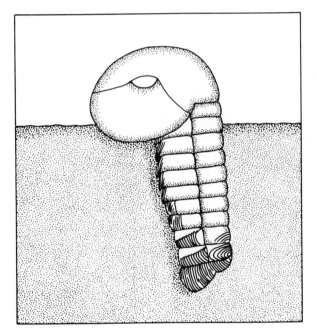

Figure 5.17 Side view of *Stenopilus pronus* in normal living position. After Stitt (1976), Journal of Paleontology, Society of Economic Paleontologists and Mineralogists.

race ridges, wedge-shaped ridges having one gentle slope and an opposite steep slope much like a topographic cuesta. The pygidium is narrow, convex upward, and securely fastened to the thorax. It is somewhat scooplike in shape and is oriented nearly perpendicular to the thorax (Fig. 5.16). When the animal is enrolled, the pygidium tucks securely into the anterior arch of the cephalon (Fig. 5.16).

Each of these features is adapted to the functioning of the trilobite (Stitt, 1976). The trilobite normally was burrowed into the substrate with the thorax oriented vertically and the cephalon lying horizontal on the sediment surface. The animal is interpreted to have been a filter feeder, bringing water with suspended food through the large anterior arch. It may have been oriented with the anterior arch facing into the prevailing current. We have no details at this stage as to the pattern of circulation of the water or the method of extraction of the suspended food. Many of the unusual morphologic features of *Stenopilus pronus* are related to its burrowing habit. The shape and orientation of the pygidium appear to be adaptations to digging into soft sediment. The terraced ridges aided in the burrowing process by allowing the pygidium to easily penetrate the sediment with sediment sliding past the gentle slope but resisting movement in the opposite direction when sediment pressed against the steep slope. The telescoping thoracic segments could then pull the body down into the sediment while the terraced ridges served as an anchor. The terrace ridges are analogous to ridges on the carapace of the modern mole crab, *Emerita*, which burrows backward into the sand in a manner similar to that proposed for *S. pronus*. Several bivalve genera have a similar sculpture pattern on their shells as an aid in burrowing.

The convex posterior thoracic axial rings give the posterior part of the thorax a crudely circular cross section. This shape minimizes resistance to penetration by minimizing the surface area. This shape is used by modern burrowing millipedes. The broad axial lobe in the trilobite probably accommodated the large muscles that were needed by the large, strong, digging appendages. The reduced pleural spines on the posterior segments minimize friction with the sediment in burrowing.

One problem with the mode of life indicated for *S. pronus* is what to do when the animal needed to move or was washed out of the substrate by currents. Upon sudden exposure by currents or perhaps when uncovered by predators this trilobite could roll up into a tight, subspherical ball with the soft tissue well protected by the calcareous covering. When the animal needed to move to a new site to burrow into the sediment it could probably strongly arch its back using the telescoping segments (Fig. 5.18) and crawl, probably in a rather slow, awkward manner, to a new location.

Stenopilus pronus was adapted to life in an area of moderately firm substrate with sufficient suspended food for filter feeding. It probably required at least a weak current to bring in new food, and its ability to burrow partially into the substrate suggests that it was adapted to areas with moderately strong currents. One advantage to being partially buried could have been to increase stability in currents. Another might be for protection of the buried part of the body. All reported occurrences of *S. pronus* seem to be in limestones. Those occurrences that Stitt (1976) was able to document in some detail were associated with stromatolites. Stitt suggested that *S. pronus* lived in the shallow subtidal sediment between stromatolite heads. Currents would be concentrated as they flowed between the stromatolite domes. The specimens are usually found in

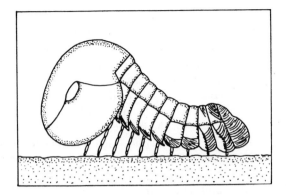

Figure 5.18 Side view of *Stenopilus pronus* in crawling position on substrate surface. After Stitt (1976), Journal of Paleontology, Society of Economic Paleontologists and Mineralogists.

biomicritic limestones, but the presence of oncolites, intraclasts, and occasional beds or lenses with a sparry matrix suggest at least periodic higher energy conditions.

Stitt (1976) also mentions several other trilobite taxa that have a morphology similar to *S. pronus* and thus probably had a similar mode of life.

Crinoids

The analysis of the adaptation of Mississippian crinoids for filter feeding by Ausich (1980) is a good example of the combination of the taxonomic uniformitarian approach and an engineering approach. Study of modern crinoids has revealed a distinctive pattern of feeding (e.g., Meyer, 1973, Macurda and Meyer, 1974). In most cases the crinoids form a parabolic filtration fan with the concave side of the fan and the mouth facing away from the current. Food carried by the currents passes through the fan into the area of reduced currents behind and is trapped on mucus secreted by the tube feet. The food is entangled in mucus strands that pass down the ambulacral food grooves in the arms and branches to the mouth. The crinoids feed on both living and dead organic material that has a size equal to or less than the width of the food groove.

The fact that many crinoid species live together in the same community suggests that different species must in some way be adapted to utilizing different types of food or feeding in different locations within the biofacies. Otherwise, by the *principle of competitive exclusion* (see Chapter 9) one species would be expected to be at least somewhat more efficient than the others and through competition would eliminate the less well adapted species. Ausich suggests that the crinoids subdivide or partition the filter feeding niche between the species so that each species is best adapted to utilize one particular type of food from a portion of the biofacies.

More specifically, he suggests that the crinoids feed at different levels above the sea floor and use food of differing sizes. Crinoids can adapt to feed at a particular height above the sea floor by growing stems of differing lengths. Ausich recognizes three arbitrarily defined feeding levels: low, intermediate, and high. Because of the ease with which the crinoid stem is disarticulated, we only know stem lengths for a few especially well preserved individuals. The data that are available for the Mississippian crinoids studied by Ausich do suggest a range of stem lengths from 0 to about 1 m or more. By feeding at these different levels the various species are not directly competing with one another but are using food from different locations with the biofacies.

The crinoids also appear to have subdivided the filter feeding niche on the basis of food size. The crinoid filter fan can be analyzed mechanically on the basis of filter theory. Filter efficiency can be determined on the basis of the number of openings per unit area of the filter. The "mesh size" of the crinoid filter is basically a function of the branching pattern of the arms of the crinoid. The larger the number of branches the finer the mesh. Crinoids with simple, unbranched arms have the coarsest mesh and those with many branches have the finest mesh (Fig. 5.19). Ausich determined the approximate mesh size in the following manner: He counted the number of branches (N) on the crinoid by counting the arms and arm branches, pinnules, ramules, and any other branches in the food-gathering system. He also determined the height of the crown (HC), that is, the distance from the base of the cup to the tip of the arms and the height of the dorsal cup (HD) (Fig. 5.20). These measurements allow the calculation of the filter fan area $A = \pi [(HC)^2 - (HD)^2]$ and the density of branches $D = N/A$. The larger the value of D, the smaller the openings in the filter fan. The D value should thus correlate with food size, large values of D being adapted for small food particles and small D values for large food. The width of the food groove, W, also varies with food size. The animal cannot use food that is larger than its food groove. W and D are thus inversely correlated.

Ausich plotted W values vs. D values for all of the species in a given community and found that, in general, each species occupied a separate position on the diagram (Fig. 5.21). This position corresponds to the particular feeding niche of that species. In some cases species did plot close together on the diagram indicating that the species used similar types of food. In practically all of these cases the species had different stem lengths and thus were not competing directly at the same locality but occupied different levels in the tiered community. A three-dimensional plot with the third dimension being stem length would serve even better to separate the feeding niches.

Vertebrates

Because of the complexity of the skeleton and the closeness of the relationship between the skeleton and the unmineralized portion of the body, the morphology of the vertebrate skeleton clearly reflects functioning of the organism. Many aspects of functioning of the organism that can be determined only with great difficulty from invertebrate skeletons are often readily apparent from the vertebrate skeleton. For example, determination of the method of feeding and

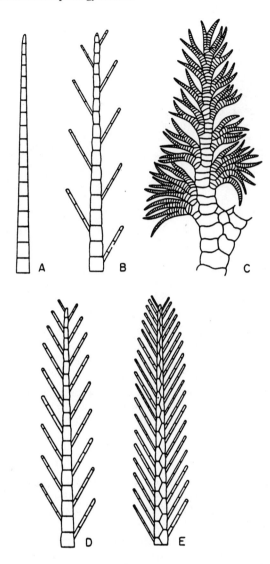

Figure 5.19 Various modes of crinoid arm branching. A—atomous or nonbranching; B—simple ramulate; C—complex ramulate; D—uniserially pinnulate; E—biserially pinnulate. After Ausich (1980), Journal of Paleontology, Society of Economic Paleontologists and Mineralogists.

the nature of the food used by an extinct bivalve or brachiopod may be highly speculative (e.g., the feeding method of rudists described above). On the other hand, the jaw structure and nature of the teeth usually readily reveal the type of food and feeding method used by vertebrates. Interpretations such as this are a routine part of the description of vertebrate fossils. More challenging to interpret are unusual structures that have no modern counterpart. The dinosaurs in-

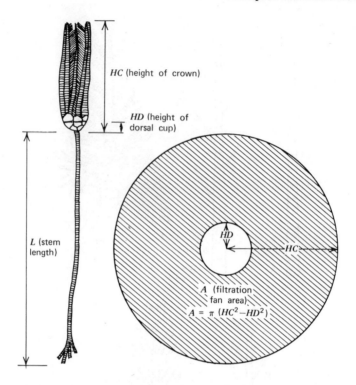

Figure 5.20 Measurements made to determine filtration fan area. The filtration fan area is used to determine the density of branches in the fan and niche partitioning among the crinoids. After Ausich (1980), Journal of Paleontology, Society of Economic Paleontologists and Mineralogists.

clude a number of examples of such unusual structures, and one of the most unusual is the series of bony dermal plates projecting along the back of the well-known genus *Stegosaurus* (Fig. 5.22). The interpretation of the function of these plates by Farlow et al. (1976) is another example of the engineering approach (both experimental and theoretical).

The *Stegosaurus* plates have usually been considered to perform a defensive function as armor, or they perhaps functioned as a means of sexual display. The plates may have performed such function, but Farlow et al. proposed an additional and probably primary function as a forced convection cooling system. The similarity between the plates and cooling fins on man-made heat exchangers suggests this.

In order to test this possibility, Farlow et al. constructed models of *Stegosaurus* which they tested in a wind tunnel by a method similar to that used to test mechanical heat exchangers. The rate of heat loss was measured on models with various types of fins in order to determine the "paradigm" for heat exchange fins on dinosaurs. They constructed scale models from aluminum cylinders to represent the body and thin metal fins to represent the plates on the

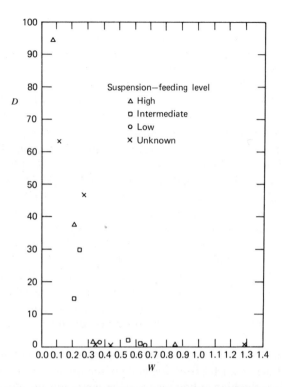

Figure 5.21 Example of a community showing a crinoid niche partitioning spectrum based on *D, W,* and feeding level. The community is from the lower Mississippian Edwardsville Formation in Monroe County, Indiana (locality IU 15109 of Ausich, 1978). After Ausich (1980), Journal of Paleontology, Society of Economic Paleontologists and Mineralogists.

Figure 5.22 The dinosaur *Stegosaurus.*

257

back. Heaters were embedded within the cylinder and thermocouples were attached to the model to determine its temperature. The rate of heat loss can be determined by monitoring the temperature of the model and of the surrounding air. Three types of fins were tested: paired, continuous; paired, interrupted; and staggered, interrupted (the latter type actually found in *Stegosaurus*). The difference in the rate of heat loss was then determined for each of the three models relative to a model with no fins.

When oriented parallel to the wind each of the finned models increased the rate of heat transfer relative to the unfinned model by about 35%. The continuous finned model was less efficient, however, in terms of heat loss per unit area because the continuous fin had a larger surface area. The three models differed considerably when oriented perpendicular to the wind direction. Paired, continuous fins increased heat transfer by only about 15%; paired, interrupted fins increased heat transfer by 27%; and the staggered, interrupted fins increased heat transfer by 33%. Thus the staggered, interrupted condition is almost equally effective in either orientation. A cooling fin that is equally effective in all orientations would certainly be an advantage to the dinosaur. Otherwise the dinosaur would have to orient itself parallel to the wind for maximum cooling effectiveness. The fact that *Stegosaurus* has a plate arrangement close to the paradigm strongly suggests that the plates did indeed function as cooling fins.

In order for the cooling fins to be effective the dinosaur had to have an effective method of heat transfer from the body to the plates. In the models this was accomplished by the high conductivity in the metal but in the animal this was probably accomplished by the circulation of blood through the plates. The highly vesicular nature of the bone suggests that it was supplied by numerous blood vessels. The dinosaur was probably able to control its cooling rate by controlling the amount of blood flowing through the plates.

Farlow et al. calculate that to work effectively as a heat exchanger a wind of at least 4.5 m/sec (10 miles/hour) would be required. *Stegosaurus* might thus be used as a crude paleowind velocity detector! *Stegosaurus* appears to have lived in a fairly open, savannah-like area where moderate to strong winds might be expected.

In recent years considerable controversy has arisen over the possible endothermic or warm-blooded nature of the dinosaurs (Bakker, 1975). A heat loss mechanism might be an advantage to a cold-blooded animal to allow it to dissipate metabolic heat. It would be even more advantageous to a warm-blooded animal which controls its temperature within narrow limits.

Vascular Plants

Vascular plants also have many morphologic features that are related to environmental conditions. Perhaps the best examples are adaptations found in desert plants for life under extreme aridity. This can be observed in modern desert plants and the results extrapolated to fossil plants by the method of homology. Solbrig and Orians (1977) present a comprehensive review of both functional morphology and physiology of desert plants.

Several morphologic characteristics of desert plants might be preserved in the fossil record. The leaves of many of the plants are small but relatively thick and leathery, minimizing the surface area and thus reducing water loss through transpiration from the leaves. The leaves of some plants have been lost entirely and photosynthesis is carried out by chlorophyll-bearing stems. Some have developed fleshy stems that store large amounts of water taken up during the infrequent desert rains. Other plants have developed extensive root systems to collect the limited moisture from a large area or to tap very deep permanent sources below the water table. The branching pattern on many plants produces a compact, globose shape, minimizing the area exposed to direct sunlight.

FUNCTIONAL MORPHOLOGY OF *DENDRASTER* FROM THE KETTLEMAN HILLS

The sand dollar (echinoid) *Dendraster* has been the subject of a detailed functional morphologic analysis (Stanton et al., 1979). It is one of the most widespread and abundant fossils in the Neogene sediments exposed in the Kettleman Hills and the surrounding area. The two *Dendraster* species are also among the most morphologically variable species found in the area. In fact, specimens of the species intergrade morphologically and may actually belong to a single highly variable species. The general morphology of *Dendraster* is typical for a sand dollar with a flattened, nearly circular, disc shape (Fig. 5.23). The mouth is approximately centrally located on the oral side of the test, the anus is at the posterior margin, and the apical system on the adoral side of the test is slightly offset toward the posterior margin. The offset nature or eccentricity of the apical system is the most distinctive feature of the genus and is the morphologic feature with which we are concerned here.

Dendraster is apparently unique among the echinoids in being able to feed both on deposited and suspended food. Observations of living specimens of *D. excentricus* reveal the feeding methods (Merrill and Hobson, 1970). When deposit feeding *Dendraster* moves horizontally on the sea floor. When suspension feeding the animal partially buries the anterior part of the test in the sediment with the oral surface vertical or slightly inclined into the current (Fig. 5.24), although orientation and inclination of the test varies depending on water turbulence. The suspension feeding mode seems to explain the function of the eccentricity of the apical system. In order to prevent the test from being flipped over by currents while it is in the suspension feeding orientation, the animal must bury a relatively large part of the anterior portion of the test. The eccentricity of the apical system allows the center of the apical system including the madreporite to be above the substrate so that respiration can take place and the water vascular system not become clogged with sediment. The stronger the currents, the more deeply the test must penetrate into the sediment for stability and

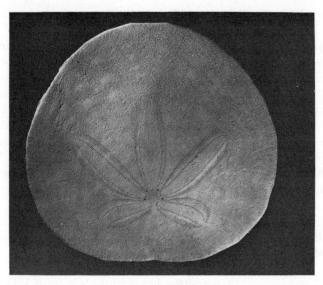

Figure 5.23 The modern sand dollar (echinoid) *Dendraster excentricus*. Collected at Morro Bay, California ($\times$ 1).

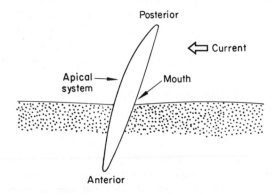

Figure 5.24 Diagrammatic side view of *Dendraster* in suspension feeding position. After Stanton et al. (1979), Lethaia, *12*.

thus the more eccentric the apical system must be to prevent its becoming buried. Greater stability in currents can also be attained by increasing the size of the test.

Raup (1956) noted that the eccentricity of living specimens of *Dendraster* varied geographically. He noted that populations living on the open coast were more eccentric than those living in bays. In fact MacGinitie and MacGinitie (1949) indicated that the difference between open coast and bay forms was distinct enough that they should be recognized as separate subspecies. Raup's data suggest that the forms intergrade and that the morphologic difference is probably environmentally controlled (ecophenotypic) rather than genetic in origin. At any rate the observation of living specimens and reasoning from the engineering approach suggest that the eccentricity of the apical system and specimen size both vary with current velocity. These morphologic features in fossils may thus be used as indicators of the velocity of ancient currents.

In the North Dome of the Kettleman Hills eccentricity is high in specimens of *Dendraster* from the lower part of the section and decreases upward. This indicates that the current velocity at the sites where the specimens lived was greater during deposition of the lower part of the section than the upper. This is in agreement with the sedimentologic and paleontologic evidence for a more open embayment during deposition of the lower part of the section than later when the basin was becoming increasingly restricted and less marine. In the upper part of the section eccentricity remains high in some localities in the Kreyenhagen and Jacalitos Hills to the west of the Kettleman Hills. This may indicate more open conditions with strong currents in those areas. They may occupy the site of the entrance to the San Joaquin Embayment from the open Pacific Ocean. See Chapter 11 for a further discussion of eccentricity in *Dendraster*.

6

Trace Fossils and the Effects of Organisms on Sediments

Body fossils have been, historically, the focus of interest in paleontology and the major source of data for the paleontologist interested in determining ancient environments. This concentration on body fossils is logical. They are obvious, easily identified and compared to one another and to living relatives, and possess morphologic and skeletal features that can be studied and interpreted in detail. Trace fossils, on the other hand, are relatively lacking in these attributes and so have been little studied until the last few decades. Even this limited study has shown, however, that they yield much information for many aspects of paleontology, and that they provide valuable information about the environment.

Trace fossils are of particular value because they are found in many sedimentary rocks devoid of body fossils, and thus in those instances provide the only record of life. They supply virtually the only record of all the organisms that have little or no mineralized skeleton and are thus unlikely to be preserved as body fossils. In sediments originally containing both skeletal material and life traces, diagenesis will commonly destroy the skeletal material by solution or recrystallization, but may enhance the traces. The diagenetic enhancement is caused by local differences in sediment porosity and permeability and in concentration of organic material in the filling or wall of the trace. In addition, and of special value in paleontology, trace fossils are almost always preserved in place. Although skeletal parts may be transported for considerable distance from the life site before burial, or be reworked from older to younger strata, the sedimentary record or trace of an organism can be redeposited under only very unusual circumstances. True fossils are also important from a sedimentologic point of view because they indicate the role of biologic processes in the deposition and penecontemporaneous subsequent modification of sediments.

Biogenic sedimentary structures include a wide range of features formed by the activity of organisms during their life but exclude the remains of the organisms themselves. The most common and thoroughly studied category consists of *trace fossils* (*ichnofossils or lebensspuren*)—the tracks, trails, burrows, and borings generated during an organism's life. Coprolites, fecal pellets, and similar evidence of organisms are not generally grouped in the broad category of trace fossils but are important though little studied biogenic sedimentary particles. The third broad category of material considered in this chapter consists of biostratification. Foremost in this category are stromatolites, formed primarily by trapping and binding of sediments by filamentous algae. Additional organisms and processes leading to distinctive stratifications are also considered.

The literature on biogenic sedimentary structures is widely scattered. However, several recent books provide comprehensive coverage of major parts of the topic, and thus a starting place for more specific information. The volume on trace fossils and problematica of the *Treatise on Invertebrate Paleontology* (Häntzschel, 1962) and its second edition (Häntzschel, 1975) is a comprehensive taxonomic treatment of trace fossils. Two volumes edited by Crimes and Harper (1970, 1977) are important collected sources on trace fossils. *The Study of Trace Fossils,* edited by R. W. Frey (1975), is a comprehensive survey of trace fossils. *Trace Fossil Concepts,* edited by P. B. Basan (1978), is less comprehensive but covers the major topics in this field. A recent book edited by Walter (1976) summarizes the knowledge of stromatolites.

TRACE FOSSIL CLASSIFICATION

Trace fossils have been categorized (1) as biologic entities, (2) on the basis of behavior represented, and (3) by their position in the sediment. Each approach yields specific and useful information for their analysis. As morphological-biological entities, they are described on the basis of their distinctive morphological features. The initial categorization is morphologic and descriptive, reflecting the geometric characteristics of the fossil. Subsequent study tends to lead to a biological classification that more or less conforms to the rules of zoological nomenclature (Häntzschel, 1962, 1975). The nomenclature that has developed is primarily restricted to the generic and specific level. A biological classification of traces is difficult for several reasons. The relation of trace fossil to causative organism is generally tenuous at best. In only relatively few cases did the animal "die in his tracks" so that the association of organism and trace fossil is certain. The study of traces of living organisms reveals that most trace-forming organisms do not leave distinctive traces. The same trace morphology may be made by several organisms and, under different environmental conditions, in different sediments, or at different stages of life, the same organism may form traces that are very different in appearance. Thus, because of the dearth of information linking trace to organisms, the names in common use largely represent morphologically distinct form taxa and do not connote a particular causative organism. Indeed, a single genus or species of trace fossil may be produced by several species of organism and a single species of organism may produce several trace fossil taxa. Differences in organism, behavior, and sediment open the way for numerous slight morphologic variants to be recognized by species names. In fact, however, the recent trend among trace fossil researchers has been to synonymize many existing specific names of this sort and to work at the generic level in environmental interpretations, using specific names only in detailed local studies.

By describing trace fossils in terms of the behavior of the organism forming the trace, the nomenclatural problems just described can be largely avoided. This important interpretative approach is based on the logic that if specific environmental conditions lead to specific behavioral responses, regardless of the kind of animal, then, conversely, the trace can be interpreted to determine the environment without having to worry unduly about the kind of animal that was responsible. The cases mentioned above of the same animal making different traces under different conditions and of different animals making similar traces under similar conditions provide the initial support for the logical assumption. They are interpretable if form alone can be related to specific conditions.

The fundamental behavioral categories are resting, locomotion, feeding, and dwelling. Many traces, however, are not the result of a single activity but the combination of such activities as locomotion and feeding or dwelling and feeding. The categories presented in Table 6.1 are based on the work of Seilacher (1953) and Frey (1978) but are modified in order to emphasize in practical terms the criteria for categorizing traces. A set of commonly used names for behavioral types of trace (Seilacher, 1953) are not used here because, derived from latin roots, they are specialized terms that only detract from the concepts we wish to emphasize.

Several classifications have been proposed that are based on the stratigraphic position of the trace, whether within a bed or, if on a bedding plane, in a position relative to that surface. From this descriptive information may be determined the stratigraphic position of the trace at the time it was formed. The complex terminology that has developed is discussed by Hallam (1975). The critical objectives are to determine whether the trace was formed on the surface, at an internal bedding plane, or within a bed; and if the trace formed below the sediment surface, to determine where the surface was and how the organism was linked to it. The trace is related geometrically and temporally, and thus genetically, to the enclosing sediment.

Table 6.1 Behavioral categories of traces

I.	Resting	The mobile animal temporarily comes to rest and more or less buries itself in the substrate. The reason is generally unknown but, for example, the organism may actually be resting or it may be a predator in hiding, awaiting prey. The trace is characterized by its temporary use, its shape as a depression on the substrate, the outline of the organism, and, occasionally, details of the ventral morphology.
II.	Dwelling	The organism constructs a dwelling place by burrowing into soft sediment or by boring into hard rock, skeletal material, or wood. The trace is characterized by (1) in contrast to feeding burrows, having a permanent nature and relatively fixed form, as indicated by being lined (for example, by sand grains, shell fragments, or organic material), or by conforming closely to the shape of the inhabitant and (2) evidence that the organism derives nourishment from outside the dwelling rather than from the sediment itself.
III.	Feeding	The organism moves on or through the substrate in search of food.
	A. Mining	Structures formed by organisms working through the sediment are distinguished from dwelling burrows by having a changing, impermanent outline, unlined walls, and by commonly being immediately back-filled so that no dwelling space is formed. Many have a high degree of symmetry indicating that the organism followed a systematic search strategy. Nevertheless, the burrow is being occupied by the organism, and the relative importance of feeding and dwelling behavior may be difficult to determine.
	B. Grazing	The organism may form a highly structured trace as it carries out an efficient search procedure in feeding on the surface. For example, the trace may resemble the pattern of a lawn mower or may radiate from the opening of a dwelling burrow. Less structured traces are more difficult to relate to a feeding activity.
IV.	Locomotion	The organism travels on or through the sediment. No purpose for the travel may be evident but may be as diverse as idle wandering, searching for a mate or food, or migration to a more suitable location. Locomotion traces are characterized by the absence of a highly structured grazing or mining pattern and by a strongly linear aspect. Many are probably non-diagnostic feeding traces.

TRACE FOSSIL ANALYSIS AND ENVIRONMENTAL INTERPRETATION

Distribution Patterns and Ichnofacies

The understanding of trace fossils has developed as information has been gathered from two sources. One has been the study of trace fossils in their geologic setting. By relating trace fossils to the associated sedimentologic and diagenetic features, their environmental significance can be established from the predetermined environmental criteria of the associated features. The other source of information has been the observation of modern traces. This approach has the advantage of combining trace and organism within the known environment, but the disadvantage of not incorporating the effects of geologic time and preservational processes. Both approaches are necessary to understand and apply trace fossils in paleoecology. The study of modern traces is handicapped by the difficulty of viewing the sediments and traces in three dimensions, but is aided by knowledge of the organism and its behavior responsible for the trace. The study of traces within the stratigraphic framework is handicapped by lack of knowledge of the animal, or its behavior in many cases, but is aided by working within the context of both time and space.

The early work with trace fossils, as with any group of fossils, was largely descriptive, and the description of new forms of both modern and fossil traces, their environmental and stratigraphic distribution, biologic affinity, and mode of origin continues to be an important task. Out of the descriptive effort has come the recognition that traces are not randomly distributed but occur in distinctive assemblages. These, and the facies (*ichnofacies*) they define, have been more clearly recognized in the fossil record than in modern sediments for several reasons. First, in sedimentary rocks, traces can be studied in three dimensions whereas in modern sediments the vertical view has been largely inaccessible until the limited use of box cores in recent years. Second, trace fossils can be studied from the full range of depositional environments as preserved in sedimentary rocks whereas many modern environments have been largely inaccessible or little studied. For example, deep marine traces have been difficult to sample, even with modern techniques, and because of the biased sampling provided by cores the assemblage of traces known from deep sea cores is very different from that recorded in photographs of the deep sea floor and from that known from deep water sedimentary rocks (Hollister et al., 1975; Chamberlain, 1975a; Ekdale and Berger, 1978; Berger, Ekdale, and Bryant, 1979). Terrestrial traces, as another example, have been poorly known until relatively recently because the sediments of modern terrestrial environments have historically been relatively little studied by the paleontologist interested in traces (Chamberlain, 1975b).

The study of the traces of living organisms has been carried on for many years, originally particularly in Germany, and primarily dealing with the traces of shallow water and tidal flats. Schäfer (1972), for example, has published an excellent compendium of organism activities and traces in the southern North Sea. Other noteworthy studies on shallow marine traces are by Howard and Frey (1973) who have described traces typical of Georgia estuaries, by Hertweck (1972, 1973) who has described the traces found in the shallow marine environment of coastal Georgia and the Gulf of Gaeta, Italy, and by Dörjes and Hertweck (1975) who synthesized the work of Hertweck with information about the physical environment and the macrobenthos of these areas and of the German North Sea area as well (Fig. 6.1). Numerous other studies of the traces in marginal marine settings are scattered through the literature, and although they may deal with a limited area or only an individual trace they may be of major significance. A case in point is the literature that has accumulated describing the distinctive burrows of the decapod *Callianassa* in the littoral and inner sublittoral, the recognition of their environmental significance, and their application in paleoecology (Weimer and Hoyt, 1964).

As in the study of modern traces, the study of fossil traces has led to the recognition of individual traces characteristic of specific environments. More importantly, characteristic assemblages or ichnofacies have been described. The *Cruziana*, *Zoophycos*, and *Nereites* ichnofacies, characteristic of relatively shallow shelf, outer shelf, and basinal environments, and the *Scoyenia*, *Skolithos*, and *Glossifungites* ichnofacies, representative of nonmarine and marginal marine environments, have been recognized in numerous studies and have become standards in the study of trace fossils (Fig. 6.2; Seilacher, 1967).

The paleoenvironmental interpretation on the basis of distribution patterns of individual traces, assemblages of traces, and ichnofacies has been the predominant form of analysis. The range of approaches and potential conclusions is illustrated by several examples. In Devonian and Mississippian strata of Montana and Utah, traces and ichnofacies can be analysed within a well established stratigraphic framework (Rodriguez and Gutschick, 1970). On the basis of regional stratigraphic relationships, sedimentologic features, and body fossils, it is clear that the sedimentary rocks had been deposited in environments ranging from supratidal and intertidal through shallow marine with associated organic mounds into deeper marine, below wave base, with sedimentation both from suspension and from turbidity currents. The distribution of distinctive traces and ichnofacies within this environmental gradient provide a standard for other studies in mid-Paleozoic strata of that region (Fig. 6.3). In addition, because the assemblages can be related to the standard ichnofacies described above, they are of general applicability in recognizing this range of depositional environments.

The description of the trace fossils within the framework of the Upper Cretaceous stratigraphy of eastern central Utah by Howard (1966) and Frey and Howard (1970) provides another example of the analysis of trace fossils in a geologic framework. The strata, formed by sedimentation in a west-to-east

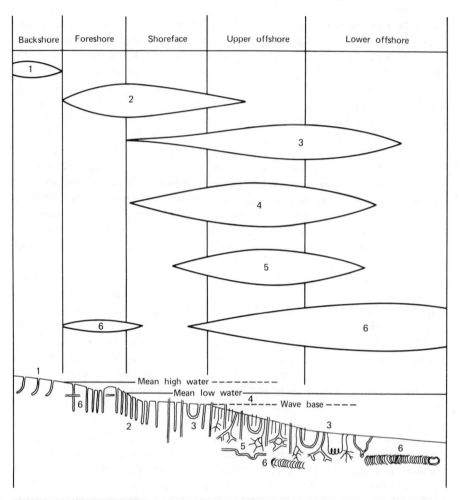

Figure 6.1 Distribution and zonation of ecologically indicative lebensspuren in a generalized beach-offshore profile. 1—burrows of terrestrial crustaceans (mostly scavengers); 2—straight vertical burrows of low-level suspension feeders; 3—burrows having two openings (basically U-shaped) of low-level suspension feeders or collectors; 4—tubes of high-level feeders; 5—dwelling burrows of intrasedimentary feeding animals; 6—crawling traces of intrasedimentary feeding animals. After Dörjes and Hertweck (1975).

pattern of alluvial, coal-bearing deltaic, sand-rich shoreline, and muddy offshore environments, are well exposed in the Wasatch Plateau and Book Cliffs of Utah (Spieker, 1949). The three distinctive lithofacies within these strata consist of (1) thin-bedded clayey siltstone, (2) very fine-grained, poorly sorted clayey sandstone with horizontal to gently inclined laminations and cross-laminations, and (3) interbedded very-fine-to-medium-grained sandstone with small and large-scale cross-bedding and poorly sorted and organic-rich shaly siltstone. These facies, which are grouped into cycles, represent re-

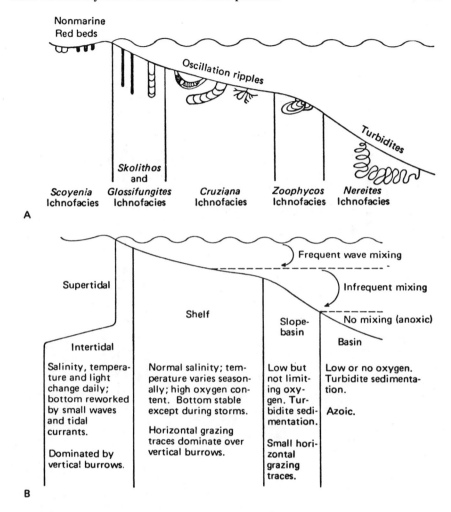

Figure 6.2 Terrestrial to offshore profile showing generalized distributions of trace fossil morphology, ichnofacies, and important environmental parameters. After Rhoads (1975).

gressive deposition successively in quiet offshore, more turbulent inner sublittoral, and littoral and shallow inner-sublittoral environments. The ichnofacies corresponding to each of the lithofacies is distinctive in taxonomic composition, in diversity and density, in general trace morphology, and in the inferred causative behavioral origin. They are most comparable to the *Cruziana* and *Skolithos* ichnofacies.

Carboniferous and Permian strata of the Oquirrh Basin of central Utah consist of 8000 m of predominantly limestone and sandstone. On the basis of lithologic characteristics and regional stratigraphic relationships, it has been inferred that these sediments were deposited in a gradually deepening sea.

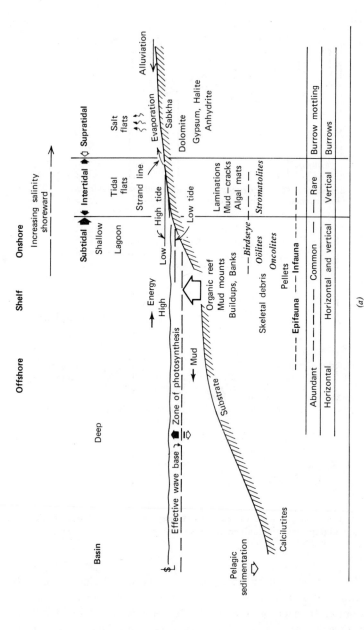

Figure 6.3 (*a*) Carbonate sedimentation model; (*b*) clastic sedimentation model and distribution patterns of organism behaviors, trace fossil morphologies, and ichnofacies. Although the ichnofacies model has been largely developed in clastic sediments, it proves to be valid in carbonate sediments as well. After Rodriguez and Gutschick (1970), Geological Journal Special Issue 3, Trace Fossils (ed. Crimes and Harper), Seel House Press, Liverpool.

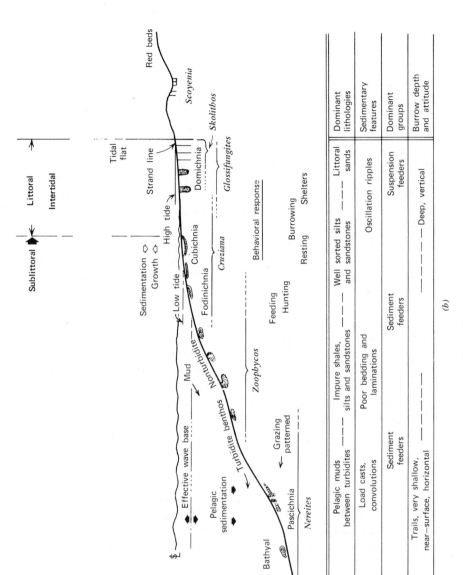

| Sublittoral | Littoral | | | |
| | Intertidal | | | |

Red beds

Scoyenia

Skolithos

Tidal
flat

Strand line

Glossifungites

Domichnia

High tide

Cubichnia

Cruziana

Behavioral response

Sedimentation
Growth

Low tide

Fodinichnia

Nonturbidite

Burrowing

Resting   Shelters

Mud

Feeding

Hunting

Zoophycos

Grazing
patterned

Effective wave base

Pelagic
sedimentation

Turbidite benthos

Pascichnia

Bathyal

Nereites

| Dominant
lithologies | Pelagic muds
between turbidites | Impure shales,
silts and sandstones | Well sorted silts
and sandstones | Littoral
sands |
| Sedimentary
features | Load casts,
convolutions | Poor bedding and
laminations | Oscillation ripples | |
| Dominant
groups | Sediment
feeders | Sediment
feeders | Suspension
feeders | |
| Burrow depth
and attitude | Trails, very shallow,
near–surface, horizontal | | Deep, vertical | |

(b)

Figure 6.3 (Continued)

271

The vertical sequence of trace fossils, which corresponds to the *Cruziana*, *Zoophycos*, and *Nereites* ichnofacies, agrees with the inferred change in bathymetry and conforms to the general bathymetric distribution established for these standard ichnofacies (Fig. 6.4).

The *Scoyenia* ichnofacies, which is characteristic of the terrestrial environment, has been poorly defined and can be interpreted in only general environmental terms. The description of traces in Miocene nonmarine strata of Nebraska and the comparison of the Miocene traces with modern traces in the sediments of the Platte River indicate the importance of beetles in generating the traces of the *Scoyenia* ichnofacies (Stanley and Fagerstrom, 1974). More significantly, however, many of the traces in these ancient and modern fluvial sediments are not easily distinguished from morphologically similar traces found in marine sediments.

Data from studies of trace fossils in both recent sediments and in sedimentary rocks have been combined in Fig. 6.5 (Chamberlain, 1978). Comparison of this figure with earlier trace fossil distribution charts such as Fig. 6.2 points out the great increase in knowledge of trace fossils that has been accumulated in just a few years. Examination of Fig. 6.5 also shows that with knowledge comes complexity: individual taxa are found in increasingly wide ranges of water depth, as perhaps best indicated by (1) *Teichichnus*, ranging from lagoonal to abyssal depths and (2) a single name, such as *Zoophycos*, applicable to a number of closely similar traces within a broad environmental range.

Construction of a chart like Fig. 6.5 that would be of universal applicability in paleoecology is difficult, and perhaps even misleading because the effort tends to oversimplify the environmental parameters that determine trace morphology and distribution and because evolutionary changes during geologic time, both in the organisms available to create traces and in their behavior itself (Seilacher, 1974), cannot be easily included.

The definition of ichnofacies and of distribution patterns of traces has been primarily related to water depth. Specific traces or assemblages diagnostic of other environmental parameters, such as salinity (Seilacher, 1963) or temperature, have not been recognized. As the studies of distribution patterns of modern and fossil traces accumulate, however, they provide the data for increasingly detailed paleoenvironmental interpretations, particularly to the extent that a broader range of environmental factors and the effects of evolution and geologic age can be taken into consideration.

Substrate characteristics such as firmness or bearing strength, Eh, and food content, play a major role in determining trace morphology by determining the kinds of organisms that may live in a given environment, their activities,and subsequently the preservability of the traces (Rhoads, 1970). Because the importance of substrate and other unrecognized environmental factors is poorly known, however, the environmental interpretation of the established ichnofacies must be in general terms at the present. The complexity of the controls of trace distribution is well illustrated by the trace

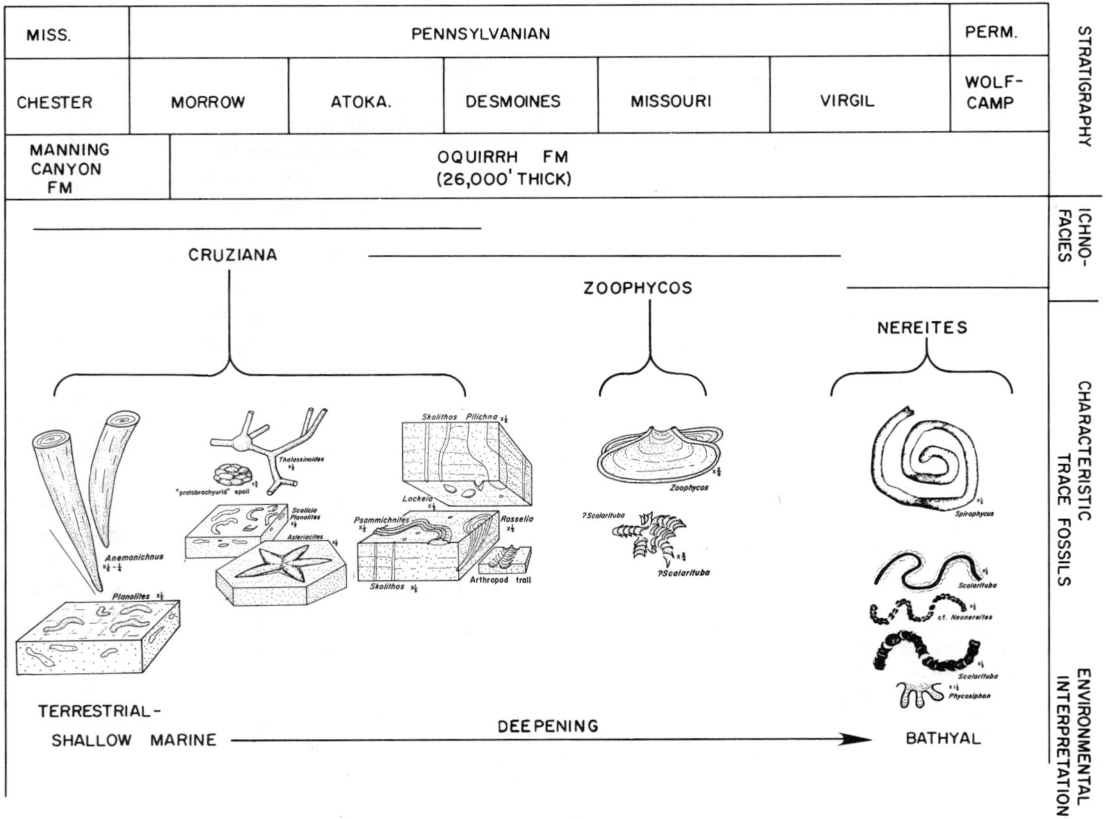

Figure 6.4 Distribution of trace fossils and ichnofacies in the Pennsylvanian and Permian Oquirrh Basin of Utah. Bathymetric interpretation of the trace fossils and also of associated conodonts indicates increasing water depth during deposition. After Chamberlain and Clark (1973), Journal of Paleontology, Society of Economic Paleontologists and Mineralogists.

273

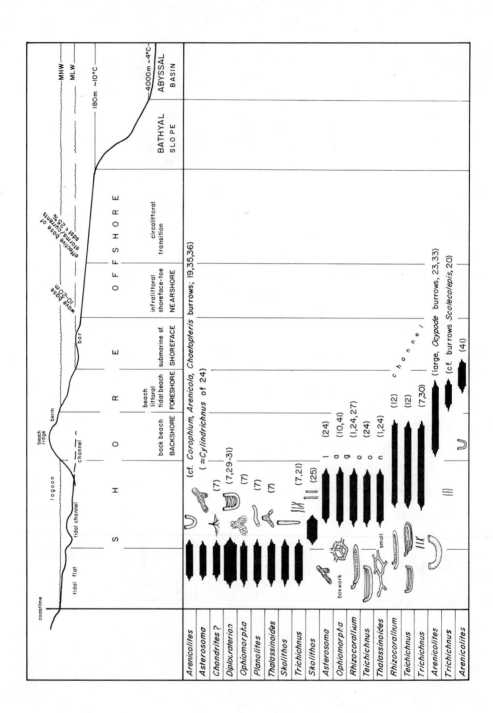

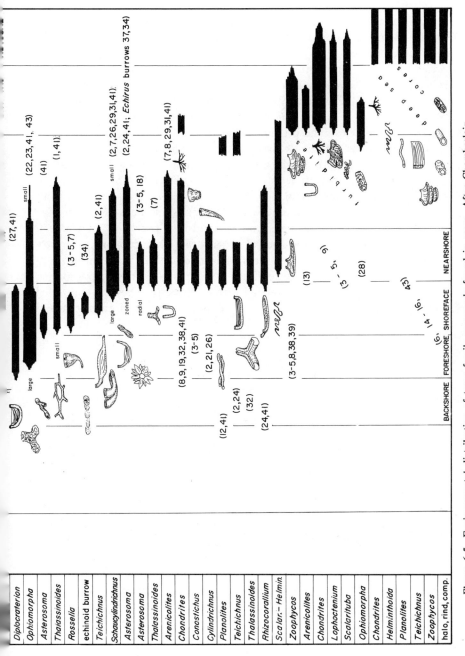

Figure 6.5 Environmental distribution of trace fossils commonly found in cores. After Chamberlain (1978), Trace Fossil Concepts, Society of Economic Paleontologists and Mineralogists Short Course.

275

fossils in Cretaceous strata at San Diego, California (Kern and Warme, 1974). The interbedded mudstone and massive ungraded sandstone of the Upper Cretaceous Point Loma Formation were deposited largely by grain-flow processes in a deep, probably bathyal, environment. Trace fossils include *Ophiomorpha* and *Thalassinoides*, usually considered characteristic of the *Cruziana* ichnofacies; *Zoophycos*, the nominate form in the *Zoophycos* ichnofacies; and *Nereites*, *Scolicia*, *Chondrites*, and other traces characteristic of the *Nereites* ichnofacies. The organism that formed the *Ophiomorpha* traces in the sandstone beds apparently lived only on a sandy substrate, but when it burrowed through the sand into the underlying mud it formed a *Thalassinoides* trace in the mud in continuity with the *Ophiomorpha* (Fig. 6.6). In addition, differences in the morphology of *Zoophycos* can be correlated with sediment texture and bed thickness: *Zoophycos* in thin mudstone beds is a short flat spiral but in thick mudstone beds and in sandstone overlying mudstone it is a high-spiral form. Thus the control of substrate character on trace morphology and the co-occurrence of traces indicative of three different environments clearly demonstrate that trace fossil assemblages cannot be interpreted simply in terms of ichnofacies correlated with bathymetry. They are controlled by, and must be interpreted in terms of, the total complex of environmental factors.

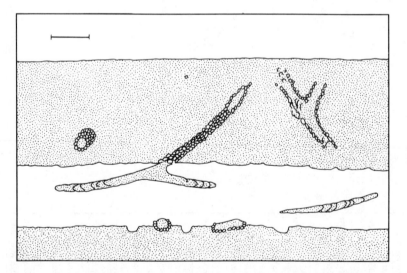

Figure 6.6 Drawing of Cretaceous strata in the Point Loma Formation, San Diego, California, showing several specimens of *Ophiomorpha* and *Thalassinoides*. Bar is 5 cm long. Typical sand-filled burrows of *Ophiomorpha* have pelletal mud walls in sandstone; unlined *Thalassinoides* occurs in mudstone. One specimen of *Ophiomorpha* in sandstone is continuous with *Thalassinoides* in underlying mudstone. Burrows at sandstone-mudstone interfaces may be lined or unlined. Crescentic backfill structures are visible in several specimens. After Kern and Warme (1974), Geological Society of America Bulletin.

Behavior and Trace Morphology

The recognition of trace fossil assemblages and their correlation with specific environments, most commonly in terms of bathymetry, is useful as a first descriptive step in paleoenvironmental analysis. It is of limited value beyond this from both theoretical and practical points of view. The approach that has proved to be most productive in understanding and interpreting trace fossils has been to analyze them in terms of the behavioral activities that produce them and to search for the relationship between environment and behavior. The assumptions made in this approach are: (1) traces, whether of living organisms or in the fossil record, represent particular behaviors of organisms during life, (2) the causative behavior is controlled by the environment, and (3) if the relationship of environment to behavior can be determined, the trace can be interpreted in terms of the environment. A basic behavioral classification of traces was presented in a previous section (Table 6.1). Trace morphology characteristic of each of the categories of behavior, and the fundamental environmental parameters that control behavior, are more fully analyzed in this section in order to develop a basis for interpreting traces in terms of the environment.

Behavior Recognition

The majority of traces are formed as the organism is constructing a dwelling or is feeding, and so it is essential to be able to distinguish between these types of traces.

If the trace is a boring in a hard substrate, the decision that the trace served a feeding or dwelling function is largely based on whether the substrate, in itself, was of any nutritive value. If the boring was of no evident feeding purpose, as is the case for excavations made by rock-boring organisms and many shell-boring organisms, the boring must have served a dwelling function. At the other extreme, borings by carnivorous gastropods into other molluscs and the tooth marks of fish feeding on coral are clearly feeding traces. Borings that are primarily for feeding but also serve a dwelling purpose, such as those made by *Teredo* into wood, should be classified as feeding traces.

If the trace is a burrow, it may be more difficult to determine if the causative function was a dwelling or a feeding activity. Dwelling burrows are relatively permanent structures; commonly those made by benthic marine invertebrates are lined by sand grains or shell fragments, as in the case of the burrow of modern callianassid crustaceans and their apparent trace fossil equivalent, *Ophiomorpha* (Weimer and Hoyt, 1964). Being more permanent, a dwelling burrow may remain open after the constructor has died, and thus be filled eventually with cement or with sediment that may be very different from the surrounding sediment (Shinn, 1968).

Organisms of many taxa undoubtedly generate only a single burrow morphology, as determined by stereotyped behavior while living within a narrow environmental range. Others, however, with wider environmental tolerances, adjust their behavior to different conditions and correspondingly generate a number of burrow morphologies. The callianassid ghost shrimp discussed above, for example, forms a distinctive relatively long and straight vertical tube, *Ophiomorpha*, in sand at the seaward edge of the beach. It forms a much more branching and complex trace in muddier sediment of bays and deeper water, and may form the irregular branching trace *Thalassinoides* in deep water mud as well as in sandy sediments of shallow water. These differences are determined by sediment character and feeding mode: the simpler burrow is primarily for dwelling as the animal feeds from food in suspension in the water above; the more complex burrow represents intensive deposit feeding in quieter water containing less food in suspension.

Tracks and trails on the sediment surface, resulting from either feeding or locomotion, are distinguished on the basis of symmetry. If the pattern has a high degree of symmetry, it is assumed to reflect an efficient feeding behavior as the organism systematically searched for food. If the trace is more or less linear or erratic, the organism may have been feeding or perhaps was just passing by for some unknown reason—it is virtually impossible to determine which.

Dwelling Traces

The objective of a dwelling burrow is to provide protection for the inhabitant from predators and from the environment above the substrate. The role of predation on burrow construction must be important but is difficult to evaluate because generally both predator and prey organisms are not preserved, and thus the interaction between them cannot even be guessed. Even if the trace maker is preserved, the role of biologic interaction is difficult to evaluate. For example, the deep burrow of the bivalve *Panopea*, reflecting its ability to burrow rapidly and deeply, must represent a strategy that has evolved to avoid either predation or an unfavorable aspect of the environment in the overlying water. However, because in this case, as in general, the biologic interaction is indeterminate, the trace can be interpreted only in terms of the physical environment.

For each parameter of an organism's environment, life conditions consist of those that are optimal and, beyond that, of those that are tolerable for survival but not reproduction, or perhaps for growth but only at a reduced rate. Even where conditions are optimal, the environment fluctuates. Thus in order to survive, the organism must be able to cope with the excursions of the environment outside the tolerance limits. A dwelling burrow represents a survival strategy in fluctuating conditions, whether the unfavorable intervals are predictable and periodic, ranging from diurnal tidal to seasonal, or occur unpredictably and irregularly as in the case of infrequent storms or extreme temperatures.

Most organisms live at or very near the substrate surface in both the subaerial and subaqueous environment because the substrate provides support for both plants and animals and nutrients for rooted plants. Thus it is a

surface of both production and accumulation of organic material. The overlying medium contains the oxygen for respiration and, in aqueous environments, nutrients in solution and in suspension as phytoplankton and zooplankton and organic detritus in various stages of decomposition. The underlying sediment is a zone of accumulation of the organic material, but lacks
light for photosynthesis and commonly is low in oxygen for respiration. For
all these reasons, the substrate surface is the area of maximum flux of resources for respiration and nourishment. In general, the environmental
variability is high above the substrate, decreases markedly just below the
surface, and decreases less rapidly with increasing depth, approaching stable
invariant conditions asymptotically (Fig. 6.7). Thus the essential survival
strategy for many animals is to utilize the resources at and above the substrate
in which to dwell permanently or in which to retreat when conditions above
the surface are not tolerable.

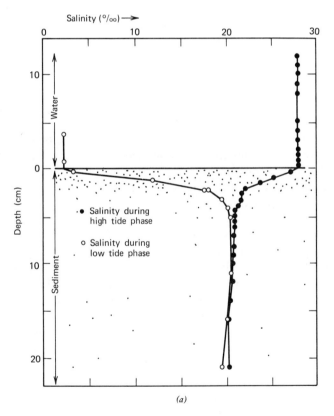

Figure 6.7 Plots of salinity and temperature ranges versus depth in the sediment. (*a*) Salinity
profiles at high and low tide of the sediment and the immediately overlying water in the Pocasset
River estuary, Massachusetts. After Sanders et al. (1965). (*b*) Temperature variation at depths of
1-cm (crossed line), 10-cm (dashed line), and 20-cm (dotted line) depths in intertidal sand, 3 July
1963, in Tomales Bay, California. After Johnson (1965b).

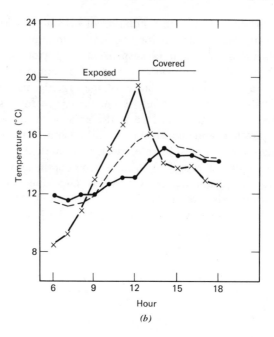

Figure 6.7 (*Continued*)

In terrestrial environments, the predictable periodic fluctuations range from diurnal to seasonal, with the magnitude largely determined by climate and thus broadly correlated with latitude. Superimposed on the global gradient, geographical conditions also control the environmental fluctuations. For example, maritime climates fluctuate less than continental ones, and variability is modified by elevation much as it is by latitude, with increasing elevation being equivalent to increasing latitude.

In marine environments, variability is broadly determined by water depth and distance from land. The daily to seasonal fluctuations of temperature and salinity in the water column that are caused by climatic fluctuations in the overlying atmosphere are greatest at the surface and decrease rapidly with depth to the thermocline at a few hundred meters or less, below which conditions are stable. Near land, the environment is affected both by conditions on land and by the local bathymetry and geography. Salinity, turbidity, and sedimentation rate fluctuate in response to variations in freshwater runoff from the land, and the magnitude of these fluctuations depends on the local geography as it restricts mixing of the coastal waters and runoff from the land with the open ocean. If restriction is sufficient under the proper climatic conditions, salinity may fluctuate widely.

The validity of the inference that dwelling burrows are primarily a means for benthic organisms to avoid fluctuations in the environment above the substrate is shown by the way dwelling-trace morphology reflects the gradients described. The characteristic traces in the shallow water *Skolithos* ichnofacies are deep burrows such as *Ophiomorpha* and *Diplocraterion*. In the *Cruziana* and *Zoophycos* facies of the open shelf, the burrows are much shallower, and in the *Nereites* facies of the deep ocean floor, surface traces are characteristic. This relation of burrow depth with environmental variability in the open marine environment was first stated explicitly by Rhoads (1966). Sanders et al. (1965) have described the same relationship in the fluctuating estuarine environment of the Pocachet River estuary (Fig. 6.7a). There, epifaunal taxa are poorly represented and are largely restricted to the most marine end of the estuary; each infaunal taxon has a preferred depth within the sediment and those that burrow most deeply are the taxa with the least tolerance for rapid salinity fluctuations. The correlation of burrow depth and environmental variability can be used in paleoenvironmental reconstruction, but caution is necessary keeping in mind that burrow morphology may also be affected by sediment characteristics, as previously mentioned, and by the nature of food resources in the environment, to be considered in a subsequent section.

For terrestrial animals, the dwelling burrow serves as a place to avoid both predators and the diurnal and seasonal climatic extremes, but, as previously discussed, the relative importance of environmental variability and predation is difficult to determine. Several kinds of seasonal climatic change cause terrestrial vertebrates to burrow. A few mammals and many reptiles and amphibians burrow during the cold season to survive by hibernation until warmer weather returns. For reptiles, the ability to burrow deeply allows some species to survive far north of what their climatically determined limit would otherwise be. Dry hot spells of the summer season are equally intolerable for many amphibians and lungfish, and they retreat into burrows to aestivate. Thus the existence in the fossil record of hibernation and aestivation burrows could indicate seasonality of temperature or rainfall in the ancient climate. Potentially, the depth of hibernation burrows could be mapped as an indication of climatic gradient and thus of paleolatitude, but they have yet to be found in the fossil record. Similarly, the occurrence of aestivation burrows indicate strongly seasonal terrestrial climate characterized by a summer period of drought. These have been described (Berman, 1976; Olson and Bolles, 1975) and have been used to infer lateral climatic gradients and temporal climatic changes during the Permian in the southwestern United States (Olson and Vaughn, 1970).

Feeding Traces

Feeding burrows are formed as the animal works through the soft sediment for food. The animal may mine the sediment, extract the food, and expel the sediment on the surface to form a volcano-shaped mound at the burrow opening and an ever larger burrow that serves also as a dwelling. On the other hand, as the animal mines the sediment, it may backfill the burrow so that an open dwelling burrow is not formed. The feeding trace morphology is determined by genetically controlled anatomy, physiology, and behavior and by environmental factors such as sediment character and kind and distribution of food in the sediment. Examples of the range of genetically determined differences in feeding behavior and in resulting trace morphology are provided by holothurians, scaphopods, and callianassid shrimp. The holothurians, which are nonselective deposit feeders, swallows the sediment and extracts the nutritive organic material as the sediment passes through its gut. Its burrow is immediately back-filled with the same sediment, unchanged except by slight solution of the carbonate grains as they pass through the gut. Repeated burrowing through the sediment would destroy any primary sediment structures and leave it thoroughly churned. The scaphopod mollusc *Dentalium* moves slowly through the sediment with tentacle-like appendages at the anterior end searching for meiofauna (Dinamani, 1964). Because *Dentalium* is very selective as it feeds, it does not modify the sediment texture; it does leave the sediment with a general churned aspect but no distinct burrows—a process aptly described as bioturbation. The Callianassid shrimp is also a selective deposit feeder like *Dentalium*, but feeds by digging in the sediment to excavate an open burrow. As the sediment is thrown into suspension and passes under its body toward the burrow opening, the shrimp picks out the bits of organic detritus upon which it feeds. The burrow may remain open long after the individual has died, and may then be filled with material different from the surrounding sediment (Shinn, 1968).

The morphology of the feeding trace of an aquatic benthic organism depends on the location of the food being used—whether it is within the water, on the substrate, or within the sediment. Most benthic organisms feeding out of the water column are relatively fixed in position and rely on food coming to them—either as active prey or as phytoplankton and zooplankton and organic detritus in suspension. Thus the burrows they build function as dwellings and would have those characteristics. Organisms feeding from inside the sediment or on the substrate range from macrophagous carnivores and scavengers to microphagous animals utilizing microscopic -plants, animals, or organic detritus. Whatever the feeding mechanism, the animal moves constantly in its search for food, leaving a path through or on the sediment. The feeding paths that different animals produce differ widely in pattern and symmetry as a result of the search strategy used by the particular animal. The strategy may have evolved as an instinct in the species and only individuals with the correct preprogrammed strategy for the local environment prosper. Alternatively, the organism may have some flexibility of action and be able to modify its behavior under different conditions. Whatever the specific mechanism, the end result is a record of feeding behavior that we can interpret as an efficient feeding strategy.

We assume that the feeding strategy must be to gain the maximum amount of food with the minimum expenditure of energy. For an animal with a nonselective feeding behavior, the problem must be like that of mowing a lawn—how to cover all the area, minimizing the overlap. For a selective deposit feeder or a predator, the search strategy must include this first step but then a choice must be made between the potential food items encountered. For example, the potential prey for a large terrestrial carnivore such as a wolf ranges from abundant but small mice to large but infrequent moose. The wolf must choose between small, yet common prey that require only a little effort to find but provide only a small payoff in food, and larger, though less common prey that provide a more adequate meal but only after much more time and effort.

The morphology of the feeding trace depends on the abundance of food and the location of the food on and in the sediment. As a broad generality, the abundance of food is greatest in shallow water near the continental margin because organic productivity is determined by the contribution of nutrients and organic material from the land, by the localization of upwelling and nutrient recycling along the continental margin, and by the restriction of benthic plants to the photic zone of the shallow shelf. Overlaid on this general pattern of decreasing organic material from shore to the deep ocean floor are local variations due to geographic and oceanographic conditions. The proportions of the available food in suspension, on the substrate, or in the sediment depend primarily on the *water energy*. (Water energy is a poorly defined term, but qualitatively describes the sum total of water motion due to current and wave action.) Where it is high, the bulk of the trophic resources is in suspension, little is on the substrate, and an intermediate amount accumulates in the sediment as organic material is buried by bioturbation and by occasional intervals of rapid deposition.

Suspension-feeding organisms should predominate in areas of high water energy because the efficient feeding strategy would be to utilize the relatively abundant food in the water column. Consequently, dwelling traces should predominate. At the other extreme, in a setting such as the deep ocean floor, both organic productivity and water energy are low so the total amount available will be less, and the amount that is present will largely be able to settle out of suspension and will accumulate on the substrate. In addition, the rate of sedimentation is low so that little of the organic material will be buried. Consequently, surface-feeding traces from animals reworking the small but consistently present food on the substrate should predominate.

The amount of food present within the sediment depends on the rates of organic productivity and sedimentation. If they are high enough, some of the organic matter may be buried before it can be consumed at the surface. Some organic matter on the substrate may also be buried by burrowing organisms as they dump excavated material on top of the organic-rich surface. If fluctuations in sedimentation result in deposition of coarser and less organic-rich sediment on top of finer and more organic-rich sediment, animals may burrow down through the coarser sediment at the surface and then feed from the buried, more nutritive stratum.

If food is abundant and being constantly replenished, no systematic search strategy is necessary, for food is available with little effort whichever way the animal turns. If food is scarce and its rate of replenishment is low, however, a "lawnmowing" strategy becomes desirable and a systematic search pattern with a high order of symmetry should be most efficient.

These general relationships between environmental conditions and both dwelling and feeding behavior lead to parallel trends of trace morphology. In shallow water, with unstable conditions, high water energy, and productivity, dwelling burrows of suspension feeders should predominate. In deeper water, with more stable conditions but with lower water energy, sedimentation rate, and productivity, dwelling burrows of suspension feeders should decrease in abundance relative to traces of both surface and in-sediment deposit feeders. With decreasing productivity and sedimentation rates, more efficient systematic feeding patterns become increasingly advantageous, and under condition of very low productivity the deposit feeders should be primarily surface grazers with highly symmetrical feeding traces.

This sequence of feeding strategies, which is broadly correlated with the bathymetric gradient and is evident in modern traces, results in the ichnofacies described earlier. It is important to note that this sequence of trace morphologies is not determined by water depth in itself but by the several factors described that happen to change fairly systematically with depth but may change along other environmental gradients as well. For example, changes in water energy and location of food resources and corresponding changes in traces, which occur from shallow to deep water, may also occur in going from shallow open and exposed marine to shallow but protected bay environments.

The general relationships between behavior and environment may not be valid at extreme levels of the environmental factors. For example, high water energy increases suspended food supply and so should promote suspension-feeding animals in general, but high water energy also increases sediment movement, which is disadvantageous for these largely sessile animals, increases turbidity due to suspended sediment, which inhibits suspension feeding animals by clogging the food gathering mechanism, and decreases light penetration and primary productivity, which limits the food supply. As another example, abundant organic content on the sea floor should favor deposit feeders, but Bader (1954) and Purdy (1964) have shown that total organic content in itself is not important in determining the abundance of deposit feeders, for much of it may be refractory and of no nutritive value. In addition, too much organic material, even if usable, may decrease the oxygen content within the sediment and thus inhibit deposit feeding.

Sediment character will also modify these general conclusions about feeding/dwelling trace morphology. The ease with which an animal can move through the sediment will determine the shape and depth of its burrow. The relative rates of deposition of organic material and sediment will determine the distribution of organic material and thus of burrows within the sediment. Burrowing may be concentrated in the upper parts of strata that are laid down infrequently but rapidly so that burrowing must subsequently begin at and work down from the upper surface. Finally, the bearing strength will determine the suitability of the substrate for benthic organisms. In very soft fluid sediment, all benthic organisms may be excluded unless they have special morphologic adaptations that prevent them from sinking into the substrate (see Chapter 2; Thayer, 1975a).

The net result is that these relationships between dwelling and feeding behavior and water energy and trophic resources provide valid paleoenvironmental criteria, but only as generalities. Analysis of behavior and trace morphology must take into account the full range of environmental factors.

Locomotion Traces

Tracks and trails that are linear, wandering, or erratic in plan are commonly attributed to locomotion. Presumably they reflect some purposeful behavior, such as feeding or migration rather than just idle wandering, but the behavior cannot be determined because the trace morphology is not diagnostic. Consequently, they are of little paleoenvironmental interpretive value. Footprints made by dinosaurs have been described that are more or less impressed in the sediment as the animal waded in water 1 m or so deep. By estimating the size of the animal from its stride, it is possible to infer the water depth (Langston, 1974). If the nature of the sediment is known, the depth of the imprint can be used to determine the weight of the animal. Conversely, it has been suggested that the depth of imprint can be used to determine the strength of the earth's gravitational field if the strength of the sediment and the mass of the animal are known. In environmental terms, the footprint sequence of the wading animal defines the location of the shoreline at that time. Vertebrate footprints are uncommon in the fossil record because they are preserved only if buried soon after being formed in damp sediment.

Resting Traces

Resting traces are rare in modern sediments and are not common in the fossil record. The best-known resting trace fossil is *Rusophycus*. This trace is made by a trilobite, as indicated by the association of *Rusophycus* with the trilobite crawling trace, *Cruziana*, and in a few cases, with the trilobite itself, dead in its tracks. Although resting traces, like locomotion traces, can be used to tell something about the animal's size, shape, ventral anatomy, behavior, and gait, they are of limited independent environmental use. They may be important as components of assemblages or ichnofacies indicative of particular environments.

TRACE FOSSILS AS CLUES TO SEDIMENTATION

The morphology of a trace is controlled in part by the characteristics of the sediments in which it forms. Consequently, traces may be of great value in describing in detail the depositional processes and the primary sedimentary characteristics. When humans first walked on the moon, views of their footprints in lunar soil were immediately transmitted to earth, and the depth of the impressions and the apparent cohesiveness of the soil in these vivid pictures formed the basis for much speculation about the nature of the soil, the depositional history, and the lunar environment. Trace fossils in the geologic record offer similar interpretive possibilities. The use of dinosaur footprints to determine sediment characteristics or the strength of the earth's gravitational field, as mentioned in the preceding section, is one example.

Sediment Strength

The original nature of a sediment soon after it was deposited and as it forms the substrate is important in several respects in determining the original conditions of sedimentation and of life. The sediment may be described in terms of the spectrum ranging from soft, flocculent, and easily eroded to firm, with high bearing strength, and with strong resistance to erosion. These original characteristics are not easily determined from the sedimentary rock but are necessary in order to predict depositional water energy, to recognize subaerial exposure or specific physical-chemical conditions that might have altered the strength of the sediment after deposition, and to explain the nature of the benthic fauna in terms of the bearing strength and cohesiveness of the substrate. An example of this last application is the recent interest in paleontology in explaining dwarfed faunas and a wide range of morphologic characteristics in benthic organisms as being adaptations for life on a soft substrate with very limited bearing strength (Carter, 1972; Surlyk, 1972; Mancini, 1978b).

Although these original sedimentary characteristics generally cannot be determined from the sedimentary rock, they can, potentially, from the morphology of the burrows the rock contains (Rhoads, 1970). Burrows can be described according to a broad spectrum of preservational conditions. At one extreme are those that are well-preserved in original shape and outline even though they had apparently been unlined and had apparently remained open to 1 m or more depth during and after the life of the inhabitant. These burrows would be indicative of a sediment with high cohesiveness and bearing strength. Burrows that are similarly well-preserved but that had a lining of organic matter, pellets, shell fragments, or sand grains may have maintained their form in somewhat less cohesive sediment. At the other extreme of the spectrum, indistinct and collapsed and deformed burrows and a general swirled wispy bioturbation structure would suggest soft and fluid sediment. In this last case, of course, if the sediment is very well sorted and homogeneous in composition, compaction may eliminate all evidence of burrowing.

In addition to providing information about the sediment strength, the trace fossils need to be considered in determining the subsequent history of the rock. The extent to which trace fossils are changed from their original shape is a useful measure of sediment compaction. The fact that traces are seldom described as having been compacted or flattened suggests that the bulk of sediment compaction occurs very soon after deposition and very near the surface. Bioturbation structures in very soft sediment may superficially resemble soft-sediment deformation. In examining such structures, both possible origins should be considered.

Rate of Sedimentations

The abundance of trace fossils in a sedimentary rock is determined by the rate of deposition and by the rate of trace formation. The rate of trace formation in turn is controlled by the complex of environmental factors that determine the abundance of organisms and level of their activity. Thus an increase in trace fossils may indicate either a decrease in depositional rate or an improvement in the environment and a consequent increase in the abundance or activity of burrowing animals. Because variations in number of trace fossils within a stratigraphic column are controlled by these two independent variables, they cannot be interpreted without additional information.

An example of integration of stratigraphic and lithologic data with trace fossil data in order to understand the sedimentation process is provided by Lower Eocene strata at a locality in east-central Texas (Warme and Stanton, 1971). The strata, in the Rockdale Formation of the Wilcox group, consist of laminated unfossiliferous claystone overlain by several 0.6 to 1.0 m beds of muddy sandstone (Fig. 6.8). Regional stratigraphic data suggest that the sediments were deposited in a delta-fringe setting. Extensive deposits of lignite only slightly higher in the section are mined within a few miles of the outcrop. Each sandstone bed is cross-bedded in the lower part but grades upward into increasingly bioturbated sandstone containing distinct *Ophiomorpha* burrows. The boundary between the cross-bedded and burrowed parts of each bed is very irregular whereas the upper contact of each bed is an uneven scoured surface with 10 to 20 cm of local relief.

The interpretation of these features is that the sandstone beds were deposited as crevasse splays, built out into the shallow water of an interdistributary or interdeltaic bay. Trace-forming organisms were probably absent from the underlying delta-fringe claystone because of the combination of rapid sedimentation and high turbidity, a soft fluid substrate, and variable low salinity. Each sand bed was deposited rapidly as individual pulses associated with storms and overlevee river flow. The rapidity of deposition of each sand unit is indicated by the scour of the underlying depositional surface and by the cross-bedding within each sandstone. Following each of these depositional events, callianassid crustaceans and perhaps other infauna colonized the sandy sea floor and began to burrow into the soft sediment. The burrowing thoroughly obliterated all primary sedimentary structures in the upper part of each bed. The burrowing may have been entirely the result of callianassid dwelling burrows with only the last generation of them being preserved as abundant *Ophiomorpha,* but other burrowers may also have been present that left no distinctive traces. A very shallow water setting is inferred

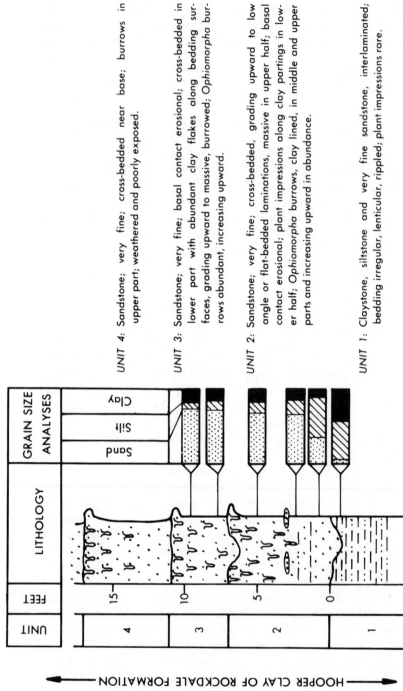

UNIT 4: Sandstone; very fine; cross-bedded near base; burrows in upper part; weathered and poorly exposed.

UNIT 3: Sandstone; very fine; basal contact erosional; cross-bedded in lower part with abundant clay flakes along bedding surfaces, grading upward to massive, burrowed; *Ophiomorpha* burrows abundant, increasing upward.

UNIT 2: Sandstone; very fine; cross-bedded, grading upward to low angle or flat-bedded laminations, massive in upper half; basal contact erosional; plant impressions along clay partings in lower half; *Ophiomorpha* burrows, clay lined, in middle and upper parts and increasing upward in abundance.

UNIT 1: Claystone, siltstone and very fine sandstone, interlaminated; bedding irregular, lenticular, rippled; plant impressions rare.

GRAIN SIZE ANALYSES

Clay
Silt
Sand

LITHOLOGY

FEET

15

10

5

0

UNIT

4
3
2
1

◀— HOOPER CLAY OF ROCKDALE FORMATION —▶

Figure 6.8 Lithologic column in the Hooper Clay Member of the Eocene Rockdale Formation, Texas. Units are individual pulses of sedimentation. During intervening time intervals, units were burrowed from upper surface, and then scoured as the overlying unit was deposited. After Warme and Stanton (1971), reprinted from Trace Fossils: A Field Guide to Selected Localities in Pennsylvanian, Permian, Cretaceous, and Tertiary Rocks of Texas and Related Papers, B. F. Perkins, Editor. By permission of the School of Geoscience, Louisiana State University. Copyright 1971 by the School of Geoscience, Louisiana State University.

from the *Ophiomorpha,* from the vertical sequence in each bed of original cross-bedding being progressively destroyed upward by increasingly pervasive burrowing, and from the scoured uneven upper surfaces cutting across the vertical *Ophiomorpha* of each bed. The depth to which burrowing extended in a bed may reflect the length of time before the next sand layer was deposited and terminated the burrowing in that bed. On the other hand, the depth may have been the natural burrowing limits for the organisms responsible, or it may have been determined by sedimentary characteristics, chemical conditions within the sand, or some other limiting aspect of the environment, inhibiting burrowers that under more favorable conditions would have gone deeper.

Sedimentation in shallow marine environments is commonly very much like that in this Eocene example, consisting of infrequent but rapid pulses of sedimentation or scour and relatively longer intervals of little activity. Thus the trace-forming organisms living within the sediment may be repeatedly forced to shift their positions up or down to reestablish themselves at the optimal depth below the sediment surface. This activity may generate distinctive sequences containing overlapping generations of traces. Goldring (1964) illustrates a number of different examples of the adjustment of organisms to scour and deposition events. Erosional and depositional events during the Pliocene in central California are well illustrated in Fig. 6.9. The presence of stable surfaces from which the underlying sediment was intensively burrowed indicates that deposition must have been very slow in general. Sometimes burrowing was abruptly terminated by subsequent deposition, although no evidence of scouring on the existing surface can be seen (horizon a in Fig. 6.9a). At other times, a period of little or no deposition and the development of an intensively burrowed surface was terminated by erosion of the upper part of the burrowed zone and followed by rapid redeposition, leaving only the lowest parts of the burrows intact (horizon b in Fig. 6.9b). Deposition in a high-energy, probably shore-face environment is suggested by the regional stratigraphic setting and the texture and sedimentary structures. The episodic nature of deposition can be recognized only from the trace fossils.

The ability of benthic organisms to move up or down during sedimentation or scour in order to maintain their preferred position at or below the depositional surface differs from one species to another (Kranz, 1974). Among benthic marine organisms, attached epifaunal individuals obviously have no migration ability whereas mobil infaunal individuals should be well adapted for vertical migration. If deposition is very rapid, the upward migration path becomes truly an escape structure (Fig. 6.10). In addition to providing information about the depositional process, vertical migration histories are helpful in reconstructing the sequence of original life assemblages of body and trace fossils out of the mixed and overlapping assemblages in the sedimentary section. If an assemblage living at the surface is buried by a sudden influx of sediment, it will be dispersed upward, scattered between the old surface and the new one, according to the escape potential of each species.

Figure 6.9 Burrows in the Pliocene Etchegoin Formation, California. Dark sandstone is medium grained, well sorted, thin bedded to laminated. It was deposited in beach to nearshore conditions. Matchbook in each photograph provides scale. (*a*) General view of outcrop; (*b*) closeup view. Light-colored burrowed sediment is muddy sandstone.

Conversely, the benthic organisms living at a single time include those at the surface and those burrowing to different depths. Thus an assemblage of fossils at a specific stratigraphic horizon may comprise individuals that had lived at that sediment surface along with individuals that migrated upward during sedimentation and others that had burrowed down from higher and younger positions of the depositional surface.

In addition to this active mixing, shells may be concentrated passively at certain levels within the sediment through the burrowing of other organisms to that depth. Van Straaten (1952) has described an example of this process. As worms of the genus *Arenicola* burrow into the sediment, they cast it onto the surface. Consequently the sediment to a depth of 20 to 30 cm is being constantly turned over. In the process, shells of *Hydrobia* and other molluscs settle to the lower limit of burrowing. If the rate of sedimentation is slow, a shell layer will accumulate that combines many generations of individuals that had lived within the actively burrowed interval. This assemblage represents the external marine environmental conditions corresponding to the depositional surface 20 to 30 cm higher in the section.

Hardgrounds and Depositional Hiatuses

During intervals of nondeposition, the marine substrate may become lithified, either by submarine cementation or during brief periods of subaerial expo-

Figure 6.9 (*Continued*)

sure, forming hardgrounds. In either case, no erosion may have occurred or be evident and the hardground may be evident only from the trace fossils consisting of borings extending down from the surface. The traces are recognized as borings because they cut sharply across all components of the rock, such as grains and cement, body fossils, other borings, and prelithification burrows and bioturbation. Because hardgrounds form in areas of fluctuating depositional rate, nondeposition, and local scour, a complex detailed stratigraphy may develop including both burrowed surfaces and bored surfaces. The chronology of overlapping generations of trace fossils is commonly difficult to unravel, but may be the only means of determining the sequence of depositional events. An example is discussed in Chapter 11 and illustrated in Fig. 11.6.

Intensive boring of the hardground may reduce it to an apparent rubble of clasts of the underlying rock type in a matrix of the overlying rock type (Fig. 6.11). As a result, a surface of nondeposition may appear to be an eroded,

Figure 6.10 Escape burrow in Pleistocene oolite lime grainstone, New Providence Island, Bahamas.

unconformable contact, and thus to indicate a much greater hiatus in the stratigraphic record. A surface of this type has been described from the Middle Eocene Stone City Formation of east-central Texas by Stanton and Warme (1971). These strata, overlying deltaic sandstone and underlying marine mudstone, were deposited in a shallow, delta-margin setting during the initial phase of the last major depositional transgressive-regressive cycle of the Middle Eocene in the area (Fig. 6.12). Historically, detailed correlation of exposed strata was difficult in the Gulf Coast region because of cyclic deposition and marked lateral facies variations. Consequently, regional disconformities were sought as possible time horizons to establish an improved stratigraphic framework. The upper surface of the Stone City Formation was interpreted as a regional disconformity (Stenzel, 1935) on the basis that the surface was uneven and that clasts lying upon it were composed of the underlying material. Therefore the surface appeared to represent a period of subaerial exposure, lithification, and erosion of sufficient magnitude to be of

Figure 6.11 Intensively burrowed surface, Eocene, Alabama, showing two generations of burrows (1—early; 2—late) and very irregular resulting surface.

regional extent. Analysis of these features in terms of trace fossils, however, leads to the alternate conclusion that apparent clasts are the lithified fillings of burrows or borings that are more resistant to weathering than the surrounding ground mass. It is not evident whether the surface was a hard ground or was burrowed intensively during a period of slow to nondeposition. The apparent lithoclasts and surface unevenness are clearly only the normal effects of biologic activity, either burrowing or boring, or both.

ORGANISMS AS AGENTS OF SEDIMENTATION

The role of organisms in modifying primary sedimentary structures and the sediments themselves may be profound. Much of the discussion so far has focused on the traces formed by the activity of organisms. The stated or implied corollary has been that primary sedimentary structures were being destroyed concurrently. The activities of organisms may not leave distinctive traces but may be more pervasive and sedimentological in nature. Because they are primarily of sedimentologic interest they are discussed only briefly.

Stratification

The formation of shell lag-layers through the activity of burrowing organisms has been described in the preceding section. This is a process that may be

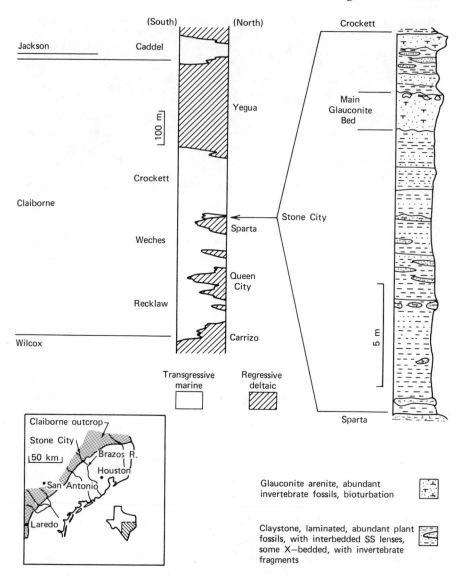

Figure 6.12 Stratigraphy of the Stone City Formation, Eocene, southeastern Texas.

common but has seldom been incorporated in the study of fossil shell concentration.

More widely recognized are the binding and trapping effects of organisms. The foremost example is the deposition and stratification of carbonate sediments as stromatolites through the action of filamentous blue-green algae (Chapter 2, and Walter, 1976). Sediments are trapped and held by a wide range of other organisms as well, however. For example, Fager (1964) has described how a marine tube-building polychaete worm and small anemone colonized a sandy sea floor, stabilizing the sediment against the force of the wave surge. Stratification may also result from biological sorting and re-deposition of a previously poorly sorted sediment. An example of this process has been described at Mugu Lagoon in California by Warme (1967). As burrowing organisms move through the sediment and redeposit the excavated material onto the surface, tidal currents sort out the finer fraction of the excavating sediment and carry it onto the higher tide flats, where it settles out and is stranded. As a result, the remaining sediment in the bay and lower tide flats becomes progressively coarser and better sorted. The resulting stratification is the typical fining-upward sequence of bay to intertidal to marsh sediments.

Texture

As just described for Mugu Lagoon, sorting can be improved by organisms resuspending sediment for subsequent winnowing by waves and tidal currents. The opposite effect of decreasing sorting is probably more common. This is accomplished by burrows cutting across and mixing beds of different textures. This mixing is evident from the fact that the filling or lining of burrows is texturally different than the surrounding sediment, and from the swirled, churned mixture of different textures in thoroughly bioturbated sediments. It is probable that many poorly sorted mud-sand rocks are the result of mixing of original thin distinct beds of mud and sand by organisms.

Sediment Destruction and Construction

Burrowing and boring both destroy and create sediments. Sediment passing through the gut of sediment swallowers may be dissolved by the acidic gastric juices so that grain size reduction and net loss occur. Mayor (1924) estimated that holothurians swallowing carbonate sediment in Pago Pago Harbor dis-solve about 400 g of sediment per square meter each year. Mechanical abra-sion in the guts of organisms may also be an important process of sediment destruction. Boring and scraping by a wide range of animals, fungi, and algae break down and recycle large amounts of loose sediment and rock (Warme, 1975; Golubic, Perkins, and Lukas, 1975). MacGeachy and Stearn (1976) have estimated that the volume of coral skeleton removed by the boring activity of microorganisms on living corals of the reefs of Barbados ranges from 3 to 60% and averages 20%.

These processes both destroy sediment and generate fine sediment from coarser sediment. They also may generate coarser material from originally finer material. For example, Ahr and Stanton (1973) have described ledges of beach rock in Puerto Rico, skeletal lime grainstone, which are being bored by the pedunculate barnacle *Lithotrya*. On the one hand, the mechanical boring produces lime mud from a lime grainstone. However, on the other hand, the boring weakens the ledges of beach rock so that they break readily into cobble-to-boulder-size clasts (Fig. 6.13). Another example of grain-size increase by biological action is provided by both deposit and suspension feeding organisms as they accumulate fine clay-size sediment into fecal pellets, which are coarser and have different sedimentologic properties than the original sediment (Pryor, 1975).

Chemical Effects

Organisms modify the chemistry of sediments, removing organic material as they feed within the sediment and adding organic material as fecal pellets, as organic burrow linings, and as carcasses from dying within the sediment. The clay minerals swallowed by crustaceans may be modified in the gut and excreted as fecal pellets of a different mineralogy (Pryor, 1975).

The more general effect of organisms is to create within the sediments a well-mixed open chemical system. As they burrow they oxygenate the sediment by casting buried sediment onto the surface where oxidation can occur. In the process, however, they bury organic material lying on the surface. The

Figure 6.13 Fragment of beachrock, composed of skeletal lime grainstone, from Isla Icacos, Puerto Rico. Sample is in original position, with borings of *Lithotrya* extending upward from lower surface. After Ahr and Stanton (1973), Journal of Sedimentary Petrology, Society of Economic Paleontologists and Mineralogists.

net effect may be difficult to predict, but chemical reactions and change within the sediment are promoted and maintained long after deposition by the constant stirring and mixing (Schink and Guinasso, 1977).

TRACE FOSSIL ABUNDANCE

Much of the analysis of trace fossils has been focused on relating particular trace fossil forms and assemblages and causative behaviors to particular environmental conditions. General aspects of the abundance of trace fossils are also important. The rate of bioturbation observed on the sea floor greatly exceeds the normal low average rate of sedimentation. Thus it is not surprising that marine sediments are generally thoroughly churned. Although to talk of an average rate of sedimentation may obscure the fact that sedimentation is generally an episodic process, as discussed earlier in this chapter, burrowing may extend so deeply and rapidly that even in areas of more rapid sedimentation intensive bioturbation throughout the sedimentary column is the rule.

In environmental terms, too, bioturbation is nearly ubiquitous. Along common environmental gradients of increasing stress and decreasing diversity, soft bodied organisms, and thus traces, tend to remain although the shelled invertebrates may have dropped out. In places where trace fossils are absent, either deposition must have been very rapid or the environment was so exceedingly unfavorable that all benthic organisms were excluded. Examples of trace abundance being controlled by rapid sedimentation would be beach deposits and turbidites. On beaches, reworking and redeposition are continuous so that even if traces are formed, they are soon destroyed. Turbidites are deposited rapidly and generally in deep water, where burrowing is less common than in shallow water. Examples of trace abundance being controlled by limiting environmental conditions are evaporites and diatomites. Evaporites are clearly deposited in an environment that is outside the tolerance for benthic organisms. The reason for the absence of trace fossils in laminated diatomites is not immediately evident because the abundance of organic material should be advantageous. It is probable, however, that the organic material has depleted the available oxygen in and perhaps on the substrate. In addition, the diatomaceous ooze probably would not have provided a substrate firm enough for macroorganisms.

Absence of oxygen appears to be the most common reason for lack of bioturbation. In the silled basins of the Southern California borderland, water circulation is limited below sill depth, available oxygen is depleted, macrobenthic organisms are lacking, and the microbenthic assemblage is of low diversity and consists of dwarfed and deformed individuals along with a normal planktonic and nectonic assemblage. Consequently, the sediments are laminated and trace fossils are absent. Harman (1964) has described how occasionally the water below sill level turns over and is replaced by normal sea water. Then a normal benthic fauna flourishes and the sediments are burrowed, but conditions gradually revert as the oxygen level decreases.

Both the presence and absence of biogenic sedimentary structures are of paleoenvironmental significance. The distribution patterns of trace fossils will increase in value for paleoenvironmental reconstruction as we are able not only to describe them but to explain them in behavioral-environmental terms. In this context they are relatively time-independent criteria for paleoenvironmental reconstruction. Yet we need to know more about the effects on the trace fossil record of evolutionary changes in trace-making organisms and in the evolution of behavior through geologic time (Seilacher, 1977).

7

Fossils as Sedimentary Particles

During life and after death until burial, organisms and their skeletal remains are subject to the fluid-dynamic forces of the environment. For marine organisms, these forces are the result of water motion. In the terrestrial realm, they are the result of water motion in lakes and rivers and of the flow of air over the surface of the earth. These forces are an integral part of the environment and exert a broad control on the biota. For example, they determine in part the location of trophic resources in the marine environment, whether organic matter is winnowed out of the substrate and maintained in suspension in the water column in high-energy settings or settles out of suspension and is on and in the sediment in quiet-water settings. Thus they determine in part the trophic structure of the community. As another example, they help to determine the behavior of organisms and thus the characteristics of the resulting traces.

The fluid-dynamic forces are particularly important in determining the morphology of organisms. In the living organism, the fluid dynamics of the environment may result in three types of adaptive morphology. (1) *Stability adaptive*—the morphology of the organism minimizes the impact of the fluid-dynamic forces so that the organism can maintain a stable living position in an exposed, high-energy setting. Examples would be the streamlining of terrestrial plants in windswept locations and the change in coral morphology along the shallow-to-deep-water or exposed-to-quiet-water environmental gradients (see Chapter 2). Organisms living in high-energy settings and thus likely to be swept away, disoriented, or buried may be firmly attached to the substrate, be able to reorient themselves, or be able to maintain the preferred living position by moving up and out of the sediment if they are buried by episodes of rapid sedimentation. (2) *Mobility adaptive*—the morphology of nektonic organisms involves a more active response, enhancing the organisms mobility, as in the streamlining of fish and cephalopods. (3) *Suspensibility adaptive*—conversely, in planktonic organisms such as diatoms, foraminifers, and pollen, morphologic structures may increase drag and thus decrease settling velocity.

The fluid dynamic condition of ancient environments has generally been reconstructed on the basis of lithologic characteristics, but the fossils may be equally or more informative for this purpose because they comprise an autochthanous population of clasts that potentially possesses predictable initial characteristics. Terrigenous clasts, in contrast, are characterized by unpredictable initial properties determined by the extraneous characteristics and prior history of provenance, weathering, and transportation. The fluid-dynamic aspect of an ancient environment, like that of a modern environment, of course was complex, consisting of not only the usual conditions, but also the relatively rare but effectively dominant high-energy events.

Three attributes of an assemblage of fossils or of individual species or specimens are more susceptible to the effects of fluid dynamics than others and consequently are most useful in inferring environmental conditions from the fossil record. They are (1) the condition of the fossils (the extent of skeletal disarticulation, fragmentation, and abrasion), (2) textural properties (sorting or size frequency distribution), and (3) fabric (orientation and spatial distribution pattern within the sediment). Much of the material covered in this chapter falls within the realm of biostratinomy, that part of taphonomy which is concerned with the depositional processes resulting in the incorporation of fossils in the sedimentary record.

The wide range of adaptive characteristics clearly demonstrates the importance to the living organism of the fluid-dynamic aspect of the environment. The analysis and interpretation of these characteristics in order to determine this aspect of the environment is a major topic in paleoecology and is discussed extensively in several chapters. Beyond the extent to which they are able to cope with the fluid dynamics by morphologic adaptations and by adaptive strategies of their life histories, however, organisms and their skeletal remains are passive sedimentary particles (Müller, 1979). The object of this chapter is to consider how fossils can be analyzed as sedimentary particles in order (1) to determine the fluid-dynamic aspect of the environment and (2) to recognize the amount of transportation, sorting, and selective shell destruction that has occurred, so that the fossil assemblage can be related to the original community.

THEORY

The interpretation of paleontologic characteristics to determine the fluid-dynamic nature of the ancient environment depends on two assumptions. (1) The characteristics of a fossil assemblage are those of the original biocoenosis, as subsequently modified by numerous processes during the life of the individual organisms, through death, burial, sedimentation, diagenesis, exposure and weathering, collection, preparation, and identification. (2) To determine the effects of the fluid dynamics we can isolate them from the effects of all these other processes. For example, in the analysis of sorting or size-frequency distribution of a marine invertebrate species, we would ideally start with a model of the size-frequency distribution determined solely by the population dynamics of the species. Then the subsequent effects on the initial size-frequency distribution of the various taphonomic processes other than fluid dynamics would be incorporated into the model. This would result in a new model of the size-frequency distribution to be expected under the specified conditions of population dynamics and taphonomic processes, but in the absence of fluid dynamics. Only then could the differences between this model and the actual observed size-frequency distribution be evaluated and interpreted in terms of the fluid dynamics of the paleoenvironment. This logical sequence is seldom explicitly stated and the model used as a basis of comparison is generally not presented in any detail. For example, the interpretation of a well-sorted population of fossils commonly assumes that a size-frequency distribution of this type is not to be expected except by some fluid-dynamic sorting (Boucot, 1953). As another example, the interpretation of fossil orientation only rarely proceeds from a rigorous description of the expected or probable orientation of the living organisms (Toots, 1965). However, an analysis that does not specify the presumed original orientation and then compare the fossil orientation to it lacks a logical basis for interpretation and may well be erroneous if not essentially trivial.

The mechanics and effects of fluid motion on skeletal material must be known in order to interpret the fossils to determine this parameter of the ancient environment. Commonly, in studies of marine invertebrate fossils, the "water energy" of the environment is described in approximate and qualitative terms. Considering the significance of this aspect of the environment, however, "water energy" is overly simplistic. Present understanding of the effects of water motion and transportation on skeletal material is based largely on tumbling mill experiments and on flume studies in which the parameters of water flow, substrate texture, and bed form have been related to the movement and stable orientation of skeletal particles of different size and morphology. These studies have clearly demonstrated the effects of unidirectional flow in sorting and orienting skeletal material as will be described in the following section.

"Water energy," however, is more complex than can be adequately modeled by unidirectional flow in a flume or by a rotating barrel. The laboratory experiments may satisfactorily duplicate conditions of a fluvial or tidal channel, the flow of a longshore current, and perhaps even the on-and-off flow of water on a beach. They are not satisfactory as models of the more-or-less gently oscillating water movement in offshore settings. They are also inadequate representations of the turbulence of the breaker zone. Ideally, an interpretation of the fossil record to determine "water energy" would be based on an understanding of the different components of fluid flow in nature, of their relative magnitudes in different habitats, and of their relative effects on skeletal material. This information is not presently available in these terms. Instead we are beginning to be able to describe characteristics of skeletal material from a range of environments. An example is the pioneering work by Lever (1958) and subsequent amplifying studies such as by Behrens and Watson (1969) on the transport of shells on a beach as it is controlled by several parameters of shell shape.

The ingredients of the ideal working model that would permit us to describe the fluid-dynamic aspects of the environment consist of (1) knowledge of populations and community characteristics in the absence of fluid dynamics and (2) knowledge of how skeletal material is modified by fluid flow. Because neither of these is presently available in detailed and quantitative terms to form a strong theoretical basis, the analysis of fossils as sedimentary particles currently follows the more pragmatic approach of accumulating information about fossil characteristics in sediments deposited under fluid-dynamic conditions as determined from other evidence. From these descriptive data are emerging the generalities described in the following sections.

CRITERIA FOR RECOGNIZING FLUID-DYNAMIC ACTION

The characteristics of the fossil population or assemblage that are most likely to be affected by the fluid-dynamic aspect of the environment are skeletal condition, texture, orientation, and fabric within the rock. These are discussed in this section in order to establish criteria useful in paleoenvironmental reconstruction. They are summarized in Table 7.1. As each criterion is discussed, processes other than fluid flow that may lead to the same characteristics are also examined.

SKELETAL CONDITION

Fluid flow, in the course of moving the skeleton or the sediment around it, will tend to break and abrade the skeleton, and to disarticulate it if it is composed of several components. The extent of abrasion, breakage, and disarticulation is generally correlated with the magnitude of the fluid forces at the site of deposition or with the amount of post-mortem transportation the skeleton has undergone during one or more depositional cycles. The correlation is only qualitative, however, with numerous exceptions, because the effect on the skeleton is determined both by the magnitude and by the duration of the fluid-dynamic process. As an example of the inconsistencies that arise if effect is simply equated with distance of transport, living molluscs eroded from the shallow sea floor off the Texas coast during a hurricane but quickly redeposited on the beach, still alive, are in perfect condition though transported into setting very different from that in which they had been living. In contrast, the bivalve *Donax* lives on the beach at the waters edge, but shells found on the

Table 7.1 Paleontologic criteria for evaluation of fluid dynamics

Characteristic	Interpretation
Fragmentation	Strongly produced by fluid-dynamic processes but fragmentation by biologic processes is so pervasive that it is of limited value
Abrasion	Rounding of edges and loss of fine sculptural detail is predominantly by abrasion. Therefore it is an important means of evaluating the fluid dynamics
Disarticulation	Biologic processes are so effective that articulated skeletons are very rare. Therefore it is of limited value. Articulated skeletons reflect no agitation and, more importantly, no disturbing biota
Differential preservation of skeletal parts	Common result of differences in hydrodynamic equivalences of skeletal parts. Therefore it is an excellent criterion of fluid dynamics if differential destruction by crushing during lithification, by predators, or by solution can be eliminated
Texture	Size frequency distribution (SFD) of individual species in assemblage is so strongly controlled by population dynamics that it is of little value in determining fluid dynamics. The size frequency distribution of the total assemblage is useful
Fabric	Original life positions and post-mortem orientations are diagnostic of the fluid dynamics. They may be destroyed by predators and scavengers and by bioturbation. Therefore their absence is not diagnostic of an absence of water energy
Lithologic data	If the sediment within a fossil is different from that surrounding it, it is strongly suggestive of reworking. Sedimentologic-stratigraphic data provide confirmation and constraints to paleontologic interpretation of fluid dynamics

beach, very close to the life site, may be disarticulated and badly worn by constant washing to and fro in the surf. In spite of such complications, the general rule is valid and skeletal condition is recognized as a basic criterion for determining if the individuals in an assemblage of fossils are "in place" (Fagerstrom, 1964).

Articulation

Skeletons are preserved in complete, articulated condition only under unusual conditions of rapid burial or of no physical or biological disturbing forces. Normally, progressive decay and decomposition of connective tissue after death is inevitably followed by disarticulation. Schäfer (1972) has described the process in detail for marine organisms as they lie on a tidal flat or float in the water. Rapid burial before the connective tissue has weakened is the principal means by which articulated skeletons are preserved. For unburied organisms, both active and passive biological processes as well as physical ones will disarticulate and scatter the skeleton.

The general tendency then is to assume that articulated skeletons have undergone little or no transport before burial. For many examples involving marine invertebrates, individuals also in living position or still attached if epifaunal confirm this conclusion. The converse conclusion, however, that disarticulated organisms have been transported, is clearly not valid. Predators and scavengers may separate and scatter the skeletal parts; bioturbation after burial may carry the process even further.

Articulated fossils may have been transported if the movement was just before or soon after death, when the skeletal parts were still held firmly together by tissue. In the example described above, of hurricane-transported bivalves, the organisms were still alive. An example of transportation soon after death has been described by Lawton (1977) for the vertebrate fossils at Dinosaur National Monument, Utah.

Preservation in articulated form, whether at the life site or after transportation, requires rapid burial as is suggested in the case of the dinosaurs, and is the common explanation. The alternative is deposition and slower burial in a quiet environment in which organisms that would scatter the skeletal parts are absent. A classic example of the latter condition is the biota of the Jurassic Solnhofen Limestone of southwestern Germany. The excellent preservation of whole fish skeletons and other organisms in the platy limestones of the Solnhofen-Eichstätt area has been explained by a cyclic model (Keupp, 1977). Deposition of the Solnhofen Limestone was in a broad lagoon separated from the open marine conditions of the Tethys of southern Europe by a more or less continuous reef sill. At times of open connection between lagoon and ocean, marine nektonic and planktonic organisms entered the lagoon and lived there. Subsequent isolation from the open ocean, however, led to hypersaline conditions and death of the marine organisms, which settled to the bottom. Their excellent preservation is due to the absence of benthic scavengers or burrowers that would disturb them and the growth of a coccoid blue-green algal mat that blanketed them on the lagoon floor. The sedimentation cycle began again when improved connection with the open ocean was reestablished and the lagoonal water column turned over.

Fish are also preserved in excellent condition in laminated diatomaceous sediments. Benthic organisms that would disturb the skeletons and sediment are apparently absent for two reasons. Because of rapid deposition of diatoms, the organic content in the sediment is very high. Thus oxygen in the sediment and in the water immediately above the substrate is lacking. The second probable cause for a lack of benthic fauna is that the diatom ooze has a very high water content and thus is very soft, with insufficient bearing strength to support scavengers.

Fragmentation

Experiments in which shells are tumbled in a barrel demonstrate that shells break as a result of the impacts that are also typical of transportation or turbulence in a surf zone (Chave, 1964). The nature of shell breakage, caused by fluid-dynamic processes and controlled by shell characteristics, is documented for recent shells on the northwest Dutch coast by Hollmann (1968). Shell fragments collected on the sandy beaches after a storm were presumably formed by the impact and grinding effects of being tumbled in the surf. By comparing the fracture lines in the different molluscs present on the beach (Fig. 7.1), Hollmann concludes that the morphology of shell fragments is independent of specific processes of breakage, but is controlled by the shell features of solidity, the concentric growth lines and seasonal rings, and sculptured detail. The crystalline microstructure and the amount of organic matrix also help determine the strength of the shell as the organic matrix decomposes, and thus the size and shape of the particles into which it breaks (Ginsburg, 1956).

Shell breakage, however, may also be the result of processes other than those of fluid dynamics. One of these is postdepositional breakage during

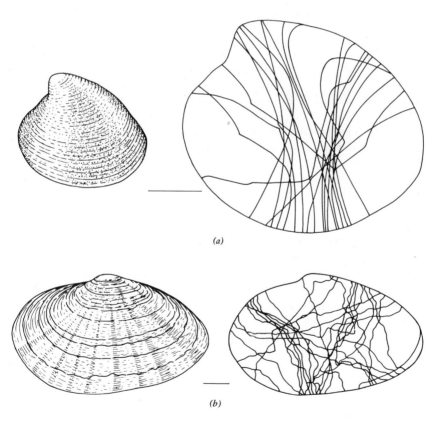

(a)

(b)

Figure 7.1 Fracture lines in two recent bivalve species from the northwest Dutch coast. Note the differences in fracture pattern and the correlated differences in shell morphology. Bar length: 1 cm. (*a*) *Venus gallina*, 20 examples; (*b*) *Mya arenaria*, 25 examples. After Hollmann (1968).

sediment compaction. Experimental work by Brenner and Einsele (1976) has provided quantitive measures of shell strength and has shown that strength is correlated with parameters of shell morphology such as solidity, and curvature. Shell strength is also determined by microstructure. From limited data, the strength in general is correlated with the size of the microstructural unit in the shell (Taylor and Layman, 1972).

Shell fragmentation by compaction should be easily recognized in the fossil record because the fragments should be together in the rock approximately in their relative position. Fragmentation of a shell floating in the sediment would probably not occur. Breakage is only likely in a coquinoid layer, in which shells are in contact and form points of impact with one another, or when the grain size of the sediment is large relative to the shell and thus cannot flow around the shell during compaction. Consequently, shell fragmentation by sediment compaction is not an important process on the basis of theoretical considerations and experimental compaction studies of shell-bearing carbonate sediment (Shinn et al., 1977).

Skeletal fragmentation is most generally caused by biologic processes. Predators are most important. Crustaceans belonging to several groups, for example, feed on bivalves by chipping away the shell margin to reach the tissue within, and feed on gastropods by chipping away the aperture margin in order to reach into the shell far enough to grasp and pull out the gastropod body (Papp et al., 1947). Numerous marine vertebrates crush invertebrate shells as they feed. For example, skates and rays crush bivalves with teeth specially adapted for that purpose. Ducks crush bivalves and gastropods in their gizzard (Trewin and Welsh, 1976) and shore birds have been observed pecking at the tests of sand dollars, breaking away the dorsal surface to expose the tissue within (Fig. 7.2). Fragmentation by predators is controlled by skeletal morphology, so that the likelihood of breakage and location of fractures are specific for different genera. In addition, morphologic features in gastropods such as low spire, narrow aperture, varices, and aperture-rim thickening will reduce the predation and shell breakage by crabs (Papp et al., 1947; Vermeij, 1978; Fig. 7.3).

A less dramatic but more pervasive biologic process is the boring by algae, worms, sponges, and a host of other organisms that weaken skeletal material and thus make it more susceptible to mechanical breakage (Warme, 1975).

Differences in fragmentation by fluid-dynamic and biologic processes should be recognizeable in the fossil record. Comparison of the fragments in Figs. 7.1, 7.2, and 7.3 is a starting point but it may be that some of the fragments in Fig. 7.1 were also caused by predation.

Under the usual conditions of slow deposition and thus of long exposure of skeletal material to biological processes such as those described, skeletal parts will almost certainly be separated, widely scattered, and selectively destroyed by predators and scavengers. Because disarticulation and fragmentation may occur from biological causes alone, they cannot be simply equated with water energy or amount of transportation. Caddee (1968), in his study of the

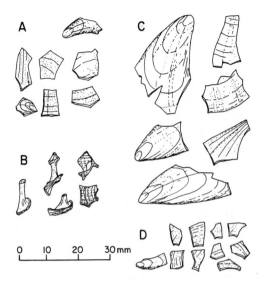

Figure 7.2 Shell fragments produced by bird predation on *Mytilus edulis* and *Littorina littorea*. (*a*) Mussel fragments; (*b*) *Littorina* fragments from eider duck excreta; (*c*) mussel fragments; (*d*) mussel fragments from gull excreta. After Trewin and Welsh (1976).

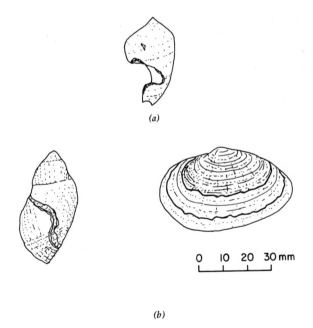

Figure 7.3 Shell fragments produced by crab predation. (*a*) Fatal predation on gastropod; (*b*) nonfatal predation on gastropod and bivalve and subsequent repair.

modern organisms of the Ria de Arosa, Spain, provides a graphic example of this fact. Shells on the beach there are entirely disarticulated and fragmented. The normal inference would be that this resulted from the constant agitation and tumbling as the shells were transported on to the beach and then remained in the surf zone. However, shells found in mid-bay, on a muddy substrate and far removed from wave agitation, are equally disarticulated and fragmented. Primarily crustacean predators were responsible for the shell condition in mid-bay and probably for the fragmentation and disarticulation of the beach material as well.

Rounding

The rounding of skeletal parts is generally explained by abrasion. As such it can be diagnostic of movement, tumbling, and wear in a high-energy environment. In contrast, a skeleton lying undisturbed where it had lived should have angular edges and fine sculpture well preserved. Semiquantitative measures of abrasion rates for different kinds of marine invertebrate skeletons are available from experimental studies using tumbling barrels (Chave, 1964) and field conditions (Driscoll, 1967). The results of these studies are that rate of abrasion is determined (1) by the relative sizes of skeletal particle and sediment grain and (2) by skeletal microstructure. The first relationship is demonstrated by the relative durability of large and small *Spisula* shells in Fig. 7.4. The shells that are smaller relative to the sediment are more likely to be broken as well as abraded. Correspondingly, as grain size increases, the rate of abrasion increases for all the kinds of skeletal material. The second relationship is illustrated by the durability of the different kinds of skeletons. The most durable skeletons are dense and fine grained—the gastropod *Nerita* and the bivalves such as *Spisula* and *Mytilus*. Intermediate forms are the corals, with hard but moderately porous structure, and oysters,with organic-rich, coarsely crystalline structure. The least durable skeletons are the very porous and organic-rich echinoderms and algae.

Although the relationship of abrasion and fluid dynamics is generally valid, a number of exceptions must be considered. As described above, in the case of shells transported during hurricanes along the Texas Coast, a storm can easily move shells several miles within a very short time and deposit them in pristine condition. In addition, shell margins may be rounded by the boring action of algae and other organisms, both by removing shell material and by weakening the shell to facilitate mechanical abrasion (Perkins and Tsentas, 1976).

TEXTURE

A transported assemblage of fossils should have the same textural characteristics as inorganic particles with a similar transportational and depositional

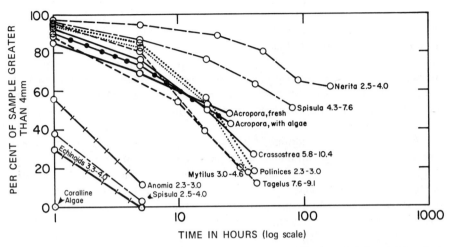

Figure 7.4 Durability of invertebrate skeletal material in a tumbling barrel also containing chert pebbles. Numbers following each name are initial size range in cm. After Chave (1964).

history. That is, the size distribution and sorting should be determined by the nature of the fluid dynamics of the environment and by the size, shape, and density of the fossils. Under particular conditions of water energy, the susceptibility of fossils to transportation or to local movement and resulting abrasion is determined by three factors: (1) size—in general the smaller fossils will be more readily moved, (2) density—less dense, more bouyant fossils should be more readily moved, (3) shape—rounded shapes are more readily rolled on the sea floor and thin flat shapes are more readily put into and maintained in suspension. Allowing for these parameters of hydraulic equivalence, a specific level of water energy operating on the individuals of a taxon should winnow out those that are most easily transported and leave behind a residual population. The fossil assemblage then should have an overall size distribution and sorting determined by the size and sorting characteristics of each taxon present. Some taxa may be relatively unmodified whereas others may be completely removed. The response of the organisms, of course, would be compatible with that of the associated sediments forming the substrate for the living organisms and forming the matrix of the final deposit. That is, the size distribution of the fossils should be in hydraulic equilibrium with that of the sediment.

The interpretation of size and sorting in a fossil assemblage depends on knowing the original size-frequency distribution and on taking into account other natural sorting processes. The size-frequency distribution in natural populations is discussed in detail in Chapter 8, and is reviewed here only briefly. In a population of living organisms the size-frequency distribution depends on the rates of recruitment (birth), growth, and mortality through time. Because of the wide range of potential values for these parameters, a wide range of size-frequency distributions is possible for populations. The composite of these for all the species present in the community will probably encompass an even wider range.

Although the size-frequency distributions of the original populations and community generally cannot be described rigorously, many studies have assumed that a typical or standard size-frequency distribution exists and have analyzed populations relative to it. This assumed size-frequency distribution is negatively skewed, with a predominance of small individuals because of high juvenile mortality. In contrast, it is assumed that this initial distribution should be modified by fluid dynamics toward a normal distribution because of preferential winnowing of the small individuals (Fagerstrom, 1964). Initially this criterion appeared to be useful for recognizing fluid dynamic facies. However, both field studies of natural assemblages and computer simulations (Craig and Oertel, 1966) have demonstrated that differences in the recruitment, growth, and mortality parameters of population dynamics in themselves can generate a wide range of size-frequency distributions. Many additional size-dependent factors can modify the initial distribution: (1) predation, (2) mineralization and solubility, (3) strength to resist crushing during sediment compaction, and (4) collection, preparation, and identification procedures. The size-frequency distribution of the fossil assemblage also depends on whether the fossil assemblage was formed by some catastrophic means and represents a census assemblage or by the gradual accumulation of individuals as they died and represents a normal or cemetary assemblage. Finally, size and sorting may differ because of general environmental conditions. "Dwarfed" or micromorph faunas, composed of smaller-than-normal individuals, are well-studied examples of this. They may be due to transportation, representing the winnowed-out part of a normal community, or they may be due to evolutionary or environmental factors that have limited their size or longevity. Mancini (1978a) has described the analysis necessary to determine the cause of the dwarfing of the fauna and then to interpret the fauna.

DIFFERENTIAL PRESERVATION

All parts of a skeleton should have an equal likelihood of being incorporated into the fossil assemblage in the absence of any post-mortem processes that preferentially destroy or remove some parts more than others. In general, smaller and more fragile skeletal parts are relatively more susceptible to the winnowing, fragmenting, and abrading effects of fluid dynamics. This relation has been refined and applied particularly in the study of vertebrate fossil assemblages in order to determine (1) the impact of predators on the fossil assemblage and (2) the depositional setting (Voorhies, 1969; Behrensmeyer, 1975).

Sorting of bivalve shells on the beach has been described by many workers. The extent of this sorting is indicated by data (Table 7.2) from the Dutch North Sea Coast collected by Lever (1958). The sorting is primarily caused by the mirror-image assymmetry of the two valves. Because of a longshore component to the waves, water flows onto the beach at an angle, but then flows off the beach directly down the slope with a reduced longshore component. The two valves are affected differently by this directional aspect of the water flow. The one that is able to be oriented so that a relatively gentle slope faces both directions of water flow will remain on the beach. The one that is not will be continuously rotated and moved with each wave and be preferentially winnowed and abraded (Fig. 7.5; Behrens and Watson, 1969). Although differences in gross morphology of the two valves are of primary importance, differences in fragility, weight, ornamentation, gastropod borings, and hinge structures projecting out from the valve are also important.

Differential preservation may be determined by selective predation on a particular part of the prey. Naticid gastropods, for example, feed on bivalves by boring a hole through the shell to get to the tissue of the animal. The hole is commonly situated near the umbo, and will be preferentially on one valve rather than the other. The valve generally not bored, being stronger, will be more likely preserved in the fossil record. Birds, too, may cause differential preservation of bivalve shells. The oyster-catcher (*Haematopus ostralegus*) feeds on clams exposed during low tide by breaking the shells with its beak. Results of a study on the north coast of Northern Ireland indicated that the left valves of individuals of *Venerupis, Spisula,* and *Mytilus* were preferentially destroyed (Carter, 1974), so that about two thirds of the remaining valves were right valves.

ORIENTATION

Each organism living on or in the substrate has a specific and preferred orientation. This orientation is likely to be preserved if the organism is firmly attached to the substrate. It is less evident or preservable among unattached or weakly attached organisms. If the life orientation is known, and if it can be recognized in the fossil record, it is evidence that the fossil was a member of the original community at that place and that disturbing and transporting processes were absent or less than those necessary to displace the individual. These processes may be fluid dynamic, biological, or mechanical. Original orientation is rarely preserved unless the organism is firmly attached to the substrate or lives in a more or less permanent burrow well below the sediment surface. If it is preserved, the life orientation may be interpreted to tell about the environmental parameter to which the organism was responding. An illus-

Table 7.2 Percentages of left and right valves of bivalves at three locations on the Dutch North Sea Coast, July–August, 1955[a]

	Texel			Den Helder			Petten		
	%L	%R	Total	%L	%R	Total	%L	%R	Total
Spisula subtruncata	39.7	60.3	1616	65.3	34.7	1948	52.0	48.0	1681
Spisula solida	44.3	55.7	149	60.7	39.3	56	45.6	54.4	57
Macoma balthica	47.9	52.1	328	43.7	56.3	213	48.3	51.7	149
Donax vittatus	41.0	59.0	178	63.0	37.0	370	46.3	53.7	121

[a]From Lever (1958).

OCEAN   BEACH

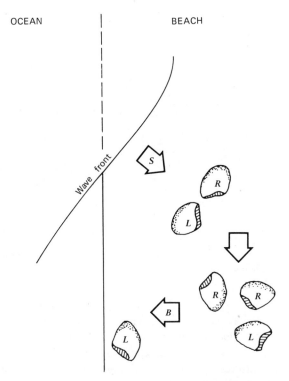

Figure 7.5 Differential sorting mechanism. The stippled portion of each valve is the broad, gently sloping edge that faces the current when the valve is in a stable orientation. In this position the valve is moved only by strong currents. However, when the steep-sided beak (vertically lined) faces a current the valve is unstable and may be rolled or rotated and transported by a relatively weaker current. Thus the oblique swash (*S*) approaching from the right as one faces the sea orients the valves so that the *R* valve is relatively stable with respect to the backwash (*B*), but the *L* valves tend to be carried back down the foreshore. Furthermore, as a result of backwash orientation the beak of the *R* valve points into the next oblique swash which will thus be able to carry it further up the foreshore. Oblique waves approaching from the left would sort the valves in the opposite manner. After Behrens and Watson (1969), Journal of Sedimentary Petrology, Society of Economic Paleontologists and Mineralogists.

tration is provided by Oligocene oysters, *Crassostrea gigantissima*, in North Carolina (Lawrence, 1971b). Living oysters of the species, *Crassostrea virginica* are preferentially oriented so that the plane of commisure is approximately parallel to the tidal currents flowing over the oyster bank. Measurement of the plane of commissure of fossil oysters, then, should indicate the direction of current flow in the ancient environment. This is confirmed by the Oligocene occurrence because the orientation of the tidal channel can be determined from independent geologic evidence. The direction of current flow based on the oysters closely approximates the channel direction (Fig. 7.6).

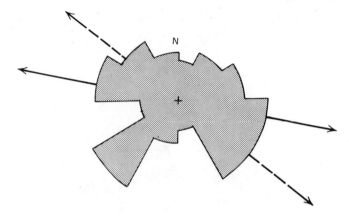

Figure 7.6 Orientation of plane of commissure of 86 Oligocene oysters. These are in a bank that formed in a channel. The dotted arrow is orientation of channel. The solid arrow is vector mean of oyster orientations. After Lawrence (1971b), Journal of Paleontology, Society of Economic Paleontologists and Mineralogists.

The more common phenomenon is not for original orientation to be preserved, but for skeletal parts to be oriented by fluid dynamics whether they had or had not been oriented during life.

Numerous examples have been described from the fossil record of bedding plane exposures scattered with skeletal material more or less clearly aligned. The usual way to display these data is by plotting the azimuth of a linear axis of the fossil on a rose diagram. Reyment (1971) describes the quantitative statistical procedures for analyzing this type of data. In some cases, features of the fossils or traces they made on the substrate indicate clearly the direction of flow. For example, the starfish illustrated in Fig. 7.7 has been shoved toward the top of the slab. This is suggested by the bent-under upper arm tip and the downward bend of the right arm. If only the starfish were used in the interpretation, it might also be suggested that the starfish had been deformed by a current flowing from top to bottom. In this case, however, the drag marks leave no doubt as to the current direction (Seilacher, 1960).

More generally, only the oriented fossils provide information about the fluid dynamics. The correct orientation of these has required flume studies and field observations of present-day material. A notable study in which all of these approaches have been combined is in Devonian strata of Pennsylvania (Nagle, 1967). Typical orientations of the fossils are illustrated in Fig. 7.8 for two brachiopods, an elongate bivalve and pelmatozoan stem segments at the Deer Lake locality in the Mahantango Formation (Nagle, 1967). The interpretation of these fossils was based on a general model developed by experimentation in flumes and natural settings (Fig. 7.9). The bimodal rose diagrams indicate that the fossils were oriented by wave action rather than a unidirectional current. The predominant wave movement was from south-southeast to north-northwest.

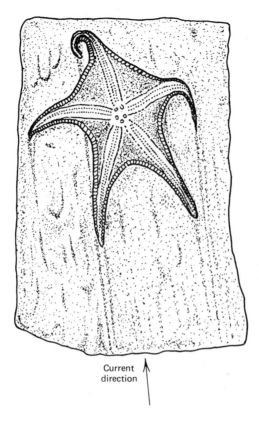

Current direction

Figure 7.7 Drag marks on substrate and bending of starfish indicate current direction was from bottom to top of illustration. After Seilacher (1960), copyright Hessisches Landesamt für Boden-forschung, Wiesbaden.

The movement and orientation of skeletal material has been carried out in a great many flume studies. Through this work, the parameters of skeletal shape and density, substrate texture, and water depth and flow conditions have been investigated. The role of current velocity is demonstrated by the orientations of *Scrobicularia* valves at different velocities (Fig. 7.10). When several shells are near each other on the sea floor they interact and become oriented in the accumulations (Fig. 7.11).

Skeletal parts generally occur on a substrate of soft sediment. Conse-quently, the sediment, as well as the shells, is in motion and plays a dominant role in the orientation of the shells on the substrate and as buried. The topic of fossil orientation is large and complex because of the wide range of factors that may be important. The work of Futterer (1978a and b) provides a com-prehensive survey of results of work in this field.

Orientation of fossils is generally interpretable in terms of the fluid dynamics of the depositional environment. The absence of orientation cannot

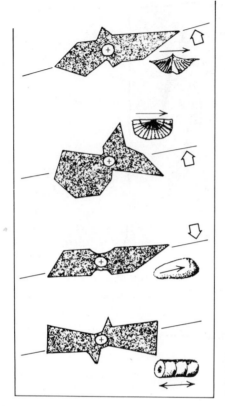

Figure 7.8 Orientations of two brachiopods, an elongate bivalve, and pelmatozoan stem fragments in the Devonian Mahantango Formation, Pennsylvania. Arrow by each fossil is direction measured. Double arrow is inferred current direction. After Nagle (1967), Journal of Sedimentary Petrology, Society of Economic Paleontologists and Mineralogists.

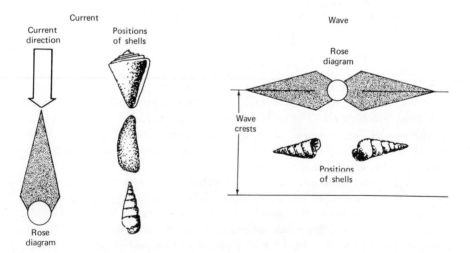

Figure 7.9 Diagnostic orientations of bivalve and gastropods by current and wave energy. After Nagle (1967), Journal of Sedimentary Petrology, Society of Economic Paleontologists and Mineralogists.

316

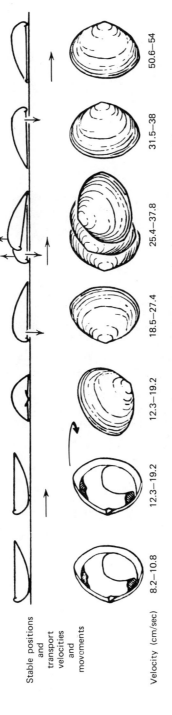

Figure 7.10 Behavior of single valves of the bivalve *Scrobicularia plana* in a water current at different velocities. The velocity is measured 1 cm above the substrate. The arrows indicate the velocity ranges and patterns of motion for shell transport. After Futterer (1978a).

317

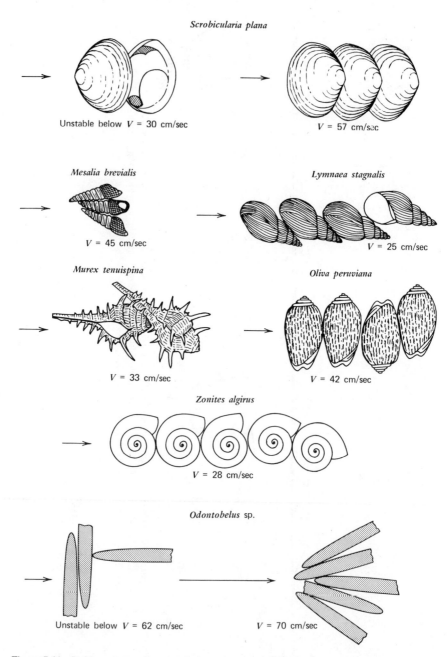

Scrobicularia plana

Unstable below V = 30 cm/sec

V = 57 cm/sec

Mesalia brevialis

V = 45 cm/sec

Lymnaea stagnalis

V = 25 cm/sec

Murex tenuispina

V = 33 cm/sec

Oliva peruviana

V = 42 cm/sec

Zonites algirus

V = 28 cm/sec

Odontobelus sp.

Unstable below V = 62 cm/sec

V = 70 cm/sec

Figure 7.11 Stable patterns of accumulations of shells of different morphology. The arrow indicates current direction. The current velocity in each case is that at which the figured pattern is attained. The velocity is measured 1 cm above the substrate. After Futterer (1978a).

318

be considered indicative of lack of fluid dynamics. On a soft bottom, scour of the sediment and disorientation during burial may obscure the orientation of shells lying on the surface. Initially oriented shells may be scattered by predators and scavengers after deposition. Skeletons that were originally strongly oriented at the substrate surface are likely to be "randomized" by bioturbation within the substrate. On the other hand, compaction of the enclosing sediments subsequent to both preburial orientation by water currents and to postburial disorientation by bioturbation will tend to rotate inclined fossils into a more horizontal position that might resemble a current-formed planar orientation.

APPLICATIONS

The use of the criteria described in the preceding section is common in paleo-environmental reconstructions. These criteria are important in determining not only the depositional environment but also the extent to which the fossil assemblage is in place and represents the original community (Fagerstrom, 1964). It is important to apply these paleontologic criteria in conjunction with sedimentologic criteria. The results from use of the two types of criteria should be compatible. Thus extensive transport or winnowing of a fossil assemblage is not likely to have occurred if the assemblage were in a very fine-grained sediment. Similarly, a fossil assemblage in a conglomerate bed that clearly required considerable energy for its deposition is probably transported. These examples, of course, may be so obvious as to be trivial. Another approach involving joint evaluation of fossil and sedimentologic data is to analyze the fossils in the context of the regional stratigraphic framework. This provides a strong limitation on the range of possible environmental conditions. The latter approach is particularly emphasized in Chapter 11.

Each of the criteria is suggestive, but not definitive, and therefore not independently diagnostic. Used together, the criteria provide valuable information about the fluid dynamics of the environment and about the taphonomic processes that molded the original community into the fossil assemblage. A detailed and comprehensive use of these criteria has not been made in the analysis of the Kettleman Hills fossils in order to determine the fluid-dynamic conditions of the Pliocene environment there. The case studies described below, from other areas, provide examples of the comprehensive application of these criteria in paleoecology.

Mammal Assemblage of the Arikaree Group (Miocene), Nebraska

Mammal bones are concentrated in a calcareous ashy sandstone several feet thick in localized outcrops in northwestern Nebraska. The Tertiary vertebrate fossils from this region have been important in establishing a mammalian biostratigraphy. To relate the assemblage to the previously established stratigraphic framework, it is desirable to determine the depositional conditions of the assemblage. The fluid-dynamic characteristics are important because they reflect the amount of transportation and the degree of mixing of disparate habitat life assemblages within the fossil assemblage. Hunt (1978) has carried out a comprehensive study of the assemblage that includes many of the criteria described above.

The assemblage consists of several hundred bones, 120 of which are identifiable. They represent a minimum of 13 or 14 individuals belonging to 11 species and 10 genera. None of the bones are articulated, although two individuals may each have contributed several bones to the assemblage as indicated by proximity and similarity of the bones in size and condition. The orientation of the bones are described by three parameters—the bearing, in which the ends are not differentiated; the polarity, in which the azimuth of one end (the larger in this study) is measured; and the inclination. The rose diagram of the bearing of the long axis of 141 bones (Fig. 7.12) indicates a strong but diffuse northwest orientation and a lesser broad maximum in the northeast quadrant. The plot of the azimuth of the large end of each bone provides a much clearer picture of northwest and northeast maxima (Fig. 7.13). The inclination of all the bones is low and not diagnostic. Flume studies and observations in nature have shown that elongate particles tend to be aligned either parallel or perpendicular to water movement (Fig. 7.8; Voorhies, 1969). The equal sizes of the northeast and southwest peaks in the polarity diagram suggest that the direction was normal to current flow. Hunt concludes that the predominant water flow was toward the northwest because flume studies by Voorhies have shown that bones aligned in the flow direction have a strong tendency to lie with the large end downstream.

Bone abrasion was estimated on a subjective scale. In general, the amount of abrasion is intermediate. A decrease in abrasion upward in the section suggests that the water energy and/or amount of transportation decreased with time. In addition to the general level and to the vertical gradient in abrasion,

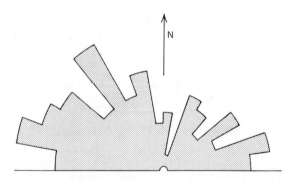

Figure 7.12 Bearing of long axis of 141 bones from Harper Quarry, Nebraska, Miocene. After Hunt (1978).

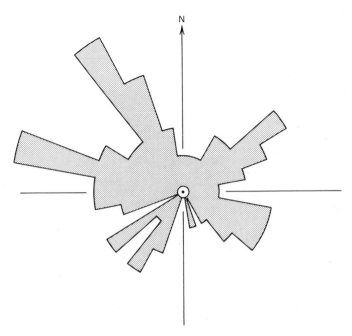

Figure 7.13 Rose diagram of polarity (bearing of large end) of 74 mammal bones from Harper Quarry, Nebraska, Miocene. After Hunt (1978).

the bones of some kinds of animals are more abraded than of others. This type of evidence might be used to distinguish autochthonous and introduced elements in the assemblage.

The different parts of a skeleton, whether invertebrate or vertebrate, have different preservation and transportation potentials. Consequently, the proportion of skeletal parts in a fossil assemblage compared to that in the organisms is a clue to the extent of winnowing and transport, abrasion, and diagenesis. The proportions of skeletal parts in the vertebrate assemblage is portrayed in Fig. 7.14. Ungulate bones constitute 71% of the identifiable bones; 29% are carnivores. No complete skulls are in the assemblage. Ribs are relatively abundant. Most ungulate bones are broken and fragmented.

The complete interpretation of the fossil assemblage by Hunt includes sedimentologic and stratigraphic data as well as paleontologic. The best interpretation in any case is that based on the widest range of information. The orientation of bones in two perpendicular directions and the moderate to slight abrasion indicate that the bones were deposited in shallow water with generally weak currents. The relative abundance of easily transported vertebrae and ribs is also indicative of weak current action. With stronger currents, they would have been winnowed out. The decrease in abrasion upward in the section suggests decreasing water energy as the bed was being deposited.

The conclusion that water energy was low is confirmed by the small amount

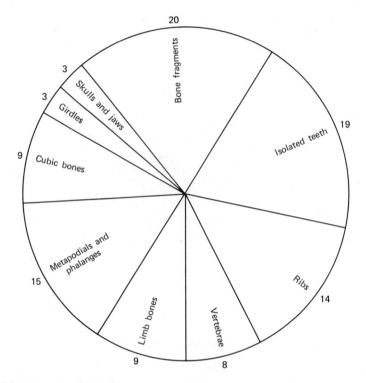

Figure 7.14 Percentages of skeletal elements in 198 mammal bones from Harper Quarry, Nebraska, Miocene. Cubic bones include podials (astragali, calcanea), sesamoids, sternebrae. Girdle elements include scapulae, innominates. After Hunt (1978).

of abrasion of the small and compact bones. Another explanation than water agitation is necessary, then, for the generally fragmented condition of skulls, lower jaws, limb bones, and the other more fragile skeletal parts. The limitation of water energy leads to the conclusion that much of this fragmentation was caused by carnivores.

Permian Bivalves of the Park City Formation, Wyoming

Boyd and Newell (1972) have described in detail a thin bivalve-rich limestone and determined the taphonomic and diagenetic processes that have produced the characteristics of the fossil assemblage. Shell condition and orientation are major clues to the fluid dynamics of the depositional environment.

The fossil assemblage is contained in a limestone layer 13 cm thick. Many of the fossils are silicified, others are preserved as sand casts or coarsely crystalline calcite. The fossils have been subjected to a complex diagenetic history in which formation of molds of the fossils was an important step. The present siliceous and sand casts are fillings of these molds. The replacements of the original fossils faithfully preserve fine details of the original shells so that their condition can be interpreted. Bivalves, represented by more than a dozen species dominate the assemblage; other fossils present include bryozoans, gastropods, scaphopods, trilobites, crinoid columnals, nautiloids, and brachiopods.

The bivalves are largely disarticulated (Table 7.3) and scattered, so that pairs of valves cannot be found. The relative abundance of paired specimens of *Nuculopsis* sp. may be attributed to the numerous interlocking teeth in this form, which might have helped to keep the valves paired after death. At the same time, however, the proportions of right and left valves for the most common bivalve species are not significantly different (Table 7.3). Also, the size range encompassed by populations of each bivalve species is wide and generally normally distributed.

Fragmentation of shells of two of the most common bivalves, *Scaphellina bradyi* and *Oriocrassatella elongata,* were analyzed. Breakage is common in these shells and occurs in all parts of the shell. It is not concentrated in the thin and relatively fragile parts, but in fact commonly is across the umbo and hinge plate. In spite of the high frequency of broken shells, abrasion is slight—generally fracture surfaces are sharp and angular.

The shells appear to be randomly oriented. No preferred orientation in terms of concave-up or concave-down, flat-lying or inclined is evident. Shell alignment as described by Nagle (1967) and illustrated in Fig. 7.8 and 7.9 for invertebrates and by Hunt (1978) in Fig. 7.12 and 7.13 for vertebrates is also lacking.

Boyd and Newell conclude from these data that some effective agent was operating to disarticulate the bivalves and to scatter and break the valves on the sea floor. However, the poor sorting by size and right-left differences in morphology suggest that current or wave energy was too slight to be that agent. The lack of a preferred orientation reenforced this conclusion as did the lack of abrasion. This interpretation of the fossil characteristics to arrive at a description of the fluid dynamics of the ancient environment leads, then, to the

Table 7.3 Articulated and single valves of most common bivalve species in samples from a fossiliferous bed in the Park City Formation (Permian, Wyoming)[a]

Species	Left Valves	Right Valves	Articulated Specimens
Kaibabella sp.	757	747	2
Scaphellina bradyi	557	575	1
Oriocrassatella elongata	390	364	2
Schizodus sp.	225	213	6
Astartella aueri	91	100	1
Nuculopsis sp.	66	68	16

[a] From Boyd and Newell (1972).

suggestion that the fossil characteristics described were the result of predators that disturbed and scattered the shells lying on the sea floor as they searched for and crushed living bivalves. Boyd and Newell propose that fish were the major group of predators. This is based (1) on arcuate "bites" along the edges of some shells and (2) on the common pattern of breakage across the hinge area, indicating a predator large enough to crush the whole articulated living bivalve. Holocephalan fish are likely candidates because their teeth are common in the Park City Formation.

Eocene Molluscs of the Main Glauconite Bed, Stone City Formation, Texas

The fauna of the Main Glauconite bed of the Stone City Formation has been studied in order to attempt to reconstruct the original life community (Stanton and Nelson, 1980). The stratigraphy is illustrated in Fig. 6.12; the paleontology is described fully in Chapter 9.

Analysis of the fauna to assess the fluid dynamics of the environment was necessary in order to derive as much as possible from the fossil assemblage about the original community. In particular, the extent to which the composition and relative abundances of species in the fossil assemblage were the result of transportation or winnowing needed to be determined.

The Stone City Formation consists of two distinct facies: (1) interbedded lenticular cross-bedded and rippled sandstone and brown laminated mudstone with common small chunks of lignite but with few invertebrate fossils and (2) olive-gray glauconite arenite with abundant invertebrate fossils.

The Main Glauconite bed is 1.5 m thick and consists predominantly of ovoidal glauconite grains 0.5 mm long and 0.2 mm in diameter. The grains are authigenic replacements of fecal pellets. The term glauconite is used in a general sense; chamosite is the probable actual composition. The pelleted sediment was thoroughly mixed by bioturbation so that original primary sedimentary structures are not preserved. The matrix of the glauconite sandstone is made up of about equal proportions of clay and fine to very fine, well sorted, angular quartz sand. The basal contact of the bed is sharp, burrowed, and slightly irregular. The upper contact is also irregular because of intensive bioturbation. Continuous bioturbation during deposition formed a patchy churned structure throughout the bed that reflects local differences in pellet and matrix content. In addition, distinct trace fossils in the upper 25 cm consist of abundant *Thalassinoides* and rare *Gyrolithes* that were probably formed after the bed had been deposited (see Chapter 6).

Body fossils are abundant throughout the unit comprising 96 genera and 120 species in a sample of 6616 individuals. The fauna is primarily gastropods and bivalves. Bryozoans, octocorals, fish otoliths, scaphopods, crustaceans, and nautiloids are minor elements.

There are eight characteristics of the fauna that can be used to determine the fluid dynamics of the paleoenvironment. (1) Fragile skeletons and fine sculptural detail are generally excellently preserved. (2) Less than 1.0% of the bivalves are articulated. (3) No fossils are found in a recognizeable living position. (4) Many of the fossils are fragmented, but rounded, worn shells suggestive of abrasion are rare. (5) The total assemblage is poorly sorted, with shell size ranging from microscopic to several inches in length. (6) There is no preferred orientation among the scaphopods (Fig. 7.15). (7) Bivalves are predominantly inclined or concave-down, but all orientations are present; a plot of the inclinations of the shells does not suggest any preferred orientation (Fig. 7.16). (8) The right-left ratios for only 5 of the 21 species are significantly different at the 95% level using the Chi-square test. *Cubitostrea, Crassostrea,* and *Anomia,* three species of "oysters," are represented by an overabundance of the right valve, which is smaller and thinner, and so could be transported more easily than the thicker and heavier left valve. *Notocorbula* and *Vokesula,* two corbulids, are represented, in contrast, by an overabundance of the heavier, thicker right valve.

Current-formed sedimentary structures such as cross-beds and ripples are absent. Terrigenous sand is a minor component of the sediment and is not coarser than fine sand. The glauconite grains were originally fecal pellets and

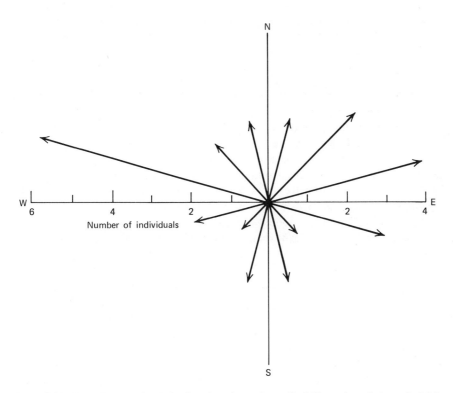

Figure 7.15 Rose diagram of polarity (bearing of anterior end) of 30 scaphopods from the Main Glauconite bed, Stone City Formation, Texas, Eocene. After Nelson (1975).

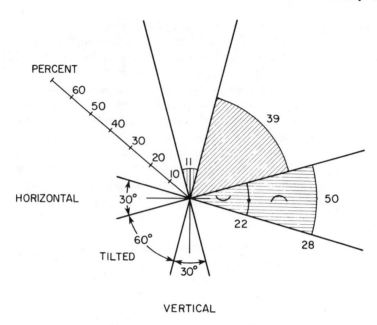

Figure 7.16 Orientation of bivalves in the Main Glauconite bed, Stone City Formation, Texas, Eocene. Radial distance in each sector is percentage of shell with vertical, tilted, horizontal concave-up, or concave-down orientation. After Nelson (1975).

were probably not mechanically strong enough to have survived more than local transportation.

The fluid-dynamic interpretation of these data is at first contradictory. The scarcity of articulation and the abundance of fragmented shells suggest water agitation sufficient to have broken and separated the shells. The right-left proportions suggest that water currents were sufficiently strong to winnow out the more easily transported valves of *Notocorbula* and *Vokesula* and to bring in the lighter valves of *Cubitostrea, Crassostrea,* and *Anomia,* but were not strong enough to affect the more robust species. In contrast, the poor sorting of the assemblage and the absence of shell orientation, of current-formed sedimentary structures, of shell abrasion, and of coarse terrigenous sediment argue for very little water energy.

The resolution of these divergent interpretations must take into account, as in the preceding Permian example, the effects of biologic processes in the creation of the fossil assemblage. Evidence of predation by crustaceans is abundant: both the outer lips of gastropods and the ventral margins of bivalves are chipped. Many individuals bear multiple scars, indicating repeated unsuccessful attack. Thus much of the fragmentation may be the results of predation rather than of agitation by water movement. In addition, this activity of predators would have been effective in disarticulating and disorienting the shells. The intensive bioturbation that occurred probably also disoriented and perhaps disarticulated the shells. We conclude that water energy during deposition of the Main Glauconite bed was slight. Positive evidence for this is found in the fossils and in the sediment texture. The lack of sedimentary structures and fossil orientation provides negative evidence for the same conclusion but is not so strong for it cannot be proved that these features were not present but subsequently obliterated by bioturbation. The magnitude of the currents that were present could potentially be determined by flume experiments like those illustrated in Fig. 7.10 by Futterer (1978a), in which the hydraulic characteristics of the different shells would be studied.

Storm Deposits

Present-day rates of sedimentations are very different from one place to another. Observations over only a very few years also indicate that sedimentation rate is highly variable at a single location. This fact has been emphasized by Wolman and Miller (1960) and by Gretener (1967). They point out that the bulk of sedimentation, like the effects of many other geologic processes, is accomplished during rare and infrequent periods. Ager (1973) has generalized this concept in his broad evaluation of the nature of the stratigraphic record. The Eocene beds described in Chapter 6 as a sequence of crevasse-splay deposits provide an excellent illustration of episodic sedimentation (Fig. 6.8). In this example, each bed was laid down rapidly with sufficient intervening time for the upper part of the bed to be intensively burrowed.

The rare but large geologic event is important in paleontology because strata in which fossils are relatively abundant may form by the normal slow accumulation of shells on the sea floor during times of slow sedimentation or, in contrast, by some rapid process of concentration during infrequent times of elevated fluid dynamics. The latter deposits are described as storm deposits or tempestites. They are probably important in the fossil record but have been overlooked. The paleontologic features discussed in this chapter, however, should be particularly useful in their recognition.

The first distinction to be made is between shell accumulations resulting from storms and those resulting from gradual accumulation during periods of slow deposition of sediment matrix. The paleontologic criteria described above should be sufficient to differentiate the former, high-energy case from the latter, low-energy case.

Shell concentrations may form by storm action in two different ways. If the water energy is not too great, the sediment and shells on the sea floor will be disturbed and the sediment will be put into suspension and winnowed out, leaving behind a lag concentration of essentially autochthonous shells. These have been referred to as shell-lag (Brenner and Davies, 1973). An example studied by Aigner (1977) is described below. Under conditions of greater water energy, the shells may be transported and deposited as allochthonous concentrations. Two examples of this process (Ball, 1971, and Stanton, 1967) are also described.

Shell Beds in Lower Hauptmuschelkalk, Middle Triassic of Southwest Germany

The sedimentary rocks in the Triassic Lower Hauptmuschelkalk in the vicinity of Crailsheim include cyclic sequences of shale and marl at the base, dense lime-stone, shelly limestone, and a shell bed at the top (Aigner, 1977). The shale and marl beds are 5 to 20 cm thick; the limestone beds are 5 to 10 cm thick (Fig. 7.17).

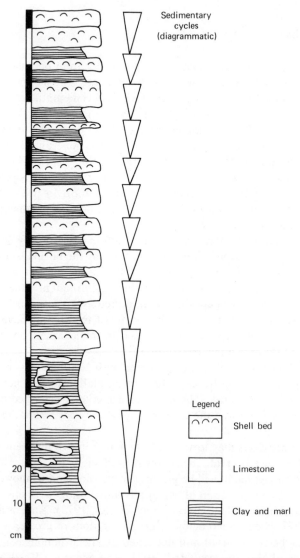

Figure 7.17 Sedimentary cycles in the Lower Hauptmuschelkalk, southwest Germany, Middle Triassic. After Aigner (1977).

The shale and marl beds contain few fossils, which are generally articulated and scattered in the rock as single individuals. The dense limestone overlying the shale-marl beds is characterized by the absence of body fossils but abundant trace fossils. The trace fossils occur along the lower surface of the limestone, and within it as living traces and pellet-filled feeding traces or spreiten. Body fossils may make up 50% of the shelly limestone; lime mudstone forms the matrix of the rock. The fossils are commonly articulated and lithoclasts are rare. The fossils are more concentrated in the shell layer at the top of the limestone than in the underlying rock and may be so concentrated as to form a coquina. The top of the shell layer is an omission surface, representing an interval of nondeposition before the cyclic sedimentation was renewed with deposition of shale and marl.

The interpretation of the lithologic and paleontologic features by Aigner (1977) is illustrated in Fig. 7.18. The overall cyclic pattern is explained by changes in water salinity and temperature. The shell bed at the top of the cycle is considered to be an essentially autochthonous storm deposit (shell-lag) concentrated from the underlying shelly lime mudstone. The evidence that the fauna is autochthonous consists of the bivalves, which are largely articulated, closed, unbroken, and unabraded; the scaphopods, which are presumably autochthonous because they would be difficult to transport; and juvenile ceratite ammonites, which would appear to have lived in the inferred depositional setting.

The cyclic pattern of sediments culminating in the shell bed is attributed to

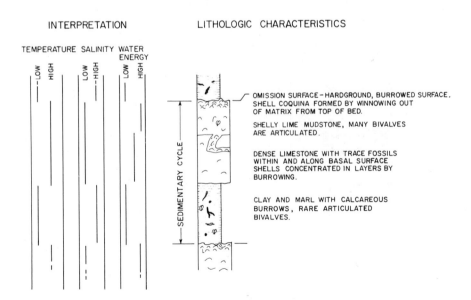

Figure 7.18 Detail of a sedimentary cycle in the Lower Hauptmuschelkalk, southwest Germany, Middle Triassic. Lithology of cycle and shell layer at top is caused by increasing water energy associated with changes in water characteristics and depth. After Aigner (1977).

the environmental changes in temperature and salinity. The increase in water energy at the end of the cycle is presumably the result of a storm. The position of the storm in the cycle may have been determined by the same climatic factors responsible for the change in salinity and temperature. The level of water energy is indicated by the presence of the juvenile ceratites, which could be easily transported, by the moderate degree of scaphopod lineation, and by the general concave-down orientation of bivalve shells. These suggest that the level of water energy was moderate as it suspended and winnowed the lime mud sediment of the shelly sea floor and left a lag pavement of shells.

Storm Deposits in the Westphalia Limestone, Pennsylvanian, of Kansas and Oklahoma

The Westphalia Limestone is a thin (0 to about 5 m) but extensive Late Pennsylvanian unit that crops out in a belt more than 200 miles long between Oklahoma and Missouri and occurs in the shallow subsurface to the west for at least 100 miles (Ball, 1971; Fig. 7.19). The Westphalia Limestone consists in large part of three very different facies. The mixed-fossil lime wackestone is the most extensive of the three, present in outcrop and throughout the subsurface area studied. It is lime mudstone with a diverse assemblage of invertebrates. The

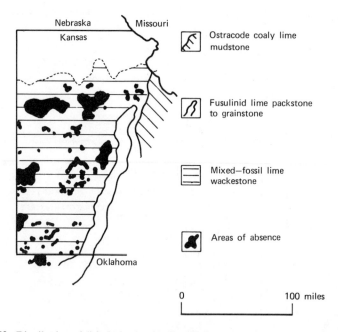

Figure 7.19 Distribution of lithofacies in the Westphalia Limestone, Pennsylvanian of eastern Kansas and vicinity. After Ball (1971), Journal of Sedimentary Petrology, Society of Economic Paleontologists and Mineralogists.

ostracode coaly lime mudstone is a lenticular unit less than 0.5 m thick and limited in distribution to a part of the outcrop belt and the nearby subsurface. The low diversity of the fauna and close association with coal deposits suggest that it is a nearshore, brackish facies. The fusulinid lime packstone to grainstone is a lenticular to sheetlike unit as much as 1.5 m thick. Fusulinids are the dominant fossils, associated with a diverse assemblage of other marine invertebrates.

The fusulinid lime packstone to grainstone is interpreted for several reasons as a storm deposit laid down in high-energy conditions. The fossils are commonly broken and abraded, and are locally current lineated. Single valves of the bivalve *Myalina* are commonly concave-down and concentrated locally in layers interpreted as shell lag deposits. Lime-mud lithoclasts, apparently derived from the immediately underlying strata, are also locally common.

Evidence that the fusulinid lime packstone to grainstone facies is not a shell lag but consists of transported, allochthonous material is derived from the fossils found in it. Most of the fossils are present in lesser concentration in the underlying mixed-fossil lime wackestone. However, the fusulinids and *Myalina* do not occur there or in Pennsylvanian strata in general, and therefore apparently did not live together. It is believed that myalinid clams lived in nearshore to bay, brackish-water habitats whereas fusulinids lived in open-marine, shallow-shelf habitats. These taxa plus the other associated invertebrates suggest that skeletal material from several distinct and geographically separate habitats were scoured from the mixed-fossil lime wackestone facies covering the shallow sea floor, were concentrated, transported, and mixed together by the storm currents, and were finally deposited in localized lenticular bodies of the fusulinid lime packstone to grainstone.

Shell Beds in the Castaic Formation, Upper Miocene, of Southern California

The Castaic Formation consists of sedimentary rocks deposited in the Soledad and Ridge Basins, east of the San Gabriel Fault (Stanton, 1967; Fig. 7.20). In the Ridge Basin the formation is approximately 7000 ft thick and unconformably overlies Paleocene and Eocene strata. In the Soledad basin, it is much thinner because of subsequent erosion as well as probably less deposition, and overlies with slight angular unconformity nonmarine Upper Miocene strata. The San Gabriel fault was active during the deposition of at least the upper half of the Castaic Formation and formed an abrupt southwestern boundary to the basins.

The sediments of the Castaic Formation were deposited along the margin of a transgressing sea, both on an open coast and within an embayment. They consist of three lateral facies (Fig. 7.21): (1) the lower part of the Violin Breccia, composed of poorly sorted sediments deposited along the western side of the basin, adjacent to the active San Gabriel Fault, (2) interbedded mud and sand deposited in mid-basin, (3) a complex of sand and gravel beds deposited as basal sediments along the eastern margin of the basin. The basal strata along the northeastern margin of the basin consist of lenticular and areally restricted sand-

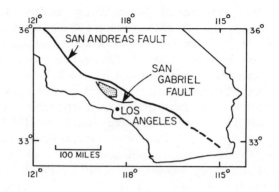

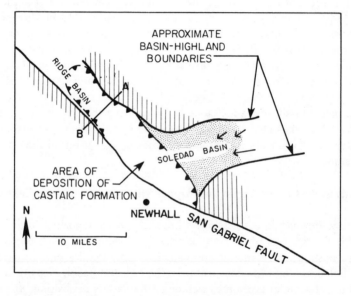

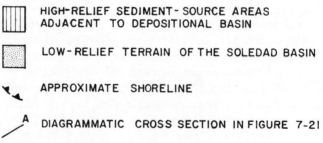

Figure 7.20 Location and local paleogeography of the Castaic Formation, southern California, Upper Miocene. After Stanton (1967).

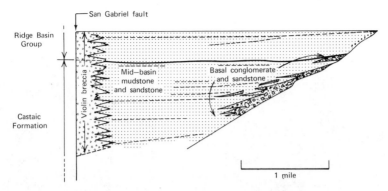

Figure 7.21 Diagrammatic cross-section showing facies relationships within the northern part of the Castaic Formation, southern California, Upper Miocene. Subsequent deformation and erosion are not portrayed. After Stanton (1967).

stone and conglomerate beds that lap against the basal contact and extend as tongues from the contact half a mile or more out into the mudstone facies.

The fauna in the basal strata consists primarily of molluscs with brachiopods, barnacles, echinoderms, and bryozoans as minor components. Two distinctive assemblages can be recognized and form the basis of this example. One is in the sandy beds adjacent to the basal contact, which were located in very shallow water near shore. The other assemblage is in beds of sandstone and conglomerate that extend basinward from the basal contact and must have been located offshore as much as a half-mile or more.

The fossils in the nearshore assemblage are commonly in growth position, are whole, unworn, articulated if bivalves, and are in sediments like those in which their modern relatives live (Fig. 7.22). The assemblage is composed of species that either lived on a sand or rocky substrate, intertidally or in shallow water. Some of the species are restricted to the intertidal habitat. The rest range from the intertidal or base of the intertidal into water as much as a few tens of meters deep. As a whole, the assemblage appears to be essentially in living position in sediment deposited in water not more than about 10 m deep.

The assemblage in the sandstone and conglomeratic sandstone beds extending offshore from the basal contact contains the same taxa as in the nearshore sandstone. However, the shells are generally disarticulated, broken, and abraded and occur in a matrix that is much coarser and less well-sorted than the substrate in which present day representatives of the species normally live (Fig. 7.23). In contrast to the fossils in the nearshore sediment, which are widely scattered in the rock, those in the offshore beds are closely packed and commonly oriented concave-down.

Several species are restricted to the offshore beds. The most common of these are *Turritella cooperi* and *Nemocardium centifilosum*. These species are most commonly found living today at depths of 60 to 100 m. The specimens of these two species are generally unbroken and unabraded in contrast to the specimens

Figure 7.22 Articulated and single valves of *Lyropecten crassicardo* in shoreline basal pebbly sandstone in the Castaic Formation, southern California, Upper Miocene. Scale is 6 in. long. After Stanton (1967).

with which they are associated and commonly are found in adjacent strata rather than being mixed with the disarticulated and broken shells.

The relative proportions of right and left valves of the bivalve *Chione elsmerensis* are also useful in the reconstruction of the fluid dynamics of the paleoenvironment. Individuals of this species lived in the shallow sublittoral environment along the shore. Storm as well as normal wave action would have tended to transport the shells onto the beach or out into deeper water below wave base. Those carried onto the beach would have been sorted by long shore currents as described earlier. A northward long-shore current would have concentrated the left valves on the beach and transported the right valves for possible inclusion in the offshore fossil assemblage. Correspondingly, if the long-shore current had been flowing southward, the right valves would have been concentrated on the beach and the left valves transported away. Few valves of *Chione elsmerensis* are found in the shoreline sand beds of the Castaic Formation. This may be the result of abrasion on the beach. It may also be the result of solution by water flowing through the porous and permeable sand. The high intensity of solution of fossils in beach sands is well documented by McCarthy (1977). The probability that this form of fossil diagenesis was important is strengthened by the relative predominance of the oyster *Crassostrea titan,* of the pecten *Lyropecten crassi-*

Figure 7.23 Shell fragments in offshore conglomerate of the Castaic Formation, southern California, Upper Miocene. After Stanton (1967), Short Contributions to California Geology, Special Report 92, California Division of Mines and Geology.

cardo, and the sand dollar *Astrodapsis fernandoensis*, forms with robust calcite skeletons, and least likely to be dissolved.

The scarcity of *Chione* in the shoreline sands and the preponderance of left valves of *Chione* in the sublittoral deposits indicate that there was a strong southward flowing long-shore current along the northwest margin of the basin. The interpretation of all these types of evidence leads to the conclusion that the beds of sandstone and conglomerate extending from the basal contact out into the basinal mudstone consist of sediment and shells reworked from shallow water near shore into adjacent deeper water, where the substrate was normally populated by a different molluscan fauna. Thus the reconstructed depositional environment would be one in which, during most of the time, mud was being deposited in mid-basin and to within a few hundred feet of the shore. Along the basin edge, the rocky shore and the substrate of sand and pebbly sand in water a few fathoms deep supported a distinctive fauna. The abundance in this fauna of *Crassostrea* specimens as much as 46 cm long and *Lyropecten* specimens as much as 20 cm across indicates that these conditions were not greatly disturbed for long periods of time. At infrequent intervals, no doubt associated with heavy storms, the nearshore substrate and organisms were in large part eroded, mixed with pebbles and cobbles from along shore or newly introduced into the marine environment, and then redeposited along shore and as tongues extending basinward for distances up to a mile and into water 60–100 m deeper. Immediately afterward, a distinctive deeper-water fauna would inhabit the off-shore sandy substrate afforded by the sand and conglomerate tongues and would persist until this substrate was buried by the normally deposited mud.

8

Populations in Paleoecology

Organisms do not function entirely as isolated individuals but are part of a *population* of many interacting members of the same species living together in an area. Members of a population interact with one another (especially in interbreeding and in utilizing a common food source) and with other populations (especially in utilizing food resources and space). Also, the population as a whole is affected by many physical aspects of the environment. To this point we have mostly been concerned with the impact of the environment on individual organisms, especially their distribution and morphology. In this chapter we consider the effects of the biological and the physical environment on population properties, such as (1) population growth rates and growth patterns, (2) age and size distribution of individuals within the population, (3) long-term variation in population size, (4) spatial distribution of population members, and (5) amount of morphological variability within the population. Because these properties are influenced by environmental parameters, we should be able to determine the nature of paleoenvironments by measuring these properties in fossil populations.

A simple definition of population is individuals of a species living together in an area. A more rigorous definition would state that the individuals live closely enough together so that each individual in the population has an equal opportunity of interbreeding with all members of the opposite sex within the group. Thus a population is a group of freely interbreeding organisms. In genetic terms, a population shares a common gene pool and there are no geographic or other barriers to gene flow. As is so often the case in ecology and paleoecology, boundaries are difficult to define. Where does one population stop and another begin? How complete must the isolation be between two areas before we should recognize members of the same species in the areas as belonging to different populations (e.g., corals on two adjacent reefs)? Should all individuals of a uniformly distributed, widespread species be considered as belonging to the same population (e.g., a planktonic foraminifera species in the open seas)? In such a case, clearly all individuals do not have an equal opportunity to interbreed, but no natural boundary exists within the overall distribution. In practice the nature of the geologic record usually helps us with these boundary problems. We deal with only a sample of the population and that sample, because of preservation and exposure of the strata, represents only a limited portion of the geographic distribution of the population. We discuss the properties of the population, but we can really only measure the properties of our limited sample and hope that they are representative. We define the boundaries of the population on the basis of the exposures or cores from which our samples came. We need not be concerned with boundaries because the boundary areas are conveniently not exposed or preserved.

As will be apparent in the ensuing discussion, some attributes of fossil populations are difficult to study because of the nature of the fossil record. A population consists of individuals living at a given instant in time. Ideally, a population should be studied by observing individuals living at the same time. Rarely does the fossil record allow us to do this. A great deal of the effort expended in studying fossil populations is in attempting to unravel or extract this instant-in-time view from the time- and space-averaged fossil record.

Ecologists have been very active in studying the properties of populations and the influence of environment on populations. Good reviews and general discussions of populations are given by Slobodkin (1962), MacArthur and Connell (1966), Odum (1971), and Hutchinson (1978). Paleoecologists have increasingly applied the approaches of population ecology, particularly the study of population structure or dynamics, in analyzing fossil populations. Useful general discussions of the application of population ecology to the fossil record are given by Hallam (1972a) and Valentine (1973a). Much of the work on fossil populations has been done to interpret the structure and evolution of the population rather than to use population properties to interpret paleoenvironments. Our emphasis is on using the properties of fossil populations in interpreting ancient environments.

POPULATION GROWTH

The change in the size of a population is a fundamental attribute of the population that is strongly related to environment. Under favorable environmental conditions a population of a given species may increase rapidly in size to a high level. Under less favorable conditions the population may grow more slowly and remain relatively small. Variable environmental conditions will likely cause population size to vary. Unfortunately, population growth patterns on a short term scale can seldom be studied in the fossil record because of the time-averaging effect. Invertebrate populations normally grow to full size in a few weeks or at most a few years. Most fossil assemblages probably accumulate over a period of many years so that short-term changes in population size are obliterated in the mixing process (Walker and Bambach,

1971; Fursich, 1978). The best opportunity to study short-term population growth in fossils would be in a section that was deposited very rapidly in an environment of low turbulence and minimal mixing. Nevertheless, the concept of population growth is important in helping us to understand many aspects of population and community structure.

Intuitively what might one expect to be the pattern of change in population size? At least two types of patterns might be expected. If we assume that the population starts small it might grow rapidly, perhaps even exponentially, to some value and then decrease abruptly. This pattern might be repeated over and over for a given population. This is a common pattern among modern populations and has been called the *J-shaped pattern* of population growth (Odum, 1971). It results from the presence of initially favorable but changeable conditions for the population. The changing parameter may be physical (e.g., temperature), chemical (e.g., salinity), or biological (e.g., food organisms). The cause for the abrupt decrease in population size may be *density independent*, that is, not dependent on the population size, as for example a sudden decrease in temperature; or the cause may be *density dependent*, that is, the overutilization of food resources. Many examples of the *J*-shaped population growth pattern are familiar to us from our daily experience—grasshoppers, mosquitoes, crab grass (Fig. 8.1A).

The other pattern we might expect is for the population to first increase and then to gradually level off at some upper equilibrium value. This pattern is also common in nature and has been called the *S*- or *sigmoidal-shaped pattern* (Odum, 1971). Any given area can support only a certain-size population for an indefinite time period because some environmental factor, commonly the food source, is maintained at a certain level that places an upper limit to the population size. This upper population limit is called the *carrying capacity* of the environment. The control on the maximum size of the sigmoidal-shaped population growth pattern is thus density dependent, and the closer the population size approaches the carrying capacity (i.e., the greater the density) the slower will be the rate of increase in population size. In a perfectly balanced ecosystem the population ideally should remain at the carrying capacity unless some outside force disturbs the system. Examples of populations showing the sigmoidal-shaped pattern would include the bass in a balanced lake or pond, the deer population in a well-managed herd, or corals on a stable reef (Figs. 8.1B and 8.2).

The most common patterns may be more complex than a simple *J*-shaped or sigmoidal-shaped one. Sometimes observed patterns are a combination of the two basic patterns or even more complex patterns (Fig. 8.1). These may, in some cases, be caused by the population temporarily exceeding its carrying capacity. In other cases the carrying capacity itself may change as a result of environmental fluctuations. Density-independent factors may occasionally be superimposed on the density-dependent control of the sigmoidal-shaped pattern.

Mathematical models have been developed to describe population growth

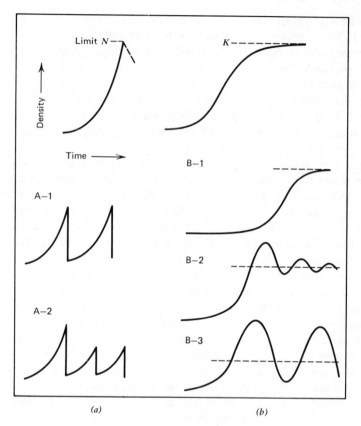

Figure 8.1 Population growth patterns (changing population density with time). (*a*) *J*-shaped pattern and variants; (*b*) sigmoidal or *S*-shaped pattern and variants. After Odum (1971).

(See Hutchinson, 1978, for an interesting and entertaining discussion of the history of development of these models). These are useful in allowing us to more rigorously test hypotheses that explain population growth by comparing the models with actual data. They also allow us to quantitatively predict the population growth rate and pattern.

The simplest mathematical model describes population growth with no upper limit to population size with the equation

$$\frac{dN}{dt} = rN \tag{8.1}$$

where N is population size, t is time, and r is a rate constant that describes the rate of increase in population size (Fig. 8.3). In words, this model simply says that the rate of change in population size is equal to the size of the population times a population growth constant. An equation of this type should be

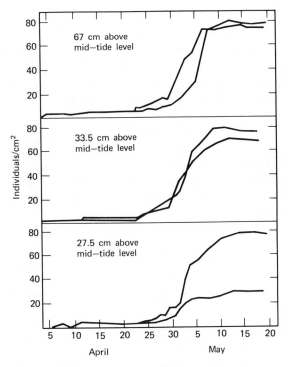

Figure 8.2 Sigmoidal population curves for populations of the barnacle *Balanus balanoides* on rocks in the intertidal zone of the Firth of Clyde, Scotland. The two curves in each box represent 2 samples at the same tide height. After Connell (1961). Reprinted from "Effects of competition, predation by *Thais lapillus* and other factors on natural populations of the barnacle *Balanus balanoides* by J. H. Connell in Ecological Monographs, Vol. 31, p. 65, by permission of Duke University Press. Copyright 1961 by the Ecological Society of America.

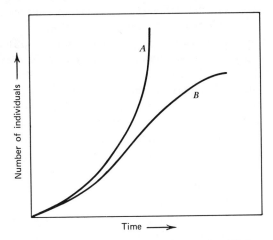

Figure 8.3 Exponential growth curve (*A*) and logistic equation (*B*) for modeling population growth.

familiar to geologists because it is identical to the radioactive decay equation used in age dating. In the decay equation N is the number of atoms of the radioactive element and the decay constant, λ, is used instead of r. λ is negative indicating a decrease in the population size of radioactive atoms, whereas r is positive indicating a growth in the number of organisms.

The integrated form of the differential equation is

$$N_t = N_0 \, e^{rt} \tag{8.2}$$

where N_t is the population size at any given time, N_0 is the population size at the start, and e is the base of the natural logarithms. The value of r can be calculated from data on the population size increase by arranging the equation in the form

$$r = \frac{\ln N_t - \ln N_0}{t} \tag{8.3}$$

The r value is a function of the interaction between *natality* (birth rate), b, and *mortality* (death rate), d.

$$r = b - d \tag{8.4}$$

Each population has a characteristic natality that is basically controlled by the physiology of the species but may vary depending on environmental conditions. Likewise, the population has a characteristic mortality that is also controlled by the organism's physiology and environmental factors. Clearly the population will grow most rapidly under those conditions that maximize natality and minimize mortality. The maximum value of r for a given population is called the *intrinsic rate of natural increase* or the *biotic potential* of the population. The actual value of r for a given set of environmental conditions is normally less than the biotic potential, and the difference between the two is a measure of the *environmental resistance* to maximum growth rate (Odum, 1971).

The biotic potential depends on the organism being investigated. Some species have an extremely high natality rate and if the mortality rate is not too high will give high r values, hence rapid rates of population growth. Diatoms, bacteria, foraminifera, and mosquitoes would be examples of species of this type. Other species have much lower biotic potential and thus much slower population growth rates. Elephants, humans, and probably dinosaurs fall into this category.

Of course the exponential growth model cannot explain the whole story of population growth, for it predicts populations of astronomical size within relatively short time periods. It is this model that led the Rev. Thomas Malthus to his predictions of disaster for the human race due to geometric population increase (see discussion in Hutchinson, 1978). Clearly something not accounted for in this model happens to restrain population growth. The exponential growth model is useful in introducing the concept of biotic potential and does fairly accurately depict the growth of many populations in their early stages. Some examples of the *J*-shaped pattern could be compared to the exponential growth pattern that is abruptly terminated at a certain point. The main shortcoming of the exponential growth model is that it does not take into account any factor limiting growth rate even though density-dependent factors will always operate and should be accounted for in the population growth model.

The *logistic equation*, which was originally developed by P. F. Verhulst (Hutchinson, 1978), takes into account the density-dependent factors and gives a reasonable approximation of the sigmoidal-shaped population growth pattern. In this model

$$\frac{dN}{dt} = rN \frac{(K - N)}{K} \tag{8.5}$$

where K is the upper equilibrium limit or carrying capacity that, in theory, the population size approaches but does not exceed and the other symbols are as previously defined (Fig. 8.3). As can be seen from the equation, as N approaches K, that is, as the population size nears the carrying capacity, the term $(K - N)$ approaches zero and thus the rate of population growth must approach zero. This models the effect of density-dependent factors acting on population growth. The fact that this model generates curves that closely approximate actual data does not necessarily mean that it is the most accurate model, but it does have the advantage of simplicity.

The two most important concepts shown by the logistic model are the parameters of intrinsic rate of natural increase (r) and carrying capacity (K). Both parameters are sensitive to the environment and thus can contain environmental information for the ecologist and paleoecologist. As we discuss later in this chapter, the ability to rapidly increase population size when environmental conditions are favorable may be a definite advantage for a species. This is especially true in those environments that are not always favorable for the species. Thus the environment selects species with high r values (*r-selection*). In other cases the species with the largest carrying capacity will have an advantage. By definition, such a species will be able to maintain a larger population size than a competing species with a smaller carrying capacity. Environments that allow the species to maintain a stable population size are especially likely to select species with high K-values (*K-selection*).

POPULATION STRUCTURE

The distribution of ages of individuals within the population, or the *population structure*, is determined by the intrinsic properties of population growth and the extrinsic properties of the environment. A number of papers and books on population structure discuss this topic. The classical work of Deevey (1947) on life tables and survivorship curves is especially useful and Slobodkin (1962) and Odum (1971) have good reviews of population dynamics. The paper by Hallam (1972a) is brief but gives a good introduction from the paleontologic point of view, as do earlier papers by Craig and Hallam (1963), Kurtén (1964), and Craig and Oertel (1966). More recently, Thayer (1975c and 1977) and Richards and Bambach (1975) have made important contributions to the application of population dynamics to fossil populations.

An example will help to demonstrate the use of population structure in paleoecology. Some populations may contain a large number of young individuals but relatively few of intermediate or old age. In such a population the mortality rate must be very high among the juvenile forms so that few individuals live to maturity. This is either a characteristic of the particular species or it is a result of environmental conditions that prevent many of the individuals from reaching maturity. The modern suspension feeding bivalve, *Mulinia lateralis,* has such a population structure in muddy environments with a soupy substrate (Levinton and Bambach, 1970). This environment is somewhat marginal for *M. lateralis,* most individuals of which are thus unable to survive beyond the young stage (Fig. 8.4). On the other hand, the deposit-feeding bivalve *Yoldia limatula* shows a relatively larger number of adult specimens (and has a more uniform mortality rate) in the same environ-

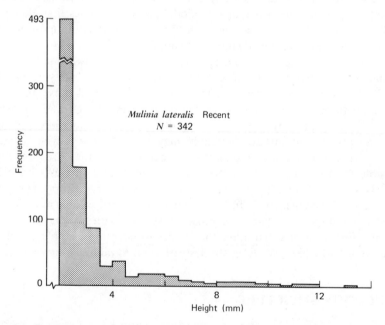

Figure 8.4 Size-frequency distribution of dead shells of modern *Mulinia lateralis.* After Levinton and Bambach (1970).

ments and thus appears better adapted for life on a muddy substrate (Fig. 8.5). By comparison, a population of the Devonian bivalve *Arisaigia placida,* which was probably also a deposit feeder, collected from a shale, has a population structure similar to modern *Y. limatula.* This suggests that substrate conditions in the Devonian seas at that locality may have been similar to the modern example to which it was compared.

Life Tables and Survivorship Curves

The structure of populations can be most readily visualized from *life tables* and *survivorship curves.* Life tables (Deevey, 1947) contain the number of survivors from a starting population, mortality rate, and mean life expectancy at regular intervals during the life of the species (Table 8.1). They have been extensively used by population ecologists but not much used directly in paleoecology. Life tables include, in tabular form, the information graphically shown in survivorship curves, which are constructed in the following way. Consider the individuals in a population that are all born or spawned at the same time (a *cohort* in ecological terminology). With time, individuals of the cohort gradually die off, leaving a certain number of survivors. A graph of the number of survivors vs. time since birth (or age of the cohort) is a survivorship curve (Fig. 8.6). The abcissa of the survivorship curve may be the age in some convenient time unit, but for ease of comparison of species with widely different longevity the percentage of the mean or maximum life span is sometimes used. Commonly, the ordinate of the survivorship curve is the number of individuals out of a starting 1000 (or some other arbitrary number), or it

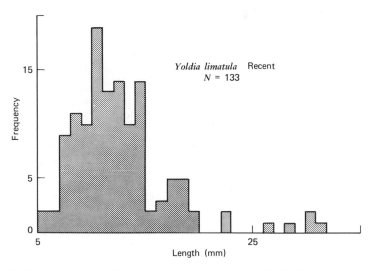

Figure 8.5 Size-frequency distribution of dead shells of modern *Yoldia limatula.* After Levinton and Bambach (1970).

Table 8.1 Life table for a population of *Tivela stultorum* from California[a]

Age (years) x	Age as % Deviation from Mean Length of Life x'	Number Dying in Age Interval out of 1000 Born d_x	Number Surviving at Beginning of Age Interval out of 1000 Born l_x	Mortality Rate per Thousand Alive at Beginning of Interval $1000q_x$	Expectation of Life, or Mean Lifetime Remaining to those Attaining Age Intervals (years) e_x
0–1	−100	550	1,000	550	2.07
1–2	−52	202	450	449	1.87
2–3	−3	72	248	290	1.99
3–4	+45	60	176	341	1.60
4–5	+93	60	116	517	1.18
5–6	+142	38	56	678	0.90
6–7	+192	13	18	722	0.75
7–8	+237	5	5	1,000	0.50

[a] After Hallam (1967).

could also be the percentage surviving. The ordinate is almost always given on a log scale. Such a scale results in a straight-line plot for the case in which the mortality rate is constant at all ages (Fig. 8.6). The rate of death of members of the cohort can be shown as

$$\frac{dN}{dt} = q_x N \tag{8.6}$$

in which q_x is the *specific death rate* and N is the number of individuals. Because q_x represents a loss of individuals, it will have a negative sign. In the case of a constant mortality rate, q_x is constant. Equation 8.6 is then a simple exponential function just as is the equation for unlimited population growth (8.1), and thus log N vs. time will plot as a straight line [see (8.3)]. In the more general case, the value of q_x will change with age of the cohort giving a curve on the survivorship plot. q_x is the slope of the survivorship curve and thus can be readily calculated. Another way of showing population structure is with a plot of specific death rate (q_x) vs. age of the individual (Fig. 8.7).

The shape of the survivorship curve differs with the structure of the population. Three general types can be recognized (Figs. 8.6 and 8.8):

1 Populations characterized by high juvenile mortality and lower mortality later in life have a curve that is concave upward. Such a population will have many young individuals and fewer older ones (Fig. 8.8a). This pattern is quite common among invertebrates that usually produce large numbers of larvae, most of which die young, with few individuals living to maturity. In most invertebrates this high mortality occurs in the larval stage which leaves no fossil record. The mortality pattern may change drastically after settling of the larvae and development of a mineralized shell. Thus the preservable portion of the record of the life of the species may not show the high juvenile mortality pattern. This has the effect of eliminating a portion of the survivorship curve.

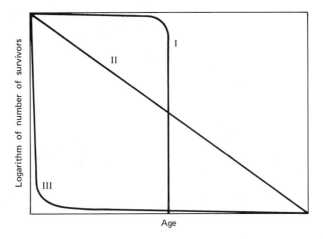

Figure 8.6 Schematic representation of three basic types of simple survivorship curves. I—mortality increasing with age; II—constant mortality rate; III—mortality decreasing with age. After MacArthur and Connell (1966).

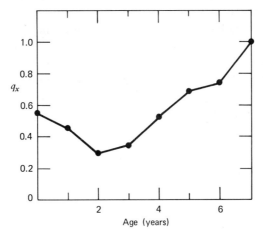

Figure 8.7 Specific mortality of *Tivela stultorum* from California. Data from Coe and Fitch (1950).

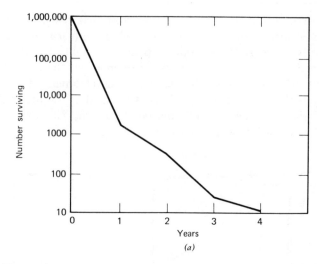

Figure 8.8 Three basic types of survivorship curves in natural populations. (*a*) Decreasing mortality in the prawn *Leander squilla*. After Kurtén (1964); (*b*) increasing mortality in the recent brachiopod *Terebratalia transversa*. Upper curves (triangles) are plotted as a function of valve length on the arithmetic horizontal axis. The lower curves (circles) are plotted as a function of valve length on the logarithmic scale, which should be approximately equivalent to age (see text for discussion). The curves with open symbols represent one sample; the curves with solid symbols represent a second sample. After Thayer (1977). (*c*) Constant mortality in the bird *Vanellus vanellus*. After Hutchinson (1978), copyright Yale University.

2 At the other extreme are convex upward curves characteristic of species with a low mortality rate early in life but an accelerating rate as maximum longevity is approached (Fig. 8.8*b*). Curves of this sort apply to some invertebrates and many vertebrates, most notably to man. Improved medical treatment and public health have resulted in the curve for man becoming increasingly convex but maximum longevity has not been changed.

3 The intermediate case is a descending straight line representing constant mortality throughout life. Few if any species would plot as a perfectly straight line but many approximate it with some irregularities. The post larval stages of some invertebrates, small mammals, and some birds are examples (Fig. 8.8*c*).

As can be seen from the examples shown here, survivorship curves contain irregularities and are not the perfect hypothetical curves shown in Fig. 8.6. A common example is a crude sigmoidal shape which indicates a higher mortality rate at the juvenile stage with a lower intermediate rate followed by a higher rate later in life. A stair-step pattern is found in some species, such as the arthropods, which have certain stages during their life history when they are more vulnerable to mortality.

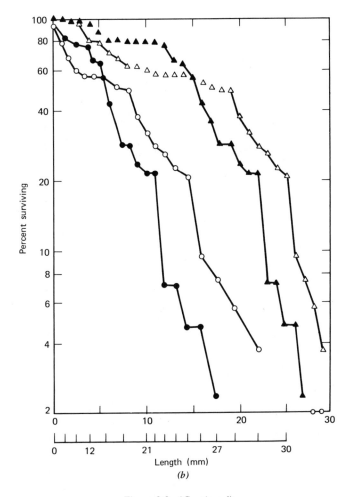

Figure 8.8 (*Continued*)

Age-Specific and Time-Specific Survivorship Curves

How can survivorship curves be constructed from fossil populations? Survivor-ship curves are often difficult to construct, even for modern populations. A true survivorship curve, sometimes called an *age-specific* survivorship curve, traces the survivorship of a single cohort and can be constructed only from records kept over the life span of the entire cohort. This is difficult and time consuming to do for modern populations and almost invariably impossible for fossil populations. The best approximation of an age-specific survivorship curve is a *time-specific* survivorship curve. This curve is constructed by de-termining the number of living individuals in each age class at a given time. It is thus called an *age-frequency* curve. To convert an age-frequency curve to

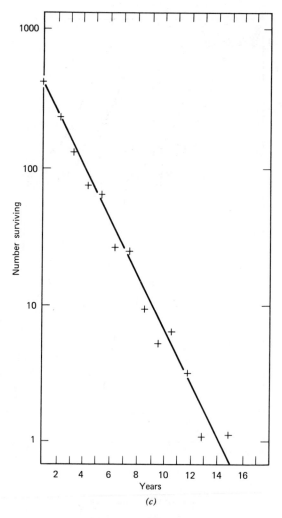

Figure 8.8 (*Continued*)

the form of a standard time-specific survivorship curve, the ordinate must be changed from frequency to percent surviving. In order to do this, the youngest class is presumed to have undergone no mortality and so is adjusted to equal 100% survivors. All other classes are then multiplied by this factor in order to normalize the curve.

$$l_x = N_x \frac{100}{N_1} \tag{8.7}$$

where l_x is the percent survivors, N_x is the number of individuals in class x, and N_1 is the number of individuals in class 1.

In a stable population which is adding new individuals at a constant rate while mortality continues at a constant rate within each age class, the time-specific survivorship curve will be identical to the age-specific curve. Because absolute stability probably never exists, the two types of survivorship curves will never be absolutely identical and may bear only a remote resemblance to one another. Commonly, age-frequency and size-frequency curves are polymodal (Fig. 8.9) with each mode being composed of individuals from a different spawning season. Usually the number of individuals in each mode will decrease with increasing age. The general survivorship pattern can perhaps be approximated from such data by drawing a curve connecting the peaks. This may not yield accurate results, however, because conditions may vary from year to year so that some years are more favorable for spawning than others. This has especially been documented in fish populations (Hutchinson, 1978). The height of the peaks in the age-frequency plot is thus a function of the success of the spawning as well as the mortality pattern. Polymodal curves may not always be the result of seasonal spawning because irregular recruitment due to any cause will produce polymodality, which is the rule rather than the exception in age-frequency or size-frequency curves. Thayer (1977) emphasizes the risk in interpreting polymodal curves as indicators of seasonal spawning.

A common way to present data on the structure of populations is to show plots of size-frequency rather than age-frequency. In polymodal size-frequency plots the width of the individual peaks is a function of differences in growth rate between individuals in a given spawning class. The peaks get closer together at larger sizes because of the usual slowing of growth rate with age. Finally the peaks may begin to merge as growth rate differences cause an increasing overlap of spawning classes. Such a plot is useful in indicating approximate growth rate (from the spacing of the peaks) and the approximate maximum age (from the number of peaks).

Census Populations and Normal Populations

Rarely does a fossil population correspond to a population of organisms that were all living at the same time, but a survivorship curve can be directly constructed only from such a population. Populations of organisms all living at the same time may occasionally occur when some catastrophic event causes the sudden killing of the entire living population and allows its burial separate from preexisting specimens. Hallam (1972a) has called such fossil assemblages *census populations* to distinguish them from *normal populations* that result from the gradual accumulation of shells over a period of time. Ager (1963) has given a colorful and insightful comparison of the two types of populations. He compares the census population to the human population of the city of Pompeii near Rome which was abruptly covered by an ash fall from Mt. Vesuvius in 79 B.C. The living population of the city was suddenly killed and preserved at the spot where they were at the time of the ash fall.

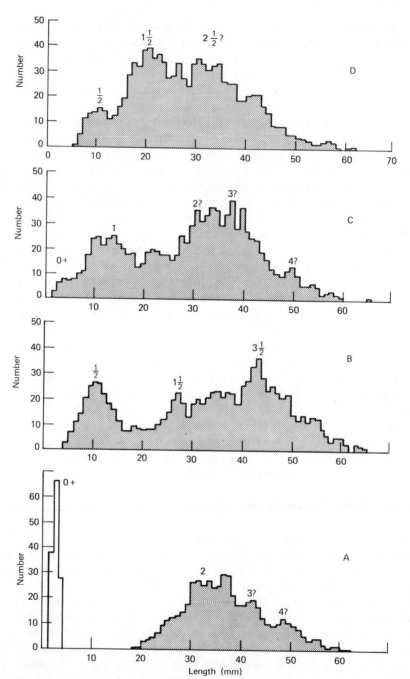

Figure 8.9 Polymodal size—frequency histograms of *Mytilus edulis* from Ferny Ness, Scotland. The numbers above the histograms are the presumed age of specimens in that mode in years. Collection dates: A—April 1961; B—November 1961; C—April 1962; D—November 1962. After Craig and Hallam (1963).

The normal population is like the bodies in the cemetery outside the city walls which had accumulated over a period of many years as the population died of various causes. A number of examples of census populations have been described in the paleontological literature. For example, Hallam (1961) describes assemblages of brachiopods from the lower Jurassic of England and Waage (1964) describes molluscan assemblages from the upper Cretaceous of South Dakota which appear to consist of census populations. There are few definite rules for distinguishing census and normal populations, although Hallam (1972a) gives some characteristics of census populations. A census population would be expected to occur in a very narrow stratigraphic interval, perhaps on a single bedding plane, and usually consists of only a few species. It would most likely be found in a quiet water environment where mixing with fossils immediately above or below is minimized. The fossils might occur in a very thin, fossiliferous unit with relatively fewer fossils above and below. A study of growth rings or skeletal structure should reveal that all specimens died in the same season. Some sedimentologic or stratigraphic evidence of a catastrophic event would further support the census population interpretation. An example might be a turbidite bed immediately above the fossiliferous bed. All of these criteria are, at best, suggestive and do not prove that the fossils in question represent a census population.

A time-specific survivorship curve (or age-frequency curve) can be obtained directly from a census population provided the age of the fossils can be determined (such as by growth lines or skeletal structure). To the extent that the size of the fossil is a function of its age, a size-frequency curve will approximate the survivorship curve in a census population. An age-frequency or size-frequency plot for a normal fossil population will *not* usually approximate a survivorship curve, however, because it does not represent the organisms living at one time.

Under favorable conditions, survivorship curves can be constructed from normal fossil populations because the age distribution in a normal fossil population is a function of the survivorship pattern. For example, in the Pompeii cemetery there was probably a larger portion of old and middle-aged bodies than young adult bodies. This is a reflection of the mortality pattern in humans with its convex upward curve indicating that relatively few younger people die to produce bodies. On the other hand, an oyster or clam with a high juvenile mortality would produce a normal population with many young shells and relatively few older shells. A population with a constant rate of mortality would produce a distribution with a regular decrease in number of individuals of increasing age.

One way of viewing the relationship between the survivorship curve and age-frequency curves for normal populations is in terms of body production rates. The biologist is interested in living organisms and so measures the rate of survival, but the paleontologist studies the dead bodies and so is interested in the body production rate. The survivorship curve shows the rate of changing size of the cohort. That rate is always zero or less because individuals are constantly dying off, and the more rapidly they die, the steeper is the descending curve and the more negative the slope. But a loss in terms of the living population is a gain for the population of dead bodies and thus potential representatives for the normal fossil population. The survivorship curve viewed upside down (or plotted upside down) is a measure of body production rate with a steep negative slope on the survivorship curve becoming a steep positive slope on the body production curve. When the slope is steep, many bodies are being produced and the age-frequency curve will be high, and when the slope is low, few bodies are produced and the age-frequency curve will be low. A plot of the slope of the body production curves vs. age (i.e., q_x vs. age, Fig. 8.7) shows the pattern of the age-frequency curve for the normal fossil population.

Before a survivorship curve can be reconstructed for a fossil population a method must be found to determine the age (life span) of the fossil specimens. In most cases this is a difficult if not impossible task. Some organisms produce growth rings on their skeletons or growth bands within their skeletons which may be annual (see Chapter 4). This has been rather convincingly shown for some molluscs, corals, fish, and echinoids, but the technique must be used with caution. The usual assumption is that slowing of growth during the winter produces the growth ring, but storms or other events that disturb the animal may also produce a growth ring. In some cases, by careful study, such disturbance rings can be distinguished from seasonal rings, but care is required in using this technique. In vertebrates, growth rings and evidence of wear in teeth and various other skeletal features have been used to determine age (Kurtén, 1964). The age of arthropods, especially ostracodes, have been determined from molt stages or *instars* (Fig. 8.10).

The more usual approach is to assume a relationship between size and age. This is at least approximately true for most invertebrates which continue to grow throughout life but at a decreasing rate in old age. Levinton and Bambach (1970) concluded that the relationship between growth rate and age for bivalves can be approximated by the equation

$$D = S \ln(T + 1) \tag{8.8}$$

where D is size, T is time, and S is a constant characteristic of the particular population. For purposes of constructing generalized survivorship curves, this equation is probably as good as any. A simpler approach is to plot size on a logarithmic scale (Fig. 8.8b) as was done by Thayer (1977). This takes into account the exponential decrease in growth rate which is included in the equation above so that the scale should be approximately arithmetically equivalent to age. The relationship between size and age may be more complex than this in detail. Many species seem to show a sigmoidal shape with a somewhat slower growth rate very early in life, a higher but gradually decreasing rate later, and a slow rate late in life. Perhaps an even greater problem is that growth rate varies between individuals so that specimens of equivalent size are not necessarily of

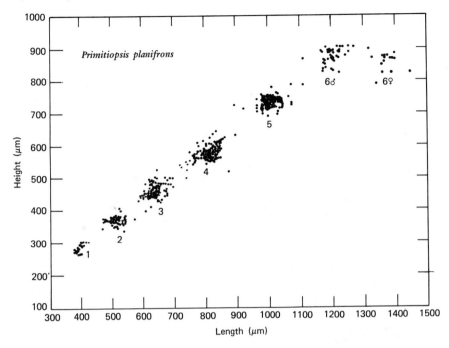

Figure 8.10 Variation of length with height of shells of the ostracode *Primitiopsis planifrons*. The numbers below the point clusters are molt stages. After Kurtén (1964).

equal age. This especially makes difficult the interpretation of polymodal curves (Thayer, 1977).

A size-frequency plot is an approximation to an age-frequency plot. Several studies of population structure have used size-frequency plots and not attempted the refinement of converting size into actual or relative ages.

In general terms the mortality pattern of a normal population can be interpreted directly from age-frequency (or crudely from size-frequency) curves (Figs. 8.11 and 8.11A). A population having high juvenile mortality will have a strongly positively skewed curve with a large number of young (small) individuals. A population with a constant mortality pattern will also have a positively skewed curve but less so than in the case of the high-juvenile mortality population. A population with mortality rate that increases with age will have a more symmetrical or even negatively skewed curve (see Richards and Bambach, 1975).

Constructing a Survivorship Curve from a Normal Population

An approximation of a survivorship curve can be constructed from a size-frequency distribution for a normal fossil population by the following procedure:

1 Establish a series of age or size (linear dimension) classes for the population. The age classes should be of equal size. The number of age classes established will probably be determined by the precision with which these age classes can be recognized from growth lines or skeletal structures in the fossil. Size classes can be set at the convenience of the researcher, but for most cases 8 to 20 classes should be sufficient. Numerous finely

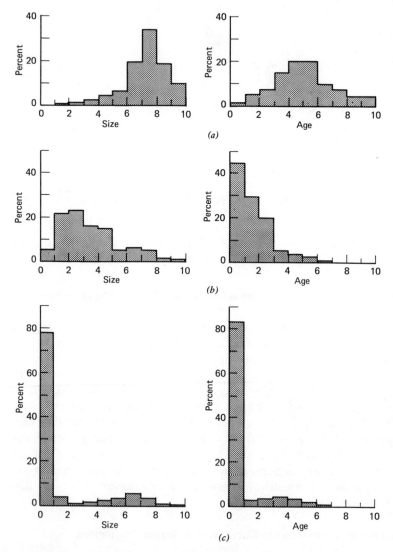

Figure 8.11 Representative size-frequency distributions and their corresponding age-frequency distributions. Growth rate (and thus size) is presumed to decrease logarithmically with age. (*a*) Population with increasing mortality with age; (*b*) population with approximately constant mortality; (*c*) population with decreasing mortality with age. After Richards and Bambach (1975), Journal of Paleontology, Society of Economic Paleontologists and Mineralogists.

subdivided classes require more specimens in the sample. Age classes are necessary to construct a true time-specific survivorship curve and they obviously should be used if the age can be determined.

2 Determine the age or measure a linear dimension for all specimens in the population and assign them to the classes established in step 1.

3 If size rather than age is used for the x-axis of the survivorship plot it may be converted to age or a time factor by use of an equation such as (8.8). The x-axis may also be a logarithmic scale to more closely approximate the age-size relationship. If an arithmetic scale is used, the

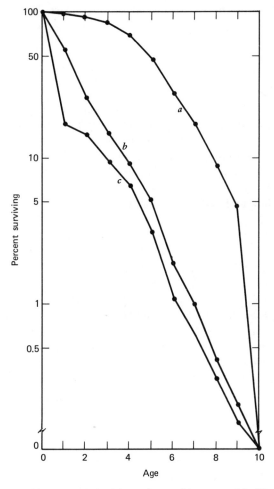

Figure 8.11A Survivorship curves derived from the age histograms (*a*), (*b*), and (*c*) shown in Fig. 8.11. After Richards and Bambach (1975), Journal of Paleontology, Society of Economic Paleontologists and Mineralogists.

probable nonlinear relationship between size and age should be considered in interpreting the results.

4   The ordinate should be logarithmic, with the scale most conveniently expressed as percent of specimens surviving.

5   At age zero or size zero, 100% of the specimens are surviving. At the end of the age or size of the first class, subtract all the members of that class from the total number of specimens. These are the specimens that did not survive beyond that class. Determine the percent remaining (or surviving) after the subtraction and plot the percentage at the boundary between class 1 and 2. Repeat this procedure for each class, adding each time the members of that class to all younger or smaller fossils which did not survive to that age or size. A logarithmic scale has no zero so the plot must stop at the lower end of the last age or size class. The equation for calculating points on the survivorship is

$$l_x = \frac{N_t - \Sigma_x^1 N_i}{N_t} 100 \qquad (8.9)$$

where l_x is the percentage of the population surviving at the end of class x, N_t is the total number of fossils in the population, $\Sigma_x^1 N_i$ is the sum of the number of fossils in all classes between 1 and x. Figure 8.12 shows a hypothetical example of a survivorship curve constructed from an age frequency histogram by this procedure.

Modifying Factors on Fossil Survivorship Curves

Differential preservation and modification of the size distribution by current winnowing will modify the survivorship curve. Small, thin shells are more subject to solution and other diagenetic processes than are larger, more robust shells, and currents are more likely to winnow small shells from the assemblage. Some of the earliest interest in size-frequency plots was as a method of investigating whether or not the fossils had been transported (e.g., Menard and Boucot, 1951; Boucot, 1953; and Fagerstrom, 1964). The hypothesis was that current transport or sorting would produce normal bell-shaped curves whereas *in situ* populations would be positively skewed. This may often be correct, but clearly *in situ* populations may produce normal or even negatively skewed size-frequency curves (Craig and Hallam, 1963). A common condition seems to be for small (juvenile) individuals to be under-represented in fossil populations because of physical and chemical destruction. This results in the common bell-shaped distributions in fossil populations. Kurtén (1964) suggests that although such distributions make impossible the study of juvenile mortality, the adult mortality can still be studied if the larger individuals are preserved. The effects of transportation and mixing may be less important than once thought (Walker and Bambach, 1971).

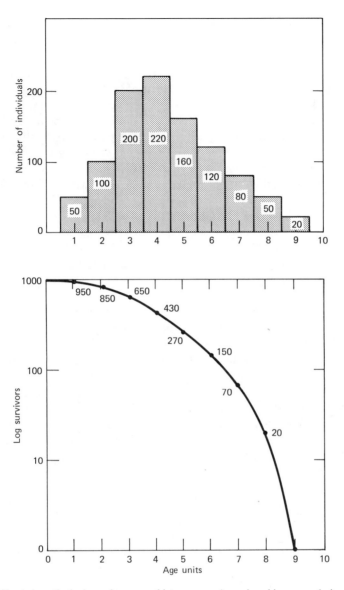

Figure 8.12 A hypothetical age-frequency histogram and survivorship curve derived from it. The numbers on the histogram refer to the number of individuals in each age class. The numbers on the survivorship curve refer to the number of individuals surviving to the indicated age. After Hallam (1972a). From Models in Paleobiology, T. J. M. Schopf, Ed., Freeman, Cooper and Co.

Accordingly, Hallam (1972a) believes that the evidence suggests that size-frequency distributions are largely accounted for by growth and mortality rates.

In some respects the time-averaged nature of the fossil record is useful in the study of survivorship because the effects of seasonal spawning and of good years vs. bad years for spawning are, to a certain extent, averaged out. Likewise, short-term environmental effects are minimized. Fossil size-frequency curves may nevertheless be complex; they may be polymodal despite the averaging effect because of a combination of seasonal spawning and high seasonal mortality. For example, a species that spawns in the spring may have an unusually high mortality rate in the late fall or early winter. Mortality is thus highest when the individuals are at a certain age and approximate size, so a size-frequency plot will show a maximum at that size. Such plots yield useful information about the species population. Polymodal populations result from species that spawn seasonally, a useful bit of information in itself. The spacing between the modes should represent one year of growth. Thus the curves can indicate the longevity of the population, and the spacing will also tell the growth rate.

Arthropods are very suitable for population structure studies because their relative age can be determined from molt stages. However, because arthropods grow by molting, not all the fossil specimens may actually represent dead bodies. Thus the normal rules for survivorship curves do not apply without modification. If the arthropod fossils were all derived from dead bodies, the procedure for normal populations described above could be used. If the fossils are all molts, their abundance is a measure of the number of living organisms that successfully lived through that molt stage. Thus, in a sense, the molts represent "live bodies" rather than dead. In this case the fossils are equivalent to a census population and can be used to directly construct a time-specific survivorship curve (or an age-frequency curve). In most cases the actual fossil arthropod assemblage is a mixture of molts and actual bodies so neither approach will yield a true survivorship curve. Kurtén (1954) constructed survivorship curves for a fossil ostracode population based on each of these assumptions (Fig. 8.13). In the example he used the curves do not differ much, so he concludes that either method can be used without much error. In the more general case, however, the results could be quite different; hence the problem should not simply be ignored.

Information from Population Structure

In summary, six categories of information that potentially can be obtained from studying the size-frequency distribution of a fossil species are:

1 **Survivorship Curves** This curve contains useful information in terms of understanding the biology of the species and can also yield useful environmental information. Variation in the survivorship curves between populations of the same species may reflect differences in environmental

conditions, such as substrate hardness. For example, the study of Levinton and Bambach (1970) discussed above uses this approach.

2 **Seasonal Spawning Patterns** These can be detected from census populations and perhaps less reliably from normal populations. This would suggest that environmental parameters vary seasonally.

3 **Seasonal Mortality Patterns** These can be detected from polymodal normal fossil populations. They also suggest seasonally variable environments.

4 **Growth Rate** Spacing of modes in polymodal distributions can, under ideal conditions, yield this information. As indicated in Chapter 4, growth rate may be a function of environmental parameters, especially temperature.

5 **Maximum Age** The number of modes in polymodal distributions can give an estimate of the maximum age of individuals in the population.

6 **Current Sorting** Normal, bell-shaped distributions may suggest current sorting, particularly if other current indicators are present. Care must be taken in making this interpretation because populations with low juvenile mortality may produce bell-shaped distributions.

Fossil Examples of Population Dynamics Studies

A number of studies of the population dynamics of fossil populations have been conducted. We only discuss a small sampling here.

Richards and Bambach (1975) studied 16 brachiopod assemblages of Ordovician, Silurian, and Pennsylvanian age in one of the most extensive studies of population dynamics in fossils. They found two types of populations in their study: one with a strongly positively skewed size-frequency distribution

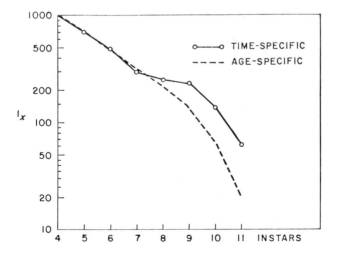

Figure 8.13 Survivorship curves for the Silurian ostracode, *Beyrichia jonesi* calculated by the age-specific and time-specific methods (see text for details). After Kurtén (1964).

(with an abundance of small specimens) and another with a negatively skewed distribution (with an excess of large individuals and few small ones). The survivorship curves for the first group showed a high juvenile mortality rate and a decreasing mortality rate with age (Fig. 8.14). The survivorship curves for the second group show a mortality rate that is low in juveniles and increases with age (Fig. 8.15). All the populations with high juvenile mortality are from assemblages that the authors interpret to have lived on a soft, muddy substrate. The juveniles had difficulty coping with the turbidity and resuspended

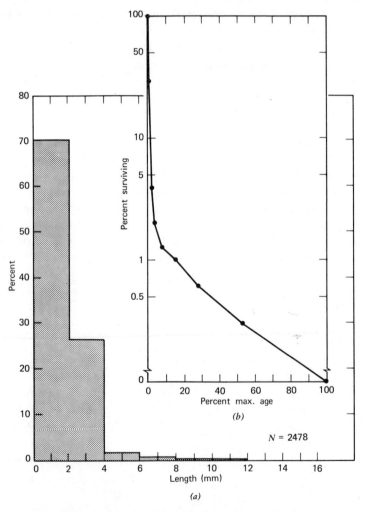

Figure 8.14 Size-frequency distribution (*a*) and survivorship curve (*b*) for a population of the brachiopod *Composita* cf. *subtilita* from Pennsylvanian strata at LaSalle, Illinois. After Richards and Bambach (1975), Journal of Paleontology, Society of Economic Paleontologists and Mineralogists.

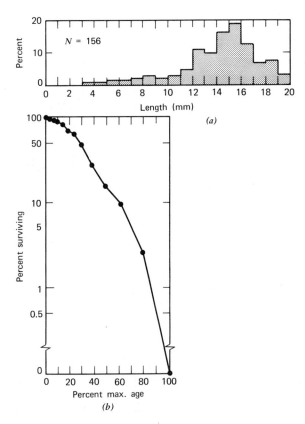

Figure 8.15 Size-frequency distribution (*a*) and survivorship curve (*b*) for a population of the brachiopod *Lepidocyclus capax* from Ordovician strata at Abington, Indiana. After Richards and Bambach (1975), Journal of Paleontology, Society of Economic Paleontologists and Mineralogists.

sediment as well as the unstable substrate; consequently the mortality rate was high. The second group lived on firmer substrates, and consequently the mortality rate among juveniles was low. The higher mortality rate in larger individuals simply reflects the gerontic mortality.

In an earlier study Surlyk (1972 and 1974) reached similar conclusions concerning the depositional environment of the upper Cretaceous (Maastrichtian) chalks of northern Europe. Several species of brachiopods that attach to small, hard objects in the substrate show a positively skewed size-frequency curve indicating high juvenile mortality. Survivorship curves indicate that mortality was fairly constant throughout life (Fig. 8.16). Surlyk concludes that the cause of this mortality pattern was the unstable, soupy condition of the substrate. A brachiopod attached to a small piece of hard substrate had a high probability of being overturned or buried if the sediment surface were disturbed by any physical or biological process. Another group of brachiopods which lived free on the surface yielded a more bell-shaped size-

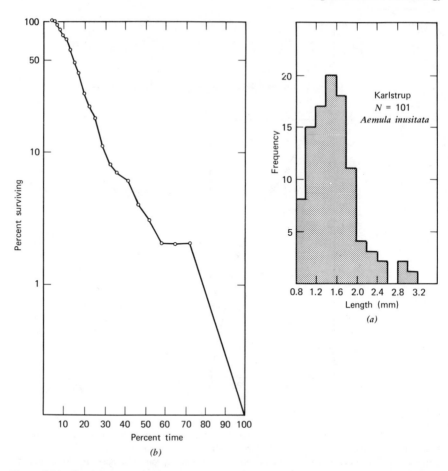

Figure 8.16 Size-frequency distribution (*a*) and survivorship curve (*b*) for a population of the brachiopod *Aemula inusitata* from Cretaceous strata at Karlstrup, Denmark. After Surlyk (1974), copyright Elsevier.

frequency distribution, indicating relatively low juvenile mortality and a higher later mortality. It may also indicate a high growth rate early in the life of the individual. Surlyk concludes that these brachiopods grew rapidly in order to raise their commissure above the sediment for feeding. Their mode of life on the substrate was more stable than that of the attached species so that they had a lower juvenile mortality rate.

Mancini (1978b) constructed size-frequency and survivorship curves for several bivalve, gastropod, cephalopod, and echinoid species from the Cretaceous Grayson Formation of Texas. Most of the survivorship curves in his study have a sigmoidal shape indicating high juvenile mortality followed by a period of reduced mortality and finally with increased mortality in the gerontic stage (Fig. 8.17). He interpreted the Grayson fauna as having lived in quiet

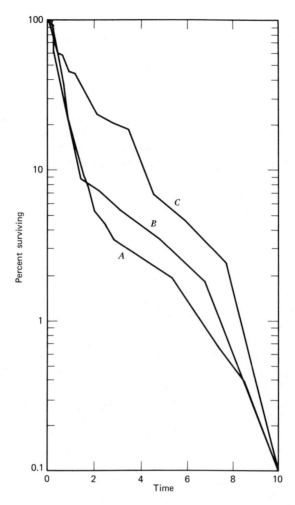

Figure 8.17 Survivorship curve for the gastropod *Turritella irrorata* (*A*), the bivalve *Plicatula incongrua* (*B*), and the echinoid *Coenholectypus costilloi* (*C*) from Cretaceous strata in McLennan County, Texas. After Mancini (1978b), Journal of Paleontology, Society of Economic Paleontologists and Mineralogists.

water on a soft mud bottom with many of the species showing adaptations for life on a soft, soupy substrate. One of the main adaptations appears to have been the early development of sexual maturity so that the body size is smaller than in related species living on a more firm substrate. This has resulted in a dwarf or micromorph fauna. The high juvenile mortality in the Grayson species may have been due to difficulty in coping with the soft substrate. Those individuals that were able to survive the early life stage in this rigorous environment were then able to continue life with a relatively low mortality rate. The rate increased again in the gerontic stage. The survivorship pattern for the Grayson species thus supports the interpretation of the depositional environment made largely on the basis of functional morphology.

POPULATION SIZE VARIATION

We have discussed how populations grow from a small starting size to a maximum. This can be viewed as a short-term fluctuation in population size. But population size may also vary on a time scale of years, decades, or longer. Many examples of population size variation have been described, especially in populations in terrestrial environments (see Odum, 1971, for a review). Marine examples are also common (Coe, 1957), although they have not been as extensively described, perhaps because of the greater sampling difficulty.

Modern Examples

Among the examples from terrestrial environments are the lynx and snowshoe hare of the Arctic which vary tremendously in number in 9 or 10 year cycles (Fig. 8.18). Every 9 to 10 years the hare becomes so abundant for some unknown reason that the demand for food exceeds the supply. The populations of lynx, which prey on the hares, also increase as their food supply grows. But when the hare population decreases dramatically (or crashes) the predators in turn suffer high mortality. The lemmings, with a 3 or 4 year cycle, show a similar variation in numbers in the North American Arctic. The migratory locust of Africa and Eurasia is likewise well known for its tremendous variation in population size.

Among the better documented marine examples are the pismo clam (*Tivela stultorum*) from the west coast of North America which has shown large year-to-

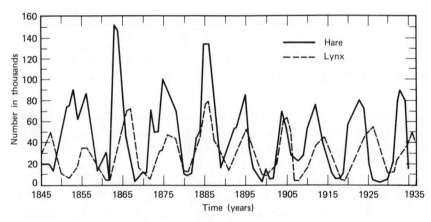

Figure 8.18 Changes in the abundance of the lynx and the snowshoe hare as indicated by the number of pelts received by the Hudson Bay Company. After Odum (1971).

year variations in abundance. Because *T. stultorum* is an important clam for human consumption, careful records of population abundance have been kept for many years by the California State Fisheries Laboratory. The abundance of young individuals especially fluctuates irregularly from year to year. The abundance of older individuals does not vary so much because they include individuals spawned over a period of many years (Fig. 8.19; Coe, 1957). Even larger variation is found in species of the genus *Donax*, the butterfly or bean clam, which may have an abundance of thousands of specimens per square meter in the intertidal zone of exposed, sandy beaches and then may almost disappear for several years. Populations of *Mytilus edulis*, the common edible mussel, often fluctuate abruptly (Coe, 1957). Phytoplankton species show marked short-term fluctuations that are largely a function of nutrient concentration. The infamous red tides are the result of such variation in population size of certain dinoflagellate species. One of the best known recent examples of a great fluctuation in population size has been the tremendous increase in abundance of the Crown of Thorns starfish (*Acanthaster planci*) on coral reefs in the Pacific (Chesher, 1969). In the early 1970s this species had become so abundant as to cause concern about survival of the reefs. The activities of man may have been involved in the case of *A. planci* as well as in some other examples of population explosions. Man has certainly had an effect in the many cases where he has introduced species into new areas (house sparrows, starlings, water hyacinths, etc.).

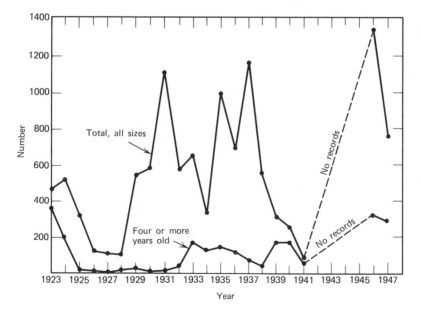

Figure 8.19 Variation in the abundance of the bivalve *Tivela stultorum* between 1923 and 1947 at Pismo Beach, California. After Coe (1957).

The cause for these fluctuations in population size can sometimes be determined but often it is not clear. The cause of the decline in population size is often much clearer than the cause of the increase. Odum (1971) has suggested that four categories of factors are likely to be involved. (1) The most obvious category consists of density-independent environmental factors. Certain physical (temperature) and chemical (salinity) environmental conditions favor the growth of the population, and when those conditions change, the population abruptly declines. Density-independent biological factors may also be involved. Disease and the production of poisonous waste products by other organisms (red tides) are often cited as causes for the catastrophic decrease in population size (Brongersma-Sanders, 1957). A change in larval-bearing currents would be another example of a density-independent factor. (2) Chance or random fluctuation in a combination of physical and biological environmental factors occasionally results in conditions favoring rapid population growth. To an extent, chance factors are probably involved in population fluctuations, but often we invoke the chance aspect to cover up our ignorance of what is really involved. The extent to which biological and geological occurrences of this sort are problablistic or deterministic is a matter of considerable debate. (3) Density-dependent population interactions will certainly affect population growth and density. Clearly the lynx in the Arctic fluctuates in abundance because its most important prey, the snowshoe hare, fluctuates. The predator may simply increase in abundance until it uses up too much of its prey and then crashes in abundance to remain at a low level until the prey has a chance to recover. (4) Complex interactions between all trophic levels of the ecosystem may have large effects on population sizes within the ecosystem. Variations in both the abundance and type of the primary producers particularly may have effects on the abundance of populations throughout the ecosystem.

Opportunistic and Equilibrium Species

All populations vary in size to a certain extent, but some, such as the examples cited above, vary much more than others. Species characterized by marked variations in population size have been called *opportunistic species* because they have the capacity to rapidly increase in abundance when the opportunity arises (Fig. 8.20). MacArthur and Wilson (1967) hypothesize that this is a characteristic evolved by these species to adapt to unstable, variable environments which provide periods of favorable conditions (with opportunities) separated by periods of unfavorable conditions (without opportunities). They have called the evolution of opportunistic species *r-selection*. The *r* refers to the logistic equation (8.5) in which *r* is the intrinsic rate of natural increase. Species populations with a large *r* value have the greatest capacity to rapidly increase in order to utilize trophic resources that have suddenly become available. They are characterized by *J*-shaped population growth curves, by high juvenile mortality and by concave upward survivorship curves (although the highest mortality record may not be preserved because it is in the larval stage).

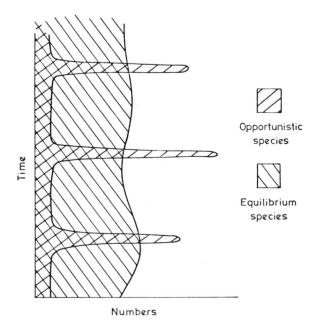

Figure 8.20 Schematic representation of variation in abundance patterns of opportunistic and equilibrium species in a stratigraphic sequence. After Hallam (1972a). From Models in Paleobiology, T. J. M. Schopf, Ed., Freeman, Cooper and Co.

Opportunistic species are generalists that are broadly adapted to a range of environmental conditions, that is, they thus occupy large ecological niches. They are likely to be common in environments dominated by physical stress, as in the case of the snowshoe hare, lynx, and lemming in the harsh climate of the Arctic, or the locust in the desert or semiarid regions. *Tivela stultorum* and *Donax gouldi* are marine examples that live in shifting sands in the turbulent, intertidal zone. Levinton (1970) concluded that opportunistic species are most common in young habitats with high environmental stress and with environmental factors that fluctuate with a period close to the reproductive age of the species. Examples of such environments would be intertidal environments (especially at high latitudes) and estuaries with variable salinity and temperature. Examples of the least favorable situation for opportunistic species would be the stable environments of the subtidal tropics and abyssal deeps. Occasionally, opportunistic species appear in what otherwise seems to be a stable environment, such as *Acanthaster planci* on coral reefs and *Mulinia lateralis* on the stable mud bottoms. The explanation for this is not clear, but even in stable environments conditions vary enough so that food resources may suddenly increase or a predator may disappear for some reason so that a species with high population growth potential suddenly undergoes a population explosion.

Populations that show little variation through time have been called *equilibrium species*. They are in equilibrium with their environment and apparently are maintaining a population size near the carrying capacity, K, for the environment (Fig. 8.20). These species have evolved through *K-selection* (MacArthur and Wilson, 1967) and are characterized by the ability to maintain their population size near the carrying capacity.

Opportunistic species are inefficient in their use of large amounts of energy for reproduction. Neither is their feeding efficiency maximized. They are generalists that can use a broad range of food but are not adapted to maximum efficiency in utilizing any one type of food. Equilibrium species minimize these inefficiencies by producing fewer offspring and thus using less energy for reproduction. They have small ecological niches, being highly specialized for a specific kind of food which they secure with great efficiency. They have S-shaped population growth curves, and their juvenile mortality rate is lower than in opportunistic species and ideally should be constant or highest in gerontic stages. Because of the narrowness of their adaptation, they are morphologically less variable than opportunistic species. Examples of equilibrium species are more common than opportunistic because they tend to occur in diverse communities. The gastropod *Conus* with its poison dart method of food capture, the giant clam *Tridacna* with algal gardens in its siphonal tissues, and deep sea angler fish with luminescent lures are three modern examples.

An equilibrium species requires a stable, dependable environment with a constant food source. The major environmental stresses are biological, from competing populations, rather than physical. They occur in the opposite type of environment from opportunistic species, being especially common in the subtidal tropics and the deep sea.

Fossil Opportunistic and Equilibrium Species

Differentiating between opportunistic and equilibrium species in the fossil record could be a useful tool in distinguishing between physically stable and unstable environments. The great abundance of a given species in a single bed may suggest that it was an opportunist, but the concentration could be due to current sorting. The following criteria are useful for recognizing opportunistic species (Levinton, 1970). (1) The species is very abundant in the assemblage (85-100%). (2) It is widespread in a given isochronous horizon although it may be patchy in its distribution within that horizon. (3) It occurs in several horizons with barren intervals between. (4) It appears in a variety of facies. (5) It occurs abundantly with faunal associations that are otherwise easily distinguishable. (6) Its morphology is variable and suggests a generalized feeding mechanism. These criteria should be combined with evidence for transportation and sorting.

Equilibrium species should have many characteristics that are essentially the opposite of opportunistic forms. (1) The species is restricted to a definite facies. (2) It occurs relatively continuously throughout the fossiliferous section. (3) It may be relatively abundant, but not more than a few percent in a well preserved fauna. (4) It should occur in diverse faunas. (5) Its morphology shows little variation between specimens. (6) It is clearly specialized for a given mode of life.

Although some species are opportunistic and some are equilibrium, most have intermediate strategies. These merely represent the extremes in a continuum. Thus one is not justified in going through a faunal list and designating each species as either opportunistic or equilibrium. Likewise, physically unstable environments do not necessarily include 100% opportunistic forms and stable environments 100% equilibrium. Even unstable environments contain a certain portion of dependable trophic resources that can be used by nonopportunists and trophic resources in stable environments fluctuate so that there is room for some opportunistic or at least semiopportunistic species.

Waage (1968) describes at least two bivalve species from the Cretaceous Fox Hills Formation of South Dakota which appear to be opportunistic. The Timber Lake member of the Fox Hills Formation was largely deposited in an environment that could not be tolerated by readily fossilizable organisms. Occasionally environmental conditions apparently improved allowing the invasion of the area by populations of *Pteria* and *Cucullaea* which apparently increased very rapidly in abundance but were periodically killed. The nature of the fauna suggests an unstable, fluctuating environment.

Alexander (1977) describes six brachiopod species ranging in age from Cambrian to Triassic which he considers to be opportunists. He uses the criteria given by Levinton (1970) as his major basis for this conclusion. All of the species are small and thin shelled and have minimal ornamentation. Growth line analysis suggests that they grow rapidly during their first year of life. Five of the species apparently lived on a fluid substrate in a generally unstable environment, which accounts for their opportunistic characteristics. They appear to be adapted to maintain themselves on the soft substrate in an environment with low oxygen concentration. One of the species apparently lived in an area of mobile (shifting) substrate that eliminated most other species.

DISPERSION PATTERNS

Individuals within a population may be spatially distributed in three basic patterns: (1) *random*, (2) *regular* (*overdispersed*), and (3) *clumped* (*underdispersed*) (Fig. 8.21). Dispersion can be viewed on many different scales, and the type of dispersion pattern may vary depending on the scale. On the global scale all populations show clumping. For example, populations of species living in the shallow sea will be found only on the shelf areas and will be absent from the broad expanses of open ocean and from terrestrial areas. Thus they will be clumped in the shelf areas. Concentrations or clumps of species might be found on a given segment of the shelf, but these clumps may be randomly distributed. Within an individual clump the distribution may be evenly spaced or regular. Thus a single species may show all the basic patterns of distribution depending on the scale of the observation. The type of dispersion pattern may be obvious in some cases but not in others. In the latter case a statistical test must be used to determine the type of distribution. Standard statistics texts, such as Sokal and Rohlf (1969), discuss the procedure for making such a test.

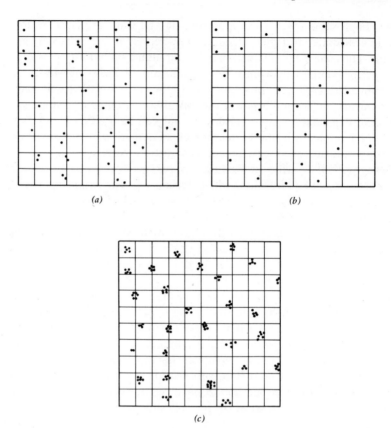

Figure 8.21 Random (*a*), overdispersed or regular (*b*), and underdispersed or clumped (*c*) distributions of points. Each of these distribution patterns can be recognized in organisms. After MacArthur and Connell (1966).

Random Distribution

In a random distribution pattern, the location of any individual is independent from the location of all other individuals. Because of the method of distribution of many species by the apparently random settling of larvae, this distribution might be expected to be very common among invertebrates. Numerous studies have shown that larval settling is not really random but that the larvae of many species can, to at least some extent, select the site where they settle. Based largely on distributions of terrestrial organisms, Odum (1971) states that random dispersal patterns are uncommon. Jackson (1968) observed a random distribution on mud flats of the suspension feeding bivalve *Mulinia lateralis*. Two-year-old individuals of the bivalve *Gemma gemma* from the same area also were randomly distributed although younger individuals showed clumping. This is probably because *G. gemma* is ovoviviparous, that is, the eggs hatch and the larvae develop within the shell of the mother. Thus the very young are concentrated around their mothers. With time they disperse enough to assume a random distribution. The opposite pattern may be expected to occur at times, that is, the very young may have a random distribution because of larval settling, but with time those individuals that settled in less favorable localities will die leaving the older individuals in a clumped distribution.

A random distribution may indicate one or a combination of several factors. (1) The environment in the area is uniform (or the species is able to tolerate a wide range of environmental variability). (2) Individuals of the population do not interact strongly with one another. Negative interactions tend to result in regular distribution and positive interactions result in clumping. (3) Currents may move the individuals about, giving them a random distribution. (Under certain conditions currents may produce clumps of shells.) (4) Reproduction (especially in sessile forms) is by random settling of larvae.

Regular Distribution

Regular distributions have been more commonly described in terrestrial environments. This distribution is the result of either an active or a passive effort to minimize competition between individuals in the population. Regular distributions may result from the dying of young individuals or larvae that have settled too close to previously established individuals. In vagile forms (or by larval selectivity) the spacing may result from individuals moving to unoccupied sites where there is a maximum of distance from other individuals. Regular distributions are most obvious and have been most actively studied in vertebrates such as birds, which often exhibit strong territorial behavior. Plants also commonly have regular distributions because of the need to minimize competition for light or moisture. Invertebrates sometimes show regular distributions (usually within clumps) because of competition for limited space. The crowded but regular distribution of barnacles or mussels on rocks in the intertidal zone is an example.

Clumped Distribution

All species are clumped on one scale or another. Indeed, the concept of the population almost implies clumping. An individual population represents a clump or aggregation that is separate from other aggregations. Clumping may simply result from inhomogeneities in the distribution of suitable habitat for the species. The clumping of organisms requiring a hard substrate on isolated rocks would be a small scale example of this. On the other hand, clumping may result from a positive interaction between individuals of the species. Extreme examples of this would be in the social insects such as bees and in colonial organisms such as corals. Perhaps the ultimate modern example of the positive interaction of individuals within a species (and the ultimate clump?!) is the coelenterate *Physalia,* the Portugese-Man-of-War. In *Physalia* the individual almost loses its identity and becomes an organ for the superorganism. Some individuals specialize for feeding, others for defense (much to the dismay of the swimmer who is stung by them), and others for flotation (Barnes, 1974), but all are clumped into a single jellyfish.

Clumping may be achieved by several processes. (1) The Larvae may postpone metamorphosis until they contact a suitable substrate. (2) The larvae may settle randomly but die out between clumps. (3) The food supply may be concentrated in certain areas so that the individuals move to those concentrations. (4) The young may be born live, have an extremely short larval stage, or develop from eggs deposited near a parent so that the young concentrate in the area of the parent. (5) Currents may physically concentrate the living specimens.

Clumping may offer a number of advantages to the individual. (1) The baffling effect of an aggregation of individuals such as crinoids or alcyonarians reduces the current velocity sufficiently to allow suspended food in the water to be captured. (2) The clumped distribution is also an obvious advantage in reproduction. (3) The clumping or schooling of individuals provides a defensive advantage to many swimming fish and invertebrates as well as many terrestrial vertebrates such as the caribou and bison. Predators that will attack isolated individuals often hesitate to attack a school or herd. (4) Many additional advantages accrue to the individual if clumping results in social and colonial animals. In this case, individuals may become highly specialized, so that the colony can operate more efficiently than separate individuals. The massiveness of a colonial skeleton such as a coral allows existence in a very turbulent environment. Coordinated beating of cilia allow the production of much stronger food-bearing currents than would be possible by the individual. The defense of a colony such as a stinging coral or Portugese-Man-of-War is much more effective than that of a single, small individual. Many additional advantages of the colonial and social existence could be cited (Wilson, 1975).

The individuals within a colony or other clump of individuals are often very evenly distributed. In some cases this may simply be a result of close packing of individuals. But at least in some cases it appears to have developed to minimize competition on the fine scale while taking advantage of clumping on the slightly larger scale.

Dispersion in Fossils

The comments on dispersion, to this point, have been concerned with living organisms. What about fossils? Several studies (e.g., Johnson, 1965; Warme, 1969; and Peterson, 1977) have compared the distributions of dead shells (potential fossils) and living organisms. The degree of correspondence depends on the species involved, the type of environment, and various diagenetic or taphonomic processes; but a few generalities emerge. Originally clumped or regular dispersal patterns will usually tend to become more random in the fossil record. Any post-mortem transport of skeletons by currents or other factors will usually tend to randomize this distribution. In some cases, however, currents may concentrate shells of certain sizes and shapes and produce clusters from originally random distributions. Jackson (1968) suggests this as a possibility for dead *Gemma gemma* shells on modern mud flats in Connecticut. Perhaps even more important is the fact that with time through successive generations the location of clumps will change so that shells in a bed that formed over a period of many generations will tend to lose their clumped pattern and become more random. Consequently, one cannot assume that all fossils that are found together actually lived together at the same time (Fürsich, 1978) and that randomly distributed fossils were not originally clumped. The diversity in an accumulation of fossils is commonly greater than that of the living organisms that actually occupied the spot at any one time if the fossil accumulation consists of elements of a number of clumps living there over a period of time. Certain clumps do tend to perpetuate themselves in the geologic record and do not migrate in time. The most obvious examples are reef or bank assemblages. The reef owes its existence to the fact that the organisms form a clump and do not migrate. Thus the fossil record may be a poor representation of the life-dispersal patterns. As is so often the case, specific examples may, however, give an excellent record of the living patterns. Perhaps the best record would be in an area of low turbulence but rapid sedimentation rate.

The paleoenvironmental interpretation of dispersal patterns of fossils, if recognized, has not been well studied, but a few speculations are possible. Clumped distributions probably indicate little post-mortem transport, especially if other indicators also suggest minimal currents. Other factors being equal, random distributions probably suggest unstable environments and opportunistic species because these species interact little with other species. In stable environments with highly specialized species, clumping should be more common because interactions between species are likely to be greater.

The following is an example of how dispersal patterns might be used to solve a paleoenvironmental problem. An accumulation of skeletal material in the rock record might be interpreted as a sedimentary accumulation produced by current transport and winnowing. Or, alternatively, it might be interpreted as an *in situ* accumulation of organisms living on an organic bank structure. A highly clumped distribution of the fossil populations within the structure might suggest an *in situ* accumulation because current transport would randomly mix the fossils. A random distribution within the populations may indicate current transport that resulted in scattering of any original clumps.

MORPHOLOGIC VARIATION

No two individuals of a population are exactly alike, and an important characteristic of a population is the morphological variability among individuals of the population. The amount and nature of this variation is in part determined by environmental parameters. Some species, particularly those that are highly specialized, have a small range of variability. Other species, especially those that are more generalized, tend to be morphologically more variable within a given population. The morphological variability may either be the result of *genotypic* (under direct genetic control) variability or of *phenotypic* (not directly controlled by genetic differences) variability.

All sexually produced organisms have some genetic variability between individuals. This is of course the basis for evolution through natural selection. Some of the morphological variability that results from the genetic variation probably has no functional significance, but some variants will be slightly better suited for a given habitat than others. Thus the genetic variability would itself have a selective value for organisms living in variable environments. Using an electrophoresis technique to determine the relative number of genetic loci that are polymorphic, Ayala et al. (1973, 1975) and others have measured the amount of genetic variability within populations. Surprisingly, the available data indicate more genetic variability in forms from stable environments than unstable. The explanation of this still is not clear. Apparently this indicates, however, that morphological variability within populations of invertebrates is due more to phenotypic than genotypic causes. Ayala et al. (1975) suggest as one explanation that genetic variability among specialists allows them to become even more highly specialized on an individual basis to their particular habitat.

Bonner (1965) has suggested two types of phenotypic effects on morphology. One, which he calls *multiple choice variability,* results in two or more morphologies being possible for the species. Each type can be called an *ecophenotype* of the species. The particular morphology that develops is determined by the environmental conditions under which the species grows. An example of multiple choice variability can be found in the coralline alga *Goniolithon* (Bosence, 1976) which may develop a crustose form under high turbulence conditions, a short branching form at intermediate conditions, and delicately branching form in quiet conditions (Fig. 2.4). This ability to develop a number of morphologies is due to the genetic constitution of the species but each morphologic type has the same basic genetic makeup.

The other type of phenotypic variation is called *range variation.* During development of the species, a certain range of morphologies is possible because of the multitude of biochemical processes involved and the effect of environment on these processes. In some cases a certain variant suits the organism to its particular habitat, and in other cases there is no obvious advantage. The variation in eccentricity of modern specimens of *Dendraster excentricus* (Raup, 1956) and in Pliocene species of that genus (Stanton et al., 1979) may be examples of range variation (see Chapter 5).

POPULATION STUDIES ON FOSSILS FROM THE
KETTLEMAN HILLS

The morphology of the genus *Anadara* has been analyzed in the Pliocene strata of the Kettleman Hills (Alexander, 1974). A number of morphologic features in

Anadara populations are correlated with environmental conditions. The distribution of mean specimen sizes in populations in the *Pecten* zone indicates that specimen size decreases with salinity of the depositional environment (Fig. 8.22). The largest specimens are found on the east flank of North Dome, an area of normal marine salinity as indicated by various lines of evidence such as taxonomic uniformitarianism, oxygen and carbon isotopic composition, and diversity. Specimens from lower-salinity areas such as the west flank of North Dome and the east flank of Middle Dome are smaller. This is probably an ecophenotypic effect. From comparison of mean size with paleotemperature, as determined by faunal (Chapters 2 and 9) and geochemical (Chapter 3) evidence, size is positively correlated with temperature. The mean value of the number of ribs on *Anadara* shells is positively correlated with salinity. Specimens collected on the east flank of North Dome from the *Pecten* zone have as many as 28 ribs on the shell (Fig. 8.23). Specimens from lower salinity areas usually have fewer ribs, and one sample from the west flank of North Dome has a mean of only 24. The positive correlation between rib number and salinity is not strong but is similar to the results of other studies. Eisma (1965) noted a similar reduction of

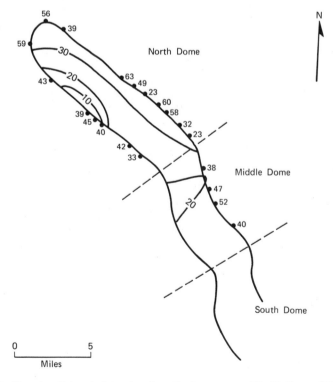

Figure 8.22 Mean length (mm) of *Anadara* from the *Pecten* zone of the Kettleman Hills Neogene. Numbered lines are salinity isopleths in per mille as determined from oxygen isotope analysis (see Chapter 3). After Alexander (1974), Journal of Paleontology, Society of Economic Paleontologists and Mineralogists.

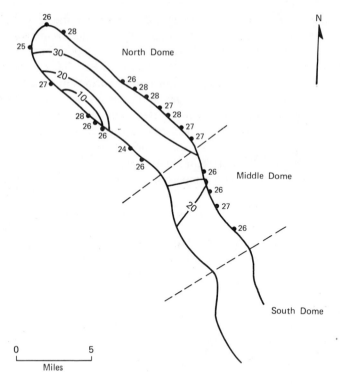

Figure 8.23 Mean rib frequency (ribs per valve) in *Anadara* from the *Pecten* zone of the Kettleman Hills Neogene. Numbered lines are salinity isopleths in per mille as determined from oxygen isotope analysis (see Chapter 3). After Alexander (1974), Journal of Paleontology, Society of Economic Paleontologists and Mineralogists.

rib frequency with salinity in populations of modern *Cerastoderma edulus* from the Netherlands (see Fig. 2.33). He interpreted this relationship as being due to the increasing undersaturation of the low salinity sea water with respect to $CaCO_3$.

Eccentricity in populations of *Dendraster* (discussed in Chapter 5) is another example of morphological variation with a functional significance in populations of Kettleman Hills fossils.

The population dynamics of *Anadara* and *Dendraster* have been studied in the Kettleman Hills (Alexander, 1974). In the *Pecten* zone, the size-frequency distributions of three of four samples that were collected from a matrix that had 30% or more of silt and clay had strongly positively skewed distributions indicating relatively high juvenile mortality. One sample, from a matrix with 49% silt and clay, had a nearly symmetrical distribution suggesting low juvenile mortality. One sample from a matrix with only 4% fines was strongly positively skewed. Thus the relationship of high juvenile mortality to fine-grained sediments found by several other authors, applies in some cases to Kettleman

Hills *Anadara* populations. Alexander did not show survivorship curves so we do not know the details of the mortality pattern.

We also have accumulated considerable size-frequency data for *Dendraster* in connection with our study of eccentricity in this echinoid species (Stanton et al., 1979). *Dendraster* is far from the ideal subject for analysis of population structure, but it can be used to point out some of the problems encountered in studies of this type. Modern sand dollars generally live in relatively coarse, well-washed sands indicating at least moderately turbulent conditions. Consequently, transport and winnowing might be important processes in generating the size-frequency distribution of the fossil assemblage. The size-frequency curves for 13 populations are approximately symmetrical (Fig. 8.24) with most being slightly negatively skewed. This suggests a low juvenile mortality. Survivorship curves were constructed from the size-frequency data (Fig. 8.24). Size

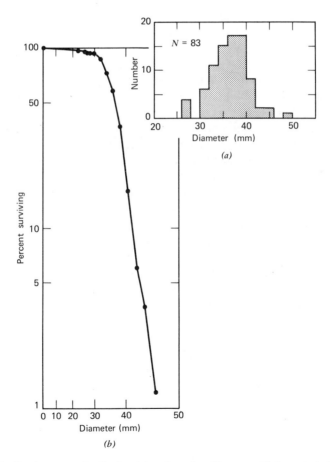

Figure 8.24 Size-frequency distribution (*a*) and survivorship curve (*b*) for a typical population of *Dendraster* from the Neogene of the Kettleman Hills. After Stanton et al. (1979).

was plotted on an arithmetic scale. It was also plotted on a logarithmic scale (as was done by Thayer, 1977) in order to correspond more directly to age. These plots suggest a low juvenile mortality (in fact a zero mortality until a diameter of 15–20 mm is reached!). Mortality rate gradually increases and then becomes approximately constant as the maximum size (age?) is approached. The survivorship curves (Fig. 8.24) have the same general shape as the brachiopod populations from firm substrate studied by Richards and Bambach (1975; Fig. 8.15). Incidentally, this is close to the shape of the survivorship curve for humans!

The apparently extremely low juvenile mortality rate shown by the survivorship curves reflects the almost complete lack of small *Dendraster* fossils. In fact, the smallest *Dendraster* specimen in the 703 specimens we measured has a diameter of 12 mm. Taken at face value, this means that young *Dendraster* specimens never die, hence we never have small fossils. We can never recall seeing very small *Dendraster* specimens in our many collecting trips in the Kettleman Hills. Almost certainly small *Dendraster* specimens occasionally die so there must be some other explanation for the lack of small specimens. The possibility that the young live in a different environment than the adults and we have not sampled that environment is also unlikely. Cadee (1968) explained the lack of juvenile shells in some mollusc species in a Spanish bay in this way. *Dendraster* might be able to migrate a short distance during its life time but probably not more than a few km at the most. We have not found juvenile specimens even over many km of outcrop.

Small specimens are mechanically more fragile than the larger ones, hence they might be physically destroyed in the rather turbulent environments inhabited by *Dendraster*. The small echinoid test is certainly more fragile than a larger one, hence more subject to physical destruction in turbulent environments. *Dendraster* skeletons are probably also chemically less stable than most of the invertebrates in the area and small specimens may be dissolved. Peterson (1976) conducted an experiment in which he measured weight loss in a number of molluscs and in modern *Dendraster* due to solution after burial in the sediment in Magu Lagoon, a modern bay in California. *Dendraster* lost weight more rapidly than any of the molluscs. In fact, Peterson estimated that *Dendraster* specimens would last for only four years in the experimental location. The rapidity of solution is probably due to the high porosity of the skeleton and its high-Mg content. Specimens certainly did not dissolve this rapidly in the Kettleman Hills Pliocene, but small specimens would dissolve more quickly and may have been destroyed by this process. However, specimens only a few mm in diameter of another echinoid, *Merriamaster*, are found in the strata in lithologies which are apparently identical to those containing *Dendraster*. Any physical or chemical process that would eliminate small specimens of *Dendraster* should also eliminate *Merriamaster*. Therefore we conclude that destruction of small specimens is minimal. We believe the most likely explanation is low juvenile mortality combined with growth rate in young specimens that is fast enough that individuals do not remain small very long. Hence few small specimens are found. There is certainly an advantage to an echinoid living in a turbulent environment to grow quickly and thus increase stability as rapidly as possible.

Other population parameters have not been systematically investigated in the Kettleman Hills but certainly could be applied. Some of the species are more opportunistic and others more of the equilibrium type. In the *Pecten* zone and the upper *Mya* zone, oysters (*Ostrea*) and associated barnacles (*Balanus*) are quite common and often occur in great abundance associated with a fauna of very limited diversity. This suggests the rapid population growth characteristic of *r*-selection. The oysters and barnacles occur in a reduced-salinity, low-stability environment. Equilibrium species in the Kettleman Hills are more difficult to identify, but many of the species in the higher, more stable salinity environments of the east flank of North Dome probably should be classified as equilibrium species (e.g., the corals *Astrangia* and *Rhizopsammia,* the brachiopod *Terebratalia,* or the bivalve *Chama*). In general, the Kettleman Hills environments were not highly stable, hence very specialized, *K*-selected species are not common.

Patchy population distributions are common in the Kettleman Hills. The oysters and barnacles, especially in the upper *Mya* zone, are good examples of patchiness. They occasionally occur in great abundance but at some areas are completely lacking. On the other hand, the fairly regular distribution of fossils in the area at given horizons is expressed by the naming of the zones. Almost everywhere throughout the hills there is a fossiliferous horizon at the base of the *Pecten* zone. No study has been made to see how regular the distribution is within this or other horizons.

9

Ecosystems and Communities in Paleoecology

Much of the material of paleoecology and its interpretation is concerned with single taxa and individuals. Examples of this comprise the bulk of the preceding part of this book. An example would be the analysis of the chemistry, microstructure, and morphology of a fossil as these attributes were determined by the particular environment in which the fossilized organism lived (Chapters 3 and 4). Other examples would be the study of population characteristics of a species as they reflect the adaptations and survival of the population as an enduring unit within its physical and biological environment (Chapter 8). The method of taxonomic uniformitarianism, in which the tolerances and preferences of living organisms are applied to fossil representatives of the same taxon in order to determine paleoenvironmental conditions (Chapter 2), is another example.

The general approach exemplified by these techniques, in which the focus is on the characteristics of the individual organism or population, is referred to as *autecology*. The contrasting approach, of *synecology*, focuses on describing, understanding, and interpreting organisms in the context of the other organisms with which they coexisted. This type of analysis can take place at two levels: (1) study of the organism in relation to associated organisms emphasizing interactions, interdependencies, and coevolution within the framework of a common environment and (2) analysis of the assemblage of co-occurring organisms, the *community*. The second approach, the main topic of this chapter, depends on those aspects of the community that are controlled by the environment and are preservable in the fossil record, and thus are potentially useful in the reconstruction of ancient environments.

The motivation for this approach has been the early and continuing recognition that the world is subdivided geographically into biological units ranging in size from broad regions encompassing several continents or large parts of

382

oceans to areas so small that they could be viewed at a single glance. Regardless of the scale, each is characterized (1) by compositional and structural attributes by which it can be readily distinguished from other units at that scale, (2) by an internal homogeneity such that the differences between samples or localities within the unit are less than between samples from different units, (3) by boundaries that can be readily defined, although they may be gradational and difficult to locate in detail, and (4) by temporal persistence and geographical recurrence. These characteristics are meaningful although expressed qualitatively; they are developed more precisely and quantitatively in this and following chapters.

Ecologic units of a wide range of size have been described. For any area, they can be arranged in a nested hierarchy with those at each level encompassing several of the next lower level. The broader, regional units are essentially geographic, and determined by the physical environment. The analysis of these larger biotic units, such as *realms, biomes,* and *provinces,* is reviewed in Chapter 10, on *paleobiogeography.* At the other end of the scale, structural characteristics and interactions between species become increasingly important in characterizing and recognizing the small units. In this chapter, these smaller units, the *ecosystem* and community, are discussed.

The community has been an exceedingly popular topic of study in ecology. As a result, the biological foundation for the paleoecologic analysis of communities is well established. However, many of the ecosystem concepts and models that have been developed by theoretical ecologists appear to be of limited value in paleoenvironmental reconstruction. This is, in part, because community characteristics essential to ecological theory are generally not preserved in the fossil record. However, the incorporation of ecologic concepts and methods into paleoecology is a slow and continuing process. Thus concepts that presently appear to be of little use in the reconstruction of ancient environments may become useful as our knowledge increases. A recent collection of papers providing an overview of current theory and understanding of community ecology is *Ecology and Evolution of Communities* (Cody and Diamond, 1975). Community concepts and applications in paleoecology are surveyed in *Structure and Classification of Paleocommunities* (Scott and West, 1976).

THE COMMUNITY CONCEPT

The description by Möbius (1877) of the ecosystem of a North Sea oyster bank provides a valuable starting point in the discussion of the community concept. This paper presented for the community of the modern oyster bank the essential point that, when the total biota and the physical environment are considered as a unit (the *ecosystem* in present terminology), it is clear that the diverse elements of the ecosystem strongly interact. As any one element of the environment changes, other elements, such as the abundance of a certain species, will change. Consequently, still other aspects of the ecosystem, both physical and biological, will be modified. In spite of such short-term fluctuations, however, the gross aspect of the ecosystem remains constant because of the long-term stability of the physical environment and because of the cybernetic interactions of the taxa that are present.

The definition of community presented at the beginning of the chapter is very general. Several specific terms have been used in ecology and paleoecology for more precise definition of the community. The total biotic component of the ecosystem is the *biocoenosis*. The biocoenosis is seldom studied in ecology because of the difficulties inherent in so comprehensive an analysis. The biocoenosis cannot be described or analyzed in paleoecology because many of the original taxa are not preserved in the fossil assemblage. Community analysis in ecology is generally based on only one or a few taxonomic groups, such as birds, corals, or molluscs. Community analysis in paleoecology is also based on one or a few taxonomic groups, such as foraminifers, brachiopods, or terrestrial vertebrates. Commonly, the community analysis of marine faunas in paleoecology is the total preservable part of the biocoenosis, comprising, for example, the shelled benthic macroinvertebrates. These partial representations of the biocoenosis have been named *organism communities* by Newell et al. (1959). The general term community is useful because of the broad concepts associated with it. The distinction between biocoenosis and organism community is important because it emphasizes the difference in information content and therefore the potential difference in completeness and precision of interpretation.

Inherent in the community concept, as epitomized by the oyster bank community of Möbius (1877), is the notion that organism interactions are critical. They determine the composition and structure of the community at any instant, as well as the changes that take place during community succession. The extreme expression of this notion is the analogy of a community as a superorganism, with the component species representing different organs that all interact for the benefit of the body or community (Clements and Shelford, 1939). The alternate extreme notion is that communities do not exist as real entities but are only statistical constructs devised by man to subdivide the biological continuum (Muller, 1958). According to this idea, organisms that characteristically cooccur are associated primarily by common physical-environmental tolerances and preferences rather than by biological interdependencies, and lateral biotic changes are gradational rather than sharp discontinuities bounding distinct and internally homogeneous communities. Each of these contrasting points of view has been championed in numerous papers during the past 50 years. As in most controversies, examples of either extreme point of view may be cited, but intermediate examples are even more common. Consequently, modern concensus would tend to avoid either extreme (Fager, 1963). The heuristic approach is to note that communities with the general characteristics noted above are commonly recognized in the living world and that, although the relative importance of the physical environment and organism interaction as the control on community composition and structure is generally indeterminate, the community is a worthwhile subject of study in attempting to understand the nature of the ecosystem.

In paleoenvironmental studies, the theoretical question of environment vs. biologic interaction is of real significance if we express the situation in the terms of the equation: community characteristic $= f$ (environment, biologic interaction). It is evident that, to determine the ancient environment from the community characteristics, we need to know the interaction term. (This is true, as well, in interpreting individual taxa by the taxonomic uniformitarian approach of Chapter 2). This is largely impossible, however, because most of the biocoenosis, and therefore most of the interactions, are not preserved in the fossil record. In fact, in much of paleoenvironmental reconstruction based on community analysis, the role of the interaction term is not discussed, and it is presumably assumed that the interaction term is 0 in the equation above. (In contrast, much current literature on the potential environmental impact of the extinction of endangered species considers that interactions are a major term in the equation—that extinction of only a few species may so upset community structure that the ecosystem will become unstable and collapse.) By and large, the application of the community concept in paleoenvironmental reconstruction has been on a very meager theoretical foundation. In particular, the use of the community has been very pragmatic without being concerned about whether community boundaries are abrupt or gradational and whether a community is a real, integrated structure or only the confluence of overlapping ranges of individual taxa, perhaps bounded by locally somewhat steeper environmental gradients (Hoffman, 1979).

Many of the theoretical aspects of community ecology remain unresolved; others are not of immediately apparent application in paleoenvironmental analysis. We put aside these uncertainties and attempt to develop applications of the community as an operational unit of potentially great value in a wide range of paleoecologic problems.

The use of the term community in paleoecology must be made clear because of the wide range of possible meanings that can be derived from ecology. First, the basic paleontologic unit, the *fossil assemblage,* clearly is quite different from the original biocoenosis because of the meager preservation of the original biota. Taphonomic analysis of the fossil assemblage may improve our knowledge of the biocoenosis but generally cannot go far toward reconstructing it in any detail (Stanton and Nelson, 1980). Thus reference to the assemblage as a community should generally not imply any knowledge about the biocoenosis. If defined as a representation of the organism community consisting of the skeletonized and readily preserved organisms in the original biocoenosis, the fossil assemblage may be comparable to the organism communities studied by the ecologist. Consequently, for example, the ecologic information gained from the study of the skeletonized organisms in modern coral-reef communities is applicable to coral-reef communities in the fossil record.

As the community concept has been carried over from ecology to paleo-ecology, use of the term community has become fashionable but in many cases has been used merely as a synonym for assemblage. This is worthwhile if it stimulates visualization of the fossils as part of an ecosystem. However, the term community is most accurately used when the relations between the fossil assemblage, organism community, and biocoenosis are clearly spelled out and form an integral part of the analysis.

The community has been used in paleoecology to study three types of problems. One is paleoenvironmental reconstruction, the major concern of this book. The second is the evolution of biological organization or ecosystem structure during earth history. The third is the role of the interactions of the organisms in the community, the community structure, on the evolution of individual species.

Two distinct aspects of the community have been most studied and used in paleoenvironmental interpretations. One is the taxonomic composition. This has most generally been in terms of presence or absence of taxa. Although the relative abundances and population dynamics of the component taxa should no doubt also be considered (Hoffman, 1979), this is difficult to do and the results may not be meaningful anyway because of the imperfect preservation of the biocoenosis. The other aspect of the community is its structure, as expressed by diversity and dominance on the one hand and trophic relationships on the other. Analysis of the first aspect is essentially analogous to the taxonomic uniformitarian method in autecology. Analysis of structural characteristics independent of the taxonomic composition is analogous to the analysis in autecology of characteristics such as functional morphology and population dynamics of individual taxa, which are more or less independent of the taxonomy of the organism being analyzed, and thus, it is hoped, less time dependent.

The term community has been used for entities of a range of magnitudes. In addition to the usage here, it has also been applied to geographically broader units determined essentially by uniform environmental conditions. This type of community has been used most commonly to characterize depth zones (Natland, 1933). We believe that these two types of communities are different in concept as well as scale and should be distinguished in nomenclature. Therefore, we would concur with Watkins, Berry, and Boucot (1973) and refer to the broader units as *life zones*.

COMMUNITY RECOGNITION

The distribution of species in nature is generally complex. The area in which each species is found may have irregular boundaries, be discontinuous, and fluctuate. The abundance of the species may differ widely and irregularly. If all the species are considered together, few will have coincident boundaries and abundance distributions. The problem, then, in community recognition is to identify the natural groupings of organisms that reflect both the distribution pattern of environmental parameters and the biological structural characteristics.

Because the number of organisms may be great, the tendency, particularly in the days before large-capacity computers, was to use only the most abundant species, the *dominants,* in the analysis. Now, however, even the rarest species can be included in the analysis. Two basic approaches, Q-mode and R-mode analysis, have been used in community recognition. Both are valid but lead to different results and interpretations.

Q-Mode Analysis

Samples are compared, and those that are relatively similar are grouped together into communities so that the average sample similarity, or homogeneity, is greater within than between communities. This has been the more common approach in ecology and paleoecology. Petersen (1915) used Q-mode analysis in a qualitative way in his classic study of the living biota of the Kategat. The subsequent restudy of his original data by computer based Q-mode analysis (Stephenson, Williams and Cook, 1972) provides an excellent comparison of the results of community recognition before and after the advent of computers. Q-mode analysis is the approach used in quadrat sampling in plant geography and it is the approach used by the marine biologist in observing successive grab or dredge samples brought on board the ship. Boundaries drawn between adjacent communities form a map description of the distribution of the communities. In geologic terms, the rock unit containing a community is a *biofacies.* Any time horizon through the biofacies represents the mapped distribution of the community on the earth's surface at that time and is equivalent to the present-day distribution pattern of a community. Comparison of samples in Q-mode analysis is largely on the basis of taxonomic composition but it may be based on other attributes such as morphologic characteristics, as has been the case in the description of plant associations.

R-Mode Analysis

Individual taxa are compared in terms of their distributions in the samples. Those taxa that cooccur are grouped together whereas taxa with distributions that are mutually exclusive are not strongly correlated and are placed in different communities. The communities resulting from either Q-mode or R-mode analysis are characterized by the species that are common and have a high degree of *fidelity* (are largely restricted to a single community). Associated species in a community would be those less abundant but still with high fidelity, and common species with low fidelity. The latter group grades into those that are nondiagnostic, even though they may be abundant, because they are ubiquitous in the samples studied.

The advantage of the R-mode approach is that it emphasizes patterns of cooccurrence and mutual exclusion among species and thus points out possible biological interactions. On the other hand, the assemblage in an individual sample is commonly difficult to relate to the communities established by R-mode analysis because the samples may contain species characteristic of more than one community and, at the same time, lack some of the species considered diagnostic of any one community. As a result, R-mode communities are more difficult to map and thus to use to define biofacies than Q-mode communities. Because the Q-mode communities as defined form mappable units, which can be readily compared to sedimentary facies, the Q-mode approach has been the more common method in paleoecology.

By either method of definition, the community is based on the organisms and their distributions. Communities may also be defined on the basis of characteristics of the physical environment. Without knowing about the biota, we may recognize habitats such as bay or ocean, lake or stream, and shelf or abyssal, and then describe as habitat communities the biota of each (Newell et al., 1959). Paleontologic examples would be the "red-bed community" or the "black-shale community." These are valuable as generalized descriptive entities. They are conceptually deficient, however, because they are based on the assumption that the human observer can recognize the significant environmental gradients and tolerances that control the distributions of the organisms and communities. Because the biota reflects the complex and multidimensional environment as it is, rather then how we imagine it might be, community recognition based on the organisms themselves results in more real and precise community boundaries and characteristics than if based on perceived habitats.

An essential criterion used in the recognition of biocoenoses and organism communities is that sample-to-sample differences (*heterogeneity*) within a community are less than between communities. Heterogeneity within a community is the result of patchy, uneven, distributions of taxa. The extent to which patchiness in a community is confused with between-community heterogeneity depends on sample size and density relative to patch dimension. General guidelines for designing the sampling program for community paleoecology are given by Stanton and Evans (1971) and by Harper (1977).

TAXONOMIC-UNIFORMITARIAN ANALYSIS

The taxonomic-uniformitarian approach in paleoecology has historically been the major method of determining ancient environmental conditions and is discussed comprehensively in Chapter 2. The essential steps in the method are: (1) list the fossil assemblage, (2) list the comparable living taxon for each of the fossil taxa, (3) list for each environmental parameter, such as depth or salinity, the tolerances of each of the modern taxa, and (4) allowing for taphonomic biasses in the fossil assemblage, assume that the overlap in tolerances of the assemblage of modern taxa defines the probable range of values for that parameter for the fossil assemblage. The method relies on the substantitive uniformitarian assumption that an environmental tolerance for a fossil taxon is like that of the closest living relative. This becomes increasingly dubious with geologic age and with departure in relatedness between fossil and modern analogue. In addition, the method depends on the abundance and precision of environmental data available for the modern analogues. These data are generally sparse. For example, temperature, salinity, and water depth are the most basic parameters of the marine environment but the limiting values of these for a species are generally not known. Instead of actual established data, tolerances are generally estimated from limits of geographical distributions. Other potentially significant parameters such as the ranges of variability of temperature, salinity, and depth are seldom available, and very little is known about the effect of one parameter on the tolerance of an organism to another parameter.

The use of the community rather than the species as the entity with which to transfer modern environmental information into the geologic record results in a more integrated and comprehensive interpretation. Instead of separating the fossil assemblage into component taxa and estimating, then, for the assemblage, the value of each individual environmental parameter from the overlap of the values for each of the modern related taxa, the assemblage as a whole is compared with modern communities; and the modern environmental conditions of the most comparable community are applied to the fossil assemblage. The uniformitarian assumption basic to the method imposes the same limitations as when the individual taxa are used as transfer entities. However, the advantage of the community approach is that when a modern community has been identified as most similar to the fossil assemblage, as many parameters of the modern environment as desired, beyond the normal few available for individual taxa, can be analyzed. The significant aspects of the environment leading to the differences in the modern communities can be determined and this knowledge can then be applied to the fossil assemblage to provide a broader and more comprehensive reconstruction of the paleoenvironment.

In addition to comparison of fossil communities with modern ones, comparisons between fossil communities on the basis of taxonomic composition are also common. This may lead to inferences about similarities or differences in the environment and about the course of evolution within the community either by changes in community structure or by the replacement of one species within a niche by another.

COMMUNITY STRUCTURE ANALYSIS

Using the community, as defined and compared by its taxonomic composition, as an information transfer unit is most effective in the Neogene. The age difference between modern and older fossil communities results in decreasing taxonomic similarity because of evolution, extinction, and changes in range. Thus the information that can be transferred by this method becomes progressively more general, potentially to the point of being trivial. The alternative approach that somewhat gets around the problem of decreasing taxonomic similarity with increasing geologic age is to analyze those structural characteristics of the community that are determined by the environment, that are independent of the taxonomic composition of the community and thus, that are relatively independent of age. The possibility of utilizing such characteristics is reinforced by the observation that structural characteristics in different communities from similar modern environments are similar. This is true even though the communities may be from different parts of the world and may be taxonomically very different, at least at the species or genus level. This leads to the working hypothesis that each environment requires certain physiologies and life histories of individual taxa and thus certain structural characteristics at the community level. These environmentally determined attributes are inherent in the community although the species making up the community may have evolved independently, from different preexisting faunas, in the different parts of the world.

The community-structural approach is readily applied in the study of Cenozoic communities, in which direct comparison with the structure of similar modern communities is possible. In the study of more ancient communities, comparison of the fossil community with modern communities is less precise because of the possibility of evolution of community structure itself. In this case, gradients and relative differences in structural measures are most useful because absolute values of the measures may have changed as individual taxa and the community evolved. Two principal community characteristics have been used in paleoenvironmental reconstruction. One is *diversity* and the related measures, *dominance* or *equitability*. The other is *trophic structure*. It is important to note that, although both of these characteristics are community attributes, they can be measured by means of a variety of indices and analyzed without necessarily determining the community composition and other attributes.

DIVERSITY TERMINOLOGY

Patterns of diversity among living organisms have been thoroughly studied by ecologists. Numerous measures of diversity have been proposed and the physical and biological determinants of diversity have been analyzed. Thus a strong ecological background is available for the paleoecologist in measuring and interpreting diversity in the fossil record.

The general concept of diversity encompasses both the number and the relative abundance of taxa present within a sample. The terms diversity and *species diversity* have been used to denote either the number of taxa (Whittaker, 1972) or a combination of the number of taxa and their relative proportions

(MacArthur, 1972; Pianka, 1978; Pielou, 1969). Because of this lack of uniformity in usage, the term diversity is suitable for general purposes but the specific concept intended must be stated when a more precise meaning is desired. The term *richness* is preferable to species diversity when referring to number of taxa.

The relative abundance or importance of taxa within a sample has been described by terms such as dominance, relative importance, evenness, and equitability. *Evenness* and *equitability* measure the uniformity of abundance of the taxa. Lloyd and Ghelardi (1964) have suggested that evenness is expressed in absolute numerical terms whereas equitability is relative to some hypothetical abundance distribution. This distinction, however, has not generally been made in the paleontologic literature and equitability has been the commonly used term. Terms such as *dominance* and *relative importance* refer to the same characteristics of the sample, but on an inverse scale, that is, high dominance corresponds to low equitability and vice versa. A number of measures combining richness and equitability have been widely used and referred to simply as diversity or as dominance diversity. To avoid the confusion resulting from the use of diversity in this way, Peet (1974) favors *heterogeneity*. The term *dominance diversity* would seem to cause even less confusion, however.

Ideally, to maximize the relationship of diversity to the ecological structure of the community, ecological units rather than taxonomic units should be used to measure diversity. The use of taxonomic units in calculating these measures implies that all individuals of a taxon are ecologically uniform, and that all taxa are equally distinct from one another and occupy niches of equal size. Because these assumptions are clearly not valid in detail, units based on other attributes such as productivity or biomass have been proposed as alternatives to numerical abundances of taxa. These are perhaps usable in ecologic analysis but are difficult to apply in paleontology.

DIVERSITY MEASURES

Publications by Sanders (1968), Whittaker (1972), Fager (1972), Hill (1973), and Peet (1974) provide an introduction to the extensive literature on the measurement of diversity and on the indices that have been proposed, their attributes, and relative merits. Diversity indices can be grouped into those that measure richness, those that measure a combination of richness and equitability (dominance diversity), and those that measure equitability. They can also be grouped into those that are or are not based on some frequency distribution pattern assumed to be valid, in general, for organisms in nature. In measuring diversity, the sample size and the taxonomic level to be used (whether species, genus, etc.) depend on the nature of the problem. Because sample size and taxonomic level strongly influence the calculated diversity value, they must be taken into account when comparing values from different studies or from samples of different size within a study. Figure 9.1 illustrates the dependence of several diversity measures on sample size.

To yield significant interpretive results, a diversity index should be ecologically sound and statistically rigorous. However, the large literature on diversity definition and measurement shows that these two requirements are difficult to satisfy simultaneously. The measurement of diversity in paleontology must also take into account the partial and biased preservation due to taphonomic processes. Because of generally low but variable preservation of the original biocoenosis, differences in diversity between samples may be more the result of preservational differences than of original environmental differences (Lasker, 1976). Diversity measures that have been most commonly used in paleoecology are described briefly below.

Richness

Richness, S, the number of species or taxa, is the simplest measure of diversity and has been widely used in comparing lists of fossils. However, it is strongly dependent on sample size. $(S - 1)/(\log N)$ is one of several attempts to correct for the sample-size dependency of S. This index is based on the commonly observed fact that if cumulative number of species is plotted against cumulative number of individuals, the equation of the curve is approximately logarithmic. The relation of richness to sample size can also be expressed by the equation $S = CA^z$, in which A is sampling area. In a study

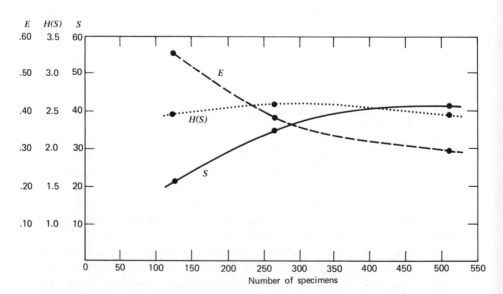

Figure 9.1 Relation between sample size and diversity and equitability measures. *S*—richness; *H(S)*—entropy; and *E*—equitability as measured by relative entropy. Curves based on a foraminiferal sample from the continental shelf off North Carolina. After Gibson and Buzas (1973), Geological Society of America Bulletin.

by MacArthur and Wilson (1967), C depends on the taxa counted and the value of Z has been determined empirically to lie within the range of 0.20 to 0.35. Richness measures that are adjusted for differences in sample size using assumed distributions are more easily compared with one another than unadjusted ones. On the other hand, they are imprecise to the extent that actual diversity in nature does not universally correspond to assumed, theoretical distributions. The *rarefaction* method (Sanders, 1968) attempts to compensate for the effect of sample size by graphically determining for a sample what the diversity would be at smaller sizes sampled. In the example of Fig. 9.2, the total number of taxa in each sample is indicated by a point at the right-hand end of each line. Because of differences in sample sizes, however, these S values are not easily compared. If the relative proportions of taxa in a sample are known, however, and assuming that these proportions would be constant in samples of different sizes from the same locality, the rarefaction curve is easily constructed. Thus the richness of samples of different size can be compared from values on the respective rarefaction curves at some common sample size. Simberloff (1972) has pointed out that the method proposed by Sanders is not mathematically correct, so that in previously published curves richness is generally overestimated. Simberloff has described the correct procedure to follow.

Dominance Diversity

Measures of entropy, based on information theory, are widely used in paleontology. Mathematically, they measure the uncertainty in predicting the identity of a randomly selected individual from a population. The *Shannon-Weaver equation*,

$$H = - \sum_{i=1}^{S} P_i \log P_i$$

is the most commonly used form because it is easily calculated if n_i/N, the proportion of the ith species in the sample, is substituted for P_i, the proportion of that species in the population. The value of H is greatest when equitability is highest. That is, for a given number of species, the maximum value of H is attained when all the species are equally abundant, and is equal to log S. Thus H is determined both by S and by equitability, and consequently is more difficult to interpret than richness measures, determined only by S. The use of sample values for population values introduces a bias into the results but this is small considering the inherent inaccuracies of the paleontologic data. The relationships between the different entropy equations have been thoroughly analyzed by Pielou (1969). The probability of two individuals, randomly drawn from a sample, being of the same species is

$$\sum_{i=1}^{S} P_i^2$$

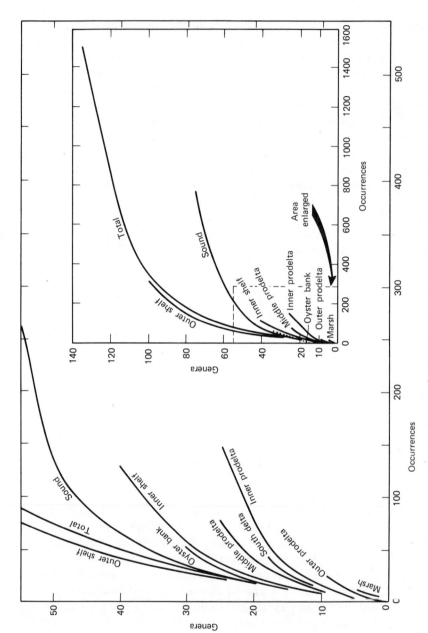

Figure 9.2 Rarefaction curves for total living fauna and modern communities on south and east of Mississippi Delta. Curves are based on shelled macroinvertebrates. Occurrences are the summation for a community of the number of samples from which each genus was collected. See Fig. 9.5 for location of communities. After Stanton and Evans (1972b), Journal of Paleontology, Society of Economic Paleontologists and Mineralogists.

Because the value of this index varies inversely with dominance diversity, other forms, such as its reciprocal, have been proposed. Neither the variants nor the basic index have been widely used in paleontology.

Equitability

The most commonly used index of equitability is

$$E = \frac{H}{H_{max}}$$

which is the ratio of the calculated entropy to that which the sample should have if all the species were equally abundant. The value of E ranges from 1, if all the species are equally abundant, toward 0 as dominance increases and equitability decreases. Other similar ratios based on entropy indices, which have also been used, have been described and evaluated by Sheldon (1969).

DIVERSITY CAUSES AND PATTERNS

A valuable approach to the understanding of diversity is through considera-tion of the environment as a multidimensional hyperspace (*environmental hyperspace*) that is partitioned into hypervolumes representing individual niches (Hutchinson, 1957; Valentine, 1969). According to this viewpoint, the axes of the hyperspace represent the factors that are important or limiting for the organism. Although the physical parameters, such as temperature or temperature variability, are most readily established, the hyperspace should also include the biologic parameters of the ecosystem. Richness, then, is equivalent to the number of niches that coexist in the hyperspace. The number of niches is determined by the size of the hyperspace, by the size of the niches or hypervolumes, and by the extent to which the niches overlap. Visualizing this for a single axis or environmental parameter or resource (Fig. 9.3; Mac-Arthur, 1972),

$$S = \frac{R}{\bar{U}} \left(1 + 2\frac{\bar{O}}{\bar{H}} \right)$$

where S = richness or number of species
$\quad\quad R$ = axis length, or dimension of the resource or parameter; for example the total temperature range within which life was possible at a locality
$\quad\quad \bar{U}$ = mean total niche width
$\quad\quad \bar{O}$ = mean overlap
$\quad\quad \bar{H}$ = approximately the mean niche width

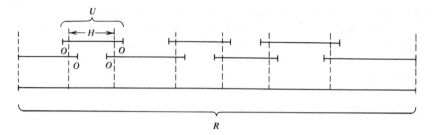

Figure 9.3 Relation of resource and niche characteristics. See text for description. After Mac-Arthur (1972), figure 7-1 (p. 171) from Geographical Ecology: Patterns in the Distribution of Species, copyright by Harper & Row, Publishers, Inc. Reprinted by permission of the publisher.

Thus richness is proportional to resource breadth and niche overlap and is inversely proportional to average niche size. Equitability, in this context, is a measure of the relative utilization of energy flux through the ecosystem by the organisms occupying the different niches. Diversity patterns, geographically and during geologic time, can readily be analyzed and described in terms of niche width and resource breadth. Niche overlap is not easily determined for living organisms and has been little studied in paleoecology.

Diversity varies both in time and space. It may vary on a relatively short time scale, such as during community succession, on a longer time scale of environmental change, such as that caused by the environmental changes associated with transgressive-regressive cycles, and on the long time scale of geologic time as a result of very broad environmental and evolutionary changes. The former two conditions are discussed in Chapter 11; the latter condition has been the subject of a voluminous literature (e.g., Valentine, 1969; Tappan and Loeblich, 1973; Bambach, 1977). In this chapter we concentrate on spatial variations in diversity.

In the modern world, richness of the total biota and within individual higher taxonomic units decreases from the equator toward the poles (Stehli, 1968). This latitudinal gradient is evident within each of the full range of environments that might be studied, whether marine, terrestrial, or lacustrine, or a subdivision of these, such as marine shallow shelf or rocky shore (Fig. 9.4). At regional and smaller scales, diversity gradients are ubiquitous and measureable at all taxonomic levels. Both richness and equitability of marine invertebrates decrease from the stable environmental conditions of the open-marine shallow shelf shoreward toward the more rigorous and variable conditions near shore and in estuaries and lagoons. Typical examples of this trend are from the Baltic Sea for both vertebrates and invertebrates (Segerstråle, 1957), from the vicinity of the Mississippi Delta for benthic shelled macro-invertebrates (Stanton and Evans, 1971), from numerous estuaries and lagoons for invertebrates (Emery and Stevenson, 1957), and from the Atlantic coast in the Cape Hatteras area for ostracodes (Hazel, 1975). Richness also increases from the shelf into deeper water of the ocean basins for benthic invertebrates (Hessler and Sanders, 1967) and foraminifera (Gibson and Buzas, 1973). The distribution of communities in the Mississippi Delta region is illustrated in Fig. 9.5. The progressive diversity increase away from the delta is illustrated by the rarefaction curves of Fig. 9.2. The diversity relationships for benthic foraminifera with depth off the eastern United States are illustrated in Fig. 9.6 (Gibson and Buzas, 1973). The understanding of modern diversity and gradients can be approached theoretically in the terms of the preceding equation in which richness is shown to be proportional to resource dimension and inversely proportional to average niche width. In less abstract terms, the possible causes of modern diversity patterns have been comprehensively reviewed by Sanders (1968), MacArthur (1972), Valentine (1973), and Pianka (1978). The wide range of explanations put forth indicates that there is no simple and universal explanation, but that diversity at any site is determined by a complex interplay of historical, biological, and physical factors within the ecosystem. The factors that appear to be of prime importance are time, stability, and resource.

Time

Diversity is determined by evolution as organisms accomodate themselves within the ecosystem to the physical environment and to other organisms. Hutchinson (1959) has proposed that the general course of this evolution is toward increased specialization, which corresponds to decreasing niche size, or niche subdivision. This results in increased biological interactions that provide opportunities for even further specialization and increase in diversity. Other time-dependent processes, in addition to decrease in niche size, may have resulted in changes in richness during geologic time. Hyperspace size has increased during earth history when biological advances opened up first terrestrial and later aerial habitats to be colonized.

Stability

Stability plays an important role in determining diversity in three ways. (1) It determines the amount of geologic time available during which evolution can proceed in any particular environment toward the attainment of the optimum degree of specialization, and thus of niche subdivision and diversification. A short-lived, unstable environment will not allow such specialization. In an unstable environment, specialized organisms cannot survive conditions outside their narrow tolerance range. Consequently, the organisms tend to be generalists, with large niches. This is commonly referred to as r-selection—that is, the best-adapted organisms are those with rapid growth and maturation and large reproductive potential (see Chapter 8). In contrast, the adaptiveness of specialized organisms with slower growth rates and lower reproductive potential in stable environments is referred to as K-selection in reference to the control by the carrying capacity of the environment on diversity. (2) Sta-

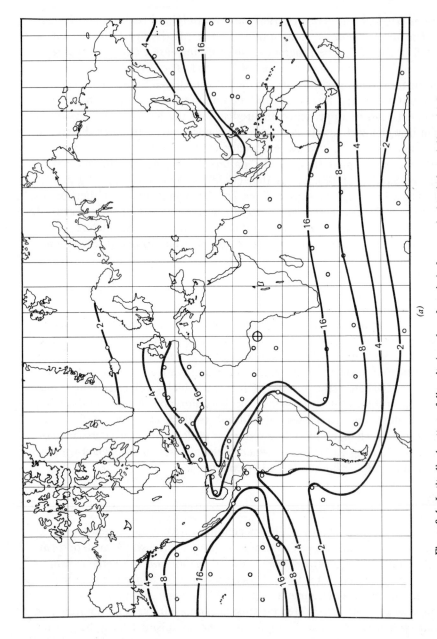

Figure 9.4 Latitudinal pattern of diversity (number of species) of recent planktonic foraminifers. (*a*) Contoured data points; (*b*) quadratic surface fitted to data points in (*a*). After Stehli and Helsley (1963), copyright 1963 by the American Association for the Advancement of Science, Science *142*:1057–1059, text figures 1 and 2.

(*a*)

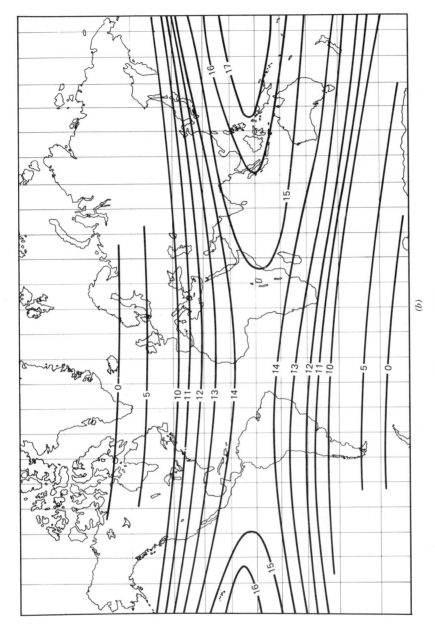

Figure 9.4 (Continued)

399

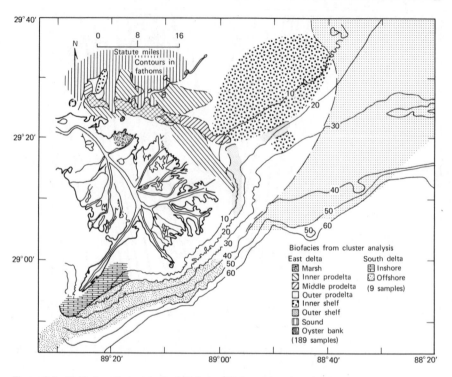

Figure 9.5 Biofacies adjacent to the Mississippi Delta, determined by *Q*-mode cluster analysis of 189 samples east of the delta (Parker, 1956) and 9 samples south of the delta. After Stanton and Evans (1972b), Journal of Paleontology, Society of Economic Paleontologists and Mineralogists.

bility determines the extent to which succession progresses, with increasing diversity, toward the climax stage. In these cases of evolution and succession, diversity is determined by a long-term or a short-term process, and the time span during which each process can continue is controlled by the stability of environmental conditions. In the first case, we are dealing with the evolutionary process during perhaps millions of years; in the second case, the successional process may last only a few months or even less. (3) Stability determines diversity by functioning as a resource, or dimension of the hyperspace. Although the maximum or average value of an environmental parameter, such as temperature, is commonly thought of when specifying conditions for life, the fluctuations in that parameter diurnally, seasonally, or during longer periods of time may actually be much more important as a limiting factor for the presence of a species or for the level of biotic activity in the ecosystem. In general, equable, stable conditions are optimal for most organisms, whereas few organisms are adapted for highly variable conditions. This is true if the environmental fluctuations are predictable, as, for example, seasonal. It is even more true if the fluctuations are unpredictable.

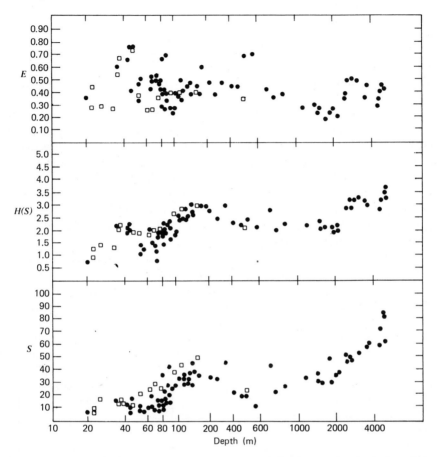

Figure 9.6 Graph of S, $H(S)$, and E versus depth for foraminifera from stations in the Cape Cod to Maryland area of the continental shelf off the eastern United States. Circles are stations north of Maryland, open squares are stations off Maryland. After Gibson and Buzas (1973), Geological Society of America Bulletin.

Resource

Any characteristic of the environment, such as space, salinity, or food, for which organisms compete or which may be limiting, may be represented as an axis of the hyperspace and be considered as a resource. Resources determine diversity because they establish the total volume of the hyperspace that potentially may be utilized. A narrow range for an environmental parameter at a locality is equivalent to little heterogeneity in the environment and thus to low diversity. This is most readily apparent to us from our experience of living in terrestrial environments. A grassy plain will contain a certain limited compliment of organisms. The addition of a clump of trees greatly expands the resources along several axes of the hypervolume—most evidently in the ranges

of temperature and sunlight and in types of shelter and food for animals—
and correspondingly increases both plant and animal diversity.

Resource and stability are of primary importance in understanding and
applying diversity in paleoecology. The effect of these two parameters on
diversity is summarized in Fig. 9.7 and analyzed in more detail by Valentine
(1971b).

As a summary example, the combined effects of time, resource, and
stability can be illustrated in the low diversity of a desert. Deserts are typically
characterized by low and sporadic precipitation and by temperature that is
high during the summer daytime but may fluctuate over a wide range diurnally and seasonally. After an infrequent but perhaps torrential rain, dormant
seeds will germinate and plant life will flourish, but only for the short time
until arid conditions return. The low environmental stability results in fluctuations in primary productivity and thus limits time for succession to lead to
a complex and diverse community structure.

Diversity is also low in deserts because of limited resources. Critical environmental conditions, such as low and fluctuating moisture, high and fluctuating temperature, and low and fluctuating primary productivity at the base
of the food pyramid, can each be represented by a range of values on an axis
on the hyperspace that is either very narrow or falls largely outside the tolerable range for most organisms. Thus the resource values in a desert environment define a small hyperspace. Because the number of niches is generally

Resources	Stable	Unstable
Poor	**Box 1** *K*—selection Small stable population, highly specialized Highest diversity	**Box 2** Compromise selection Moderate population size and specialization minimizing fluctuations Low diversity
Rich	**Box 3** *r*—selection and *K*—selection Mixed population size and specialization High diversity	**Box 4** *r*—selection Fluctuating population, often large, unspecialized Lowest diversity

Figure 9.7 Evolutionary process, population characteristics, and diversity (richness) in communities formed under different conditions of resource stability and quantity. After Valentine
(1971b), Lethaia *4*.

correlated with hyperspace volume, diversity is low. In addition, because of the variable environment, the organisms that can live in the desert must be generalists, adapted to a wide range of conditions, rather than specialists. Consequently, individual niches are large, subdividing the hyperspace coarsely and resulting in low diversity.

TROPHIC STRUCTURES

The manner in which energy flows through a community is a distinctive structural attribute, unique to each community. It is determined by the physical parameters of the ecosystem that control the composition and relative abundances of the organisms present, as well as the rate and time distribution of productivity and the spatial distribution of nutrients and food for the different organisms present. It is determined by the biological parameters of the ecosystem that reflect niche differences and overlaps, food requirements and preferences, and trophic interactions between organisms.

The trophic structure of a community may be described in two ways. The organisms may be categorized according to *trophic level* and arranged in an ecologic pyramid based on numbers of individuals, biomass, or energy. Alternatively, the food web within the community may be constructed so that the flow of energy along the multiple and interacting food chains is charted. The first approach is a more general one; the second is more detailed and contains much more information but is more difficult to carry out in a paleoecologic study. Both approaches are well surveyed in the ecological literature and in basic texts (e.g., Pianka, 1978) and in a paleoecologic context by Scott (1978).

The most fundamental way to describe the trophic structure of a community is in terms of *energy flow*. Energy input to the ecosystem is primarily as sunlight. This results in a certain amount of plant growth (*primary production*). The plants support, in turn, herbivores, carnivores, scavengers, and parasites. Ultimately, the organic material is buried in the sediment or decomposed to the original inorganic elements. The efficiency of converting organic material at one level to organic material at the next level is on the order of 10 to 20%. This is evident as we individually try to balance our caloric intake with the weight we wish to gain or lose. It is commonly portrayed in the *ecologic pyramid* (Fig. 9.8). The width at each level reflects the decreasing energy available at successive levels and the corresponding general decreasing numbers of herbivores and successive levels of carnivores.

If the ecosystem were thoroughly understood, energy flow could be described in detail. Such an attempt is illustrated in Fig. 9.9. A comprehensive *food web* would indicate the amounts of energy resulting in the primary production of each type of plant. It would indicate the extent to which each plant was utilized by the different herbivores or was decomposed back to the original elements, and the food utilizations, in similar fashion, at the successive carnivore levels.

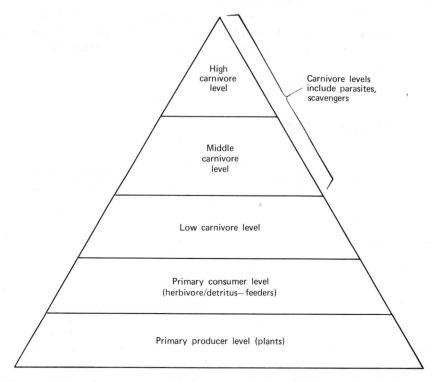

Figure 9.8 Ecologic pyramid. Decreasing width at successively higher trophic levels reflects 10 to 20% efficiency of flow of energy and corresponding decrease in number of individuals.

A comprehensive food web would require complete knowledge of the ecosystem. Consequently, none exists and even generalized ones of the type illustrated in Fig. 9.9 are uncommon. The food web is difficult to establish for two basic reasons. (1) The food utilized by the different animals is difficult to determine. For example, an organism may feed on several levels at the same time, it may feed on different levels through time, shifting its preferences with age and changing availability, or it may have very narrow food preferences that differ from locality to locality. In addition, detailed analysis is necessary to be sure that an organism is actually utilizing what it eats— amphipods, for example, feed on plant detritus but derive nutritive value from only the microorganisms attached to the detritus (Fenchel, 1970), and molluscs may generally feed on detritus in the same way, as indicated for example by Newell (1965) for the gastropod *Hydrobia* and the bivalve *Macoma*. (2) Energy and material flux into, within, and out of the ecosystem is difficult to measure.

Analysis of community structure by reconstruction of the food web is difficult for modern communities and quickly leads to complex flow charts with multiple pathways and feedback loops. It is even more difficult for fossil

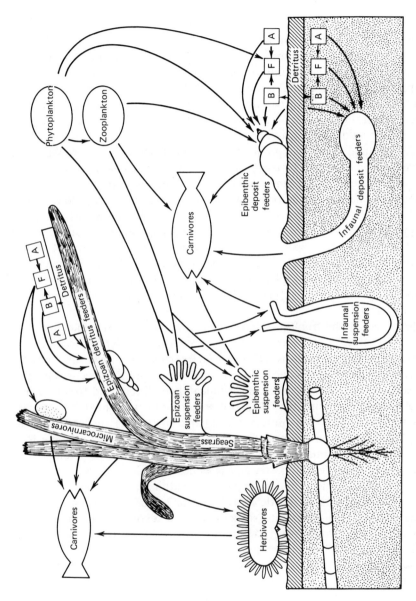

Figure 9.9 Generalized diagram of the energy flow in a seagrass community. B—bacteria; F—foraminifera and other microherbivores; A—microscopic algae. After Brasier (1975).

communities because much of the necessary information is not preserved. Consequently, analysis of the trophic structure of fossil communities is best approached through attributes that are more easily described and measured. Analysis of the fossil community in terms of levels of the trophic pyramid is one approach that has several advantages. (1) Community analysis in terms of trophic levels permits the grouping of the fossil taxa into a few trophic-level categories and thus provides a general view of the trophic structure when the more specific food chains may not be reconstructable. (2) Construction of the ecologic pyramid from the proportions of organisms at each trophic level provides a simple means of comparison of the fossil community with other fossil and present-day marine communities. Using the proportions of organisms may be a gross approximation, of course, to energy flow, but is considered reasonably accurate. More sophisticated approaches that, for example, take into account size differences and population dynamics of the individual species are probably not worth while at our present level of understanding for two reasons. (1) The increase in accuracy is small compared to the inherent errors due to taphonomic processes. (2) Size differences probably do not reflect differences in energy flow as much as they do differences in "standing crop."

The aspect of the trophic structure of marine communities that has been most widely used in paleoecology is the proportion of deposit-feeding and suspension-feeding organisms at the primary consumer level, the *trophic proportion*. The use of trophic proportions in determining paleoenvironments is based on correlations between environmental conditions and trophic structure of modern communities. The two major sources of food for primary consumers in marine communities are organic matter (1) suspended in the water and (2) on or in the sediment. The relative proportions of these food resources depend primarily on water turbulence. Because particulate food material behaves as detrital particles, suspended particles will settle out of the water onto the sea floor under very quiet conditions, whereas under more turbulent conditions food particles will remain in suspension and those on the sea floor will be swept into suspension. The texture of the substrate is determined by the same energy conditions, higher-energy conditions producing coarser sediment and lower-energy conditions, finer sediment. Dissolved organic compounds are preferentially absorbed onto clay particles from which they can be removed by organisms, further reinforcing the correlation between deposit feeders and finer sediment (Sanders 1956; Driscoll 1969).

The proportion of organisms feeding on food in suspension versus food on or in the substrate has been correlated with water energy in modern environments by Sanders (1956, 1968), Savilov (1957), and Driscoll and Brandon (1973). As a first approximation in paleoecology, water energy might be correlated with depth, but protected shallow settings might also lead to a high proportion of deposit-feeding organisms. A preponderance of suspension-feeding organisms has also been found concentrated locally in deeper water at the shelf edge (Savilov, 1957).

Other factors of secondary importance in affecting the relative abundances of deposit- and suspension-feeding organisms include substrate stability, which may be affected in part by bioturbation; oxygen content, which may be controlled in part by the amount of organic material settling onto the substrate; and the amount and diversity of the particulate food resources (Rhoads and Young, 1970; Rhoads, Speden, and Waage, 1972; Aller and Dodge, 1974).

Studies by Turpaeva (1957) and other Russian fisheries scientists on the trophic structure of fish and the associated benthic invertebrates have also added to the understanding of trophic proportions. They have shown that the primary-consumer level of a community is generally dominated by organisms of one feeding type (trophic group), that a community is usually dominated by a few most abundant species, and that, if the dominant species are ranked by abundance, successive species are of different feeding types. Furthermore, they recognized that these tropic characteristics of a community were determined by environmental conditons (Savilov, 1957).

Attempts to reconstruct in any detail the trophic web of fossil communities is a recent development in paleoecology. At first glance, the results are particularly useful in defining the preservational completeness of the biocoenosis by indicating levels or components of the ecologic pyramid or food chain that are apparently inadequately represented in the fossil assemblage. This approach is also valuable in exploring the possible evolution of community structure.

APPLICATIONS OF THE COMMUNITY IN PALEOENVIRONMENTAL RECONSTRUCTION

The full range of compositional and structural characteristics described above have been used in the analysis of fossil assemblages. Numerous examples are in the current literature. A small representative sample of approaches in community paleontology is described in this section.

Taxonomic Uniformitarian Analysis

Ancient communities have been recognized and compared with one another and with modern communities on the basis of taxonomic composition in numerous studies. The degree of taxonomic similarity is generally considered proportional to the degree of similarity in the environments in which the communities existed. The interpretation based on the communities has varied widely. In many papers, biofacies are recognized in the strata being studied and the biota characteristic of the biofacies is referred to as a community. Spatial distribution of the communities within the stratigraphic framework, and associated lithologic criteria, may lead to environmental inferences for the fossil example without making any comparison with modern communities. An example of a study of this type is the description of late Ordovician communities of the central Appalachian region by Bretsky (1969). Within more or less contemporaneous strata, Bretsky was able to define the depositional environment in general terms of bathymetry, distance from shore and deltaic areas of sediment influx, and substrate texture and composition. The three communities recognized in these strata, the *Sowerbyella-Oniella, Orthorhyn-chula-Ambonychia,* and *Zygospira-Hebertella* communities, then, define distinct tracts on the sea floor (Fig. 9.10) and, by inference, bathymetric zones (Fig. 9.10A).

Similar studies in the Lower Paleozoic (e.g., Ziegler, Cocks, and Bambach, 1968; Walker, 1972; Tipper, 1975) have led to the awareness that the communities can be compared and related (1) to environmental similarities and (2) to evolution of community composition and perhaps structure (Bretsky, 1968). A comparison of Ordovician and Devonian communities of New York has shown that the communities of different age but from presumably the same environment are very similar, reflecting the ecologic control of community composition. The similarities, as indicated by the subtidal communities (Fig. 9.11), are at a high taxonomic level but reflect the presence of equivalent niches (Walker and Laporte, 1970).

Thus the potential has been established to define communities diagnostic of the full range of environments and to use them in environmental interpretation. The subsequent extension of this approach is the collection of cartoons as in Fig. 9.11 for a wide range of environments throughout the Phanerozoic (McKerrow, 1978).

A Pliocene flora from the Mount Eden Beds of Southern California provides an example of paleoenvironmental analysis in which the plant community is used as an entity to transfer information from the Recent to the past (Axelrod, 1938). The flora consists of 30 species representing 21 genera and 16 families (Table 9.1). From a knowledge of modern plant communities occurring in the very diverse habitats of Southern California, it is evident that the Pliocene flora from the Mount Eden Beds is a composite of several plant communities (Table 9.2). These are determined by present-day differences in temperature, soil moisture, and humidity as controlled by differences in elevation and slope exposure. The recognition of analogous modern plant communities from the Pliocene flora has two significant payoffs. The first is that the mixed nature of the flora can be established. The second is that the environments of the modern analogous communities can be described in much greater detail and in terms of more parameters than would be possible from considering the individual taxa one by one. Axelrod is able to then describe the depositional setting as a low basin of shallow lakes and marshes with adjacent alluvial fans, rolling hills and plains, and nearby higher hills. Each of these areas contributed a distinctive community to the flora. In addition, the climate can be described in terms of summer and winter rainfall and temperature ranges and seasonality. Finally, Axelrod is able to illustrate with photographs the modern analogous plant communities as they occur in Southern California.

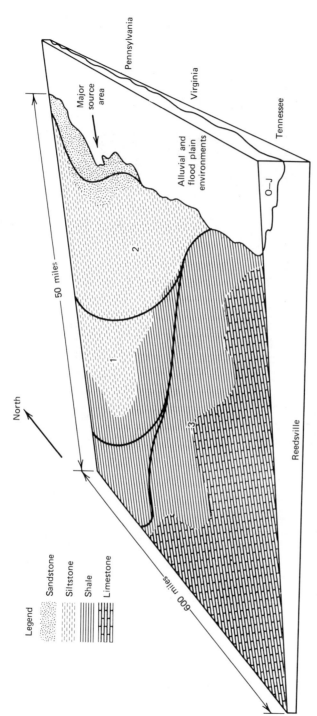

Figure 9.10 Distributions of sediments and biofacies during deposition of the Late Ordovician Oswego barrier-lagoonal deposits along the northeastern portions of the shoreline in the central Appalachians. The communities characteristic of each biofacies are: (1) *Sowerbyella-Onniella* Community; (2) *Orthorhynchula-Ambonychia* Community; (3) *Zygospira-Herbertella* Community. See Fig. 9.10A for the composition and inferred environment of each community. After Bretsky (1969), Geological Society of America Bulletin.

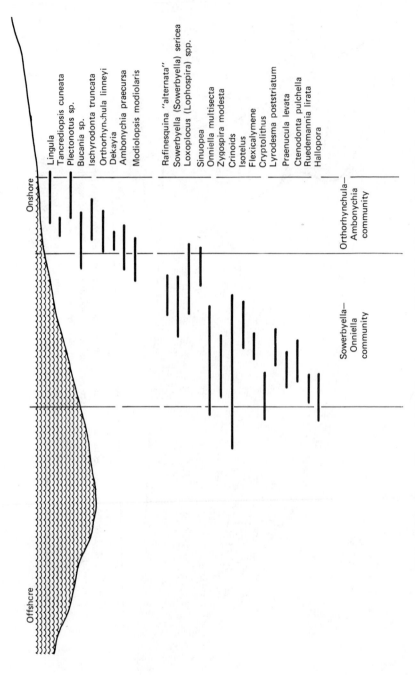

Figure 9.10A Composition and inferred environment of Late Ordovician communities in the central Appalachians. Geographic distribution of the communities is illustrated in Fig. 9.10. (*a*) Northern communities; (*b*) southern community. After Bretsky (1969), Geological Society of America Bulletin.

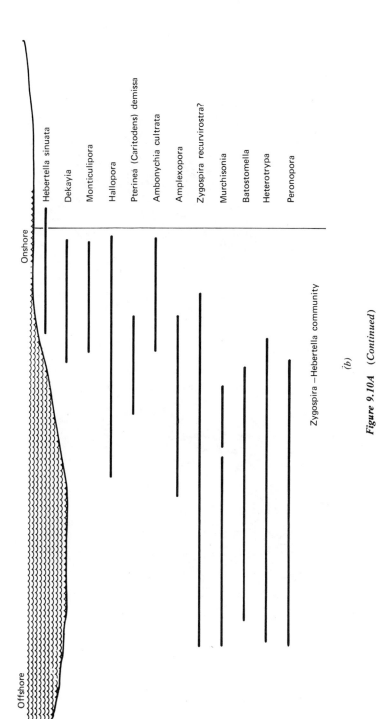

Offshore

Onshore

Hebertella sinuata

Dekayia

Monticulipora

Hallopora

Pterinea (Caritodens) demissa

Ambonychia cultrata

Amplexopora

Zygospira recurvirostra?

Murchisonia

Batostomella

Heterotrypa

Peronopora

Zygospira – Hebertella community

(b)

Figure 9.10A (*Continued*)

411

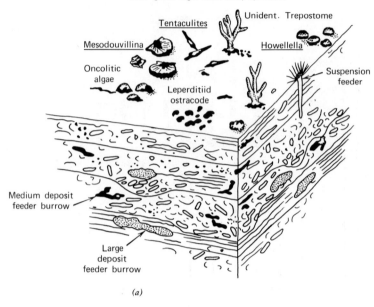

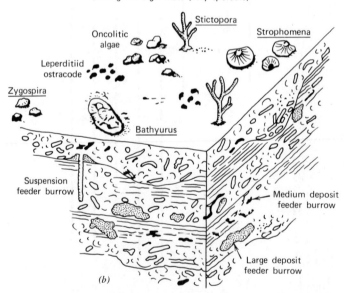

Figure 9.11 Reconstructions of communities in living position and as they accumulate as fossil assemblages. (a) Devonian Manliuss Formation; (b) Ordovician Black River group. Both communities lived in a low-intertidal environment, which presumably accounts for the high similarity in composition and form, although they are of very different age. After Walker and Laporte (1970), Journal of Paleontology, Society of Economic Paleontologists and Mineralogists.

Table 9.1 Mount Eden flora: Systematic list of species[a]

Pteridophyta	Dicotyledonae—*continued*
Equisetales	Rosales
Equisetaceae	Platanaceae
Equisetum sp.	*Platanus paucidentata*
Spermatophyta	Rosaceae
Gymnospermae	*Cercocarpus cuneatus*
Coniferales	*Prunus preandersonii*
Pinaceae	*Prunus prefremontii*
Pinus hazeni	Leguminosae
Pinus pieperi	*Prosopis pliocenica*
Pinus pretuberculata	Sapindales
Pseudotsuga premacrocarpa	Anacardiaceae
Cupressaceae	*Rhus prelaurina*
Cupressus preforbesii	Sapindaceae
Gnetales	*Sapindus lamottei*
Gnetaceae	Rhamnales
Ephedra sp.	Rhamnaceae
Angiospermae	*Ceanothus edensis*
Monocotyledonae	*Ceonothus* sp.
Pandanales	Ericales
Typhaceae	Ericaceae
Typha lesquereuxi	*Arbutus* sp.
Dicotyledonae	*Arctostaphylos preglauca*
Salicales	*Arctostaphylos prepungens*
Salicaceae	Gentianales
Populus pliotremuloides	Oleaceae
Salix coalingensis	*Fraxinus edensis*
Salix sp.	Asterales
Juglandales	Compositae
Juglandaceae	*Lepidospartum* sp.
Juglans beaumontii	
Fagales	
Fagaceae	
Quercus hannibali	
Quercus lakevillensis	
Quercus orindensis	
Quercus pliopalmeri	

[a]The Mount Eden flora contains 30 species, representing 21 genera and 16 families. Twenty-two of the species are dicotyledons, of which 7 are arborescent, 3 are normally small trees, and 12 definitely shrubby. Of the remainder 6 are conifers, and the monocotyledons and pteridophytes are both represented by single species.

Table 9.2   Mount Eden flora subdivided into habitat communities[a]

Desert-border element	Savanna and woodland
Ephedra sp.	*Arbutus* sp.
Lepidospartum sp.[b]	*Juglans beaumontii*[b]
Prosopis pliocenica	*Pinus pieperi*
Prunus preandersonii	*Quercus hannibali*
Prunus prefremontii	*Quercus lakevillensis*
Quercus pliopalmeri[b]	*Quercus orindensis*
Sapindus lamottei[b]	Chaparral
Lake-border or marsh[c]	*Arctostaphylos preglauca*
Equisetum sp.	*Arctostaphylos prepungens*
Typha lesquereuxi	*Ceanothus edensis*
Riparian	*Ceanothus* sp.
Fraxinus edensis[b]	*Cercocarpus cuneatus*
Juglans beaumontii[b]	*Fraxinus edensis*[b]
Lepidospartum sp.[b]	*Quercus pliopalmeri*[b]
Platanus paucidentata	*Rhus prelaurina*
Populus pliotremuloides	Coniferous associations
Salix coalingensis	*Cupressus preforbesii*
Salix sp.	*Pinus hazeni*
Sapindus lamottei[b]	*Pinus pretuberculata*
	Pseudotsuga premacrocarpa

[a] After Axelrod (1938).

[b] May normally occur as a dominant in more than one habitat.

[c] Most of the riparian genera may also be present about the borders of lakes or marshes.

A similar analysis by Shotwell (1964) of Late Tertiary mammalian faunas of the northern Great Basin in the western United States is illustrated by Fig. 9.12. The changes in the community representation during the Late Tertiary are interpreted to reflect changes in habitat and vegetation (Fig. 9.13) caused by climate changes. The climatic change up to the Hemphillian was one of increasing continentality, expressed by decreasing rainfall and increasing seasonality of temperature range and rainfall distribution. The climate became somewhat moister from the Hemphillian to Blancan.

The environmental reconstruction of the Pliocene in the Kettleman Hills region has relied heavily on the recognition of Pliocene communities and their comparison with modern communities (Stanton and Dodd, 1970). The modern analogue in this study has been the communities of the San Francisco Bay area, California, as established by Q-mode cluster analysis of the molluscan fauna (Fig. 9.14). Eight major biofacies are derived from the cluster analysis. Six of these are within the bay and are of particular value in the comparison with the Pliocene communities. The physical conditions characteristic of each biofacies are illustrated in Fig. 9.15, and the faunal composition of each, the molluscan community, in Fig. 9.16.

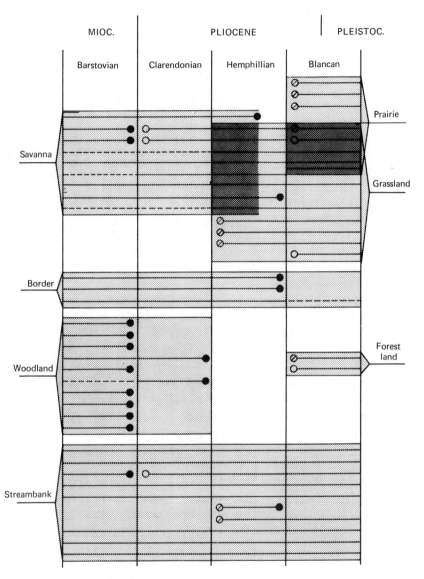

Figure 9.12 Composition by habitat of Late Cenozoic mammalian communities of the northern Great Basin, western United States. Each horizontal line represents a phyletic line made up of a number of descendent genera. Solid circles and open circles indicate terminations and beginnings of phyletic lines. The shifts in proportions of habitat communities are correlated with environmental changes in Fig. 9.13. After Shotwell (1964).

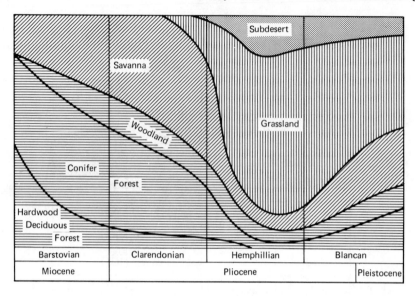

Figure 9.13 Changes in proportions of plant communities during the Late Cenozoic in the northern Great Basin of the western United States were caused by changes in climate, that is, increasing aridity until late Hemphillian followed by increasing humidity during the Blancan. After Shotwell (1964).

The pattern of biofacies is symmetrical with the bay geography. Biofacies C, D, and E are concentrically nested around the bay entrance; F, the oyster bank biofacies, is largely restricted to the south arm of the bay; G and H are restricted to the inner part of the north arm.

Going from the bay entrance into either of the arms, the environment changes from equable and normal marine to progressively less stable because of increasing isolation from the buffering effects of the open ocean, and increasing responsiveness to the adjacent terrestrial environment. Faunal diversity decreases along this environmental gradient, from the outer bay through the middle and inner bay biofacies; the total number of genera increases only slightly as the fauna from more restricted settings is added. The diversity, therefore, reflects the maximum "marineness," and not the average environmental conditions within the bay nor the environmental heterogeneity. The diversity gradient is comparable in both arms of the bay from the entrance through biofacies E.

Considering only the soft bottom, infaunal, component of the San Francisco Bay fauna, community C represents the basic marine open outer bay fauna, and the fauna of the other communities is derived from the fauna of community C and consists of those genera that are tolerant to the various stresses present in different parts of the middle and inner bay. Most of the abundant and diagnostic taxa of the inner bay infaunal community are also present and even common near the bay entrance. Exceptions are *Mya*, which

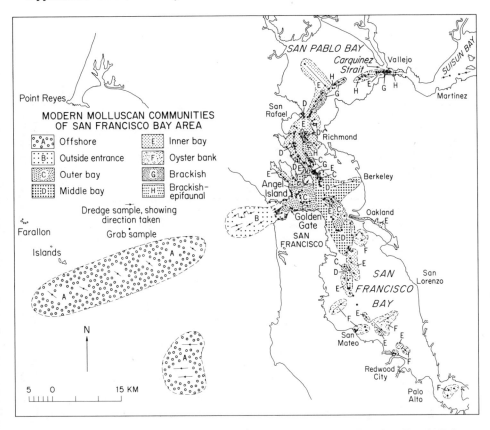

Figure 9.14 Distribution of modern molluscan communities in San Francisco Bay and the adjacent Pacific Ocean. After Stanton and Dodd (1976b), Lethaia *9.*

is rare in the outer bay but dominant in the inner bay, and *Solen,* which is confined to the mid-bay.

The gradual elimination of taxa without replacement, going from the bay entrance back into the inner bay, is not as obvious within the epifaunal component of the fauna. The oyster bank community represents, in the inner bay, the hard bottom habitat that has been sampled at only a few localities nearer the bay entrance. Consequently, the restriction of a number of epifaunal genera to community F in the south arm of the bay (Fig. 9.14) may be due to the lack of sampling of the suitable substrate as well as to actual scarcity of sublittoral shell bottoms elsewhere in the bay.

The environmental parameters that are listed in figure 9.15 seem to be those of major importance in determining the faunal distribution patterns, and thus are most applicable to paleoecology. Water depth is a major independent variable but the whole of San Francisco Bay is shallow enough that probably few of the sublittoral species have particular adaptations that would limit them to narrower depth ranges. Depth exerts a strong influence on

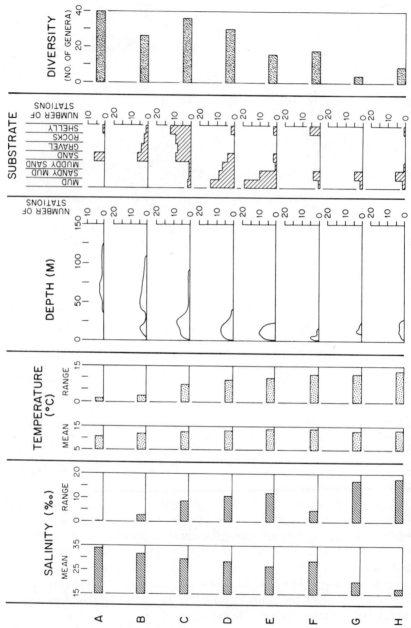

Figure 9.15 Environmental and diversity characteristics of modern molluscan communities in the San Francisco Bay area. Communities: A—offshore; B—outside entrance; C—outer bay; D—middle bay; E—inner bay; F—oyster bank; G—brackish; H—brackish-epifaunal. After Stanton and Dodd (1976b), Lethaia 9.

current and wave action and consequently on substrate texture; therefore, apparent fauna-depth correlations probably reflect the substrate-depth correlation. The plot of bottom temperature shows that average temperature is fairly constant in the bay but that the temperature range increases greatly with distance from the bay entrance into either arm. Salinity range is small near the bay entrance and throughout most of the bay except in the inner part of the north arm (communities G and H) where freshwater inflow is large.

The environmental homogeneity within individual biofacies, particularly evident in terms of substrate, confirms that the biofacies reflect discrete environments. Although the substrate may be homogeneous within a biofacies, localities with the same substrate that are grouped on faunal composition in other biofacies indicate that the fauna reflects more than substrate or any other single environmental parameter.

Comparison of the Pliocene communities with these modern communities is possible because the fossil fauna is similar to that of the model at the generic level, because few of the fossil genera have become extinct, and because few of the modern genera are new arrivals. Beyond taxonomic similarity, however, community structure may change as the role of individual species within the community changes with evolution. It is assumed that significant changes have taken place in relatively few genera. The generalized distribution of depositional environments during deposition of the Pliocene sediments is illustrated in Fig. 1.9. These conclusions based on the modern communities are cross-checked by their distribution within the cyclic stratigraphic framework.

A more detailed study of the *Pecten* zone has recognized eight biofacies by *Q*-mode cluster analysis of 93 samples from 54 measured sections. The distribution of these biofacies in the east and west flanks of the Kettleman Hills is illustrated in Fig. 9.17. The taxonomic composition of these biofacies define the communities in Fig. 9.18. The inferred paleoenvironment is again based in large part on comparison of these Pliocene communities with those in the San Francisco Bay area. In both the general study and that of the *Pecten* zone, the initial environmental description is in terms of geographical position within the bay setting. This provides the basis for detailed environmental description in terms of specific parameters as they can be discriminated in the various bay settings.

Community Structure Analysis

Diversity

Diversity is an integral feature of community structure and is essential for its description and analysis. It is also commonly analyzed in paleoecology independently of the community. Because of the wide range of factors that may control diversity, it has been used to analyze an equally wide range of paleontologic problems. The temporal patterns of diversity change through geologic

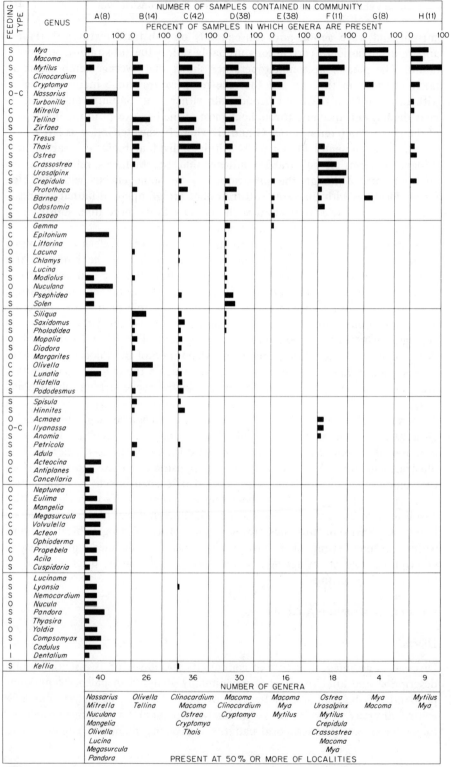

time, during the shorter intervals of regressive-transgressive cycles, and during succession, are discussed in Chapter 11.

Lateral diversity gradients have been thoroughly studied in modern environments but have been less commonly applied in paleoecology. An example of the use of diversity in the study of very-large-scale problems is Stehli's (1970) work with global diversity patterns. In this work the modern latitudinal diversity gradient is considered to be determined by the earth's rotation and the latitudinal gradient of solar energy incident upon the earth. Because these characteristics have been constant during earth history, the dependent diversity gradients should also have the same general latitudinal pattern. As a consequence, the position of the earth's rotational pole during the Permian is determined from paleontologic data and is compared with that based on paleomagnetic data. Although the conclusions may be questioned because the samples used are relatively few in number and unevenly distributed over the world, the study is an excellent example of the power of diversity as a paleoenvironmental indicator.

At a regional scale, Hallam (1972b) has analyzed the diversity of Lower Jurassic mollusc and brachiopod faunas of western Europe and Greenland. The patterns of diversity for the different mollusc classes and for the brachiopods are more complex then the simple latitudinal pattern used by Stehli. In fact, overlaid on the global pattern are the more local factors such as depth, salinity, substrate, and environmental stability.

Diversity has been an important parameter in the paleoenvironmental studies of the Pliocene strata of the Kettleman Hills. Diversity in the Recent communities of San Francisco Bay (Fig. 9.14) is plotted in Fig. 9.15 and is discussed in the preceding section. In the total Pliocene section (Fig. 1.9) diversity is high in the Jacalitos and Pancho Rico Formations (74 and 80 genera respectively), intermediate in the lower part of the Etchegoin Formation (33 to 47 genera in the *Siphonalia* zone and underlying units), and low in the upper part of the Etchegoin and throughout the Tulare and San Joaquin Formations (4 to 32 genera per unit) except for the *Pecten* zone with 59 genera (Table 9.3). Diversity fluctuates widely from unit to unit in the San Joaquin Formation and in the Etchegoin Formation above the *Siphonalia* zone. The primary interpretation of these differences of diversity is that they reflect differences in stability, correlated with degree of isolation from the open-marine environment.

Diversity is equally important in the interpretation of the fauna of the *Pecten* zone. The diversity of the communities (Fig. 9.18) is portrayed by the

Figure 9.16 Composition of the modern molluscan communities of the San Francisco Bay area. The bar length represents the frequency of occurrence of the genus within each community. Communities: A—offshore; B—outside entrance; C—outer bay; D—middle bay; E—inner bay; F—oyster bank; G—brackish; H—brackish-epifaunal. Feeding types: S—suspension feeder; O—deposit feeder utilizing resources on the substrate; I—deposit feeder utilizing resources in the substrate; C—carnivore, parasite, or macrophagous scavenger. From Stanton and Dodd (1976b), Lethaia 9.

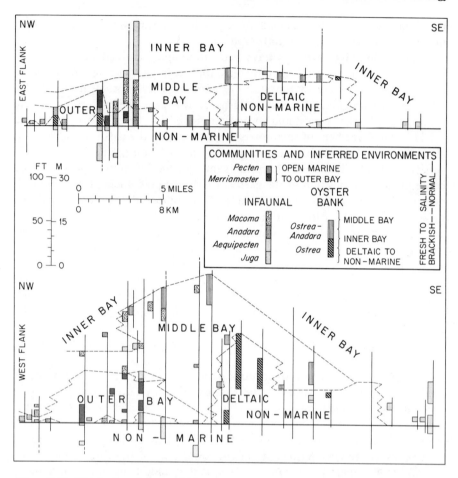

Figure 9.17 Distribution of communities and depositional environments in the *Pecten* zone along the east and west flanks of the Kettleman Hills. The datum line is the base of the *Pecten* Zone. Vertical lines indicate the extent of stratigraphic section examined. The patterned bars indicate the distribution of communities. After Dodd and Stanton (1975), Geological Society of America Bulletin.

rarefaction curves of Fig. 9.19. In general, the diversity values agree well with the interpretation based on taxonomic composition. The exception to this generality is the *Merriamaster* community, which is indicative of an open marine to outer bay environment on the basis of the composition, for it contains abundant individuals of the sand dollar *Merriamaster*. The community is found in well sorted and commonly cross-bedded sandstone, indicating a current-swept habitat comparable to that in which modern sand dollars are found. Lithology and bedding geometry suggest sand patches and bars on the sea floor, comparable to those outside San Francisco Bay (Yancey and Wilde, 1970) but on a smaller scale, rather than to those in tidal channel or beach

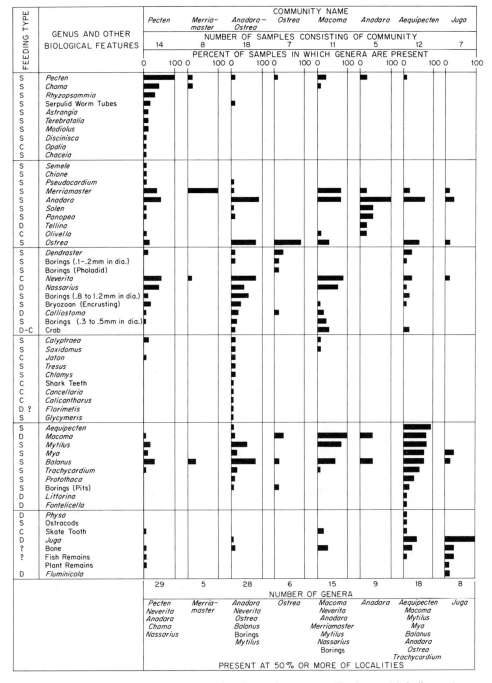

Figure 9.18 Composition of the communities in the *Pecten* zone. The bar length indicates the percentage of samples of the community which contains the taxon. Feeding types: *S*—suspension feeder; *D*—deposit feeder; *C*—carnivore, parasite, or macrophagous scavenger. After Stanton and Dodd (1976b), Lethaia 9.

Table 9.3   Diversity in the Pliocene strata of the Kettleman Hills, California[a]

Formation	Zone of Woodring et al. (1940)	Number of Genera/unit
Tulare		2
	Upper *Amnicola*	11
	Lower *Amnicola*	17
		6
	Upper *Mya*	14
San Joaquin		7
	Acila	29
		4
	Pecten	59
		11
	Neverita	21
		9
	Cascajo Cgl.	32
Etchegoin		7
	Littorina	10
		6
	Upper *Pseudocardium*	28
	Siphonalia	47
	Macoma	33
	Patinopecten	38
Jacalitos/Pancho Rico		74/80

[a] After Stanton and Dodd (1970).

settings. The *Merriamaster* community is closely associated with the *Pecten* community, always overlying it where the two occur in the same section. The *Merriamaster* community probably existed in essentially the same conditions of normal salinity, good communication with the open ocean, and environmental stability as did the *Pecten* community. It is concluded that the low diversity reflects stress conditions of the mobile sand substrate. Thus, although local diversity differences have been largely explained in the literature in terms of geographical gradients in environmental stability and restriction from the uniform, well-mixed, open-marine environment, localized stress conditions must also be taken into account in interpreting patterns of diversity.

Trophic Proportions

The study by Walker (1972) of Ordovician to Mississippian communities in the western United States provides a convenient starting point for our description of the application of trophic structure in paleoecology. Walker showed

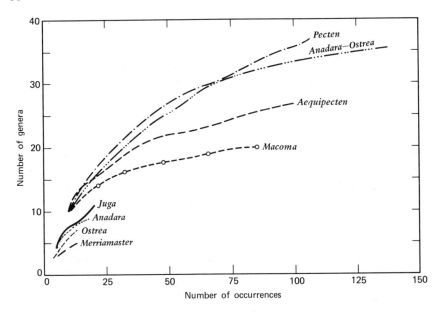

Figure 9.19 Rarefaction curves portraying the relative diversity of communities in the *Pecten* zone. After Dodd and Stanton (1975), Geological Society of America Bulletin.

that the fossil communities were like the communities described by Turpaeva in that each was dominated by one trophic group, each trophic group was dominated by one species, and the feeding types of the ranked dominants within a community alternated. Subsequently, Rhoads et al. (1972) and Wright (1974) have shown that Cretaceous and Jurassic communities of Wyoming and South Dakota are also characterized by apparently distinctive trophic structure and are each restricted to particular depositional environments. These Mesozoic communities differ from the modern communities described by Turpaeva in two respects, however: (1) suspension feeders are much more common as dominants and (2) alternation of feeding types among the dominants is much less common. In fact, many of these fossil communities are homogeneous, containing only suspension-or only deposit-feeding organisms, suggesting unusual, extreme environments in which trophic resources were limited to either the water column or the substrate.

As a result of the apparent relationship of trophic proportions and the environment, trophic proportions have become a standard attribute to be considered in the study of fossil communities. The Cretaceous example from the western interior of the United States demonstrates the value of the approach. In this study, Scott (1974) first established biofacies by cluster analysis of the paleontologic data. Each biofacies is characterized by a distinctive assemblage of fossils, or organism community. The trophic proportions for each community are illustrated by the triangular diagrams of Fig. 9.20. The depositional environment in which each community lived is deter-

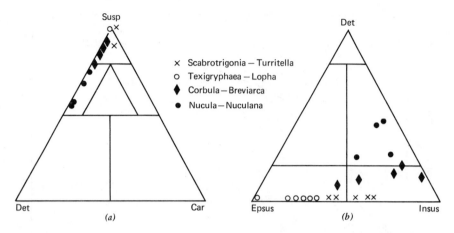

Figure 9.20 Feeding-habit (*a*) and substrate-niche (*b*) ternary diagrams for fossil assemblages in the Cretaceous Tucumcari Formation in northeastern New Mexico section. Percentages are based on numbers of taxa. Susp—suspension feeder, Det—detritus feeder, Car—carnivore, Epsus—epifaunal suspension, and Insus—infaunal suspension. After Scott (1974), Lethaia 7.

mined from the combination of the lithologic, taxonomic, and trophic information, within the stratigraphic constraint. The stratigraphic framework is important, as emphasized in the introduction, because the postulated individual depositional environments must fit into a coherent internally consistent model, as illustrated in Fig. 9.21.

A major problem in the application of trophic structure in paleoecology is caused by the generally very incomplete preservation of the original biocoenosis. Because the environment is reflected in the trophic proportions of the biocoenosis, it is necessary to assume that (1) each fossil community is a sample drawn from a corresponding original biocoenosis, and boundaries and distribution patterns based on the fossil community coincide with those based on the biocoenosis and that (2) the trophic structure of the biocoenosis is preserved in the fossil community.

These assumptions must be evaluated before the trophic proportions of fossil communities can be used with confidence in paleoecology. The living macrobenthic invertebrate communities of the Southern California shelf have been investigated for this purpose (Stanton, 1976). Communities based on the total biota were compared with those based on that component of the biota that would potentially be preserved in the fossil record. A high degree of correspondence between the total and the "fossil" communities indicates that the first assumption is valid. Strong divergence between the trophic proportions of the total and "fossil" communities, however, indicates that the second assumption is not valid (Fig. 9.22). This conclusion has been further tested and confirmed by analyzing the trophic proportions of the Pliocene communities described previously for the Kettleman Hills (Stanton and Dodd, 1976b), and independently by Bosence (1979a, 1979b). Scott (1978) has re-

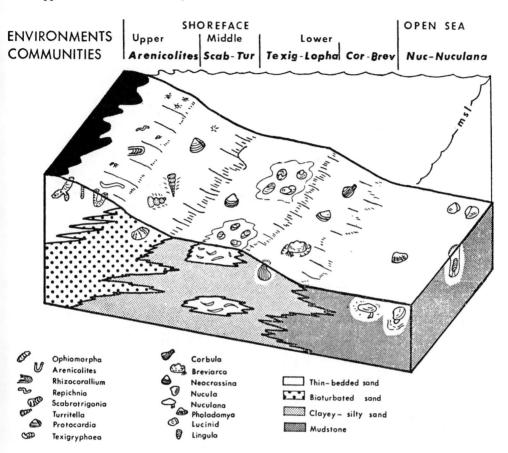

ENVIRONMENTS

COMMUNITIES

SHOREFACE				OPEN SEA
Upper	Middle	Lower		
Arenicolites	*Scab-Tur*	*Texig-Lopha*	*Cor-Brev*	*Nuc-Nuculana*

Ophiomorpha	Corbula			
Arenicolites	Breviarca			
Rhizocorallium	Neocrassina	Thin-bedded sand		
Repichnia	Nucula	Bioturbated sand		
Scabrotrigonia	Nuculana	Clayey- silty sand		
Turritella	Pholadomya	Mudstone		
Protocardia	Lucinid			
Texigryphaea	Lingula			

Figure 9.21 Environmental model for Cretaceous Tucumcari and Purgatoire Formations of the southern Western Interior of the United States showing relations between substrate and benthic communities. The model combines sedimentologic and stratigraphic data with the community characteristics of Fig. 9.20. After Scott (1974), Lethaia 7.

examined the available trophic data for both fossil and modern communities and has emphasized the problems involved but also proposed a heuristic model relating trophic proportions, diversity, and environmental conditions (Fig. 9.23). The status of trophic proportions as a paleoenvironmental tool is not well established at present. It seems to work in some cases. On the other hand, the basic assumptions may be invalid to the point that the technique has little predictive value.

Trophic Structure

Reconstruction of the trophic web of a fossil community involves an integrated analysis of the community. Minimal requirements are a complete census of the community and a thorough knowledge of the feeding habits of the orga-

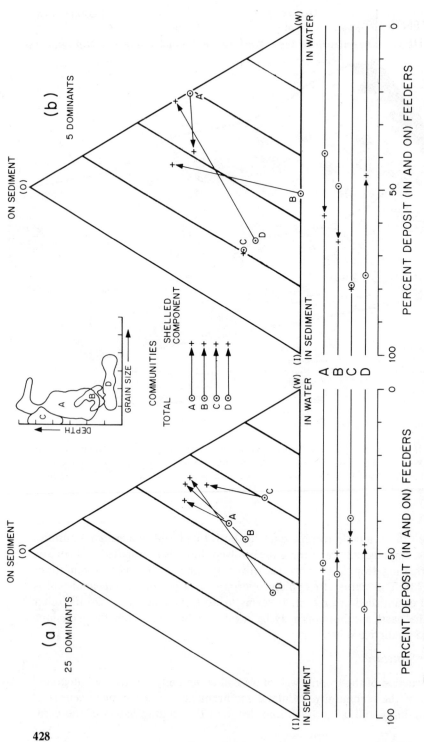

Figure 9.22 Trophic proportions of communities of benthic macroinvertebrates on the southern California shelf. Ternary diagram (*a*) is based on the 25 most abundant species in each community; ternary diagram (*b*), on the 5 most abundant species utilizing trophic resources in suspension, in the sediment, or on the sediment. Values circled are for total communities; valued indicated by cross are for shelled communities derived from total communities. Depth-texture characteristics of communities are generalized. The bar scales below the ternary diagrams represent the data in terms of proportions of deposit vs. suspension feeders. After Stanton and Dodd (1976b), Lethaia 9.

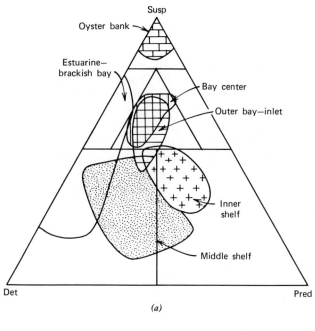

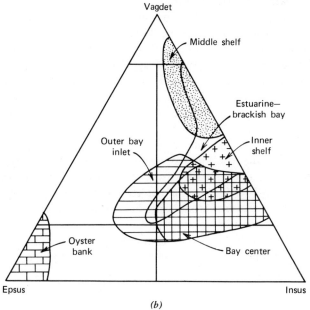

Figure 9.23 General model for trophic and habitat/mode-of-life characteristics of Cenozoic macrofaunal communities. (*a*) Feeding habits. (*b*) Substrate niche. Abbreviations as in Figure 9.20. After Scott (1978), Lethaia *11*.

nisms present. These are generally not attainable for modern communities and are virtually impossible for a fossil community. As a result, the trophic web that can be developed for a fossil community is incomplete. Nevertheless, it is valuable because it makes evident the taphonomic processes that create the biased record as preserved in the fossil assemblage, because it helps to establish organism interactions within the community, and because it provides information both about the unpreserved component of the original community and about the original environment.

The study of mammalian evolution by Olson (1966) provides an example of the value of community structure and the food web as modes of analysis in paleontology. By reconstructing the Permian communities as far as they are preserved and then incorporating other necessary but unpreserved organisms, a fuller understanding of the original community is possible. From this it becomes evident that the origin of early mammals was linked by food-chain requirements to aquatic environments and to environmental changes during the Permian and Triassic.

The argument that dinosaurs may have been warm-blooded is based, in part, on trophic evidence. Because warm-blooded carnivores have a higher metabolic rate than cold-blooded carnivores, the shape of the ecologic pyramid should differ in the two cases. The shape of the pyramid (in terms of relative abundance of carnivores vs. herbivores) can potentially be determined on the basis of the fossil record of Cretaceous vertebrates, and thus the warm blood-edness of the carnivorous dinosaurs (Bakker, 1975).

A systems analysis approach to shallow water benthic ecosystems results in the flow chart in Fig. 9.24 (Hoffman, Pisera, and Studencki, 1978). This flow chart provides a useful starting point for community analysis by presenting the interactions of biologic and physical materials, their quantities, and rates. Two examples of the reconstruction of the trophic web component of the chart are illustrated in Fig. 9.25 by Hoffman (1977) and in Fig. 9.26, by Stanton and Nelson (1980). The trophic web in Fig. 9.25 is for one of the communities in Middle Miocene Korytnica Clays of Poland. The composition of the community is illustrated by list and relative proportions in Fig. 9.27. The trophic web in Fig. 9.26 is for the community of the Main Glauconite bed in the Middle Eocene Stone City Formation, Texas (Fig. 6.12). The composition of the community and relative proportions of taxa are illustrated in Table 9.4, along with the feeding characteristics of modern relatives of the fossils. The two trophic webs are similar in construction and are based on the same assumptions. They differ only in the extent to which all the components of the Miocene community are plotted individually, and in the extent to which the different pathways are quantified in the Eocene trophic web.

The following steps are necessary for the reconstruction of the trophic web.

1   The fossil assemblage must be thoroughly sampled and identified and the relative proportions of the different taxa determined. The proportions in

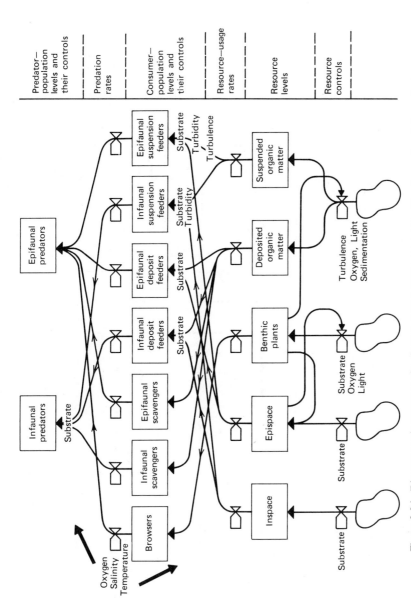

Figure 9.24 Diagrammatic structure of a shallow water benthic ecosystem. *Clouds*—system-independent states, *rectangles*—levels, *arrows*—rates (or flows), *faucets*—rate controls. After Hoffman, Pisera, and Studencki (1978).

431

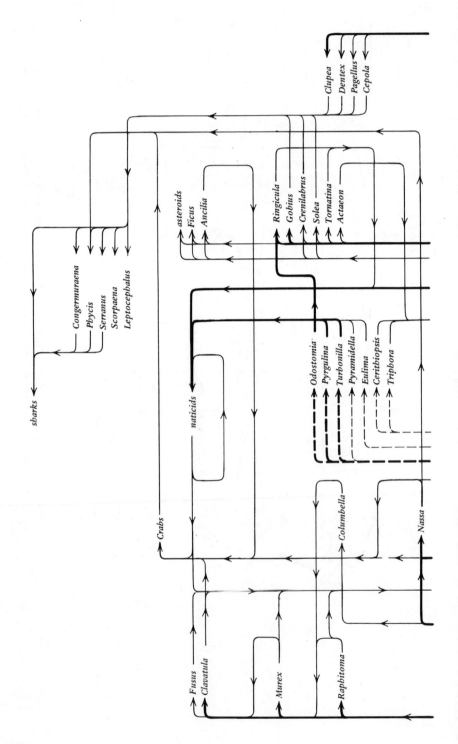

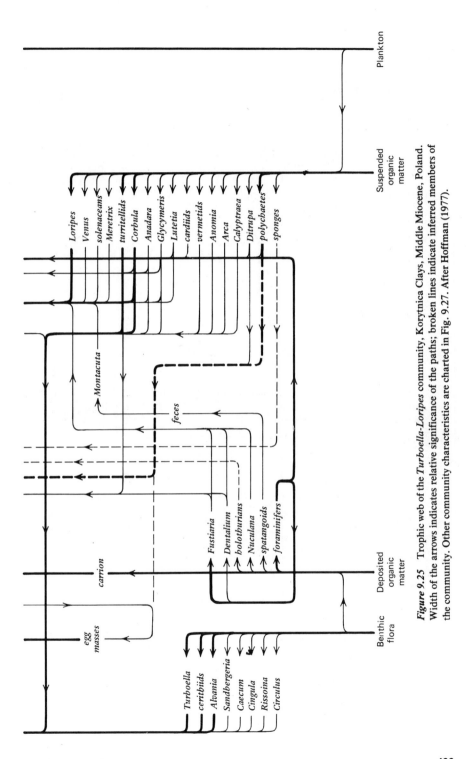

Figure 9.25 Trophic web of the *Turboella-Loripes* community, Korytnica Clays, Middle Miocene, Poland. Width of the arrows indicates relative significance of the paths; broken lines indicate inferred members of the community. Other community characteristics are charted in Fig. 9.27. After Hoffman (1977).

433

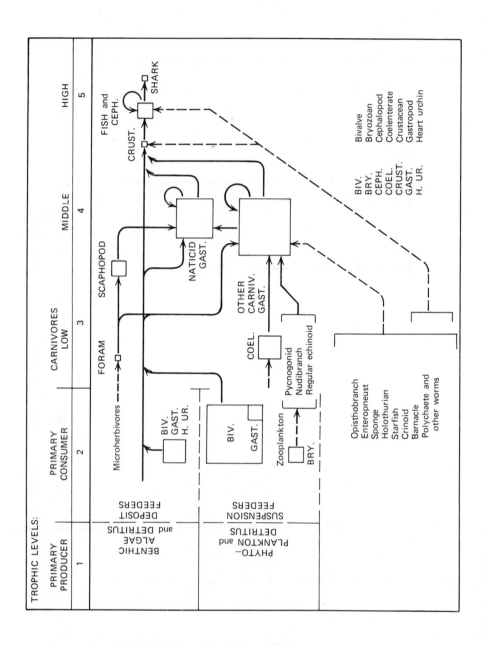

Figure 9.26 Trophic web of the community in the Main Glauconite bed of the Stone City Formation, Middle Eocene, Texas. Box sizes are proportional to numbers of individuals at each position. Solid lines and capital lettering indicate components present in the fossil assemblage and feeding relationships documented in the fossil assemblage or based on modern relationships. Dashed lines and lowercase lettering indicate inferred components and relationships in the original assemblage, based on modern trophic data involving components not preserved. After Stanton and Nelson (1980). Journal of Paleontology, Society of Economic Paleontologists and Mineralogists.

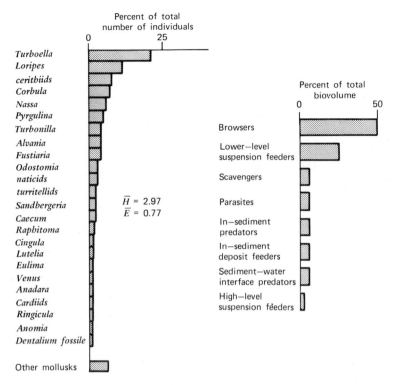

Figure 9.27 Taxonomic composition and relative abundance and trophic proportions of the *Turboella-Loripes* community, Korytnica Clays, Middle Miocene, Poland. The trophic web for the community is portrayed in Fig. 9.25. After Hoffman (1977).

Table 9.4 are based on numbers of individuals. A more precise procedure would be to take into account also the growth rates and longevity of the different species. In this way the relative trophic importance of each species could be estimated more accurately. Probably in general, this refinement is not necessary because it is small in comparison to the uncertainties caused by preservational biases and the difficulty of estimating numbers of individuals of taxa, such as crinoids, that produce numerous skeletal fragments.

2   The taphonomic processes that have formed the fossil assemblage from the original biocoenosis must be analyzed. This is necessary in order to determine, as far as possible, the organism community from the fossil assemblage. To identify the effects of transportation into or winnowing out of individuals in the Stone City assemblage, shell condition, orientation, texture, and right-left ratios were studied as well as sedimentologic features (see Chapter 7).

3   Food preferences of the members of the community are determined from three lines of evidence: (1) indications of predation, such as gastropod

Table 9.4 Composition of the Main Glauconite bed community[a]

	1 G-S	2 %	3 T.L.	4 Bor.	5 Chip.	6 Feeding habits
Coelenterata		6.5				
Scleractinia	2-2	6.2	3	—	—	Carnivorous
Alcyonaria	1-1	0.3	3	—	—	
Bryozoa	6-7	2.9	2	—	—	Phytoplankton
Scaphopoda	2-2	1.9	3	C	C	Carnivorous on forams, small bivalves
Gastropoda		49.3				
Fissurellidae	1-1	0.1	2-3	—	R	Detritus and carnivorous on small sponges
Turbinidae	1-1	0.1	2	—	R	Benthic diatoms, filamentous algae, eaten by Fasciolariidae
Vitrinellidae	2-2	0.1	2-3	A	P	Algae, parasitic on worms
Architectonidae	1-4	0.4	4	R	C	Carnivorous on anemones, corals
Turritellidae	2-6	2.2	2	A	A	Suspended detritus and phytoplankton
Caecidae	1-1	0.2	2	—	C	Interstitial diatoms
Scalariidae	2-3	0.1	4	—	C	Carnivorous and parasitic on coelenterates
Eulimidae	2-3	0.2	4	P	—	Parasitic on echinoderms (starfish, holothurians, regular echinoids), polychaetes
Pyramidellidae	1-1	0.1	4?	C	—	Parasitic/carnivorous on polychaetes, coelenterates, molluscs, starfish
Naticidae	3-3	16.1	3-4	A	A	Carnivorous on bivalves, gastropods, scaphopods
Ficidae	1-2	0.7	3	—	P	Carnivorous on urchins and other echinoderms
Cymatiidae	1-1	0.2	3-4	R	P	Carnivorous on molluscs, asteroids, echinoids
Muricidae	1-1	0.1	3	—	—	Carnivorous on bivalves, barnacles, gastropods
Pyramimitridae	1-1	0.1	?	—	P	???
Buccinidae	2-5	3.0	3	R	C	Scavengers; carnivorous on bivalves, crustaceans, worms

436

Family						Feeding
Nassariidae	1-1	0.6	2	A	A	Nonselective deposit feeder: diatoms, detritus; scavenger
Fasciolariidae	2-2	2.6	3-4	P	P	Carnivorous on gastropods, bivalves, polychaetes, barnacles
Volutidae	1-2	1.1	3	R	P	Carnivorous on bivalves
Olividae	1-1	0.3	3	R	C	Carnivorous on small molluscs, foraminifers
Marginellidae	1-1	0.4	4?	—	P	Carnivorous on ???
Mitridae	1-1	0.1	4?	—	R	Carnivorous on crustaceans, sipunculid worms
Cancellariidae	2-3	2.0	3	R	C	Carnivorous on soft bodied interstitial microorganisms
Conidae	1-1	0.1	4	—	P	Carnivorous on herbivorous polychaetes, fish gastropods
Terebridae	2-2	2.5	4	C	P	Carnivorous on worms, enteropneusts
Turridae	12-14	12.9	3-4	P	C	Carnivorous on annelids, nemerteans
Retusidae	1-2	2.9	3-4	R	C	Carnivorous on other opisthobranchs, foraminifers
Mathildidae	2-3	0.2	?	—	C	???
Ringiculidae	1-1	0.2	3	C	—	Carnivorous on polychaetes, foraminifers
Bivalvia		36.6				Feed from suspension or sediment surface. Dietary preferences generally not known, probably largely microflora and detritus, but nonselective, including bacteria, microfauna
Nuculidae	1-1	2.1	2	P	P	Deposit feeder
Nuculanidae	2-4	3.0	2	—	C	Deposit feeder
Arcidae	1-1	0.2	2	—	R	Suspension feeder
Noetidae	1-2	2.6	2	C	R	Suspension feeder
Ostreidae	2-2	1.1	2	R	P	Suspension feeder
Anomiidae	1-1	1.6	2	R	C	Suspension feeder
Carditidae	1-1	2.4	2	R	C	Suspension feeder
Diplodontidae	1-1	0.5	2	—	C	Suspension feeder
Semelidae	1-1	0.3	2	C	R	Suspension feeder
Tellinidae	1-2	0.2	2	—	P	Deposit feeder
Mactridae	1-1	0.1	2	R	P	Suspension feeder
Veneridae	1-1	0.3	2	—	R	Suspension feeder
Corbulidae	3-3	23.3	2	C	C	Suspension feeder

Table 9.4 (*Continued*)

	1 G-S	2 %	3 T.L.	4 Bor.	5 Chip.	6 Feeding habits
Cephalopoda (*Aturia, Belosepia*)	2-2	0.1	3-5	—	—	Carnivorous on crustaceans, fish, molluscs
Echinodermata (heart urchin)	1-1	0.1	2	—	—	Nonselective deposit feeder
Arthropoda (crustacean)	1-1	0.1	3-4	—	—	Carnivorous, scavenger
Foraminifera	—	—	3	—	—	Diatoms, bacteria
Chordata						
Elasmobranchii	3-3	0.1	5	—	—	Carnivorous on fish, cephalopods
Congridae	1-1	0.1	3-5	—	—	Carnivorous on bottom-living fish, crustaceans, cephalopods
Beryciformis	1-2	0.8	3-5	—	—	Carnivorous on benthic crustaceans
Serranidae	1-1	0.1	3-5	—	—	Carnivorous on benthic crustaceans and polychaetes as juveniles; on small fish as adult
Scianidae	3-3	0.8	3-5	—	—	Carnivorous on benthic polychaetes and crustaceans, molluscs?; planktonic crustaceans, fish and squid
Ophididae	3-3	0.8	3-5	—	—	Carnivorous on benthic crustaceans (shrimp, crabs, stomatopods); juvenile fish, polychaetes
Soleidae	1-1	0.1	4-5	—	—	Carnivorous on benthic crustaceans

[a]Col. 1: number of genera and species within the taxon; Col. 2: percent of individuals in macrofossil assemblage belonging to taxon; Col. 3: trophic level of taxon in ecologic pyramid, as in Fig. 9.26—1: primary producer; 2: primary consumer; 3-5: low to high levels of carnivores; Col. 4: abundance of bored individuals in taxon—A: greater than 20% of individuals; C: 10–20%; P: 5–10%; R: less than 5% of individuals are bored; Col. 5: abundance of crustacean-chipped specimens in taxon—percentage ranges as in Column 4; Col. 6: feeding information for the taxon. After Stanton and Nelson (1980).

borings into shells of bivalves and other gastropods, (2) food preferences of the living relatives of the fossils, (3) functional-morphologic indications of the mode of life of the fossil.

The two trophic webs portrayed in Figs. 9.25 and 9.26 illustrate different analytical approaches. In the Miocene example, the role of each taxon is indicated, as determined from specific predator-prey evidence or as inferred from living organisms. Because many of these inferences are not strong, the trophic web presents a large amount of detail but the accuracy of much of it may be relatively low.

The Eocene trophic web (Fig. 9.26), on the other hand, has been kept simple by grouping together taxa with similar trophic position. Specific information, such as the predation by naticid gastropods and crustaceans on each molluscan taxon (Table 9.4 and Fig. 9.26, e.g.), could be included in the web by separating out each individual genus or species. This has not been done, however, because most of these interrelationships are speculative for the Eocene community, being based on modern feeding information, which is itself incomplete. For example, the inferred predators on bryozoans and the prey of carnivorous gastropods could each add a whole new dimension to the web. In the process, however, we would only be tracing out the modern story, more and more tenuous as far as the Eocene community is concerned, and by doing so we would be denying evolution of trophic structure during the intervening time.

The Eocene trophic web also simplifies the multiple and changing roles of many taxa within the community. For example, the crustaceans are placed at a single trophic level in Fig. 9.26, but from several lines of evidence it is inferred that they were scattered throughout the food web at a wide range of positions. Modern representatives of the Eocene fish taxa have varied diets consisting, in large part, of soft-bodied benthic invertebrates (Fig. 9.28). Worms and crustaceans, identified above from several lines of evidence as major components of the Main Glauconite bed community, thus also fit into the food web as the major food source for the fish; molluscs were possibly also utilized as food, by the scianids. As in the case of the crustaceans, it is simplistic to group the fish and cephalopods into one position in the trophic web; each had distinct requirements, and feeding interactions between them were common. The input category of "Fish" consists of these fish themselves and was used in order to avoid confusing feedback loops in the diagram. To complicate the picture further, many fish change their diets as they grow, decreasing, for example, the consumption of benthic invertebrates and increasing that of other fish.

Both of these food webs are based on incomplete data and include much speculation. Although only pale shadows of the trophic structure that must have characterized the original communities, they can provide much valuable information about the original community and environment. They give information about the unpreserved components of the original community and

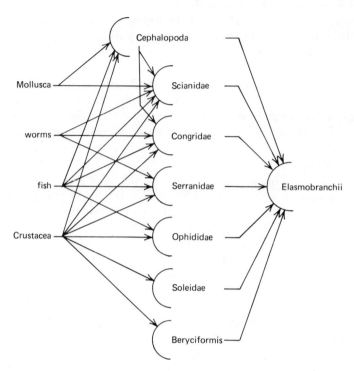

Figure 9.28 Detailed feeding interactions at the top-carnivore level in the trophic web of Fig. 9.26. Main Glauconite bed of the Stone City Formation, Middle Eocene, Texas. After Stanton and Nelson (1980), Journal of Paleontology, Society of Paleontologists and Mineralogists.

thus about the extent to which the original community structure is evident in the fossil assemblage and about the taphonomic aspects of the environment. For example, the Eocene fossil assemblage is strongly weighted toward low- and middle-level carnivores (level 3 and 4), but even this part of the original community is poorly preserved, for crustaceans, which are rare in the assemblage but abundant in the original community, are at these levels of the ecologic pyramid. The primary consumers of level 2 (inferred to be deposit-feeding worms) are even less well represented than the low- and middle-level carnivores.

The reconstruction of the trophic web can provide information about community evolution. In the Eocene community, for example, the food preferences of the crustaceans and the naticid gastropods appear to be different from those at present. Most importantly from the perspective of this book, reconstruction of the trophic web is a step toward reconstructing the paleoecosystem and so provides a basis for paleoenvironmental reconstruction.

10

Paleobiogeography: The Provincial Level

Paleobiogeography, the geographic distribution of ancient organisms, is of basic concern in paleoecology. Much of the work in applied paleoecology is concerned with describing the distribution of fossils in space and time and interpreting what that distribution means in terms of environmental patterns in space and time. Although the terms paleobiogeography and biogeography do not carry any specification of scale, they are usually used in the context of relatively large-scale distribution of organisms such as large regions of continent, ocean, or total worldwide distribution of a species or other taxonomic group. In other words, paleobiogeography is usually considered as an aspect of large-scale, "big-picture" paleoecology. Biogeography does not have to be studied in the context of ecology but simply in terms of the descriptive geographic distribution. Much of the early work in biogeography was of this nature, which Valentine (1973) has termed *geographic biogeography*. Our primary interest in this chapter is in terms of the environmental aspects and controls on biogeography or *ecological biogeography*.

Paleobiogeography is of interest to the paleoecologist for a number of reasons. Although much of paleoecological work is of a local or regional nature, the ultimate goal of applied paleoecology is to determine worldwide environmental conditions for all of geologic time. A combined study of paleoecology and paleobiogeography may help to solve certain large-scale geologic problems in other geologic subdisciplines. This approach has been extensively used, for example, to help in our understanding of the processes involved in plate tectonics (Ziegler et al., 1977). The geographic distribution of organisms is obviously strongly influenced by environmental factors so that a study of geographic distribution helps in interpreting the geographic variation of environmental parameters. For example, one of the regions of greatest change in the biota is between the tropical (or subtropical) and temperate zones. A study of paleobiogeography may help to locate the position of that transition at various times in the geologic past. Finally, the geographic distribution of organisms is strongly influenced by geologic processes and geologic history, particularly tectonic history. The paleobiogeography of vertebrate fossils and land plants in particular have long been used in interpreting the development and breakdown of barriers between continents through time. The distribution of the well-known *Glossopteris* flora is an especially well known example (Chaloner and Lacey, 1973).

In addition to the problems associated with any work in paleoecology, paleobiogeography has some special problems. Because of the large areas involved, usually one person cannot hope to do all the basic data gathering on distribution of the biota and so must depend in part on the work of others. However, each worker may have a somewhat different concept of what constitutes a particular species and even genus, so that published lists of fossil biotas will not be directly comparable. Comparing illustrations will help, or, even better, examining the specimens used by earlier workers, but these are not always available. Long-distance correlation is also difficult so that time equivalences of biotas may be imprecise. One of the purposes of paleobiogeographic studies is to look for areas with different biotas; thus by definition correlation is difficult between areas with different biotas. Because of the irregular distribution of outcrops or even subsurface samples and the unevenness of preservation, fossils may not be available from critical areas for the determination of biotic distributional boundaries. Actually, at times this may be a blessing: boundaries can easily be drawn through an area of no data whereas they are often difficult to place when data are gradational!

CONCEPTS

Definitions

The basic data of biogeography are the distributions of species and other taxa. The basic unit of interpretation is the *biotic province*. A great deal of controversy could be avoided (or more realistically perhaps simply kept in perspective) if the interpretive, subjective nature of the province were borne in mind and it were not considered as an objective reality. Many definitions of the biotic province have been suggested as have several alternative terms for the concept. Perhaps the most meaningful one is that of Valentine (1963) who defines a province as a region in which communities maintain characteristic compositions. This definition is a good expression of the concept of the province as it is now widely understood, but it is rather general to be used in practice for recognizing provinces and boundaries between them. The most important aspect of recognizing provinces is in recognizing their boundaries. These are clearly areas in which the composition of the biota of many communities change markedly and in which many species end their ranges. The areas between such zones of rapid transition constitute the provinces. The location of these transition zones changes with time so that provinces are ever changing in distribution. One of the earliest definitions of a province was an area in which 50% of the species are *endemic*, that is, restricted to that area (Woodward, 1856). This definition has the advantage of being quantitative but is still difficult to apply because many boundaries could be chosen that would set off areas in which the biota would be 50% endemic. Provinces have also been defined as groupings of similar or related biotopes or facies (*megafacies*), for example the coral reef province.

Terms other than province have been used for biogeographic units. Some are equivalent in meaning to province and others are for larger or smaller units. *Realm* is frequently used for a large unit, on a continental or multicontinental scale. For example, four realms are commonly recognized for the late Paleozoic: Gondwanan, Euramerican, Angaran, and Cathaysian (Chaloner and Lacey, 1973). Wallace (1876) used the term *region* for units of the same general scale. Schmidt (1954) proposed a hierarchy of units which, arranged from largest to smallest, are realm, region, subregion, and province. Others have also recognized subprovinces. This may be more complexity than is needed, but at least realm and province imply a convenient distinction of scale. Sylvester-Bradley (1971) gives a thorough treatment of the problems of the definition of provinces.

Despite the difficulty in finding an adequate definition of province, most workers in the field seem to recognize one when they see it. The original provinces recognized in Europe during the mid-nineteenth century by Forbes and Godwin-Austen (1859) and worldwide by Wallace (1876) have not been greatly modified by later workers.

Recognition

Ideally, a province would be recognized on the basis of the entire biota. This is rarely done, and indeed is probably rarely practical, because a worker is likely to be familiar enough with only one phylum, class, or even smaller taxonomic unit to use it for recognizing provinces. Thus much of the work on recognizing modern shallow marine provinces has been based on the distribution of a single phylum, the Mollusca. Most of the examples from the fossil record are also based on particular fossil group. Fortunately, as we discuss below, the factors controlling biotic distribution for one group will likely affect other groups in a similar manner, although certain groups, such as pelagic forms or those with a long pelagic larval stage, are likely to be less confined to provinces than are benthonic forms. In any event, recognizing biotic provinces on the basis of limited taxonomic groups is the standard practice. For clarity, however, such provinces should be prefixed to indicate their data base, as molluscan provinces, trilobite provinces, and so on.

Terms

The following is an abbreviated list of terms that are commonly used in biogeography.

Biome A group of geographically associated communities such as the coral reef biome. This term is more extensively used in terrestrial than marine ecology. Biome has occasionally been used in the sense of realm or province.

Barrier An impediment to the migration of a species, often marking the end of its range. An ocean might be a barrier for a terrestrial species or a continent or an isthmus might be a barrier for a marine species.

Cosmopolitan A species or other taxon with a broad geographic range across two or more provinces.

Endemic A species or other taxon with a limited geographic range, being confined to a single province.

Eurytopic Tolerant of a wide range of environmental conditions. This is a frequent attribute of cosmopolitan species.

Outlier An area in which a species or other taxon is found which is separated from the main part of its range.

Phytogeography The geographic distribution of plant life.

Relict A species or other taxon that once had a more extensive geographic range.

Stenotopic Limited to a narrow range of environmental conditions, an attribute of many endemic taxa.

Vagility The ease with which a taxon migrates or extends its range within its potential range. It is related to such factors as the mobility of the adult or larval stage.

Zoogeography The geographic distribution of animal life.

History

People have undoubtedly been aware of the geographic patterns of plant and animal distribution since ancient times. However, the earliest attempts to formalize these observations into biogeographic regions were made in the nineteenth century. Woodward (1856) first defined marine provinces based on molluscan species distribution. In 1859 Forbes and Godwin-Austen published *The Natural History of European Seas* which included a description of biogeographic provinces in Europe. Some of the most basic work in biogeography was done by Charles Darwin and especially A. R. Wallace (1876), both of whom worked with terrestrial biota. Landmark publications in the field of biogeography include the books of Hesse, Allee, and Schmidt (1951), Ekman (1953), and Darlington (1957), each of which contains a wealth of information of value to present-day workers in this field. Hedgpeth (1957a) published an especially useful review for paleoecologists. Until relatively recently few reviews have dealt specifically with paleobiogeography. Perhaps the first such publication was a volume on the proceedings of a symposium on faunal provinces held at the University of London in December of 1969 (Middlemiss, Rawson, and Newall, 1971). This volume includes general discussions of paleobiogeography as well as several specific studies in rock strata ranging from Cambrian to Cenozoic. The *Atlas of Paleobiogeography* (Hallam, 1973a) and *Paleogeographic Provinces and Provinciality* (Ross, 1974) contain numerous examples of provinciality among various fossil groups. A symposium on paleontology and plate tectonics held at the North American Paleontology Convention in 1977 included several papers dealing with paleobiogeography (West, 1977). A very good review of this subject, especially its theoretical aspects, is that of Valentine (1973) who has long been active in research in this field. Part A of the *Treatise on Invertebrate Paleontology* (Robison and Teichert, 1979) includes discussions of the paleobiogeography of each of the Phanerozoic Periods.

FACTORS CONTROLLING GEOGRAPHIC DISTRIBUTION OF SPECIES

Each species has a potential geographic range that is determined by its habitat requirements, but few if any species actually occur throughout their potential range. There are two basic reasons for this. First, barriers of one sort or another prevent their expansion into separate areas of suitable habitat; and, second, the species may not have had sufficient time to spread to all suitable areas, especially to overcome barriers. This is not to imply that there are vast areas of suitable habitat available for occupation that have unoccupied niches. The usual pattern is for similar species or similarly adapted species to occupy a given niche in areas separated by barriers. If the barriers were to disappear the species would compete for the same niche; and, as a result of competitive exclusion, the less well adapted species would become extinct or evolve and become adapted to a different niche. On the other hand, if a barrier should appear and divide a once continuous area of suitable habitat, the separated populations would gradually evolve into different species, each with a more restricted geographic range than the parent species. Obviously, biogeography is closely tied to evolution, a fact recognized at an early date by Charles Darwin.

Biogeography and biotic provinces are closely related to barriers and the geologic history of the barriers. Any feature of the environment that limits the distribution of a species is a barrier. Valentine (1973) uses the phrase habitat failure to describe such barriers. The types of environmental parameters that will be of greatest importance in determining provinciality will be those that affect many species at the same or approximately the same locality. Also of importance is how systematically the environmental parameter varies geographically and how effective is the barrier produced by changes in that parameter. Sometimes rather than being due to change in a single parameter, barriers may result from the interaction of two or more factors. Although practically any environmental parameter can be important in specific cases, the two most important are depth-elevation and temperature.

Depth-elevation

As indicated previously, depth is not a pure environmental parameter because many parameters change with depth. The same could be said of the related concept of elevation. However, depth-elevation is a convenient concept in discussing barriers in biogeography. All marine organisms have a limited range of depths over which they can survive; hence water that is too deep or too shallow forms a barrier. A broad expanse of deep water is a very effective barrier to the distribution of shallow water benthonic organisms, and terrestrial organisms, too, have certain elevation limits. The most effective barrier to distribution of either terrestrial or marine organisms is sea level. Terrestrial organisms cannot exist below it (except obviously in terrestrial basins that go below sea level) and marine organisms cannot exist above it. Another way of expressing the same idea is the presence or absence of the life media, be it air or water. Even a relatively narrow expanse of elevation above sea level, such as the Isthmus of Panama, is an effective barrier to marine organisms. The boundaries of all modern marine provinces are in part determined by the depth-elevation parameter.

The depth-elevation parameter can be used to illustrate the importance of the geographic extent and effectiveness of barriers. Barriers obviously may differ in effectiveness, and effectiveness will not be the same for all species. One measure of the effectiveness of a barrier is the ease with which species circumvent it. Simpson (1940) classified dispersal routes around or through barriers into three categories on the basis of their effectiveness. The *corridor* is an open route for migration with many species easily passing through the barrier. The broad shelf connecting the coastal waters of Siberia with those of Alaska would be a corridor for shallow marine organisms. A *filter bridge* is a more limited dispersal route available to some species but not all, at least in geologically short time periods. A relatively narrow strait of deep water between shoal water areas such as the Mozambique Channel between Africa and Madagascar would be an example. A *sweepstakes route* is a low-probability route that few species are able to take. The broad expanses of deep water around the Hawaiian Islands would be an example both for the dispersal of shallow marine as well as terrestrial forms.

Temperature

The boundaries of all modern biotic provinces are in part temperature controlled. All species, even eurythermal ones, are sensitive to temperature variation. But the reason temperature is so important in determining provincial boundaries is its systematic global variation. Year-round warm oceanic temperatures are restricted to the area relatively near the equator. Cooler, seasonally variable temperatures on either side of this zone form barriers. Not only is this low-temperature barrier effective, but there simply is no other side of the barrier! Hence warm water taxa are restricted to the tropical zone.

Likewise, the tropical zone forms an effective barrier to cold water taxa that might otherwise migrate from one temperate belt to that across the equator. However, the cold water taxa have two alternatives not available to warm water forms: First, there is cold water at depth even in the tropics; so a cold water species that can tolerate depth can find suitable habitat. Many species do this, exhibiting the phenomenon of *submergence*, that is, occurring at greater depth in the low-latitude end of their range (Fig. 10.1). Surprisingly, a few species also submerge at the high-latitude end of their range because the surface waters may chill more during the winter than water slightly below the surface. The second alternative for cold water forms is that, if they can break through the tropical barrier, there is cold water at the surface in the other hemisphere.

A few species or closely related species do occur on either side of the equator giving rise to *bipolar distributions* (Hesse et al., 1951), that is, the species occurs in middle or high latitudes in both hemispheres but not at low latitudes (Fig. 10.2). The method by which the species accomplish this distribution has

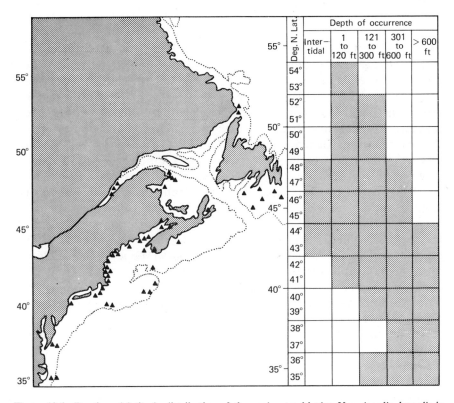

Figure 10.1 Depth and latitude distribution of the cool-water bivalve *Venericardia borealis* in the northwest Atlantic. Triangles on the map show the localities of living specimens as determined from published reports. The graph on the right summarizes the water depth of all known occurrences for each two degrees of latitude. After MacAlester and Rhoads (1967).

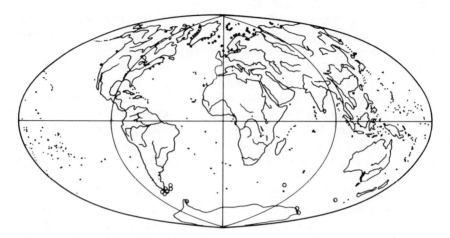

Figure 10.2 Geographic distribution of the species pair of priapulid worms *Priapulus caudatus* (dots) and *P. tuberculato-spinosus* (circles) showing a bipolar distribution. After Hedgpeth (1957a), Geological Society of America Memoir 67.

been the subject of some controversy (Fig. 10.3; Hedgpeth, 1957a). One theory is that the species may submerge into deeper water in the tropics and occur at the surface in higher latitudes. No tropical deep water forms of bipolar species have been described, however. Another theory is that in the past the temperature in the tropics was lower so that the species ranged from one hemisphere to the other. Warming of the tropics caused elimination of the equatorial forms of the bipolar species. There is no evidence for such a cold period in the equatorial regions in the recent geologic past, however. A third theory states that the species once lived throughout the tropics and into both temperate zones but that a strictly tropical species evolved that was adapted for the same niche. The original species was eliminated from the tropics by competitive exclusion but remains in the temperate zones that the new species cannot invade. This theory may be reasonable but so far is not proved by paleontologic evidence. The generally geologically younger nature of most tropical species is compatible with the theory. The tropics are a barrier that temperate species have a certain probability of penetrating albeit perhaps a very low probability. But as Simpson's term sweepstakes route implies, the chance is there that given enough time the species may be able to cross the barrier and become bipolar.

Of the temperature-related barriers in the marine environment, none is more effective than the tropical-temperate barrier. A very high proportion of the species cannot cross the tropical-temperate boundary, which has a mean temperature of about 15 to 18°C for the coldest month of the year. Wherever that temperature boundary occurs in the shallow seas there is a provincial boundary. If adequate data were available, the tropical-temperate boundary should be readily identified in the geologic record on the basis of distribution of ancient biotic provinces.

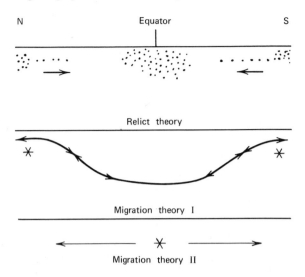

Figure 10.3 Schematic representation of three theories proposed to explain bipolarity. These three theories are discussed in detail in the text. Place of origin is indicated by an asterisk. After Hedgpeth (1957a), Geological Society of America Memoir.

Provincial boundaries often occur at other temperatures also, although the biotic change is usually not as great (Hall, 1964). Biotic changes usually occur in areas where there is a rapid temperature change over a relatively short distance. Such areas often correlate with changes of direction in the coastline, especially insofar as these are related to current patterns. For example, on the California coast one of the biggest changes in biota occurs at Point Conception (Fig. 10.4) south of which the coast runs approximately E-W for many miles and north of which it runs N-S. North of Point Conception the water is uniformly cool as the result of the southward flowing California Current immediately offshore and extensive upwelling of deep water caused by the surface waters moving offshore. South of Point Conception the temperature is warmer because of a northward flowing countercurrent, the more offshore location of the California Current, and less extensive upwelling. Outliers of the more northerly fauna occur on the Channel Islands and at places far south into Baja California where intense upwelling brings cold water to the surface.

A final point should be made in connection with temperature control of species distribution. The nature of the biota is controlled not only by the absolute temperature but also by the amount of temperature variation and the durations of low or high temperatures. The more stenothermal forms may not be able to survive in areas where the temperature is highly variable. This partly explains why some aspects of the shallow water biotas of opposite sides of the same ocean are so different (Hedgpeth, 1957a). Generally, the major oceanic circulation patterns produce less temperature variation on the eastern sides of ocean basins

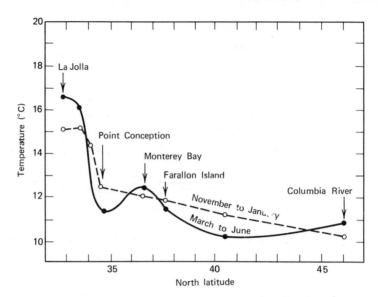

Figure 10.4 Surface temperatures along the coast of California and Oregon in March to June and November to January. Note the thermal break at Point Conception. After Sverdrup et al. (1942), *The Oceans,* copyright 1942, renewed 1970, p. 724. Reprinted by permission of Prentice-Hall, Inc., Englewood Cliffs, N.J.

than on the western sides. This also accounts for the greater diversity of the biota on the eastern sides of ocean basins.

Other Environmental Parameters

Other environmental parameters are capable of forming barriers that could mark provincial boundaries; however, they appear to be distinctly less important than depth-elevation or temperature. Currents are at least indirectly quite important in some areas because temperature patterns are strongly determined by the oceanic current pattern. Currents aid in the dispersal of larvae (although they probably do not form an effective long-term barrier to dispersal of organisms), and they modify the odds in sweepstakes route dispersal.

The amount of rainfall is almost as important as temperature in controlling the distribution of terrestrial plants and animals (Good, 1974, and Darlington, 1957). Both plants and animals are highly adapted to the amount of water available in their environment, hence do not generally have a very broad range of rainfall tolerance. Maps of terrestrial biomes clearly reflect the precipitation pattern (Fig. 10.5). Rainfall, however, is not so important in determining the really large-scale global realms, which are likely to include a range of climatic conditions.

Salinity has been suggested as an important factor controlling provincial

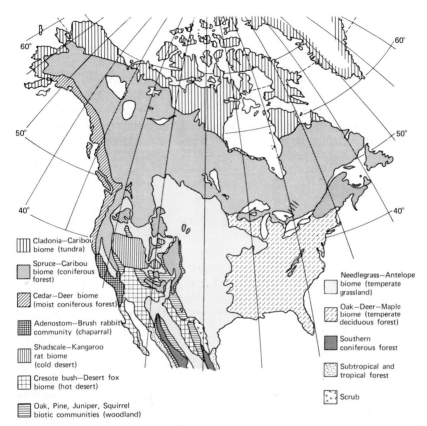

Legend on figure:

- Cladonia–Caribou biome (tundra)
- Spruce–Caribou biome (coniferous forest)
- Cedar–Deer biome (moist coniferous forest)
- Adenostom–Brush rabbit community (chaparral)
- Shadscale–Kangaroo rat biome (cold desert)
- Cresote bush–Desert fox biome (hot desert)
- Oak, Pine, Juniper, Squirrel biotic communities (woodland)
- Needlegrass–Antelope biome (temperate grassland)
- Oak–Deer–Maple biome (temperate deciduous forest)
- Southern coniferous forest
- Subtropical and tropical forest
- Scrub

Figure 10.5 Distribution of terrestrial biomes in North America. After Shelford (1963), The University of Illinois Press, copyright 1963.

boundaries. It certainly is important in marking the boundary between fresh-water and marine provinces, for few freshwater species are able to tolerate salinities above 1 to 2 ‰ (Kinne, 1971). Likewise, very few marine forms are able to live below that salinity barrier. In marine environments salinity is of minor importance in defining provincial boundaries. The reason for this is that most of the oceans have a relatively uniform salinity of 35 ± 3 ‰ (Sverdrup et al., 1942). The rather irregular, smaller areas of lower or higher salinity are not distributed in such a way as to form barriers that cannot be penetrated by normal marine species. Salinity is important in controlling the local distribution of communities but not of provinces. Low- or high-salinity communities are simply part of the assemblage of communities that makes up the province.

The same sort of statements can be made about the importance of other parameters, such as substrate or water turbulence. These parameters are highly variable at the provincial scale and, although they are important at the community level, they are not involved in forming barriers at provincial boundaries.

Water Mass

Pelagic, especially planktonic, species distributions are apparently controlled by different parameters than are shallow marine benthonic species. Major changes in biotic composition of the open ocean mainly occur at boundaries between water masses (Funnell, 1971). The water masses owe their properties largely to their characteristic density which is a function of a combination of temperature and salinity, temperature usually being the more important. Some mixing does occur at the boundaries of water masses, but they and their contained biota remain distinct with gradients in composition at their boundaries.

Geologic History

The distribution of species is in part a function of geologic history, particularly the geologic history of the environmental parameters that form barriers. The most important of these in terms of geologic history is the depth-elevation parameter or, more specifically, the relative position of land and sea. The most pronounced and best known examples of the effect of geologic history is on terrestrial biotas, especially mammals. The distinctive biota of Australia is an excellent example of this effect (Keast, 1972). Before the coming of man, only a few species had been able to penetrate the oceanic barrier around Australia since Cretaceous or Early Cenozoic time. Most members of the biota of Australia seem to have evolved from Cretaceous worldwide cosmopolitan ancestors, suggesting that no effective barriers to migration existed at that time. The subsequent establishment of barriers has been effective in preventing immigration of most outside species. Of course, man has artificially broken down that barrier by the introduction of many outside species such as the rabbit and the dog.

South America had a similar history of isolation from the rest of the world beginning in Cretaceous or Early Cenozoic time (Patterson and Pascual, 1972) which lasted until Pliocene time. With the development of the Isthmus of Panama, the modern terrestrial biota is becoming less distinctive than that of much of Cenozoic time. Some South American animals such as the opossum and the armadillo have migrated to North America and many North American animals have migrated to South America.

A related marine example is provided by the biotas of the Central American shelf and adjoining areas. Before the development of the Isthmus of Panama, no land barrier existed between the Caribbean and Atlantic on the one hand and the Pacific on the other. Since Pliocene time, distinctive biotas have developed on opposite coasts of Central America and adjoining areas of North and South America. Hallam (1973b) has pointed out that often the disappearance of a barrier to terrestrial migration is accompanied by the formation of a marine barrier and vice versa (Fig. 10.6). An aspect of these faunas relating to their evolutionary history is that many species have a similar but slightly different counterpart on the opposite side of the land barrier (Ekman, 1953). This is the result of evolution of the species from the same parent stock which occurred in both

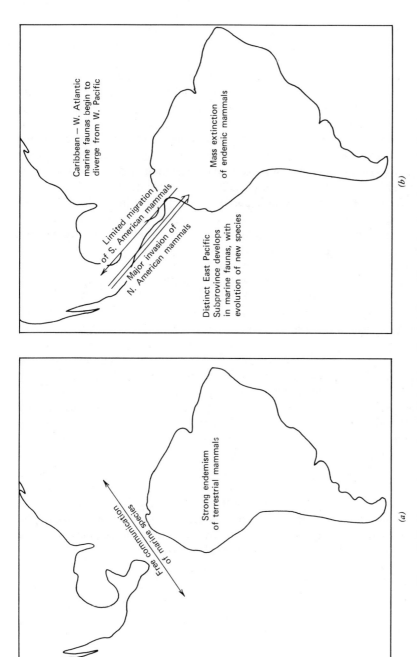

Figure 10.6 Changes in patterns of terrestrial and marine faunal distribution following creation of the Central America land bridge. (a) Pre-late Pliocene; (b) Late Pliocene-Pleistocene. After Hallam (1973b).

453

areas before the formation of the land barrier. These are classic examples of *sibling species*. One might speculate about what would happen if the Panama land barrier were, at least in part, broken by the construction of a sea-level canal. Some have predicted an ecological disaster. The long-term effect might well be significant in terms of the extinction of many species and the spread of others. However, similar events have occurred naturally many times in the earth's history and would hardly be considered disasters.

Historical changes in temperature have also had pronounced effects on the distribution of organisms. The positions of temperature barriers are constantly moving through time allowing organisms to change their distribution as the barriers move. This effect has been especially pronounced during the Pleistocene. Many effects of fluctuating climatic patterns are apparent in the distribution of modern organisms. In terrestrial biotas there are numerous relict populations of species occurring far outside the main populations of that species. Hesse et al., (1951) cite several examples of glacial relicts in Europe and North America such as the arctic ptarmigan and varying hare, which are now found in isolated populations in mountain-top localities such as the Alps, Pyrenees, and Caucasus. Many arctic plants and animals have been described far south of their normal range on Mt. Washington in New Hampshire. The populations have managed to continue to live in these relict locations by staying in areas such as on mountain tops, which are cooler than most of the surrounding areas.

There are many similar marine examples. Several species found in the Baltic Sea are relict populations of species that otherwise occur only far north in the Atlantic (Segerstråle, 1957). At one time, sea water temperatures were apparently low enough that the species extended through the Straits of Kattegat and throughout the surrounding area. Now the relict populations occur only in the northern portion of the sea far from the main area of distribution of the species.

Many shallow marine species on the southeastern U.S. coast show a disrupted distribution because of the high temperatures in south Florida (*disjunct endemism*). The species (e.g., *Mercenaria mercenaria*, *Mytilus edulis*, *Crassostrea virginica*) occur on the Atlantic coast and the Gulf of Mexico coast but not in south Florida where the temperature is too warm (Valentine, 1963). The general similarity of the Gulf Coast and the Atlantic faunas suggests that the temperature in south Florida has not always been too warm to act as a barrier to distribution.

These are but a few of many examples of the historical effects. One of the most important contributions of paleoecologists to the study of biogeography is a better understanding of this process, which leads to a better reconstruction of earth's history.

Tectonics

The major controlling factor on the distribution of land and sea is the tectonic activity. This is related to the historical factor, and tectonic development is really a part of it. The establishment and breakdown of barriers is often closely related to plate tectonic events. Provinciality is greatest during times when plate motion has produced a maximum number of separate continents (such as the present time). There are fewer barriers and thus less provinciality when plate motion has welded together continents, such as was apparently the case at the end of the Paleozoic (Valentine, 1971a).

Several specific relationships between plate tectonics and biogeography can be stated (Fig. 10.7; Table 10.1). (1) When spreading ridges lie parallel to continents they produce deep and ever widening ocean basins and thus barriers to migration of terrestrial or shallow marine biota (Mid-Atlantic Ridge). (2) Transform faults parallel to continental margins also are usually associated with deep water barriers (San Andreas fault, California). (3) Subduction zones parallel to and dipping toward the continent form deep water barriers (Peru-Chili Trench). (4) Subduction zones dipping away from continents may have island arcs that aid in breaking down barriers (subduction zone from Burma to New Hebrides). (5) Mid-plate volcanoes, perhaps related to mantle plumes, help break down deep water barriers (Hawaiian Islands). (6) Subduction zones, spreading ridges, and associated island arcs at high angles to continents may provide migration pathways breaking down barriers (Aleutian Islands) (Valentine, 1971a).

Many studies have used paleobiogeography along with plate tectonic theory to attempt to determine the relative position of the continents on the globe for various times in the geologic past. One of the more ambitious of such efforts has been that of Ziegler et al. (1977) and Scotese et al. (1979) who have prepared maps for seven time periods during the Paleozoic ranging from Upper Cambrian to late Permian. They have combined paleomagnetic pole locations, sedimentologic data on climates (evaporites, tillites, etc.), and petrologic data (andesites, ophiolites, etc.) with data on paleobiogeography to get the best fit for continental positions. They show the continent of Gondwana as a unit throughout the Paleozoic whereas the northern continents were more widely separated but began to coalesce later in the Paleozoic and eventually joined Gondwana in the Permian to produce a single supercontinent (Figs. 10.8 and 10.9). Similar maps have been prepared by Smith et al. (1973) based largely on paleomagnetic data.

HISTORY OF SPECIES RANGE

The biogeography of a species, that is, its geographic range, changes during its existence (Valentine, 1973b). Species that develop through geographic speciation of a small, isolated population would obviously start with a small geographic range. Ideally the geographic range will increase to a maximum and will then decline to eventual extinction (Fig. 10.10A). Species that form by phyletic evolution in large populations or groups of populations by the slow accumulation of genetic changes may start with large ranges and eventually decline to extinction (Fig. 10.10B). A species may start from a small population, expand geographically and then gradually evolve phyletically into a new species (Fig.

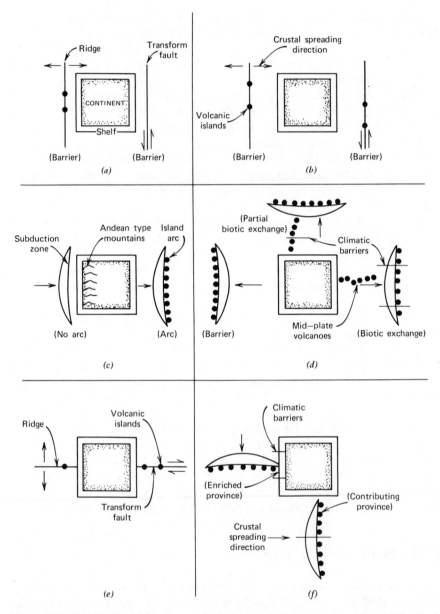

Figure 10.7 Relationships between crustal plates and continents as they affect biotic distributions. See Table 10.1 and text for a detailed explanation of each case. After Valentine (1971a). Reprinted by permission from Systematic Zoology, *20:* 261.

Table 10.1 Relationship of crustal plates to continental margins and its effect on biogeography

Geometry of Relation	Character of Margin	Distance	Biogeographic Implications for Continental Shelf	Fig. 10.7
Parallel	Ridge or	Near	Barrier but with depauperate provincial outliers on isolated islands	a
	transform	Far	Barrier	b
	subduction	Near	Barrier if truly marginal with no island arc; source of rich biota and dispersal route if arc present	c
	zone	Far	No effect unless intervening region bridged by mid-plate volcanoes, then source of rich biota and dispersal route if no climatic barriers intervene	d
High angle	Ridge or	Near	Little effect, with depauperate provincial outliers on isolated islands	e
	transform	Far	Not a case	
	subduction	Near	N-S shelf, E-W arc: arc system a source of rich biota for local province. E-W shelf, N-S arc: proximal province of arc system a source of rich biota for entire shelf	f
	zone	Far	Not a case	

From Valentine (1971a).

10.10C). In detail this pattern may be more complex, perhaps with several expansions and contractions of range, but in the idealized, simplified case the range will first expand and then contract. The rate at which the range will increase is a function of the vagility of the species. Pelagic species and those with long pelagic larval stages will spread much more rapidly than benthonic taxa with no or a short pelagic larval stage. Although the time required for the species to spread can not be predicted, the species will continue to spread if there are no barriers. This is a deterministic occurrence, that is, the species is certain to spread through the operation of normal biotic processes. The presence of barriers will affect the rate of expansion by at least temporarily stopping it. The penetration of a barrier by a species is in part a probabilistic occurrence, that is, whether the species penetrates the barrier is a matter of chance. The species has a certain probability of penetrating the barrier in a given time, so the longer the time available the greater will be the chance that the species

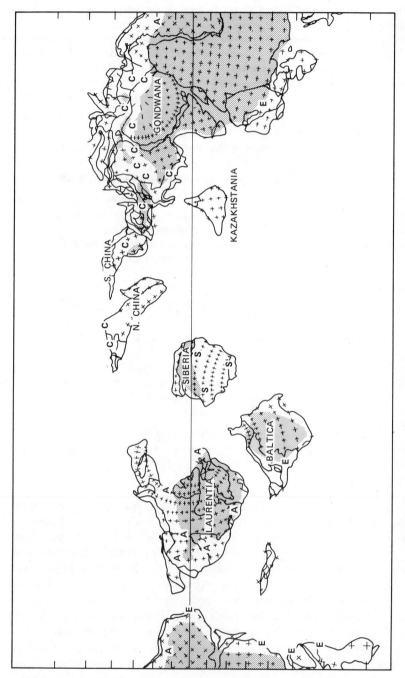

Figure 10.8 Upper Cambrian trilobite provinciality. Faunal realms are A—North American; C—Chinese-Australian; S—Siberian; and E—European. The shaded areas are inferred land areas. The X's on the continents are modern latitude-longitude intersections. After Ziegler et al. (1977).

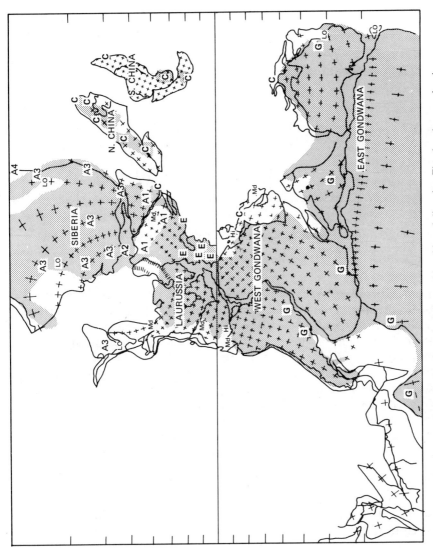

Figure 10.9 Upper Permian floral provinciality and brachiopod diversity. Floral realms are A—Angaran with four subdivisions; E—Euramerican; C—Cathaysian; and G—Gondwanan. Hi, Md, Lo - High, medium, and low brachiopod diversity. Other symbols are as in Figure 10.8. After Ziegler et al. (1977).

459

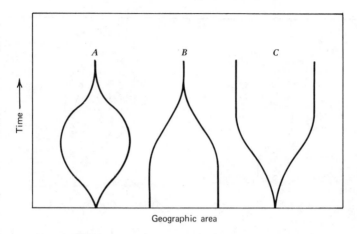

Figure 10.10   Three models of variation in geographic distribution of species through time. More complex patterns are also possible.

will manage to overcome the barrier (Gretener, 1967). Thus the general trend will be for the species range to increase with time. The location of the barriers may change during the life of the species, further complicating the details of distribution. An example of such fluctuating barriers would be temperature fluctuations during the Pleistocene. This increase in range with time led Willis (1922) to propose the hypothesis that the size of the range of a species is proportional to the age of the species.

Eventually the species may evolve phyletically into another species or it may decline in abundance and ultimately become extinct. The decline may be the result of competition with a better adapted species that is expanding its range. During the declining phase, small populations may become isolated to form relict populations. The species may continue to live for some time in a limited area called a *refuge* for the species. The Southwest Pacific might be considered a refuge for the once widely distributed *Nautilus* and indeed all of the nautiloids (Hesse et al., 1951). The contraction of range led Rosa (1931) to hypothesize an inverse correlation between species age and range. In some cases this hypothesis surely applies just as in others the Willis hypothesis of a direct relationship applies. The processes of range expansion and contraction are too complex, however, for either hypothesis to have much predictive value.

The processes of species migration and range changes are well suited for theoretical and statistical treatment. MacArthur and Wilson (1967) have published a much quoted study of this sort based largely on data on the populating of islands. Although this work is based on the biota of actual islands, it is more generally applicable. Areas of suitable habitat for a given group of species can be considered as a *habitat island* with the intervening unsuitable area as an ocean barrier. Freshwater lakes surrounded by a sea of land would be such an analogy.

MacArthur and Wilson reason that the number of species living on an island will be determined by a balance between immigration rate of new species and local extinction of species already on the island. This is shown graphically in Fig. 10.11, in which the descending curve represents immigration rate. The greater the number of species already on the island the slower will be the rate of addition of new species. Ultimately if all available species are on the island the rate of immigration of new species must be zero. The ascending curve in Fig. 10.11 represents the local extinction rate. Of necessity that curve must start at zero because if there are zero species, there are none to become extinct. When there is a large number of species the rate at which they are becoming extinct will be greater. At some point, s, immigration rate and extinction rate will be equal. The number of species corresponding to s is the equilibrium value for that particular island.

Each island will have its own characteristic immigration and extinction curves, but certain relationships should exist between the curves. For example, the immigration rate for distant islands should be lower than that for near islands because of the greater width of the barrier to be overcome (Fig. 10.12). Consequently, far islands should have fewer species than near islands. Small islands should have less space for species and also a lesser variety of habitats and thus the extinction rate should be higher than for larger islands (Fig. 10.12). As a result, large islands should have more species than small islands. In fact, MacArthur and Wilson have noted an excellent correlation between the size of an island and the number of species it supports. Limited data also in-

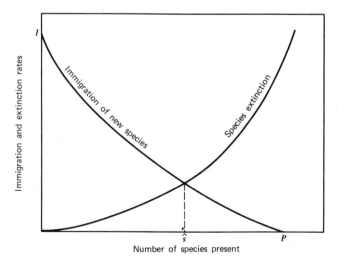

Figure 10.11 Variation in the rates of immigration, I, and extinction of species on an island as a function of the number of species on the island. P is the number of species available for immigration. The equilibrium species number is reached at the intersection point, ŝ, of the curves. See text for a detailed discussion of this model. After MacArthur and Wilson (1963), Evolution.

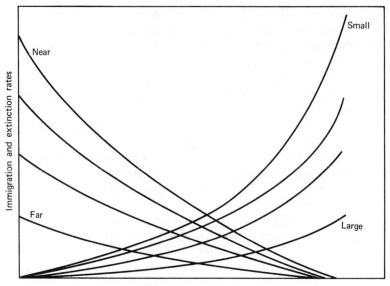

Figure 10.12 Equilibrium model as in Fig. 10.11 for biotas of islands of varying size and distance from the principal source area. After MacArthur and Wilson (1963), *Evolution*.

dicate that far islands do have fewer species than near islands (Fig. 10.13). Perhaps the main value of this approach is that it gives a conceptual framework for considering the process and results of colonization of islands, either real or habitat islands.

DEFINING PROVINCIAL BOUNDARIES

Several methods of defining provincial boundaries have been suggested, but no one method has been generally accepted. Recognition of provincial boundaries has been largely a subjective process somewhat like the recognition of boundaries between species by taxonomists. The original definition of a province by Woodward, which specifies that a province should have 50% endemic species, may seem on the surface like a fairly objective method for determining provincial boundaries. However, there are several difficulties with using this definition, as already noted briefly. First, there is nothing natural about the figure 50%. Zones of rapid species change may define areas with any percentage of endemic species, and a province characterized by low-diversity communities may even have no endemic species (Valentine, 1973b). Rigorous application of the 50% rule would often require placing the boundaries in areas other than those with the greatest biotic change. The boundary would have to go wherever the 50% point happened to occur even if it is in an area of a relatively homogeneous

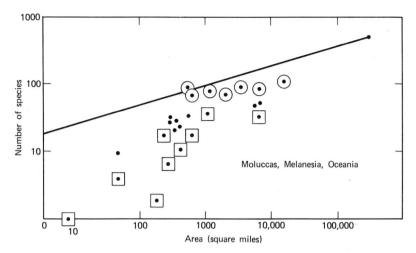

Figure 10.13 Number of land and freshwater bird species on Pacific islands as a function of the area of the islands. Islands which are near (less than 500 miles) a major land mass are circled. Islands which are far (more than 2000 miles) from a major land mass are in squares. Islands at intermediate distances are dots. After MacArthur and Wilson (1963), Evolution.

biota. Finally, the boundaries could be placed in a large number of positions to include the 50% endemic species (Fig. 10.14), so the method is still subjective. In the hypothetical example shown in Fig. 10.14, *A, B,* and *C* are just three of many possible choices for provinces. Province A probably comes closest to what most workers would recognize as a natural province. Subdivision at the break in shoreline into areas *D* and *E* does not satisfy the Woodward definition because it does not produce areas that are 50% endemic.

Schenck and Keen (1936) proposed that shallow marine provinces can be defined on the basis of the midpoints of the range of species along the coast. They used the Pacific Coast of North America as their example. They defined provinces as centering in areas where there is a concentration of range midpoints. The main difficulty of this method is that it attempts to find the center of the provinces rather than the boundaries (Newell, 1948). The midpoint of a range has no special natural significance whereas the end points do by ideally showing the position of barriers for the species. Defining boundaries on the basis of a concentration of range end points would seem to be a reasonable approach, and indeed, the more subjectively defined provincial boundaries are tacitly based on this method. Areas of greatest faunal change are areas of many species range end points.

Valentine (1966) used a numerical approach to defining provincial boundaries which is also closely related to the range end point method (Fig. 10.15). He used a *Q*-mode cluster analysis in which he clustered Jaccard coefficients calculated by considering the area bounded by each degree of latitude along the coast as a separate sample. Thus each 1 degree segment of coastline was compared to

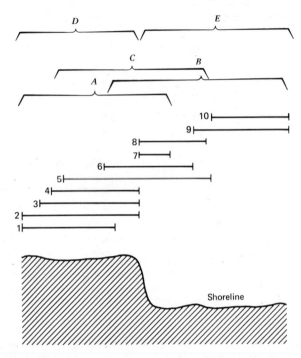

Figure 10.14 Schematic representation of various ways of defining biogeographic provinces so that they contain 50% endemic species. The horizontal lines represent the geographic ranges of species and the lettered brackets represent possible provinces. In the hypothethical example shown, many shallow water species have ranges that end at a break in the shoreline direction. See text for details.

all other 1 degree segments on the basis of presence or absence data on the oc-currence of species (he used only molluscan species). Two segments that had many species in common of course had much higher Jaccard coefficients than segments with dissimilar biotas. Usually the most similar coastal segments are those that are adjacent to each other. Widely separated segments have low similarities. Side-by-side segments in areas where marked biotic changes occur have lower similarity than side-by-side segments in areas with uniform biotas. Provinces can thus be recognized on the basis of these breaks in similarity as shown by the Jaccard coefficient. They can also be recognized by forming clusters in the cluster analysis (Fig. 10.16). This method has the advantage of objectively showing breaks in the biotic distribution. It is still subjective, however, because one must decide at what level of similarity a cluster (or prov-ince) will be recognized.

Simpson (1960) suggested a somewhat different approach to measuring pro-vinciality. He does not attempt to define formal provinces by any quantitative measure but simply compares the degree of similarity of the biota between two

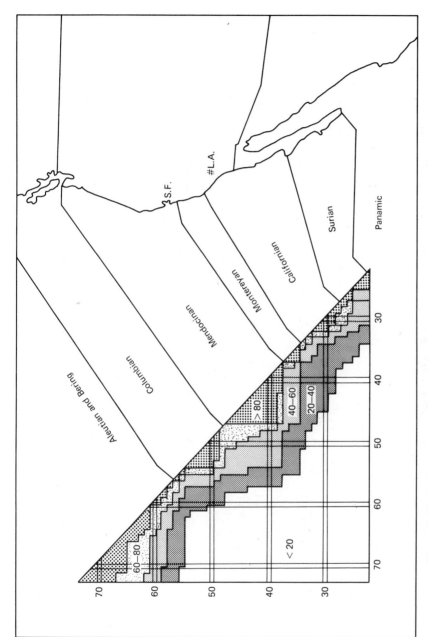

Figure 10.15 Modern northeast Pacific molluscan provinces and matrix of Jaccard coefficients of similarity for one degree latitude increments, based on shelled benthic gastropods and bivalves. After Valentine (1966).

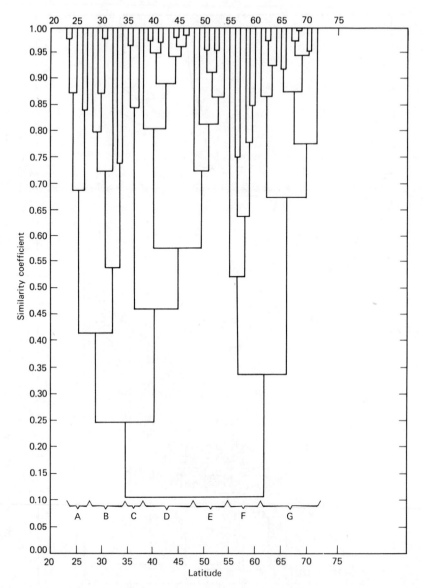

Figure 10.16 Dendrogram showing clustering of molluscs of each degree of latitude with those of other degrees of latitude with similarity determined by the Jaccard coefficient. The weighted pair-group method of clustering was used. A—Surian; B—Californian; C—Montereyan; D—Mendocinian; E—Columbian; F—Aleutian; G—Bering. After Valentine (1966).

areas. Johnson (1971) used this technique to compare brachiopod faunas. He did this by determining a provinciality index (PI) by the formula

$$PI = \frac{C}{2E_1}$$

where C is the number of taxa in common between two areas and E_1 is the number of taxa restricted to the area with fewer taxa. This method is especially well adapted to studying provinciality in fossils where the areas of occurrence are limited so that relatively few indices can be calculated. Johnson used this method to compare the Devonian brachiopod faunas of the Appalachian and Great Basin regions of the United States with the Western and Arctic Canada region. His analysis showed variation with time in the faunal similarity in both of these comparisons, suggesting that barriers existed part of the time but not all of the time between the regions. Simpson's method is actually quite similar to the Valentine method. He uses a different similarity coefficient but one quite similar to the Jaccard coefficient. He makes no attempt to recognize provinces and provincial boundaries but simply uses a quantitative method of comparing the degree of similarity between areas.

None of these methods attempts to directly define provincial boundaries on the basis of the definition of biotic provinces as suggested by Valentine, that is, a region in which communities maintain characteristic taxonomic compositions. Ideally, all communities should be studied and variations noted. Areas where many communities changed in taxonomic composition would be provincial boundaries. The question still remains as to how much change should be considered significant and how many communities have to show the change. Because provinces are by nature interpretive constructs and not really natural entities, the placing of boundaries will remain subjective.

Defining provincial boundaries on the basis of gross biotic changes without any particular reference to communities may often give satisfactory results but it can lead to erroneous results. For example, the biota may change from one area to another simply because of a change in habitat that eliminates a certain community or group of communities. The elimination of the community or communities would certainly result in a biotic change but it would not be a provincial change in which the same community types would be present but have different taxonomic compositions. For example, one segment of shallow shelf might have an abundance of rocky coasts and hard substrate with characteristic communities occupying these areas. An adjoining segment may have little or no hard substrate so that an entire group of communities associated with hard substrate is absent. In our example we will say that the soft substrate communities are taxonomically identical. There would undoubtedly be a large biotic difference between these areas, but they would not represent different provinces in the sense of the Valentine definition. The area without hard substrate would be a barrier to species requiring hard substrate. Perhaps further along the coast where hard substrate reappears the communities will have a different composition because some of the species were not able to penetrate the barrier. Thus the two isolated areas may represent two different hard substrate provinces although communities occupying the soft substrate may be the same. Should these areas be called separate provinces? This is another question with no definite answer. The biogeographer must make a subjective decision on that point.

The paleoecologist making a paleobiogeographic study first looks for significant faunal differences between areas. He or she then determines if the change is simply due to habitat difference. If it is, the differences may not represent provincial differences. If the same habitats are represented but the biotas are different, the areas may be recognized as separate provinces. How much difference is significant enough to justify recognizing provincial boundaries? This decision is based on comparison with the level of difference present in previously recognized provinces. Rather than naming provinces, quantitative measures of provinciality such as the similarity index of Simpson, the clustering technique of Valentine or simple percentage differences in the faunas of different areas might substitute for the formal defining of provinces.

MODERN PROVINCES

Wallace (1876) recognized five terrestrial zoogeographic regions (Fig. 10.17) and modern zoogeographers still use his basic classification. The northern regions, the Palearctic and the Nearctic, are sometimes combined as the Holarctic region because they are less distinctive than the other regions. The Neotropical and the Ethiopian regions are peninsulas extending into the southern hemisphere and are semiisolated by oceans and climates. The Oriental region is isolated by oceans on the south and mountains on the north. The Australian region is most isolated, being completely separated from other regions by water.

Floral regions which are essentially identical to the zoogeographic regions have also been recognized (Fig. 10.17; Good, 1974). Although different names are used for the floristic regions, most boundaries are nearly the same as those for the zoogeographic regions. A separate small region is recognized in South Africa. An Antarctic region is also recognized and the Indo-Malaysian region is subdivided into three major subregions.

The geographic distribution of the terrestrial biota, especially the plants, can be shown as biomes (Fig. 10.5). These are not provinces in the usual sense but complexes of related biotas. Nevertheless, they are useful in studying the geographic distribution of terrestrial life. In fact, they are perhaps more useful in paleoecologic studies than are the biotic regions because their distribution is strongly controlled by environmental parameters, especially temperature and precipitation. Comparison between modern and ancient biomes has especially been used in palynological studies of Pleistocene sediments (e.g., Davis, 1969).

Modern marine shelf provinces (Fig. 10.18) clearly show the effect of the depth-elevation and temperature factors. Boundaries between provinces are between north-south coastal segments or are continental masses or deep oceans. Valentine (1973b) has pointed out that equatorial provinces are broad because the same temperature gradient is repeated north and south of the equator. The arctic and Antarctic provinces are relatively broad because water temperatures in these regions are uniformly near or at freezing. North-south coastlines in the temperate regions have many, small provinces because of the steep temperature gradients (and concentration of biogeographers?). East-West coast lines have large provinces because they parallel isotherms (and thus have uniform temperatures).

Five shelf provinces have been recognized on the east coast of North America (Caribbean, Gulf, Carolinian, Virginian, and Nova Scotian). Five provinces have also been named on the west coast (Panamic, Surian, Californian, Oregonian, and Aleutian). The Oregonian province has been further subdivided into three subprovinces (the Columbian, Mendocinan, and Montereyan) by Valentine (1966). Six provinces are recognized in Europe and Northern Africa (Mauritanian, Mediterranean, Lusitanian, Celtic, Caledonian, and Norwegian). All of the North American and European provinces are clearly determined by temperature gradients. The depth-elevation factor accounts for the occurrence of three groups of provinces across the same temperature gradient.

The west coast North American provinces illustrate the factors involved in forming shallow marine provinces (Fig. 10.15). They occur on a north-south trending coast and so the boundaries between provinces are basically temperature dependent. The west boundary is formed by the deep water of the Pacific and the east boundary by the North American continent. The north and south boundaries occur at locations where there are marked temperature gradients between areas of more uniform temperature. The boundary between the Oregonian and Californian provinces (Point Conception) discussed earlier in this chapter is a classic example of a provincial boundary in an area of rapid temperature change due to the current pattern. The Panamic province is tropical, extending along the mainland Mexican coast on into South America. The northern part of this province has been separated as the Mexican province by some workers (Briggs, 1974). The fauna of the northern part of the Gulf of California reflects the tectonic and depth-elevation effects and perhaps also the geologic history effect. The Gulf of California has been interpreted as being formed by crustal rifting. This has produced a narrow, deep embayment with effective barriers on all sides except the narrow southern entrance. This narrow entrance has apparently been in the tropics since the origin of the Gulf. Although this area is in the tropical Panamic province, the temperature in the upper Gulf is not truly tropical. However, there is no available route for more temperate immigrants into the area; hence the biota consists mainly of those members of the tropical biota that are able to withstand occasionally cooler than tropical temperatures. Because of the rather high proportion of endemic species in the Gulf of California, Briggs (1974) recognized this area as a separate province, the Cortez province.

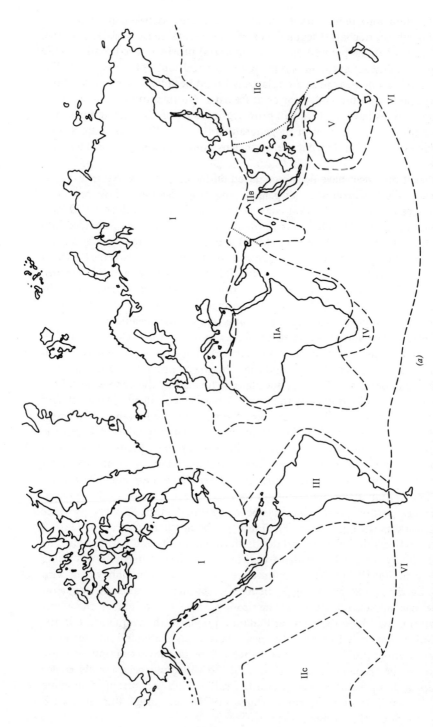

Figure 10.17 Terrestrial biogeographic regions of the world based on plants (*a*) and animals (*b*). The floristic regions are I—Boreal; IIA—Paleotropical (African); IIB—Paleotropical (Indo-Malaysian); IIC—Paleotropical (Polynesian); III—Neotropical; IV—South African; V—Australian; VI—Antarctic. The zoogeographic regions are I—Palearctic; II—Ethiopian; III—Oriental; IV—Australian; V—Nearctic; VI—Neotropical. After Good (1974) and Odum (1971).

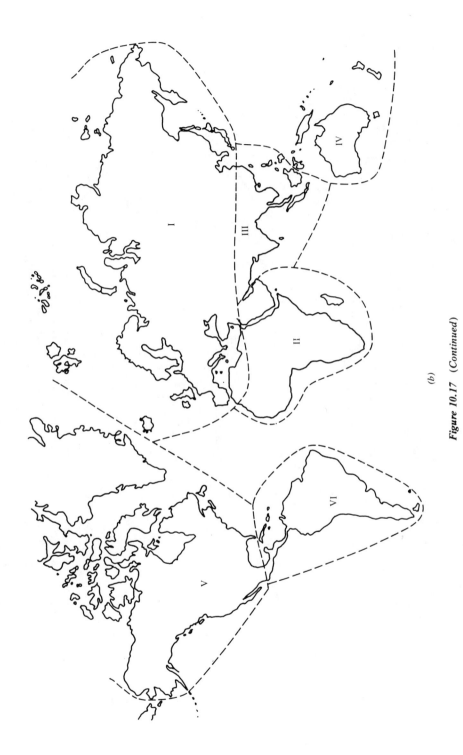

Figure 10.17 (*Continued*)

(*b*)

471

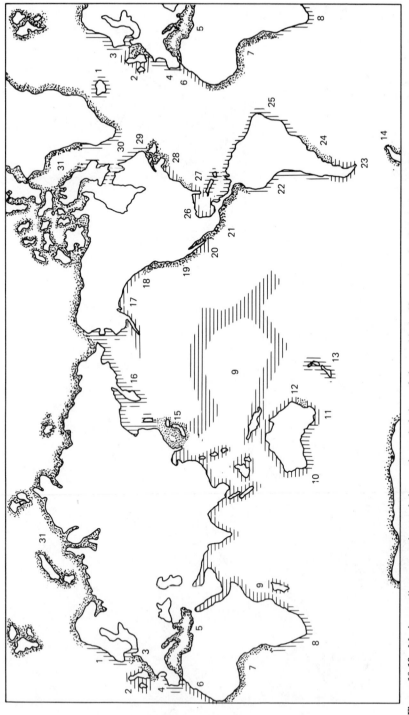

Figure 10.18 Marine molluscan provinces of the continental shelves of the world. 1—Norwegian; 2—Caledonian; 3—Celtic; 4—Lusitanian; 5—Mediterranean; 6—Mauritanian; 7—Guinean; 8—South African; 9—Indo-Pacific; 10—South Australian; 11—Maugean; 12—Peronian; 13—Zeolandian; 14—Antarctic; 15—Japonic; 16—Bering; 17—Aleutian; 18—Oregonian; 19—Californian; 20—Surian; 21—Panamic; 22—Peruvian; 23—Magellanic; 24—Patagonian; 25—Caribbean; 26—Gulf; 27—Carolinian; 28—Virginian; 29—Nova Scotian; 30—Labradorian; 31—Arctic. After James W. Valentine, *Evolutionary Paleoecology of the Marine Biosphere*, copyright 1973, p. 356. Reprinted by permission of Prentice-Hall, Inc., Englewood Cliffs, N.J.

The Surian province is transitional between the tropical Panamic and the warm temperate Californian provinces. The Surian province is complex in having a combination of protected tropical bays and much cooler outer coastal areas where upwelling of deeper, cold water occurs. It is almost like an intermingling of two separate provinces. The northern boundary of the province, at Punta Eugenia, reflects a temperature change at a geographic discontinuity on the coastline.

The Californian province contains a mixture of northern and southern faunal elements with a fair proportion of endemic species. It also contains patches of colder water, with a high proportion of Oregonian species, in areas of upwelling. Some researchers combine the Californian and Surian provinces (Briggs, 1974).

The broad Oregonian province has a rather uniform cold temperate climate (Fig. 10.4). Minor temperature breaks at Monterey Bay and at Cape Flattery, Oregon have been used to subdivide the Oregonian province into the Montereyan, Mendocinan, and Columbian.

Provinciality can be recognized in pelagic biotas of the open oceans but formal provincial names have not been generally used (Funnell, 1971). As indicated above, the pelagic biota is relatively uniform within a given water mass so that the biotic regions can simply be named for the water mass with which it is associated. Many groups (e.g., the geologically important planktonic foraminifera, diatoms, coccoliths) are much more diverse in the warmer water masses with only a few cosmopolitan genera and species at high latitudes. Many workers who have studied these groups discuss high-latitude, low-latitude, and transitional biotas. Biogeographic studies of the pelagic biota of Mesozoic and Cenozoic sediments have been especially common since the advent of the Deep Sea Drilling Project has made available numerous deep sea cores. These studies have especially emphasized the use of paleobiogeography (1) to study past climates, as in the CLIMAP program, which studied Pleistocene climates (Fig. 10.19; Cline and Hayes, 1976); (2) to study temperature fluctuations (e.g., Haq et al., 1977); and (3) to study ancient circulation patterns. Based largely on the paleobiogeography of fossil plankton, Berggren and Hollister (1977) traced the development of oceanic circulation from a relatively tranquil state in the Mesozoic to the more vigorous system of today.

Knowledge of the modern benthonic biota of the deep sea is still incomplete. A distinctive pattern of provinciality occurs below about 4000 m depth (Menzies et al., 1973). In shallower water the biota is very uniform with many species occurring in all oceans. Below about 4000 m, however, the ocean floor is subdivided by oceanic rises and ridges that act as barriers to migration; temperature differences between different bottom-water masses may also help to establish provincial patterns. Bottom water which sinks from the surface in the Antarctic area extends northward into the Atlantic, Pacific, and Indian basins. These tongues of Antarctic water support a characteristic biota that can be traced far north of the Antarctic (Fig. 10.20). Menzies et al. (1973) have recognized five deep water regions corresponding to the five major oceans (Fig.

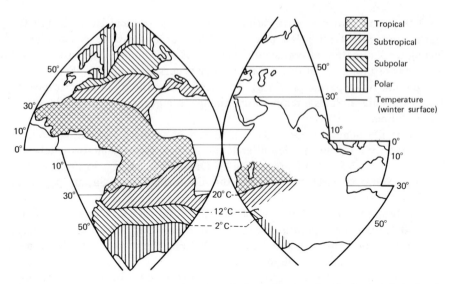

Figure 10.19 Biogeographic regions of the Atlantic Ocean based on the distribution of planktonic foraminifera on the sediment surface. After Imbrie and Kipp (1971), copyright 1971 by Yale University.

10.20). These have been further subdivided into provinces and some of the provinces have been subdivided into areas, and in some cases subareas.

ANCIENT BIOTIC PROVINCES

The paleontologic literature contains many examples of studies of probable provincialism among fossil biotas. A particularly informative example is the apparent provincialism in the Cambrian trilobite faunas of Europe and North America (Ross, 1975). Distinct differences between the trilobites of these two regions have long been recognized and attributed to provincial differences resulting from a barrier to migration between the two areas. Some anomalous localities are difficult to explain, however. Typical American genera occur in northern Scotland and European genera occur in southern Newfoundland, Nova Scotia, and eastern New England. What kind of barrier to migration could have resulted in such an uneven distribution? No convincing answers to that question were evident. One solution not based on provinciality was to recognize in North America a cratonic, shallow water realm, an intermediate depth, miogeosynclinal realm, and a deeper water extracratonic-euxinic biofacies realm (Lochman-Balk and Wilson, 1958). Each of these realms contained different trilobite genera, those of the euxinic realm being particularly distinctive. Lochman-Balk and Wilson suggested that the euxinic realm corresponded to the European province and that the presence of European genera in North America resulted from the environmental pattern and not provinciality. As

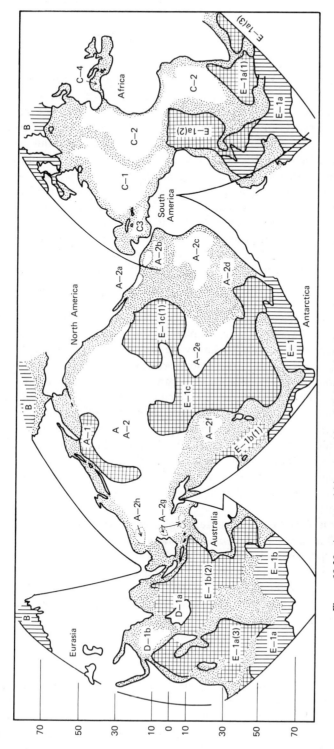

Figure 10.20 A proposed biogeographic subdivision of the lower abyssal regions of the oceans. Stippled areas are less than 4000 m deep and are thus above the lower abyssal depth. The letters and numbers refer to the regions and subdivisions as recognized by Menzies et al. (1973). After Menzies et al. (1973).

475

Valentine (1973a) points out, they explained the faunal differences in community rather than provincial terms.

Later work by Palmer (1969 and 1973) and Ross (1975) reevaluated the work of Lochman-Balk and Wilson. Palmer made an even more detailed biofacies study and recognized trilobite faunas associated with three concentric belts of sediment in North America: an interior belt of shallow water, relatively coarse, terrigenous sediments; an intermediate belt of clean, shallow water carbonates; and an outer belt of deep water, fine-grained terrigenous and carbonate sediments. Each belt contains a characteristic fauna, the inner belt having more endemic forms than the deeper water, outer belt. The shallower water environments are also found in Europe where they contain distinctive faunas. Palmer's work thus suggests that there were separate European and North American provinces, the deeper water communities being less distinctive than the shallow. The more uniform conditions of the deeper water would have allowed more easy migration in the Cambrian seas just as today. Based largely on the data of Palmer and Ross, Ziegler et al. (1977) recognizes four Upper Cambrian trilobite provinces: North American, European, Siberian, and Chinese-Australian (Fig. 10.8).

The question of the unusual distribution of the faunas between the continents is perhaps best answered in terms of plate tectonics. Southern Newfoundland, Nova Scotia, and eastern New England were probably part of the European continental mass in Cambrian times and Northern Scotland was part of North America. Later in the Paleozoic the ocean basin separating the continents closed, welding the continents together. The continents separated again during the opening of the Atlantic in Mesozoic time, but not along the old boundary, so that parts of the old European continent stayed with North America and vice versa. Tectonic and structural evidence support this interpretation (e.g., Wilson, 1966, and Dewey, 1969). The example of the Cambrian trilobites points out the necessity of including a community analysis with any study of provinciality. It also shows how studies of provinciality can be useful in interpreting geologic history.

Perhaps the best known of all examples of provinciality of fossils is the *Glossopteris* flora which lived on the Gondwanaland continent during Late Paleozoic time (Chaloner and Meyer, 1973). *Glossopteris,* a cycad, has given its name to the distinctive, rather low-diversity fossil flora found in Carboniferous and Permian rocks in South America, Africa, Australia, India, and Antarctica. Partly because of the similar flora in these now widely separated areas as well as other geologic evidence, these areas have been described as part of the supercontinent of Gondwanaland. Gondwanaland later split into smaller continents which have separated on moving crustal plates. The frequent association of the *Glossopteris* flora with glacially formed tillites, and determination of ancient magnetic pole positions, suggest that this was a high-latitude flora growing in a temperate to cold climate. Three other floral provinces have been recognized in the northern hemisphere in Late Paleozoic rocks (Fig. 10.9): the Euramerican, Angaran, and Cathaysian floras. Similar provinciality has also been noted on the basis of invertebrates (Waterhouse and Bonham-Carter, 1975) and vertebrates (Colbert, 1973). The nature of the barriers giving rise to these provinces is not clear, but almost certainly temperature as well as the depth-elevation factor are involved.

A third well-known example of provinciality among fossils is in the molluscan faunas of Jurassic age. Arkell (1956), referring extensively to earlier work, recognized three major provinces (or faunal realms) for the Jurassic: Tethyan, Boreal, and Pacific. Much work has been done in more recent years, particularly by Hallam (1969 and 1977) in delineating and explaining the Tethyan-Boreal boundary in Europe and Asia (Fig. 10.21). Hallam suggests that the boundary may not represent a true provincial boundary but a widespread facies change. That is, a different group of communities was occupying the Boreal region than occupied the Tethyan region. Hallam recognizes three lithofacies associations: (1) a terrigenous clastic facies with a relatively low diversity fauna, (2) an intermediate facies of calcareous sediments and shales, and (3) a pure carbonate facies, often with reefs or banks with a diverse fauna. The first lithofacies association is restricted to the Boreal province, the third is found only in the Tethys, and the second occurs in both. Hallam speculates that the Boreal biota occupied an area of environmental stress, such as reduced and/or variable temperature and salinity, and that only the low-diversity communities adapted for high-stress conditions could survive there. The more traditional view has been that the Boreal region was an area of lower temperature and the

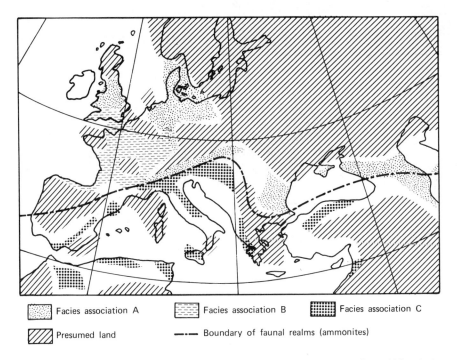

Facies association A   Facies association B   Facies association C

Presumed land   —·— Boundary of faunal realms (ammonites)

Figure 10.21 Faunal realms and lithofacies associations in Europe and northern Africa during Jurassic (Pliensbachian) time. After Hallam (1969).

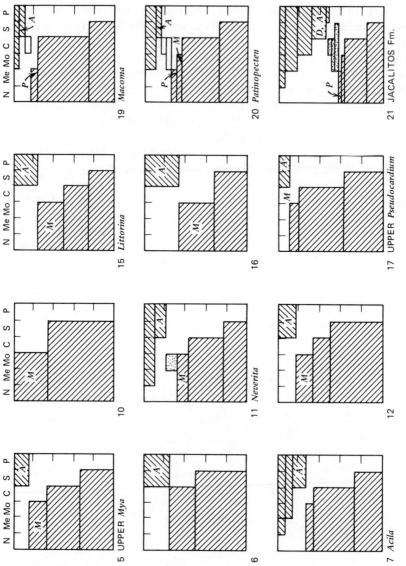

Figure 10.22 Modern provincial distribution of genera present in the Neogene strata of the Kettleman Hills. The percentage values for a stratigraphic unit are for those genera for which latitudinal distribution data are available and for which the range is less than N-P, that is, not cosmopolitan throughout the region of analysis. Stratigraphic units numbered as in Figure 1.9. After Stanton and Dodd (1970), Journal of Paleontology, Society of Economic Paleontologists and Mineralogists.

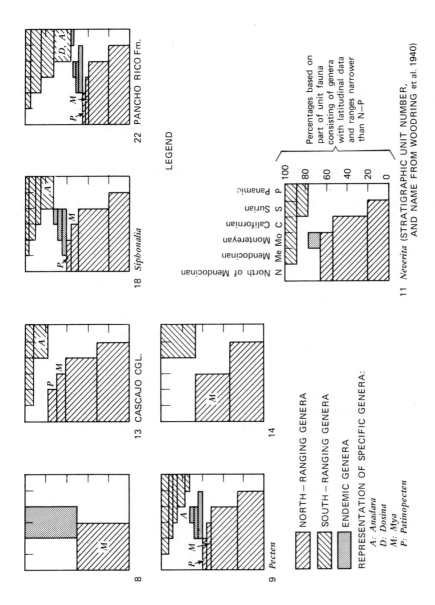

Figure 10.22 (*Continued*)

479

Boreal-Tethys boundary represented the tropical temperate boundary. The abundance of carbonate and reefs in the Tethys could be a reflection of temperature. Shallow water carbonate sediments are not common in the temperate zone today probably largely because of the temperature effect on carbonate solubility. An area of high-stress environment as large as the Boreal province also seems unlikely. Further work on comparison of the composition of similar communities from the two regions should help to settle the question.

PROVINCIAL APPROACH IN THE KETTLEMAN HILLS

The composition of the Pliocene biota of the Kettleman Hills is very similar to that of the modern biota along the Pacific Coast of North America. This allows a comparison of the fossil biota with those of the various modern provinces. The composition of the provinces has undoubtedly changed some since Pliocene time. In fact, Valentine (1961) used different terms for Pleistocene provinces that in composition are similar to but in location slightly different from the modern ones. Temperature is the main factor forming the barriers that define provincial boundaries along the Pacific Coast. Thus if the fossil biotas can be assigned to the modern province on the basis of the greatest similarity, an estimate can be made of the temperature conditions existing during the life of the fossils.

In order to determine which modern province most closely resembles that in which the fossil biota lived, we made graphs showing the modern provincial ranges of all the fossil genera for which data were available (Stanton and Dodd, 1970; Fig. 10.22). Very wide ranging genera, those that live over the entire range from the Panamic to north of the Mendocinan province, were excluded from the analysis. The width of each bar in the graphs is a measure of the proportion of the fauna in that unit which has the indicated provincial range. The province where the maximum overlap of ranges occurs should correspond to the most similar modern province. A perplexing problem of this approach is that some of the ranges of the fossils that occur together do not presently overlap. That is, some genera that lived in the same province in Pliocene time no longer do so. This must indicate that some genera have evolved in terms of their temperature requirements (see Chapter 2). Perhaps in some cases they have become extinct in a given province because of other factors. Several examples of such changing environmental requirements have been identified.

We used this technique on each of the zones recognized by Woodring et al. (1940) in the Kettleman Hills. The faunas of most of the units shown in Fig. 10.22 consist mainly of temperate species that extend southward. Warm water genera ranging northward from the Panamic province are uncommon in most units except 9, 18, 21, and 22. These units all appear to have faunas indicating depositional climates that were somewhat warmer than the rest. A relatively small proportion of the faunas consists of genera that are endemic to the Montereyan subprovince or the Californian province. Most of the zones contain faunas that are most similar to those of the modern Montereyan subprovince of the

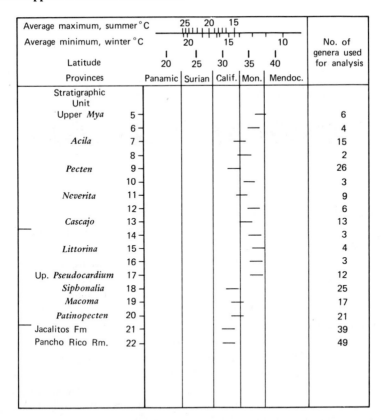

Average maximum, summer °C			25 20 15				No. of
Average minimum, winter °C			20	15		10	genera used
Latitude			20	25	30	35	40	for analysis
Provinces			Panamic	Surian	Calif.	Mon.	Mendoc.	
Stratigraphic Unit								
Upper *Mya*	5							6
	6							4
Acila	7							15
	8							2
Pecten	9							26
	10							3
Neverita	11							9
	12							6
Cascajo	13							13
	14							3
Littorina	15							4
	16							3
Up. *Pseudocardium*	17							12
Siphonalia	18							25
Macoma	19							17
Patinopecten	20							21
Jacalitos Fm	21							39
Pancho Rico Rm.	22							49

Figure 10.23 Modern provinces having fauna most similar to those of various units within the Neogene strata of the Kettleman Hills. The interpretations are based on the data summarized in Figure 10.22. Data for Jacalitos Formation derived from Nomland (1916) and Adegoke (1969). Data for the Pancho Rico Formation derived from Durham and Addicott (1965). After Stanton and Dodd (1970), Journal of Paleontology, Society of Economic Paleontologists and Mineralogists.

Oregonian province. Several of the zones have faunas most similar to those of the Californian province and some are intermediate. The paleotemperatures of individual stratigraphic units or zones were then determined by comparison with the temperature conditions found in the modern provinces (Fig. 10.23). This analysis indicates that the lower part of the stratigraphic section was deposited under slightly warmer conditions than most of the upper part of the section. A return to slightly warmer conditions occurred during *Pecten* zone time followed by a cooling trend toward the end of deposition of the section.

These results are in general agreement with those obtained from trace chemical analysis and from the taxonomic uniformitarian approach. The agreement between all of these approaches gives added confidence in all of the methods. The methods are complimentary, each giving some information not provided by the others. Our results with the Kettleman Hills biota suggest the use of multiple techniques whenever possible.

11

Temporal Patterns

The objective of this chapter is to examine paleoecologic procedures as they are determined by stratigraphy. We focus particularly on the recognition and interpretation of the temporal patterns in paleontology.

Paleoenvironmental interpretations are most commonly qualitative and relative because the absolute value of an environmental parameter in the geologic record can seldom be determined precisely. Instead, biotic gradients within the stratigraphic framework of space and time are what are usually sought and interpreted. As a result, the value of an environmental parameter at one locality is described in relative terms as being more or less than at an adjacent locality. The reconstruction of the local environment on the basis of a single fossil assemblage is difficult if the assemblage is isolated by limitations of sampling or exposure from other samples. The paleoecologic interpretation is made possible, however, by working within the stratigraphic framework, which provides a basis and constraint for the interpretation of individual assemblages and facies.

Walther's law, which states that the vertical sequence of the facies at a single locality was formed by the horizontal shift through time of a laterally arranged pattern of depositional environments, is an expression of the dominant control stratigraphy imposes on paleoenvironmental reconstruction. The most evident expression of Walther's law is provided by the lateral and vertical facies distributions during a transgressive-regressive cycle (Fig. 11.1). The potential interpretation of any specific locality in the stratigraphic succession, as shown in the idealized example of Fig. 11.1, is determined by its stratigraphic position relative to other localities. Walther's law, as it expresses this relationship, establishes a basic starting point in paleoecologic interpretation.

A classic example of the way in which this stratigraphic concept guides environmental interpretation is provided by the interpretation of deep water sediments in the years before the process of sedimentation by turbidity currents was well understood and widely accepted. Fossils in basinal mudstones had been recognized as having lived in deep water, whereas those in the sandstones interbedded with the mudstones had apparently lived in shallow water. The discordance in the environmental interpretations was recognized, but a feasible transport mechanism to move the sands and shallow water

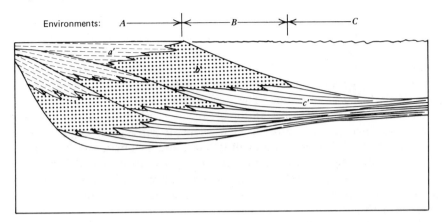

Environments: A ──────⟩|⟨────── B ──────⟩|⟨────── C

Figure 11.1 Cross-section through deltaic sediments consisting of three asymmetric cycles of rapid transgression with little deposition and prolonged regression with major deposition to form a single, primarily regressive, genetic unit. *a'*, *b'*, and *c'* are facies deposited in environments *A*, *B*, and *C*. After Selley (1976), with permission from An Introduction to Sedimentology, Fig. 128, copyright by Academic Press, Inc. (London) Ltd.

fossils into the deep basin was lacking. The standard explanation for the interbedded deep water and shallow water sediments was to invoke large-scale tectonically produced fluctuations in sea level. This was clearly unsatisfactory, and thus the discordant stratigraphic and paleoenvironmental relationships spurred the acceptance of turbidite deposition (Natland and Kuenen, 1951).

Lateral and vertical gradients may exist for any paleontologic parameter, as they may for modern biotic characteristics. Commonly observed ones, for example, are in faunal composition, in diversity, and in morphologic characteristics. The chart relating plant morphology with climatic parameters in Fig. 1.3 represents this last category. Lateral paleontologic gradients must reflect differences in the original environment and/or in the taphonomic conditions. If the taphonomic processes that modify the original community of living organisms can be identified and analyzed, in the manner suggested by Fagerstrom (1964), Johnson (1960), and Lawrence (1968), the remaining or residual paleontologic differences between samples must represent contemporaneous differences in the original environment of deposition. Vertical paleontologic differences within the stratigraphic framework are not so easily explained, however. They may represent changes in the depositional and/or taphonomic environment, as in the case of lateral patterns. In addition, however, they may be the result of successional change attributed to biological factors, or they may be the result of evolutionary causes more or less independent of the external environment in that area. To interpret the fossil record correctly these possibilities must be distinguished.

In analyzing a vertical sequence of fossils, the rate of sedimentation is an important consideration. It is important, however, not in the usual geologic sense of centimeters per million years, but in the sense of whether it is great enough to overwhelm, kill, bury, and preserve the individual organism. If sedimentation were rapid, the fossil assemblage may be essentially a census of the living community (see Chapter 8). In the ideal case of mass mortality and isolation by rapid burial, the fossiliferous horizon would contain only the individuals living at one time. The two important attributes of this census population are that it includes all the organisms, with few escaping, and that it does not include admixtures from previous or subsequent populations. A graphic example is the preservation of Pompeii and its inhabitants in the midst of life. Alternatively, if sedimentation were slow, an individual fossil assemblage may have gradually accumulated over a long period of time and consist of an incomplete and probably biased sampling of a sequence of many populations (Fürsich, 1978).

Sedimentation on the average is slow compared to the life span of most organisms that are likely to be preserved as fossils (Olsen, 1978; Glass and Rosen, 1978). Typical rates of deposition on the sea floor distant from land and a source of terrigenous sediment are on the order of fractions of a mm per year. Deposition on the continental shelf since the Pleistocene has been very slow away from shore, and in many localities has been essentially nonexistent. Even in shallow marine and terrestrial settings the rate of deposition is apparently low if sediment thickness is divided by the time interval represented by the sediments. From either this calculation or from modern rates of deposition, it would be logical to conclude that the fossil record formed by slow gradual accumulation and that the total environmental history of a stratigraphic section may be preserved without time gaps in the fossil record. If so, however, the representation of successive communities would have to be exceedingly incomplete, with only a few individuals preserved in an assemblage from each generation. If this were not so, given that rates of deposition are slow compared to rates of skeletal production, the present-day sea floor should be paved with shells, and fossils should be the dominant component of sedimentary rocks.

A more reasonable model of the formation of the fossil record is based on the fact, as Ager (1973) and others have convincingly pointed out, that apparent depositional rates are net and average rates; sedimentation actually is in pulses of relatively rapid deposition and longer intervals of nondeposition or even erosion. Consequently, the fossil record in general more nearly consists of a sequence of census assemblages than of time-averaged assemblages which accumulated gradually. Thus, in a vertical sequence, it consists of discrete, disjunct bits of the record of life at that locality. If the environment were changing through time, the time gaps in the fossil record must be short relative to the rate of environmental change in order for the change to be evident. If the environmental changes were relatively rapid and the gaps in the fossil record were relatively long, the sequence of events would be like a jerky motion picture, with large gaps in the action.

PATTERNS

Vertical paleontologic changes within a stratigraphic section are caused primarily by (1) succession, (2) environmental change, or (3) evolution of individual taxa or community structure. Paleontologic examples of these categories are listed in Table 11.1. They are only a small sampling of studies in the current literature. In many examples, the cause of change may be difficult to recognize. A principal reason is that the categories are not distinct and isolated but may cooccur and interact.

SUCCESSION

Succession has been applied most commonly and effectively to temporal changes in terrestrial plant assemblages. As indicated in Table 11.1, two types of succession have been recognized. The distinction between the two is based largely on the extent to which the plants themselves permanently modify their own physical environment.

Primary succession refers to the changes in the biota that occur as the underlying substrate develops from the original rock to a mature soil through plant growth and animal activity. The time span for this process is much longer than the life span of individual plants. Concurrently, this biotically induced change in the physical environment leads to the systematic successional changes in the biota itself. In the analysis of primary succession, changes in the local physical environment are significant in causing the biotic changes, and the overall external aspects of the environment, such as climate, are generally held constant.

Secondary succession refers to biotic changes that occur independent of, or in the absence of, any permanent modifications of the environment. During an ideal and complete secondary succession, the initial abiotic setting is colonized by a pioneer community. Because of the rigorous initial conditions, the pioneer community is composed of species that mature and reproduce rapidly, have a great many offspring, and have relatively short life spans. That is, they are *opportunistic,* and have an *r-selected* life strategy. The diversity and equitability of the pioneer community are low because few individuals are adapted to the initial conditions, but those that are present may be abundant. The pioneer community alters its local or microenvironment by damping fluctuations in the microclimate, by changing the composition and physical properties of the substrate, and by creating new niches for other organisms as it provides nutrients and new physical and biological dimensions to the environment. As these modifications continue during succession, the pioneer community and each following one is replaced by a community of new species adapted to new, biologically determined conditions. Diversity and equitability increase, the proportion of opportunistic (*r*-selected) species decreases, and the proportion of

Table 11.1 Examples of temporal paleontologic change

I. Secondary succession

Benthic marine invertebrates on soft substrate; pioneer community provides firm substrate for succeeding community
Walker and Parker (1976), Johnson (1977)

II. Primary succession

   A. Growth of Silurian reefs from deep quiet water into shallow rough water, and corresponding changes in community of reef builders and reef dwellers
Lowenstam (1957), Nicol (1962), Shaver (1974).

   B. Growth of Cretaceous rudist banks and associated changes in community
Kauffman and Sohl (1974).

III. Environmental change

   A. Substrate becomes lithified to hardground; burrowing, boring, and epifaunal benthic marine invertebrates
Bromley (1975), Goldring and Kazmierczak (1974)

   B. Water chemistry changes as Great Salt Lake fluctuates from full to dry during Quaternary climatic cycles; ostracods
Lister (1976)

   C. Environmental changes during transgressive-regressive marine cycles
Pennsylvanian-Shaak (1975), Rollins, Carothers, and Donahue (1979); Pliocene-Stanton and Dodd (1970, 1972)

   D. Terrestrial climate and habitat: Permian fish, amphibians, and reptiles
Olson (1952); Neogene mammals-Shotwell (1964);

   E. Cenozoic floral composition and plant morphology
Wolfe (1978)

   F. Long-term subtle fluctuations in the marine environment; Ordovician benthic invertebrates
Watkins and Boucot (1975), Bretsky and Bretsky (1975)

   G. Paleogeographic changes modify water roughness; echinoid morphology
Stanton, Dodd, and Alexander (1979)

IV. Evolutionary changes

   A. Ordovician brachiopod communities
Watkins and Boucot (1975)

   B. Pliocene echinoids
Stanton, Dodd, and Alexander (1979)

   C. Permian vertebrate communities
Olson (1952)

equilibrium or *K-selected* species increases. (In contrast to *r*-selected species, *K*-selected species are more specialized or have narrower niches, reproduce at lower rates, and have longer life spans.) The final or climax stage of succession is stable and in theory will continue until disturbed.

When the secondary succession is disturbed, it is commonly by a short-term, catastrophic event such as fire, storm, or epidemic. Because in theory the physical environment has not been permanently changed by the biota, as during a primary succession, the community that reestablishes itself after the disturbance will be one of those characteristic of some earlier stage of the succession. If the disturbance has been drastic, as after a severe fire, the new community might be the pioneer one. If the disturbance is mild, the existing community might be replaced by one only a stage or two earlier in the succession. If the environment is disturbed frequently relative to the time span required for the complete succession to develop, the communities characteristic of the later stages of the succession may never or only rarely occur.

The time span during which succession takes place is determined by the frequency of disturbances that terminate the succession and return it to an earlier stage. In seasonal environments, secondary succession may be of annual duration, beginning in the spring, running through the summer and fall, and terminating with the cold weather of winter. Examples of settings in which annual succession occurs would be high-latitude arctic tundra and strongly seasonal temperate bays, lakes, and terrestrial habitats with communities not dominated by perennial plants. Succession may be an even shorter-term phenomenon, as in ephemeral ponds. In all these cases, definition of a climax community, to which the succession might develop if there were more time, is not possible. The usually attained last community is as close as the succession comes to a climax condition. Alternatively, if disturbance is infrequent, secondary succession may continue to the climax stage, which will remain as the equilibrium condition for an indefinite period.

The succession leading to the climax stage should not be longer than several times the life span of the dominant species in the climax community. This is because the succession consists of several stages, and because the life spans of early-stage species tend to be shorter than those of later-stage species. Thus the duration of most secondary successions will be measured in years or tens of years. In an extreme case, such as a *Sequoia* forest, it will be measured in thousands of years.

Geologic examples of secondary succession are not likely because the duration of a secondary succession is short relative to geologic time and to the processes of sedimentation by which it would be preserved. Secondary succession in plant communities is not likely to be preserved by plant macrofossils because of their limited preservability and because of the high rate of sedimentation necessary to bury and preserve the successive communities. It is more likely to be preserved in the pollen record if sedimentation is rapid enough so that sequential events are not combined in a condensed horizon, and if bioturbation is absent so that they are not intermixed. Fürsich (1978) has described in detail the conditions of high sedimentation rate and low bioturbation rate that are necessary to ensure that temporally distinct communities will be vertically separated in the sedimentary rock. The scarcity of unequivocal examples of secondary succession in the fossil record indicates how uncommon these conditions are.

Walker and Parker (1976) and Johnson (1977) describe secondary succession in Paleozoic benthic invertebrate communities. In the Silurian Hopkinton Dolomite of Iowa (Johnson, 1977), the pioneer community was dominated by orthotetacean brachiopods that were able to populate the soft lime mud and criniodal sand substrate. The orthotetacean shells formed a suitable substrate, then, for the subsequent community dominated by pentameran brachiopods. Once the apparently necessary firm shelly substrate had been established by the orthotetacean brachiopods, the pentameran community would persist as the climax stage, and would directly be reestablished if the succession had been interrupted (Fig. 11.2). Thus this succession of brachiopods is a simple sequence of modifications determined by substrate suitability. The role of substrate in controlling succession is also well exemplified by the organisms commonly found encrusting shells on the beaches of the Texas Gulf Coast (Fig. 11.3). The encrusting bryozoan uses the empty pectinid valve; the scleractinian coral is apparently unable to attach itself directly to the bivalve but does very well using the bryozoan as a substrate.

The succession of benthic communities in the Ordovician Chickamauga Group of Tennessee was apparently more complex (Walker and Parker, 1976). It began with strophomenid brachiopods as the pioneer population on a soft muddy substrate. Subsequent stages in the succession were characterized by encrusting bryozoans, ramose bryozoans, and a diverse late stage, or perhaps climax, community dominated by a rhynchonellid brachiopod and including gastropods, bivalves, bryozoans, and pelmatozoan echinoderms. The climax community was apparently killed by an influx of argillaceous sediment. The succession was repeated many times, forming in each instance a limestone bed 2 to 6 cm thick interbedded with the mudstone. In both the Ordovician and Silurian examples, only the very local microenvironment was modified by the successive community. After each disturbance the environment returned to its presuccession condition. In that each of these successions was controlled by a permanent modification of the substrate, they could well be considered examples of primary rather than secondary succession.

Examples of primary succession in the fossil record that are most evident are associated with reef growth, in which both the substrate and local topography are permanently modified. As the reef-building organisms construct, within an otherwise uniform physical environment, a stratigraphic unit from the sea floor into shallow water they modify the local environment so that different communities of reef-building and reef-inhabiting organisms form a succession. They also may modify the adjacent environment for a considerable distance by altering oceanic circulation and patterns of sedimentation.

Reef growth as an example of primary succession has been proposed by several workers for the Silurian reefs of Illinois and Indiana. The funda-mental study was by Lowenstam (1950, 1957). Nicol (1962) phrased Lowen-stam's description of the growth of the Silurian reefs in explicitly successional terminology. Textoris and Carozzi (1964) and Shaver (1974) have made significant additions to our understanding of the reefs.

In the initial growth stage, the Silurian reefs were mounds of lime mud containing a low-diversity assemblage of benthic invertebrates. Through continuing *in situ* accumulation of biogenic carbonate sediment, the reefs grew upward from deeper, quiet water conditions into shallower and rougher water within an external environment that otherwise was apparently not changing. Because of the local environmental change to shallower and rougher water, however, a correlative sequence of distinctive changes in the fauna also took place (Fig. 11.4). Interpretation of this as a primary succession is rea-sonable because temporal biotic changes were dependent on the biologic activity of the community at each stage of the succession as it created a suitable environment for the following community, rather than on external environmental changes. The succession is clearly not a secondary succession because its duration is orders of magnitude longer than the life spans of any of the species and because the environment was being continuously and permanently modified by the organisms. Consequently, if the environment had been disturbed at some stage of reef growth the reef surface would not have been recolonized by the original, pioneer, deeper water community, but by some pioneer community of the secondary succession for that stage of the reef growth. Thus as the reef grew there always existed a climax or steady-state community and a short-term, secondary succession leading up to it after disturbances. The primary succession consists of the sequence of climax communities as the reef environment changed. The communities of the sec-ondary successions have not been described and probably would be difficult to recognize because of the slow rate of reef growth relative to the duration of the potential secondary successions at any stage in the primary succession.

The Silurian reefs formed large structures up to hundreds of meters high. A comparable pattern of succession is evident but at a very small scale in the Permian mounds described by Toomey and Cys (1979) from the Labor-cita Formation in south-central New Mexico (Fig. 11.5). Although these only grew to a height of 1 m, the succession is evident. Their small size, however, brings out very clearly the problem in describing succession in the fossil record. These mounds might be an example of secondary succession resulting from substrate modification; however, they could be an example of primary succession resulting not only from substrate modification, but also from more significant and longer-lasting environmental changes such as elevation off the sea floor or perhaps growth up into what was already very shallow water. Finally, the vertical biotic sequence may not be an example of either primary or secondary succession, but was caused by external environmental changes (Toomey and Cys suggest deepening water) and thus was not self-induced and controlled by the biota.

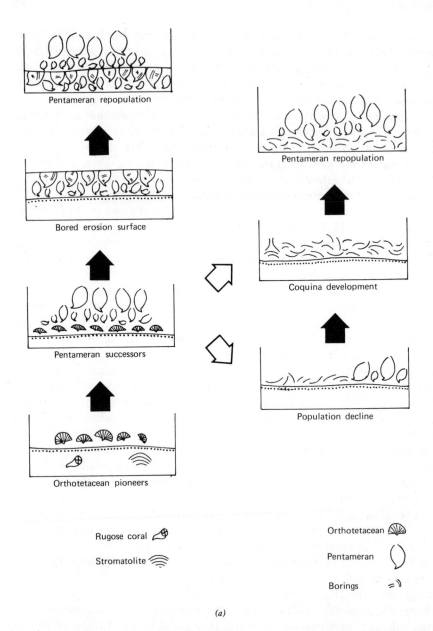

Pentameran repopulation

Bored erosion surface

Pentameran successors

Orthotetacean pioneers

Pentameran repopulation

Coquina development

Population decline

Rugose coral

Stromatolite

Orthotetacean

Pentameran

Borings

(a)

Figure 11.2 Succession of brachiopod communities in Silurian strata, Iowa. (a)—Orthotetacean pioneers form hard shelly substrate for pentameran successors, which remain dominant even after interruptions as long as substrate is suitable. (b)—Variations in successor communities during a long period of time as a result presumably of environmental differences. After Johnson (1977), Lethaia 10.

Pentameran successors

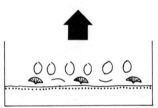

Orthotetacean repopulation

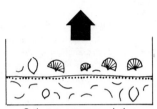

Stricklandian successors

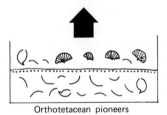

Orthotetacean pioneers

Orthotetacean

Stricklandian

Pentameran

(b)

Figure 11.2 *(Continued)*

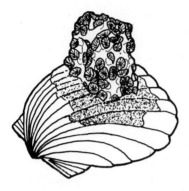

Figure 11.3 Substrate-determined succession. The bryozoan encrusts the *Pecten*, the coral then encrusts the
bryozoan. Corals do not attach to *Pecten* shells directly,
but only where bryozoan substrate is present. Length of
shell: 3.8 cm.

ENVIRONMENTAL CHANGE

A temporal change in the biota or community is commonly a response to
change in the external environment. The response may take any form but, as
indicated by the examples in Table 11.1, it has generally been described as
a change in the morphology of an individual taxon or as a change in taxonomic
composition or in diversity.

The distinction between the two environmental causes of vertical biotic
change—change in the environment caused by the organisms (primary succession) and change in the environment caused by extrinsic factors—is generally
not easily made. In both cases, the environment changes, but primary succession is caused by the organisms themselves rather than being external and
independent of them. Although difficult, the distinction between these two
causes must be made because only by recognizing and taking into account
primary succession can the external environmental changes be recognized and
interpreted from the fossil record. In general, environmental interpretations of
vertical paleontologic changes have not taken succession into account. The
correlation of environmental change and paleontologic change has been
direct. We do not believe that precision of terminology or thought is well
served by using the term succession for biotic changes caused by externally
determined changes in the environment. Nevertheless, some recent publications, such as Walker and Alberstadt (1975), have used succession in this
broad sense. As the examples in Table 11.1 clearly show, when, as is common,
succession in our restricted sense is not an apparent factor, the paleontologic
data are analyzed and interpreted by the methods and criteria outlined in the
other chapters and result in explanations involving a changing environment.
The important aspect of these interpretations is that the environmental changes
follow a limited number of temporal sequences, and the distinctive time-
stratigraphic framework of each of these sequences provides an aid to the
interpretation.

As an example, the time framework of interpretation is clearly important

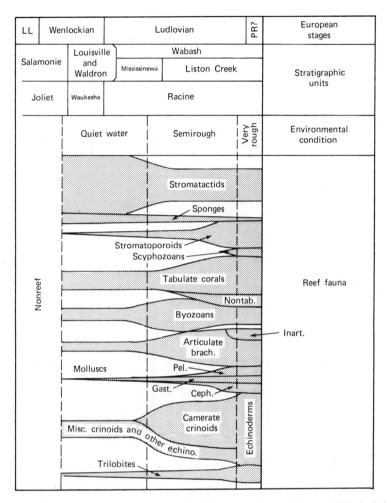

Figure 11.4 Successional changes in community of Silurian reefs of Indiana and Illinois. Inferred cause of change was growth of the reefs from quiet, deeper water into shallow, rough water. Relative abundances are based on volumetric contributions to the reef sediment. Stromatactids are enigmatic structures that are largely probably inorganic. Abbreviations: PR—Pridolian; LL—Llandoverian. After Shaver (1974), published with permission of The American Association of Petroleum Geologists.

in the analysis of organisms associated with surfaces inferred to be hardgrounds. The *sequence* of (1) infaunal organisms and burrows in the underlying stratum, (2) lithification and erosion, (3) epifaunal organisms on the upper bed surface, and (4) subsequent borings may be very complex and difficult to unravel (Fig. 11.6; Bromley, 1975). In this example it can be shown by detailed analysis that the sequence of events consists of the following. (1) Deposition and lithification of the Carboniferous sediments and subsequent planation to form hardground I. (2) Boring into the hardground

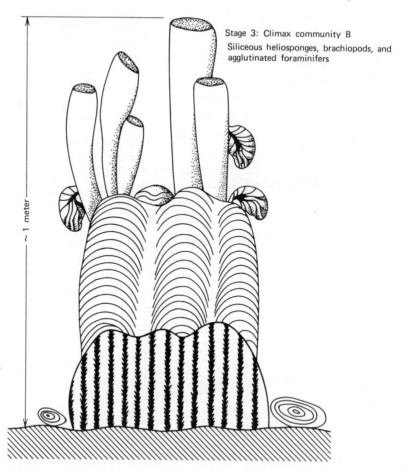

~ 1 meter

Stage 3: Climax community B
Siliceous heliosponges, brachiopods, and
agglutinated foraminifers

Figure 11.5 Biotic changes in Lower Permian sponge-algal mound in New Mexico. After Toomey and Cys (1979), Lethaia *12*.

by three kinds of organisms (A, B, C) and encrustation by oysters (D). From cross-cutting relationships, the boring and encrusting occurred concurrently over some period of time. (3) Deposition of Jurassic sediments, covering hardground I and filling the borings into it. On the basis of the Jurassic sediment within the borings, it is probable that the planation and borings described above occurred during the Jurassic. The age of the oysters of course would pin down the age of the borings. (4) Lithification of the Jurasic sediments and their planation to form hardground II. (5) Encrustation of epifaunal organisms on hardground II and borings into the Jurassic sediments and through them into the underlying Carboniferous sediments. Truncation of the bivalve borings at the upper hardground (as at the left side of B) indicates that planation of this surface may have taken place at several times while it formed the sea floor.

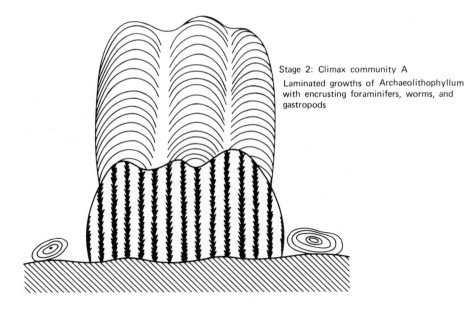

Stage 2: Climax community A

Laminated growths of Archaeolithophyllum with encrusting foraminifers, worms, and gastropods

Stage 1: Pioneer community

Digitate—shaped colonies of plumose blue—green algae, encrusting foraminifers, crypostome bryozoans with sedentary polychaete worms, and pedunculate brachiopods

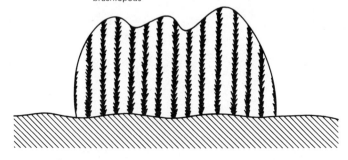

Figure 11.5 . (*Continued*)

The recognition and timing of events in this sequence are largely determined by cross-cutting and superpositional relationships. The sequence, more than the individual features, is essential for the significant, interpretive consequence of the analysis, which is the recognition of the sedimentologic events such as variations in depositional rate, including nondeposition, subaerial exposure and erosion or submarine corrosion, and subaerial or submarine lithification. Characteristics of hardgrounds are in large part determined by the associated trace fossils, and so are also discussed in Chapter 6. For example, recognition of the surface as having been lithified or as having been a

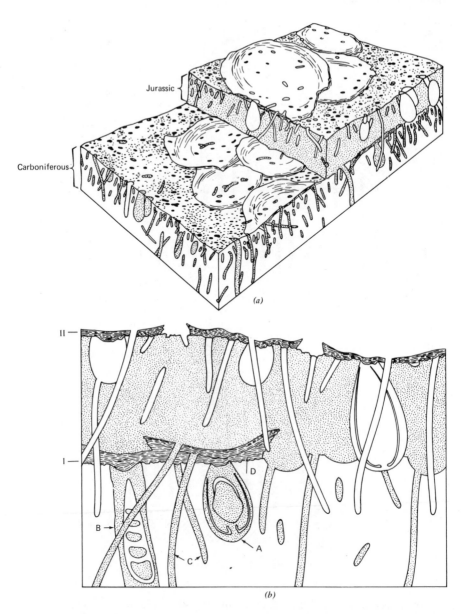

Figure 11.6 Block diagram (*a*) and cross section (*b*) through Carboniferous and Jurassic hard-grounds, England. After Bromley (1975).

soft-sediment surface of slow or nondeposition is based on whether the traces are burrows or borings. Geochemical diagenetic characteristics are also important in the recognition and interpretation of hardgrounds, as illustrated by the example from the Triassic Regensteinzone of Germany (Scherer, 1975), but have not been thoroughly studied.

Vertical paleontologic sequences are particularly useful if they are cyclic, analogous to the vertical lithologic sequences that are distinctive for specific depositional environments and so have proved to be essential in the interpretation of the physical record of sedimentation. The oscillation chart of Israelsky (1949) was a pioneer use of cyclic paleontologic distribution patterns. Correlation in thick basinal Tertiary mudstones of the Gulf Coast had proved difficult because of the lithologic uniformity and the small amount of vertical change in the foraminiferal fauna. In addition, in the cyclic transgressive-regressive deposits, first or last appearances of species were not useful for correlation because they were too strongly influenced by the depositional environment. By determining, for a series of wells along a dip section, characteristics such as relative abundances of depth-diagnostic assemblages and ratios of pelagic to benthonic, and arenaceous to calcareous foraminifers, Israelsky developed what he termed the oscillation chart, which showed from the faunal data the cyclic changes in environment and bathymetry. To construct an oscillation chart, the relative abundances of environmentally significant foraminiferal assemblages are determined for each section or well (Fig. 11.7). By arranging the wells in a dip section, as determined by the regional geology, and tying together environmental intervals from well to well, the cyclic pattern of transgression and regression becomes evident (Fig. 11.8). The oscillation chart provides a paleoenvironmental framework that can be used for further analysis of the sections. The stratigraphic significance of this procedure is that the point of maximum water depth in each well provides a reliable horizon for correlation because the point of maximum water depth is the time of reversal from transgression to regression in the cycle and as such is correlative between wells even though the water depth at that time at each site was likely to have been different. The paleoenvironmental significance in the context of this chapter is that the analysis of the faunal sequence within the stratigraphic framework reveals a cyclic pattern, and this pattern provides the means both for paleoenvironmental reconstruction and for correlation.

Another example of the value of the vertical sequence in paleoecology is provided by the Cretaceous sedimentary rocks and fossils of the western interior of the United States (Kauffman, 1977, and earlier papers listed therein). The key step in reconstructing the Cretaceous paleoenvironments has been to group both sediments and fossils into cycles like those portrayed by Israelsky's oscillation charts (Fig. 11.9). The interpretation of the cyclic pattern of rock types and fossils is in terms of a transgressive-regressive model that serves as an hypothesis to be tested and as a framework for more detailed analysis of the Cretaceous strata and fossils of the region. For example, in

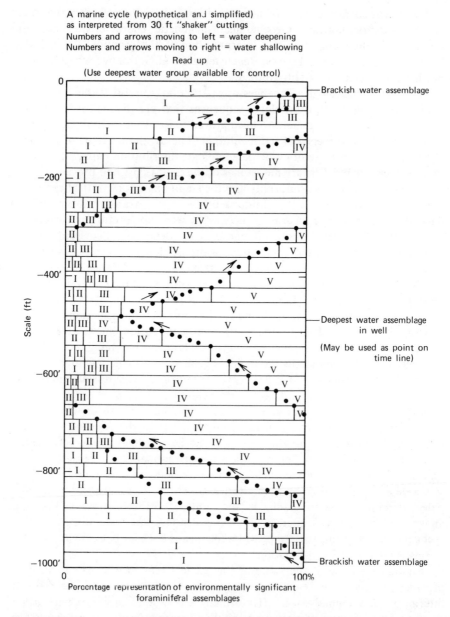

A marine cycle (hypothetical and simplified)
as interpreted from 30 ft "shaker" cuttings
Numbers and arrows moving to left = water deepening
Numbers and arrows moving to right = water shallowing

Read up
(Use deepest water group available for control)

Figure 11.7 Plot for a single section or well of the relative proportions of environmentally significant foraminiferal assemblages. I—brackish-water assemblage; II—beach assemblage; III—seaweed-zone assemblage; IV and V—successively deeper marine assemblages. After Israelsky (1949), published with permission of The American Association of Petroleum Geologists.

Columns (well or surface)

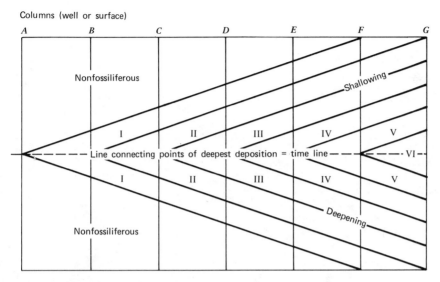

Figure 11.8 Depositional dip section of a marine cycle. Each column is simplified from the data as in Figure 11.7 by indicating at any elevation only the dominant assemblage. After Israelsky (1949), published with permission of The American Association of Petroleum Geologists.

general, cycles within a basin should be correlative as in the oscillation chart, unless the geography and bathymetry is being modified by tectonism as well as transgressive-regressive sea level changes. Thus the cyclic model puts constraints upon the interpretation of correlative strata and fossils in New Mexico or Texas, more seaward in the Cretaceous from the western interior region (Fig. 11.10).

The model presented by Fig. 11.9 must be tested, as well as used. The symmetry of the cyclic model as portrayed in Figs. 11.8 and 11.9 may be misleading, for commonly deposition during transgression may be much less than during regression, or even absent. Asymmetric cycles are the general rule in carbonate strata (J. L. Wilson, 1975), and are characteristic also of clastic deposits in tectonic areas, in basins that attain their maximum depth rapidly but then fill gradually during the regressive phase. This pattern was described in Chapter 1 in general terms for Neogene strata in the Coast Ranges and San Joaquin Valley of Central California (Fig. 1.6). A similar pattern of basin development and filling is displayed in the Humboldt Basin of northwest California (Fig. 11.11). This small basin formed in the Late Miocene and reached its maximum water depth (about 2000 m) during the Late Miocene or Early Pliocene. Sediments deposited during the period of basin formation and deepening water cannot be clearly distinguished from one another. Deposition at the time of maximum transgression was largely of diatomaceous mudstones. During the regressive phase, the vertical sequence of basin floor, slope, shelf, and finally terrestrial environments is recorded in the microfauna

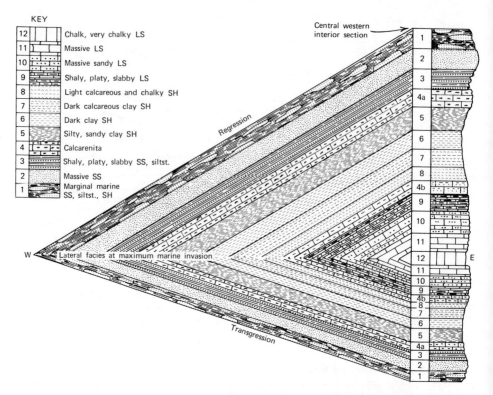

Figure 11.9 Model of sedimentation patterns within a Cretaceous marine cycle, Western Interior United States, showing: distribution and repetition of 12 major lithologic types; complete lithologic development of cycles in the central part of the basin and thinning toward strand (actual sections thicken from basin center to strand of maximum transgression); thinner transgressive than regressive sequences of rocks. After Kauffman (1969b), reprinted with the approval of the author and Allen Press, Inc., copyright 1970.

and macrofauna (Ingle, 1976; Piper, Normark, and Ingle, 1976; and Burckle, Dodd, and Stanton, 1980). The largely regressive sequence of shallowing bathymetry as recorded by microfossils in the section at Centerville Beach is illustrated in Fig. 11.12. Thus the strata that have filled the basin consist of probably not more than a few tens of meters of transgressive sediment laid down as the basin was forming, followed by more than 2500 m of strata laid down as the basin filled. This sedimentation model forms the basis for an integrated understanding of the basin history as additional sections and outcrops are incorporated into the study.

The pollen record in lake or bog sediments may reflect two aspects of the environment. The first of these is the local change as the initial lake filled with sediment to become a bog, and eventually a meadow. The transition from lake to dry ground is accompanied by changes in sediment texture,

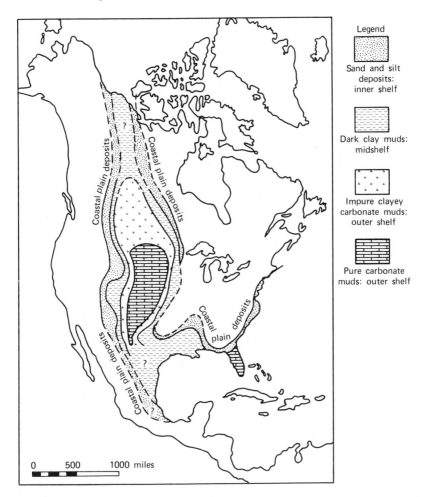

Figure 11.10 Early Late Cretaceous paleogeography of North America. After Kauffman (1969b), reprinted with the approval of the author and Allen Press, Inc., copyright 1970.

oxygenation, and fauna as well as soil moisture. Thus the presence of specific plants and the composition of the whole plant community will undergo drastic changes. Although this phenomenon has been described in the literature as an example of primary succession, we would view it instead as a simple record of environmental change. Our reason for this is that the modification of the plant community is largely externally induced rather than the result of the active growth of the plants themselves.

The second environmental aspect that may be reflected in the pollen record is the climate. This has been particularly well described for the postglacial record of the northeast United States. The Pleistocene ice sheet that

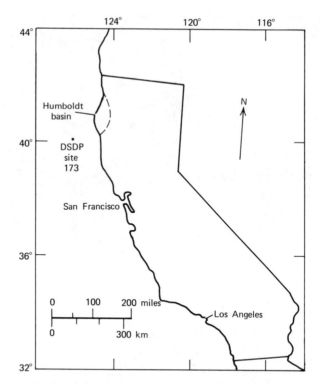

Figure 11.11   Location of the Humboldt Basin, California.

was centered over eastern Canada was bounded on the south by climatically determined plant associations arranged in more or less concentric bands, with the most arctic flora fringing the ice sheet and more temperate floras further to the south. As the climate moderated following the last glacial advance, the edge of the ice sheet moved progressively northward, and was followed in turn by the bordering floras. Consequently, the pollen record in the sediment of a lake originally at the edge of the ice sheet would reflect the northward migration of climatically determined floras and thus the climatic change itself.

   Temporal change in biota is also well preserved in a thick and well-exposed section of Upper Ordovician marine strata in the St. Lawrence Lowlands of Quebec (Bretsky and Bretsky, 1975). The pattern of changes in the marine invertebrate fauna is most evident in the lower 1500 ft. of the section. Three asymmetric cycles are evident from the sequential distribution of six assemblages within the strata (Fig. 11.13). The cyclic faunal pattern is believed to have been caused by cyclic changes in the physical environment. The parameters that are most evident as having been important are substrate texture, water depth, and the associated environmental variability. Succession in the sense we have defined it is not a factor in this example.

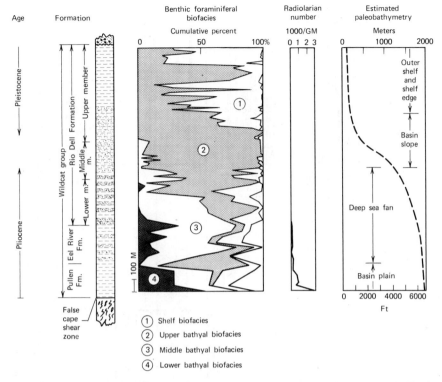

Figure 11.12 Stratigraphy of Wildcat Group at Centerville Beach, Humboldt Basin, California, and bathymetry as inferred from micropaleontologic data. After Piper et al. (1976), Blackwell Scientific Publications Limited.

EVOLUTION

Temporal changes in the fossil record may be caused by evolution as well as by environmental change and succession. The role of evolution, however, is generally difficult to recognize and to evaluate for several reasons. The first reason is that a well-developed stratigraphic framework in which biotic changes can be described is necessary but seldom available or adequate. The second reason is that the result of the evolutionary process is difficult to distinguish from that of environmental change. Evolutionary, ecophenotypic, and biogeographic phenomena of migrations due to changing environment may not be separable.

These difficulties are well illustrated by the early Tertiary Paleocene and Eocene mammals of the Big Horn Basin, Wyoming (Gingerich, 1976). As illustrated by the condylarth *Hyopsodus,* a considerable morphologic range is present among the Early Eocene specimens within a section spanning 2 or 3

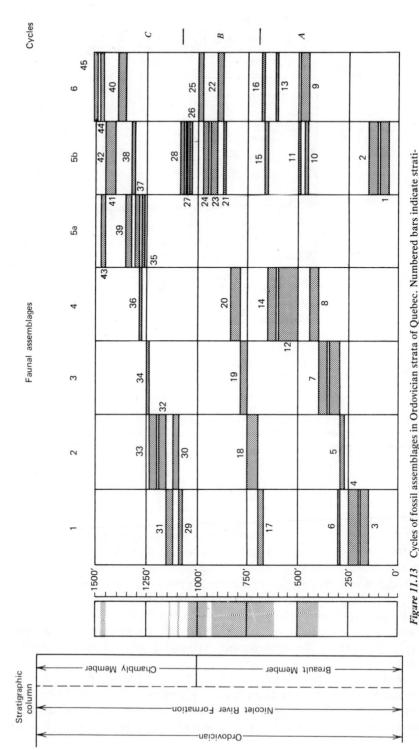

Figure 11.13 Cycles of fossil assemblages in Ordovician strata of Quebec. Numbered bars indicate stratigraphic ranges of assemblages. Stippled pattern on lithologic column represents siltstone and fine sandstone in shales and silty shale. After Bretsky and Bretsky (1975).

million years (Fig. 11.14). Species that have been described are well distributed through the section and within the range of morphology as reflected by tooth characteristics. The dashed lines of Gingerich subdivide the objective data into postulated pathways of evolution that establish a phylogeny for *Hyopsodus*. The major interpretation of these data was in terms of evolutionary mechanism. The significant point here is that some independent additional data are also necessary to be sure that these changes are evolutionary. Within a changing environment (in some aspect of climate, or resulting vegetation, e.g.) a corresponding immigration of forms with greater tolerance for the new conditions could equally well have yielded the pattern of apparent gradual change between *H. latidens* and *H. minor*. The apparent sudden appearance of *H. miticulus* could be the result of a barrier to immigration for *H. miticulus* breaking down, or it could be the result of a new subhabitat appearing as the environment changed, and being occupied by the immigration of *H. miticulus*.

The problem posed in distinguishing evolutionary change from environmentally induced change in the case of a single taxon such as *Hyopsodus* is also present in analyzing community evolution. The temporal dimension in the fossil record is an important aid in paleoecologic interpretation. It is accurate, however, only to the extent that the evolutionary, environmental, and successional components can be evaluated.

TEMPORAL PALEONTOLOGIC PATTERNS IN THE PLIOCENE OF THE KETTLEMAN HILLS

Three types of vertical paleontologic patterns are present in the Pliocene strata of the Kettleman Hills. These have been important in the paleoenvironmental interpretation and in establishing a stratigraphic framework for further study.

The broadest is an overall change from base to top in faunal composition and diversity. These changes are portrayed by the range chart of Fig. 11.15. The most evident trend is of decreasing diversity upward in the section. When the habitat preferences of the living representatives of the genera in the fauna are analyzed, it is evident that a number of taxa in the lower part of the section are essentially restricted to normal marine conditions. These include taxa such as the brachiopods *Discinisca* and *Terebratalia*, the corals *Coenocyathus*, *Astrangia*, and *Rhizopsammia*, and the bivalves *Chama*, *Patinopecten*, and *Lucinoma*. Upward, the most marine forms drop out; the taxa more tolerant of a wide range of environmental conditions, such as *Ostrea* and *Littorina*, persist. Finally, in the Tulare Formation at the top of the section, the fauna consists of the low-salinity and freshwater genera.

The interpretations of diversity as an indication of environmental stability and of the faunal composition by the methodology of taxonomic uniformitarianism have led to the general picture for the lower part of the section of an Early Pliocene sea open to the Pacific Ocean and consequently charac-

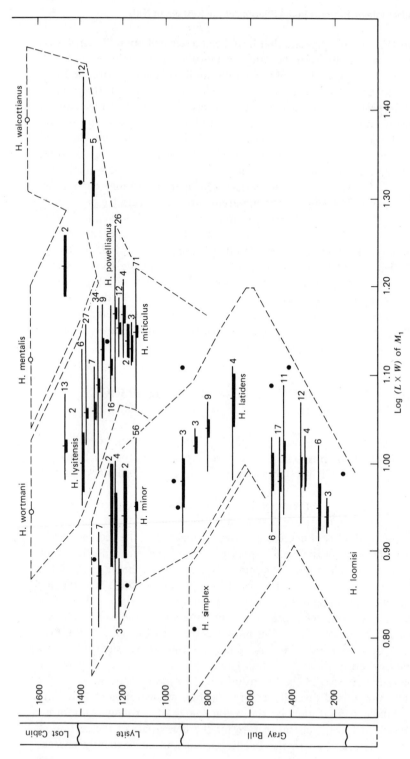

Figure 11.14 Morphology of *Hyopsodus* specimens in Lower Eocene strata of the **Big Horn Basin** Wyoming. The morphologic measure is the surface area of the first lower molar. Line length is the range of values for the number of specimens as indicated at each stratigraphic level. The central bar on each line is the standard error for that sample. Dashed lines indicate Gingerich's inferred phylogeny. Species names indicate morphologic and stratigraphic positions of established taxa. After Gingerich (1976), *American Journal of Science 276*.

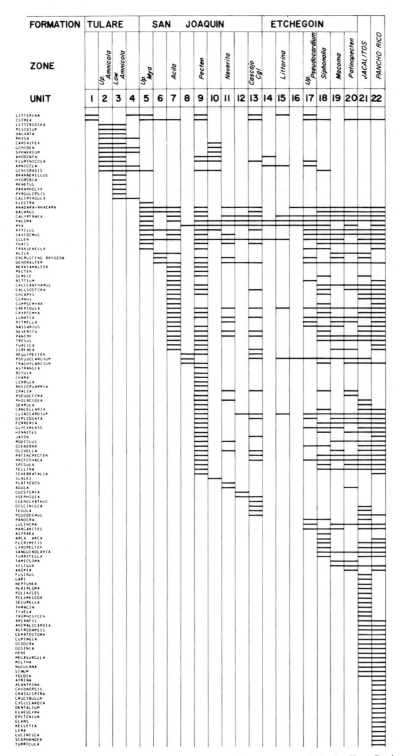

Figure 11.15 Stratigraphic distribution of genera in Pliocene strata of the Coalinga Region. After Stanton and Dodd (1970). Journal of Paleontology. Society of Economic Paleontologists and Mineralogists.

terized by stable conditions and normal salinity. Subsequently, as the communication with the open ocean became progressively reduced, the environment became less normal marine, and by the end of deposition of the San Joaquin Formation and during deposition of the Tular Formation it fluctuated between a brackish restricted embayment and a freshwater lake.

Superimposed on this overall gradient is a cyclic pattern expressed both in the sedimentary rocks and in the biota. In general, the named fossil zones (Fig. 1.9) represent the relatively marine transgressive and initial regressive parts of each cycle; the strata between zones, the less marine to nonmarine regressive parts of each cycle. Because the strata in the North Dome of the Kettleman Hills were deposited in an area of better communication with the open ocean than the strata in South Dome, the cycles there are generally more fossiliferous, represent more normal marine conditions, and grade southward into strata deposited in more nearly nonmarine conditions (Fig. 1.9).

The data on which these general interpretations are based are summarized in Fig. 11.16. Fossils are distributed vertically in a systematic and cyclical way. Diversity (number of genera) and abundance (number of specimens) of macrofossils are greatest at the base of each zone, decrease upward within the zone, and are at or near zero within the strata overlying the zone. Crab remains and fish bones, teeth, and dermal plates are most abundant in the

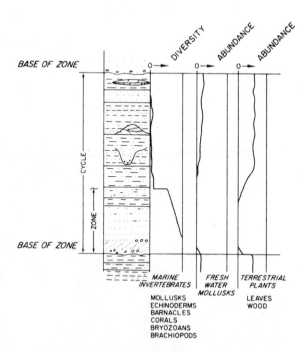

Figure 11.16 General lithologic and paleontologic characteristics of Pliocene cycles, Kettleman Hills, California. After Stanton and Dodd (1972).

upper part of the zone and in the overlying strata. Terrestrial plant remains (leaves and stem fragments) are dominant in the strata between zones, and freshwater mollusks are generally found only in the strata between zones. Fragmentary remains of terrestrial vertebrates are most common between zones or in the basal gravel of zones.

The vertical variations in lithology parallel those of the biota but are not as evident because the pattern is less uniform from the cycle to cycle. The base of each lithologic cycle coincides with the base of a zone and is generally marked by a bored and irregular erosion surface overlain by a thin basal pebble conglomerate or pebbly sandstone. The pebbles are mudstone clasts derived locally from erosion of the underlying strata, and igneous and metamorphic clasts derived from basement sources to the west. Dominant rock types within each zone consist of laminated or crossbedded sandstone low in the zone and bioturbated sandy mudstone and muddy sandstone.

Strata between zones and in the upper parts of some zones are predominantly dark green to gray claystone and thin-bedded to laminated tan siltstone with locally abundant plant remains; lenticular channel deposits of muddy, medium- to fine-grained sandstone with locally abundant plant remains are less common. The channel deposits are characterized by an erosional basal contact, by basal gravel containing clasts of the underlying and laterally adjacent rock types, and by a vertical sequence of (*a*) decreasing grain size, (*b*) increasing number of mudstone partings and thin beds, and (*c*) festoon crossbeds with clay drape replaced upward by rippled and cross-laminated beds. Most of the channels were probably fluvial but some contain shark teeth and marine fossils in the basal lag gravel and may have been tidal or estuarine. Probable aeolian dune deposits are found only in the intervals between zones and are flat-based lenses of unfossiliferous well-sorted sandstone with planar truncated crossbedding and uniform texture from base to top. Thin discontinuous beds of argillaceous dolomite also occur only in the upper parts of the cycles. The dolomite is massive to laminated; it contains floating sand grains and lacks fossils other than rare ostracods. However, adjacent mudstone may contain freshwater molluscs. Because mud cracks and stromatolitic structures suggestive of tidal flat deposition were not found in the dolomitic sediments, the dolomite probably formed in shallow lakes on, or isolated lagoons bordering, a deltaic plain.

The cyclic pattern as described above and illustrated in Fig. 11.16 is a composite derived from the study of outcrops throughout the exposed section in all three domes of the Kettleman Hills. A typical section is described here in order to add detail to the generalized description and to indicate the range of variability present in the cyclical pattern (Fig. 11.17). This stratigraphic column, which comprises the *Siphonalia* zone and small intervals of the adjacent cycles, was measured on the east side of North Dome.

Underlying the *Siphonalia* zone is siltstone and very fine sandstone of the *Macoma* zone with abundant plant fossils that probably were deposited in a very restricted marine to swamp environment. A lenticular sandstone at

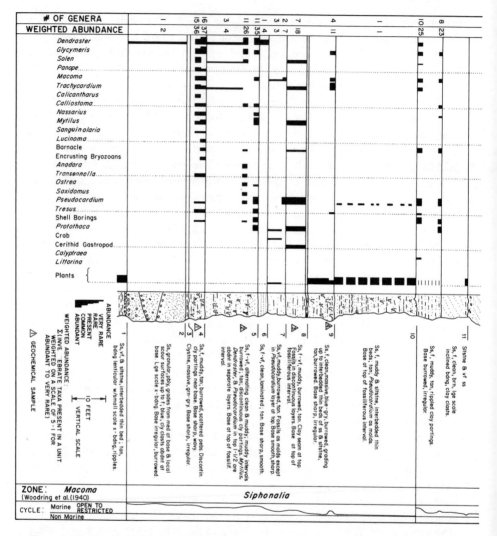

Figure 11.17 Detailed lithologic and paleontologic characteristics of the *Siphonalia* zone, a single cycle in the Pliocene Etchegoin Formation, Kettleman Hills, California. After Stanton and Dodd (1972).

approximately the same stratigraphic level 1½ miles to the northwest of the section plotted is probably an aeolian dune, in harmony with the restricted marine to nonmarine interpretation. In the *Siphonalia* zone, gradients upward include decreasing grain size, sorting, amount of inclined and crossed bedding, faunal diversity and abundance, and increasing abundance of plant fossils. Consequently, the sediments and fossils at the top of the cycle are very similar to those at the top of the underlying cycle. These overall gradients, along with trends and variations in the composition of the fauna, are interpreted as reflecting a fluctuating but general change in the environment from relatively open-marine or outer bay with normal salinity to highly restricted marine of probably reduced salinity.

The cycle overlying the *Siphonalia* zone consists of siltstone very much like that in the upper part of the *Siphonalia* zone plus lenticular bodies of blue sandstone of diverse genesis. It begins with an erosional bored basal surface of a blue sandstone, grades upward through crossbedded blue sandstone, muddier and finer brown sandstone, and into plant-bearing siltstone. Marine fossils are confined to the lower part of the cycle, associated with the relatively clean high-energy, crossbedded deposits.

The cause of the cyclic changes within the section is not known, but climatic fluctuations, tectonic movement controlling subsidence of the basin, or tectonic movement controlling the degree of communication with the open ocean would seem to be most probable. The climatic mechanisms that have been proposed are considered unlikely because they do not provide a comprehensive explanation of the sedimentologic as well as paleontologic patterns, and because paleoclimatic analysis based on both biogeochemical and faunal evidence has indicated that climatic fluctuations during deposition of the Etchegoin and San Joaquin Formations were minor.

The great number of widespread thin shallow marine to nonmarine cycles points to episodic basin subsidence under tectonic control as a prime cause of the cycles. Episodic lateral movement of the San Andreas Fault, located along the western margin of the San Joaquin Basin, may also have been important because the basin was largely land locked during the Middle and Late Pliocene with only a narrow connection across the San Andreas fault to the Pacific Ocean (Fig. 1.8). Thus changes in the environment would have resulted from variation in the width of the inlet as lateral movement occurred along the fault.

These changes would have been primarily in salinity and degree of circulation and not in water depth. If so, sediments in the upper part of some cycles would have been deposited in brackish to fresh water when connection with the open ocean was closed, and subsequent apparent transgression would have occurred when communication with the Pacific Ocean was increased or reestablished and more normal marine conditions were reintroduced to the basin.

The third type of temporal paleontologic trend in the Pliocene strata of the Kettleman Hills is in the morphology of a single taxon rather than in bulk faunal and lithologic characteristics as in the two trends described above.

The irregular echinoid *Dendraster* (Fig. 5.23) is widely distributed throughout the stratigraphic section in sandstones deposited in relatively open-marine environments. In the North Dome of the Kettleman Hills, the eccentricity of *Dendraster* (see Chapter 5 and Fig. 11.18) decreases from approximately 0.55 in the *Patinopecten* zone to approximately 0.85 in the *Acila* zone (note that the numerical value of eccentricity is 1.0 for no eccentricity and decreases as eccentricity increases). Species of *Dendraster* have been distinguished in large part on

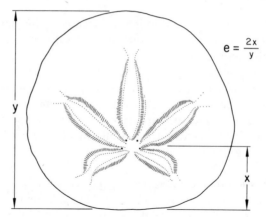

$$e = \frac{2x}{y}$$

Figure 11.18 Morphologic features on aboral surface of the irregular echinoid *Dendraster*; eccentricity = $2x/y$. After Stanton et al. (1979), Lethaia *12*.

their eccentricity, and in the Pliocene of the Kettleman Hills area the highly eccentric form in the lower part of the section has been named *D. gibbsii;* the less eccentric form in the upper part of the section, *D. coalingaensis.* Thus the temporal morphologic change may be interpreted as an instance of evolution of the phyletic gradualistic type.

On the other hand, eccentricity in the living species, *Dendraster excentricus,* on the north eastern Pacific Coast is correlated with water roughness and geography (Raup, 1956). Eccentricity there is greatest in individuals living along open exposed coasts and is less in individuals living along protected coasts and in bays. Thus the temporal change in eccentricity of the Pliocene sand dollars may have been determined by a progressive environmental change.

A detailed analysis by Stanton, Dodd, and Alexander (1979), described more fully in Chapter 5, has concluded, in fact, that the latter explanation is the correct one—that is, that the temporal morphologic change is essentially ecophenotypic and is controlled by the same overall environmental change from relatively open-marine to restricted and protected embayment that resulted in the first type of pattern described above.

Bibliography

Abbott, R. T., 1954, American seashells: Van Nostrand, Princeton, N.J., 539 p.

Adams, F. D., 1938, The birth and development of the geological sciences: Dover, New York, 506 p.

Adegoke, O. S., 1969, Stratigraphy and paleontology of the Neogene formations of the Coalinga region, California: Univ. Calif. Pub. Geol. Sci. *80*, 269 p.

Adey, W. H. and I. G. Macintyre, 1973, Crustose coralline algae: A re-evaluation in the geological sciences: Geol. Soc. Amer. Bull., *84:* 883-904.

Ager, D. V., 1963, Principles of paleoecology: McGraw-Hill, New York, 371 p.

Ager, D. V., 1965, The adaptation of Mesozoic brachiopods to different environments: Palaeogeogr., Palaeoclimat., Palaeoecol., *1:* 143-172.

Ager, D. V., 1967, Brachiopod paleoecology: Earth Sci. Rev., *3:* 157-179.

Ager, D. V., 1973, The nature of the stratigraphical record: Wiley, New York, 114 p.

Ahr, W. M., 1971, Paleoenvironment, algal structures, and fossil algae in the Upper Cambrian of central Texas: J. Sedim. Petrol., *41:* 205-216.

Ahr, W. M. and R. J. Stanton, Jr., 1973, The sedimentologic and paleoecologic significance of *Lithotrya,* a rock boring barnacle: J. Sedim. Petrol., *43:* 20-23.

Aigner, T., 1977, Schalenpflaster im Unteren Hauptmuschelkalk bei Crailsheim (Württ., Trias, mol)—Stratinomie, Ökologie, Sedimentologie: Neues Jahrb. Geol. Paläont., Abh., *153:* 193-217.

Alexander, R. R., 1974, Morphologic adaptations of the bivalve *Anadara* from the Pliocene of the Kettleman Hills, California: J. Paleont., *48:* 633-651.

Alexander, R. R., 1975, Phenotypic lability of the brachiopod *Rafinesquina alternata* (Ordovician) and its correlation with the sedimentologic regime: J. Paleont., *49:* 607-618.

Alexander, R. R., 1977, Growth, morphology, and ecology of Paleozoic and Mesozoic opportunistic species of brachiopods from Idaho-Utah: J. Paleont., *51:* 1133-1149.

Aller, R. C. and R. E. Dodge, 1974, Animal-sediment relations in a tropical lagoon Discovery Bay, Jamaica: J. Mar. Res., *32:* 209-232.

Arkell, W. J., 1956, Jurassic geology of the world: Oliver and Boyd, Edinburgh, 806 p.

Arrhenius, G., 1959, Sedimentation on the ocean floor: *In* P. H. Abelson (ed.), Researches in geochemistry: Wiley, New York, p. 1-24.

Ausich, W. I., 1980, A model for niche partitioning in Lower Mississippian crinoid communities: J. Paleont., *54:* 273-288.

Axelrod, D. I., 1938, A Pliocene flora from the Mount Eden beds, southern California: Carnegie Inst. Wash. Publ. No. 476: 127-183.

Ayala, F. J., D. Hedgcock, G. S. Zumwalt, and J. W. Valentine, 1973, Genetic variation in *Tridacna maxima,* an ecological analog of some unsuccessful evolutionary lineages: Evolution, *27:* 177-191.

513

Ayala, F. J., J. W. Valentine, T. E. Delaca, and G. S. Zumwalt, 1975, Genetic variability of the Antarctic brachiopod *Liothyrella notorcadensis* and its bearing on mass extinction hypothesis: J. Paleont., *49:* 1-9.

Bada, J. L., 1972, The dating of fossil bones using the racemization of isoleucine: Earth and Planet. Sci. Lett., *15:* 223-231.

Bada, J. L., B. P. Luyendyk, and J. B. Maynard, 1970, Marine sediments; dating by the racemization of amino acids: Science, *170:* 730-732.

Baden-Powell, D. F. W., 1955, The correlation of Pliocene and Pleistocene marine beds of Britain and the Mediterranean: Proc. Geologists' Assoc. Engl., *66:* 271-292.

Bader, R. G., 1954, The role of organic matter in determining the distribution of pelecypods in marine sediments: J. Mar. Res., *13:* 32-47.

Bailey, I. W. and E. W. Sinnott, 1915, A botanical index of Cretaceous and Tertiary climates: Science, *41:* 831-834.

Bakker, R. T., 1975, Dinosaur renaissance: Sci. Amer., *232:* 59-79.

Ball, S. M., 1971, The Westphalia Limestone of the northern midcontinent: a possible ancient storm deposit: J. Sedim. Petrol., *41:* 217-232.

Bambach, R. K., 1977, Species richness in marine benthic habitats through the Phanerozoic: Paleobiology, *3:* 152-167.

Bandy, O. L., 1960, The geologic significance of coiling ratios in the foraminifer *Globigerina pachyderma* (Ehrenberg): J. Paleont., *34:* 671-681.

Bandy, O. L., 1964, General correlation of foraminiferal structure with environment: *In* J. Imbrie and N. D. Newell (eds.), Approaches to paleoecology: Wiley, New York, p. 75-90.

Barker, R. M., 1964, Microtextural variation in pelecypod shells: Malacologia, *2:* 69-86.

Barnes, R. D., 1974, Invertebrate zoology (3rd ed.): Saunders, Philadelphia, 870 p.

Barron, J. A., 1973, Late Miocene-Early Pliocene paleotemperatures for California from marine diatom evidence: Palaeogeogr., Palaeoclimat., Palaeoecol., *14:* 277-291.

Basan, P. B. (ed.), 1978, Trace fossil concepts, SEPM short course No. 5: Soc. Econ. Paleontologists Mineralogists, Tulsa, Okla., 201 p.

Bathurst, R. G. C., 1964, The replacement of aragonite by calcite in the molluscan shell wall: *In* J. Imbrie and N. D. Newell (eds.), Approaches to paleoecology: Wiley, New York, p. 357-376.

Bathurst, R. G. C., 1971, Carbonate sediments and their diagenesis: Elsevier, New York, 620 p.

Bé, A. W. H., 1968, Shell porosity of Recent planktonic foraminifera as a climatic index: Science, *161:* 881-884.

Behrens, E. W. and R. L. Watson, 1969, Differential sorting of pelecypod valves in the swash zone: J. Sedim. Petrol., *39:* 159-165.

Behrensmeyer, A. K., 1975, The taphonomy and paleoecology of Plio-Pleistocene vertebrate assemblages east of Lake Rudolf, Kenya: Bull. Mus. Comp. Zool., Harvard Univ., *146:* 473-578.

Benson, R. H., 1961, Ecology of ostracode assemblages: *In* R. C. Moore (ed.), Treatise on invertebrate paleontology, Pt. Q. Arthropoda 3: Geol. Soc. Amer., Boulder, Colo., p. 56-63.

Benson, R. H., 1975, Morphologic stability in Ostracod: *In* F. M. Swain (ed.), Biology and paleobiology of Ostracoda: Bull. Amer. Paleont., *65:* 11-46.

Berger, W. H., 1969, Planktonic foraminifera: basic morphology and economic implications: J. Paleont., *43:* 1369-1383.

Berger, W. H., 1970, Planktonic foraminifera: Selective solution and the lysocline: Mar. Geol., *8:* 111-138.

Berger, W. H., A. A. Ekdale, and P. F. Bryant, 1979, Selective preservation of burrows in deep-sea carbonates: Mar. Geol., *32:* 205-230.

Berger, W. H. and J. V. Gardner, 1975, On the determination of Pleistocene temperatures from planktonic foraminifera: J. Foraminiferal Res., *5:* 102-113.

Berger, W. H. and J. S. Killingley, 1977, Glacial-Holocene transition in deep-sea carbonates: Selective dissolution and the stable isotope signal: Science, *197:* 563-556.

Berggren, W. A. and C. D. Hollister, 1977, Plate tectonics and paleocirculation-commotion in the ocean: Tectonophysics, *38:* 11-48.

Bergquist, P. R., 1978, Sponges: University of California Press, Berkeley, 268 p.

Berlin, T. S. and A. V. Khabakov, 1966, Analytical chemical determination of the ratio of calcium and magnesium in belemnoid rostia as a method of estimating the environmental temperatures of existence in the seas of the Cretaceous period in the USSR: Geochem. Int., *3:* 1087-1088.

Berlin, T. S. and A. V. Khabokov, 1974, Calcium and magnesium paleotemperature determinations for carbonate fossils and country rocks: Geochem. Int., *11:* 427-433.

Berman, D. B., 1976, Occurrence of *Gnathorhiza* (Osteichthyes: Dipnoi) in aestivation burrows in the Lower Permian of New Mexico with description of a new species: J. Paleont., *50:* 1034-1039.

Berner, R. A., 1975, The role of magnesium in the crystal growth of calcite and aragonite from sea water: Geochim. et Cosmochim. Acta, *39:* 489-504.

Bevelander, G. and P. Benzer, 1948, Calcification in marine molluscs: Biol. Bull., *94:* 176-183.

Bidder, G. P., 1923, The relation of the form of a sponge to its currents: Quart. J. Microscop. Soc., *67:* 293-325.

Binyon, J., 1966, Salinity tolerance and ion regulation: *In* R. A. Boolootian (ed.), Physiology of Echinodermata: Interscience, New York, p. 359-377.

Boardman, R. S., 1960, Trepostomatous bryozoa of the Hamilton Group of New York State: U.S. Geol. Surv. Prof. Pap. 340, 87 p.

Boersma, A., 1978, Foraminifera: *In* B. V. Haq and A. Boersma (eds.), Introduction to marine micropaleontology: Elsevier, New York, p. 19-77.

Bøggild, O. B., 1930, The shell structure of the mollusks: Kgl. Danske Vidensk. Selsk. Skrifter, Naturvidensk. og Mathem. Afd. (Acad. Roy. Sci. Lettre Danemark) Mém. Ser. 9, 233-326.

Boltovskoy, E. and R. Wright, 1976, Recent Foraminifera: Dr. W. Junk b. r., publisher, The Hague, Netherlands, 515 p.

Bonner, J. T., 1965, Size and cycle, an essay on the structure of biology: Princeton University Press, Princeton, N.J., 219 p.

Bosellini, A. and R. N. Ginsburg, 1971, Form and internal structure of Recent algal nodules (rhodolites) from Bermuda: J. Geol. *79:* 669-682.

Bosence, D. W. J., 1976, Ecological studies on two unattached coralline algae from western Ireland: Palaeontology, *19:* 365-395.

Bosence, D. W. J., 1979a, Live and dead faunas from coralline algal gravels, Co. Galway: Palaeontology, *22:* 449-478.

Bosence, D. W. J., 1979b, Trophic analysis of communities and death assemblages: Lethaia, *12:* 120.

Boucot, A. J., 1953, Life and death assemblages among fossils: Amer. J. Sci., *251:* 25-40.

Bourget, E., 1977, Shell structure in sessile barnacles: Naturaliste Canadien, *104:* 281-323.

Bowen, R., 1961a, Oxygen isotope paleotemperature measurements on Cretaceous belemonoidea from Europe, India, and Japan: J. Paleont., *35:* 1077-1082.

Bowen, R., 1961b, Paleotemperature analyses on Mesozoic Belemnoidea from Australia and New Guinea: Geol. Soc. Amer. Bull., *72:* 769-774.

Bowen, R., 1966, Paleotemperature analysis: Elsevier, New York, 265 p.

Boyd, D. W. and N. D. Newell, 1972, Taphonomy and diagenesis of a Permian fossil assemblage from Wyoming: J. Paleont. *46:* 1-14.

Brasier, M. D., 1975, An outline history of seagrass communities: Palaeontology, *18:* 681-702.

Breimer, A., 1969, A contribution to the paleoecology of Palaeozoic stalked crinoids: Koninkl. Nederl. Akad. Wetensch., Proc. Ser. B, *72:* 139-150.

Breimer, A. and N. G. Lane, 1978, Ecology and paleoecology: *In* R. C. Moore and C. Teichert (eds.), Treatise on invertebrate paleontology, Part T, Echinodermata 2 (v. 1): Geol. Soc. Amer., Boulder, Colo., p. 316-347.

Brenner, K. and G. Einsele, 1976, Schalenbruch im Experiment: Zentralbl. Geol. Paläontol. *2:* 349-354.

Brenner, R. L. and D. K. Davies, 1973, Storm-generated coquinoid sandstone: Genesis of high-energy marine sediments from the Upper Jurassic of Wyoming and Montana: Geol. Soc. Amer. Bull., *84:* 1685-1697.

Bretsky, P. W., 1968, Evolution of Paleozoic marine invertebrate communities: Science, *159:* 1231-1233.

Bretsky, P. W., 1969, Central Appalachian Late Ordovician communities: Geol. Soc. Amer. Bull., *80:* 193-212.

Bretsky, P. W. and S. S. Bretsky, 1975, Succession and repetition of Late Ordovician fossil assemblages from the Nicolet River Valley, Quebec: Paleobiol., *1:* 225-237.

Briggs, J. C., 1974, Marine zoogeography: McGraw-Hill, New York, 475 p.

Broecker, W. S., 1974, Chemical oceanography: Harcourt, New York, 214 p.

Broecker, W. S. and J. van Donk, 1970, Insolation changes, ice volumes, and the 0-18 record in deep-sea cores: Rev. Geophys. Space Phys., *8:* 169-198.

Bromley, R. G., 1975, Trace fossils at omission surface: *In* R. W. Frey (ed.), The study of trace fossils: Springer-Verlag, New York, p. 399-428.

Brongersma-Sanders, M., 1957, Mass Mortality in the sea: *In* J. W. Hedgpeth (ed.), Treatise on marine ecology and paleoecology, v. 1, Ecology: Geol. Soc. Amer. Mem., *67*(1): 941-1010.

Brood, K., 1972, Cyclostomatous bryozoa from the Upper Cretaceous and Danian in Scandinavia: Stockholm Contr. in Geol., *26:* 1-464.

Buchardt, B., 1977, Oxygen isotope ratios from shell material from the Danish Middle Paleocene (Selandian) deposits and their interpretation as paleotemperature indicators: Palaeogeogr., Paleoclimat., Palaeoecol., *22:* 209-230.

Buchardt, B., 1978, Oxygen isotope paleotemperatures from the Tertiary period in the North Sea area: Nature, *275:* 121-123.

Buchardt, B. and P. Fritz, 1978, Strontium uptake in shell aragonite from the freshwater gastropod *Limnaea stagnalis:* Science, *199:* 291-292.

Buchbinder, B. T., 1977, The coralline algae from the Miocene Ziglar Formation in Israel and the environmental significance: *In* E. Flugel (ed.), Fossil algae: Recent results and developments: Springer-Verlag, Berlin, p. 279-285.

Buddemeier, R. W., J. E. Maragrs, and D. W. Knutson, 1974, Radiographic studies of reef coral exoskeletons: Rates and patterns of coral growth: J. Exper. Mar. Biol. Ecol., *14:* 179-200.

Burckle, L. H., 1978, Marine diatoms: *In* B. V. Haq and A. Boersma (eds.), Introduction to marine micropaleontology: Elsevier, New York, p. 245-266.

Burckle, L. H., J. R. Dodd, and R. J. Stanton, Jr., 1980, Biostratigraphy of the Wildcat Group, Humboldt Basin, California: J. Paleont., *53:* 664-674.

Cadee, G. C., 1968, Molluscan biocoenoses and thanatocoenoses in the Ria de Arosa, Galicia, Spain: Rijksmuseum Nat. Hist. Leiden, Zool. Verhandl., *95:* 1-121.

Campbell, K. S. W., 1975, The functional morphology of *Cryptolithus*: Fossils and Strata, *4:* 65-86.

Carpenter, F. M., 1953, The geologic history and evolution of insects: Amer. Scientist, *41:* 256-270.

Carpenter, W. B., 1844, On the microstructure of shells: Rep. British Assoc. Adv. Sci., 1-24.

Carter, R. M., 1972, Adaptations of British chalk bivalvia: J. Paleont., *46:* 325-340.

Carter, R. W. G., 1974, Feeding sea birds as a factor in lamellibranch valve sorting patterns: J. Sedim. Petrol., *44:* 689-692.

Chaloner, W. G. and W. S. Lacey, 1973, The distribution of Late Palaeozoic floras: *In* N. F. Hughes (ed.), Organisms and continents through time: Palaeont. Assoc. Spec. Pap., *12:* 271-289.

Chaloner, W. G. and S. V. Meyer, 1973, Carboniferous and Permian floras of the northern continents: *In* A. Hallam (ed.), Atlas of paleobiogeography: Elsevier, Amsterdam, p. 169-186.

Chamberlain, C. K. and D. L. Clark, 1973, Trace fossils and conodonts as evidence for deep-water deposits in the Oquirrh basin of central Utah: J. Paleont., *47:* 663-682.

Chamberlain, C. K., 1975a, Trace fossils in DSDP cores of the Pacific: J. Paleont., *49:* 1047-1096.

Chamberlain, C. K., 1975b, Recent lebensspuren in nonmarine aquatic environments: *In* R. W. Frey (ed.), The study of trace fossils: Springer-Verlag, New York, p. 431-458.

Chamberlain, C. K., 1978, Recognition of trace fossils in cores: *In* P. W. Basan (ed.), Trace fossil concepts, SEPM short course No. 5: Soc. Econ., Paleontologists and Mineralogists, Oklahoma City, p. 123-183.

Chamberlain, J. A., 1976, Flow patterns and drag coefficients of cephalopod shells: Palaeontology, *19:* 539-563.

Chamberlain, J. A., Jr., 1978, Mechanical properties of coral skeleton: compressive strength and its adaptive significance: Paleobiology, *4:* 419-435.

Chamberlain, J. A. and E. G. Westermann, 1976, Hydrodynamic properties of cephalopod shell ornament: Paleobiology, *2:* 316-331.

Chave, K. E., 1954, Aspects of the biogeochemistry of magnesium I. Calcareous marine organisms: J. Geol., *62:* 266-283.

Chave, K. E., 1964, Skeletal durability and preservation: *In* J. Imbrie and N. Newell (eds.), Approaches to paleoecology: Wiley, New York, p. 377-387.

Cheetham, A. H., 1971, Functional morphology and biofacies distribution of cheilostome bryozoa in the Danian Stage (Paleocene) of southern Scandinavia: Smithsonian Contr. to Paleobiol., No. 6, 87 p.

Chesher, R. H., 1969, Destruction of Pacific corals by the sea star *Acanthaster planci:* Science, *165:* 280-283.

Chilingar, G. V., 1956, Relationship between Ca/Mg ratio and geologic age: Amer. Assoc. Petrol. Geol. Bull., *40:* 2256-2266.

Clark, G. R. II, 1968, Mollusk shell: daily growth lines: Science, *161:* 800-802.

Clarke, F. W. and W. C. Wheeler, 1922, The inorganic constituents of marine invertebrates: U.S. Geol. Surv. Prof. Pap. 124, 62 p.

Clarkson, E. N. K., 1969, A functional study of the Silurian odontopleurid trilobite *Leonaspis deflexa* (Lake): Lethaia, *2:* 329-344.

Clements, F. E. and V. E. Shelford, 1939, Bio-Ecology: Wiley, New York, 425 p.

Cline, R. M. and J. D. Hays (eds.), 1976, Late Quaternary paleoceanography and paleoclimatology: Geol. Soc. Amer. Mem., *145:* 464 p.

Cody, M. L. and J. M. Diamond (eds.), 1975, Ecology and evolution of communities: Belknap Press, Cambridge, Mass., 545 p.

Coe, W. R., 1957, Fluctuations in littoral populations: *In* J. W. Hedgpeth (ed.), Treatise on marine ecology and paleoecology, v. 1, Ecology: Geol. Soc. Amer. Mem., *67*(1): 935-940.

Coe, W. R. and J. E. Fitch, 1950, Population studies, local growth rates and reproduction of the Pismo clam (*Tivela stultorum*): J. Mar. Res., *9:* 188-210.

Colbert, E. H., 1964, Climatic zonation and terrestrial faunas: *In* A. E. M. Nairn (ed.), Problems in paleoclimatology: Wiley-Interscience, New York, p. 617-639.

Colbert, E. H., 1973, Wandering lands and animals: Dutton, New York, 323 p.

Condra, G. E. and M. K. Elias, 1944, Study and revision of *Archimedes* (Hall): Geol. Soc. Amer. Spec. Pap., *53:* 1-243.

Connell, J. H., 1961, Effects of competition, predation by *Thais lapillus*, and other factors on natural populations of the barnacle *Balanus balanoides:* Ecol. Monogr., *31:* 61-104.

Conrad, M. A., 1977, The Lower Cretaceous calcareous algae in the area surrounding Geneva (Switzerland): Biostratigraphy and depositional environments: *In* E. Flugel (ed.), Fossil algae: Recent results and developments: Springer-Verlag, Berlin, p. 295-300.

Cook, P. J., 1977, Loss of boron from shells during weathering and possible implications for determination of paleosalinity: Nature, *268:* 426-427.

Coutts, P. J. F., 1970, Bivalve-growth patterning as a method for seasonal dating in archaeology: Nature, *226:* 874 p.

Coutts, P. J. F., 1975, The seasonal perspective of marine-oriented prehistoric hunter-gatherers: *In* G. D. Rosenberg, and S. K. Runcorn (eds.), Growth rhythms and the history of the earth's rotation: Wiley-Interscience, New York, p. 243-252.

Cowen, R., R. Gertman, and G. Wiggett, 1973, Camouflage patterns in *Nautilus,* and their implications for cephalopod paleobiology: Lethaia, *6:* 201-213.

Craig, G. Y., 1952, A comparative study of the ecology and paleoecology of *Lingula:* Trans. Edinburgh Geol. Soc., *15:* 110-120.

Craig, G. Y. and A. Hallam, 1963, Size-frequency and growth-ring analyses of *Mytilus edulis* and *Cardium edule,* and their paleoecological significance: Palaeontology, *6:* 731-750.

Craig, G. Y. and G. Oertel, 1966, Deterministic models of living and fossil populations of animals: Quat. J. Geol. Soc. London, *122:* 315-355.

Craig, H., 1961, Standard for reporting concentrations of deuterium and oxygen-18 in natural waters: Science, *133:* 1833-1834.

Craig, H., 1965, The measurement of oxygen isotope paleotemperatures: *In* Stable Isotopes in Oceanographic studies and paleotemperatures: Speleto, Italy, Consiglio Nationale delle Ricerdy Lab. di Geol. Mus., Pisa, p. 3-24.

Craig, H., 1970, Abyssal carbon 13 in the south Pacific: J. Geophys. Res., *75:* 691-695.

Creer, K. M., 1975, On a tentative correlation between changes in the geomagnetic polarity bias and reversal frequency and the earth's rotation through Phanerozoic time: *In* G. D. Rosenberg and S. K. Runcorn (eds.), Growth rhythms and the history of the earth's rotation: Wiley-Interscience, New York, p. 293-317.

Crimes, T. P. and J. C. Harper (eds.), 1970, Trace fossils: Geol. J., Spec. Issue 3, 351 p.

Crimes, T. P. and J. C. Harper (eds.), 1977, Trace fossils 2: Geol. J., Spec. Issue 9, 351 p.

Cuffey, R. J., 1977, Bryozoan contributions to reefs and bioherms through geologic time: Amer. Assoc. Petrol. Geol. Studies in Geol., *4:* 181-194.

Culkin, F. and R. A. Cox, 1966, Sodium, potassium, magnesium, calcium, and strontium in sea water: Deep-Sea Res., *13:* 789-804.

Curtis, C. D. and D. Krinsley, 1965, The detection of minor diagenetic alteration in shell material: Geochim. et Cosmochim. Acta, *29:* 71-84.

Dansgaard, W., 1964, Stable isotopes in precipitation: Tellus, *16:* 436-468.

Dansgaard, W., S. J. Johnson, J. Møller, and C. C. Langway, Jr., 1969, One thousand centuries of climatic record from Camp Century on the Greenland Ice Sheet: Science, *166:* 377-381.

Dansgaard, W. and H. Tauber, 1969, Glacier oxygen-18 content and Pleistocene ocean temperatures: Science, *166:* 499-502.

Darlington, P. J., 1957, Zoogeography: the geographical distribution of animals: Wiley, New York, 675 p.

Davenport, C. B., 1938, Growth lines in fossil pectens as indicators of past climates: J. Paleont., *12:* 514-515.

Davies, T. T., 1972, Effect of environmental gradients in the Rappahannock River Estuary on the molluscan fauna: *In* B. W. Nelson (ed.), Environmental framework of coastal plain estuaries: Geol. Soc. Amer. Mem., *133:* 263-290.

Davis, M. B., 1961, Pollen diagrams as evidence of Late-glacial climatic changes in southern New England: N.Y. Acad. Sci. Ann., *95* (Art. 1): 623-631.

Davis, M. B., 1969, Palynology and environmental history during the Quaternary Period: Amer. Scientist, *57:* 317-332.

Davis, W. M., 1926, The value of outrageous geological hypotheses: Science, *63:* 463-468.

Deevey, E. S., 1947, Life tables for natural populations of animals: Quat. Rev. Biol., *22:* 283-314.

Degens, E. T., 1969, Biogeochemistry of stable carbon isotopes: *In* G. Eglinton and M. T. J. Murphy (eds.), Organic geochemistry: methods and results: Springer-Verlag, New York, p. 304-329.

Degens, E. T., D. W. Spencer and R. H. Parker, 1967, Paleobiochemistry of molluscan shell proteins: Comp. Biochem. Physiol., *20:* 553-579.

deLaubenfels, M. W., 1957, Marine Sponges: *In* J. W. Hedgpeth (ed.), Treatise on marine ecology and paleoecology, v. 1, Ecology: Geol. Soc. Amer. Mem., *67* (1): 1083-1086.

Delorme, L. D., 1971, Paleoecological determinations using Pleistocene freshwater ostracodes: *In* H. J. Oertl (ed.), Paleoecologie ostracodes: Bull. Centre Rech. Pau-SNPA, *5* (Suppl.): 341-347.

Denton, E. J. and J. B. Gilpin-Brown, 1973, Flotation mechanisms in modern and fossil cephalopods: Adv. Mar. Biol., *11:* 197-268.

Devereux, I., 1967, Oxygen isotope paleotemperature measurements on New Zealand Tertiary fossils: New Zealand J. Sci., *10:* 988-1011.

Devereux, I., C. H. Hendy, and P. Vella, 1970, Pliocene and Early Pleistocene sea temperature fluctuations, Mangaopari Stream, New Zealand: Earth and Planet. Sci. Lett., *8:* 163-168.

Dewey, J. F., 1969, Evolution of the Appalachian/Caledonian orogen: Nature, *222:* 124-129.

Dinamani, P., 1964, Burrowing behavior of *Dentalium:* Biol. Bull., *126:* 28-32.

Dodd, J. R., 1963, Paleoecological implications of shell mineralogy in two pelecypod species: J. Geol. *71:* 1-11.

Dodd, J. R., 1964, Environmentally controlled variation in the shell structure of a pelecypod species: J. Paeont., *38:* 1065-1071.

Dodd, J. R., 1965, Environmental control of strontium and magnesium in *Mytilus:* Geochim. et Cosmochim. Acta, *29:* 385-398.

Dodd, J. R., 1966, Diagenetic stability of temperature-sensitive skeletal properties in *Mytilus* from the Pleistocene of California: Geol. Soc. Amer. Bull., *77:* 1213-1224.

Dodd, J. R., 1967, Magnesium and strontium in calcareous skeletons: a review: J. Paleont., *41:* 1313-1329.

Dodd, J. R. and T. J. M. Schopf, 1972, Approaches to biogeochemistry: *In* T. J. M. Schopf (ed.), Models in paleobiology: Freeman, Cooper, and Co., San Francisco, p. 46-60.

Dodd, J. R. and R. J. Stanton, Jr., 1975, Paleosalinities within a Pliocene Bay, Kettleman Hills, California: A study of the resolving power of isotopic and faunal techniques: Geol. Soc. Amer. Bull., *86:* 51-64.

Dodd, J. R. and R. J. Stanton, Jr., 1976, Paleosalinities within a Pliocene Bay, Kettleman Hills, California: A study of the resolving power of isotopic and faunal techniques (reply to discussion): Geol. Soc. Amer. Bull., *87:* 160.

Donahue, J. G., 1970, Pleistocene diatoms as climatic indicators in north Pacific sediments: Geol. Soc. Amer. Mem., *126:* 121-138.

Donnay, G. and D. L. Pawson, 1969, X-ray diffraction studies of echinoderm plates: Science, *166:* 1147-1150.

Dorf, E., 1964, The use of fossil plants in paleoclimatic interpretations: *In* A. E. M. Nairn (ed.), Problems in paleoclimatology: Wiley-Interscience, New York, p. 13-31.

Dörjes, J. and G. Hertweck, 1975, Recent biocoenoses and ichnocoenoses in shallow-water marine environments: *In* R. W. Frey (ed.), The study of trace fossils: Springer-Verlag, New York, p. 459-491.

Dorman, F. H., 1966, Australian Tertiary paleotemperatures: J. Geol. *74:* 49-61.

Dorman, F. H. and E. D. Gill, 1959, Oxygen isotope paleotemperature determinations of Australian Cainozoic fossils: Science, *130:* 1576.

Driscoll, E. G., 1967, Experimental field study of shell abrasion: J. Sedim. Petrol., *37:* 1117-1123.

Driscoll, E. G., 1969, Animal-sediment relationships of the Coldwater and Marshall Formations of Michigan: *In* K. S. W. Campbell (ed.), Stratigraphy and paleontology—Essays in honor of Dorothy Hill: Australian Nat. University Press, Canberra, p. 337-352.

Driscoll, E. G. and D. E. Brandon, 1973, Mollusc-sediment relationships, northwestern Buzzards Bay, Massachusetts: Malacologia, *12:* 13-46.

Dudley, E. C. and G. J. Vermeij, 1978, Predation in time and space: drilling in the gastropod *Turritella:* Paleobiology, *4:* 436-441.

Duplessy, J-C., C. Lalou and A. C. Vinot, 1970, Differential isotopic fractionation in benthic foraminifera and paleotemperatures reassessed: Science, *168:* 250-251.

Durham, D. L. and W. O. Addicott, 1965, Pancho Rico Formation, Salinas Valley, California: U.S. Geol. Survey, Prof. Paper *524-A,* 22 p.

Durham, J. W., 1950, Cenozoic marine climate of the Pacific coast: Geol. Soc. Amer. Bull., *61:* 1243-1264.

Durham, J. W., 1966, Ecology and paleoecology: *In* R. C. Moore (ed.), Treatise on invertebrate paleontology, Part U, Echinodermata 3 (v. 1): Geol. Soc. Amer., Boulder, Colo., p. 257-265.

Eichler, R. and H. Ristedt, 1966, Isotopic evidence on the early life history of *Nautilus pompilius* (Linné): Science, *153:* 734-736.

Eisma, D., 1965, Shell-characteristics of *Cardium edule* L. as indicators of salinity: Netherlands J. Sea Res., *2:* 493-540.

Eisma, D., W. G. Mook, and H. A. Das, 1976, Shell characteristics, isotopic composition and trace element contents of some euryhaline molluscs as indicators of salinity: Palaeogeogr., Palaeoclimat., Palaeoecol., *19:* 39-62.

Ekdale, A. A. and W. H. Berger, 1978, Deep sea ichnofacies: Modern organism traces on and in pelagic carbonates of the western equatorial Pacific: Palaeogeogr., Palaeoclimat., Palaeoecol., *23:* 263-278.

Ekman, S., 1953, Zoogeography of the sea: Sidgwick and Jackson, London, 417 p.

Elias, M. K., 1937, Depth of deposition of the Big Blue (Late Paleozoic) sediments in Kansas: Geol. Soc. Amer. Bull. *48:* 403-432.

Elias, M. K. and G. E. Condra, 1957, *Fenestella* from the Permian of west Texas: Geol. Soc. Amer. Mem., *70:* 1-158.

Emery, K. O. and R. E. Stevenson, 1957, Estuaries and lagoons: *In* J. W. Hedgpeth (ed.), Treatise on marine ecology and paleoecology, v. 1, Ecology: Geol. Soc. Amer. Mem., *67*(1): 673-749.

Emig, C. C., J. C. Gall, D. Pajand, and J-C. Plaziat, 1978, Reflexions critiques sur l'ecologie et la systematique des lingules actuelles et fossiles: Geobios, *11:* 573-609.

Emiliani, C., 1954, Temperatures of Pacific bottom waters and polar superficial waters during the Tertiary: Science, *119:* 853-855.

Emiliani, C., 1955, Pleistocene temperatures: J. Geol. *63:* 538-578.

Emiliani, C., 1966, Paleotemperature analysis of Caribbean cores P6304-8 and P6304-9 and a generalized temperature curve for the past 425,000 years: J. Geol., *74:* 109-126.

Emiliani, C., 1972, Quaternary paleotemperatures and duration of the high temperature intervals: Science, *178:* 398-401.

Emrich, K., D. H. Ehhalt, and J. C. Vogel, 1970, Carbon isotope fractionation during the precipitation of calcium carbonate: Earth and Planet. Sci. Lett., *8:* 363-371.

Epstein, S., 1959, The variation of the O^{18}/O^{16} ratio in nature and some geologic implications: *In* P. H. Abelson (ed.), Researches in geochemistry: Wiley, New York, p. 217-240.

Epstein, S., R. Buchsbaum, H. Lowenstam, and H. C. Urey, 1951, Carbonate-water isotopic temperature scale: Geol. Soc. Amer. Bull., *62:* 417-426.

Epstein, S. and T. Mayeda, 1953, Variation of O^{18} content of waters from natural sources: Geochim. et Cosmochim. Acta, *4:* 213-224.

Erickson, D. B., 1959, Coiling direction in *Globigerina pachyderma* as a climatic index: Science *130:* 219-220.

Fager, E. W., 1963, Communities of organisms: *In* M. N. Hill (ed.), The Sea, v. 2: Wiley-Interscience, New York, p. 415-437.

Fager, E. W., 1964, Marine sediments: Effects of a tube-building polychaete: Science, *143:* 356-359.

Fager, E. W., 1972, Diversity: A sampling study: Amer. Naturalist, *106:* 293-310.

Fagerstrom, J. A., 1964, Fossil communities in paleoecology: Their recognition and significance: Geol. Soc. Amer. Bull., *75:* 1197-1216.

Farlow, J. O., C. V. Thompson, and D. E. Rosner, 1976, Plates of the dinosaur *Stegosaurus:* Forced convection heat loss fins?: Science, *192:* 1123-1125.

Faure, G., 1977, Principles of isotope geology: Wiley, New York, 464 p.

Fell, H. B., 1954, Tertiary and Recent Echinoidea of New Zealand: New Zea. Geol. Surv. Paleont. Bull., *23,* 62 p.

Fell, H. B., 1966, Ecology of the crinoids: *In* R. A. Boolootian (ed.), Physiology of Echinodermata: Wiley-Interscience, New York, p. 49-62.

Fenchel, T., 1970, Studies on the decomposition of organic detritus derived from the turtle grass *Thalassia testudinum:* Limnol. and Oceanog., *15:* 14-20.

Ferguson, L., 1963, The paleoecology of *Lingula squamiformis* Phillips during a Scottish Mississippian marine transgression: J. Paleont., *37:* 669-681.

Finks, R. M., 1970, The evolution and ecological history of sponges during Paleozoic times: *In* W. G. Fry (ed.), The biology of Porifera: Academic, New York, p. 3-22.

Fischer, A. G., 1964, Growth patterns of Silurian Tabulata as palaeoclimatologic and palaeogeographic tools: *In* A. E. M. Nairn (ed.), Problems in paleoclimatology: Wiley-Interscience, New York, p. 608-615.

Flügel, E. (ed.), 1977a, Fossil algae: Recent results and developments: Springer-Verlag, New York, 372 p.

Flügel, E., 1977b, Environmental models for Upper Paleozoic benthic calcareous algal communities: *In* E. Flügel (ed.), Fossil algae: Recent results and development: Springer-Verlag, Berlin, p. 314-343.

Forbes, E., 1843, Report on the Mollusca and Radiata of the Aegean Sea and their distribution, considered as bearing on geology: Brit. Assoc. Adv. Sci., Rep. 13, p. 130-193.

Forbes, E. and R. Godwin-Austen, 1859, The natural history of the European seas: J. Van Voorst, London, 306 p.

Foster, M. W., 1974, Recent Antarctic and subantarctic brachiopods: Antarctic Res. Ser., v. 21, Amer. Geophys. Union, Washington, 189 p.

Fretter, V. and A. Graham, 1962, British prosobranch molluscs: Ray Society, London, 755 p.

Frey, R. W. (ed.), 1975, The study of trace fossils: Springer-Verlag, New York, 562 p.

Frey, R. W., 1978, Behavioral and ecological implications of trace fossils: *In* P. B. Basan (ed.), Trace fossil concepts. S.E.P.M. short course no. 5: Soc. Econ. Paleontologists and Mineralogists, Tulsa, Okla., p. 49-75.

Frey, R. W. and J. D. Howard, 1970, Comparison of Upper Cretaceous ichnofaunas from siliceous sandstones and chalk, western interior region, U.S.A.: *In* T. P. Crimes and J. C. Harper (eds.), Trace fossils: Geol. J., Spec. Issue 3, p. 141–166.

Frost, S. H. and R. L. Langenheim, Jr., 1974, Cenozoic reef biofacies: Northern Illinois University Press, DeKalb, Ill., 388 p.

Fry, W. G. (ed.), 1970, The biology of porifera: Academic, New York, 512 p.

Füchtbauer, H. and L. A. Hardie, 1976, Experimentally determined homogeneous distribution coefficients for precipitated magnesian ealcites: Geol. Soc. Amer. Abs. with Progs., *8*(6): 877.

Funnell, B. M., 1971, Post-Cretaceous biogeography of oceans—With special reference to plankton: *In* F. A. Middlemiss, P. F. Rawson, and G. Newall (eds.), Faunal provinces in space and time: Geol. J. Spec. Issue No. 4, Seel House Press, Liverpool, p. 191–198.

Fürsich, F. T., 1978, The influence of faunal condensation and mixing on the preservation of fossil benthic communities: Lethaia, *11:* 243–250.

Fürsich, F. T. and J. M. Hurst, 1974, Environmental factors determining the distribution of brachiopods: Palaeontology, *17:* 879–900.

Futterer, E., 1978a, Studien über die Einregelung, Anlagerung und Einbettung biogener Hartteile im Strömungskanal: Neues Jahrb. Paläont., Abh., *156:* 87–131.

Futterer, E., 1978b, Untersuchungen über die Sink-und Transportgeschwindigkeit biogener Hartteile: Neues Jahrb. Paläont., Abh., *155:* 318–359.

Fyfe, W. S. and J. L. Bischoff, 1965, The calcite-aragonite problem: *In* L. C. Pray and R. C. Murray (eds.), Dolomitization and limestone diagenesis: Soc. Econ. Paleontologists and Mineralogists, Spec. Publ. 13, p. 3–13.

Gainey, L. F., 1972, The use of the foot and the captacule in the feeding of *Dentalium* (Mollusca: Scaphopoda): Veliger, *15:* 29–34.

Galehouse, J. S., 1967, Provenance and paleocurrents of the Paso Robles Formation, California: Geol. Soc. Amer. Bull., *78:* 951–978.

Garrett, P., 1970, Phanerozoic stromatolites: noncompetitive ecological restriction by grazing and burrowing animals: Science, *169:* 171–173.

Gebelein, C. D., 1969, Distribution, morphology, and accretion rate of Recent subtidal algal stromatolites, Bermuda: J. Sedim. Petrol., *39:* 49–69.

Geitzenauer, K. R., 1969, Coccoliths as Late Quaternary paleoclimatic indicators in the subantarctic Pacific: Nature, *223:* 170–172.

Gibson, T. G. and M. A. Buzas, 1973, Species diversity: patterns in modern and Miocene Foraminifera of the eastern margin of North America: Geol. Soc. Amer. Bull., *84:* 217–238.

Gilbert, G. K., 1896, The origin of hypotheses: Science, *3:* 1–14.

Gill, G. A. and A. G. Coates, 1977, Mobility, growth pattern, and substrate in some fossil and recent corals: Lethaia, *10:* 119–134.

Gingerich, P. D., 1976, Paleontology and phylogeny: Patterns of evolution at the species level in Early Tertiary mammals: Amer. J. Sci., *276:* 1–28.

Ginsburg, R. N., 1956, Environmental relationships of grain size and constituent particles in some south Florida carbonate sediments: Bull. Amer. Assoc. Petrol. Geol., *40:* 2384–2427.

Glass, B. P. and L. J. Rosen, 1978, Sedimentation rates, deep sea: *In* R. W. Fairbridge and J. Bourgeois (eds.), The encyclopedia of sedimentology: Dowden, Hutchinson, and Ross, Stroudsburg, Pa., p. 692–694.

Goldberg, E. D., 1957, Biogeochemistry of trace metals: *In* J. W. Hedgpeth (ed.), Treatise on marine ecology and paleoecology, v. 1, Ecology: Geol. Soc. Amer. Mem., *67*(1): 345–358.

Goldring, R., 1964, Trace fossils and the sedimentary surface in shallow marine sediments: *In* L. M. J. U. Van Straaten (ed.), Deltaic and shallow marine deposits: Developments in sedimentology, v. 1: Elsevier, Amsterdam, p. 136–143.

Goldring, R., and J. Kazmierczak, 1974, Ecological succession in intraformational hardground formation: Palaeontology, *17:* 949-962.

Golubic, S., R. D. Perkins, and K. J. Kulas, 1975, Boring microorganisms and microborings in carbonate substrates: *In* R. W. Frey (ed.), The study of trace fossils: Springer-Verlag, New York, p. 229-259.

Good, R., 1974, The geography of the flowering plants: Longman Group Ltd., London, 557 p.

Gordon, C. M., R. A. Carr, and R. E. Larson, 1970, The influence of environmental factors on the sodium and manganese content of barnacle shells: Limnol. and Oceanog., *15:* 461-466.

Goreau, T. F., 1959, The physiology of skeleton formation in corals. I. A method of measuring the rate of calcium deposition by corals under different conditions: Biol. Bull. Mar. Biol. Lab., Woods Hole, *116:* 59-75.

Goreau, T. F., 1963, Calcium carbonate deposition by coralline algae and corals in relation to their roles as reef-builders: Ann. N.Y. Acad. Sci., *109:* 127-167.

Goreau, T. F. and N. I. Goreau, 1973, The ecology of Jamaican coral reefs II: Geomorphology, zonation, and sedimentary phases: Bull. Mar. Sci., *23:* 399-464.

Gould, S. J., 1965, Is uniformitarianism necessary?: Amer. J. Sci. *263:* 223-228.

Grant, R. E., 1972, The lophophore and feeding mechanism of the productidina (Brachiopoda): J. Paleont., *46:* 213-249.

Graus, R. R., J. A. Chamberlain Jr., and A. M. Baker, 1977, Structural modification of corals in relation to waves and currents: Amer. Assoc. Petrol. Geol. Stud. Geol., no. 4, p. 135-153.

Graus, R. R. and I. G. Macintyre, 1976, Light control of growth form in colonial reef corals: Computer simulation: Science, *193:* 895-897.

Greiner, G. O. G., 1969, Recent benthonic foraminifera: Environmental factors controlling their distribution: Nature, *223:* 168-170.

Greiner, G. O. G., 1974, Environmental factors controlling the distribution of Recent benthonic foraminifera: Breviora, Mus, Comp. Zool., Harvard University, no. 420, p. 1-35.

Gretener, P. E., 1967, Significance of the rare event in geology: Bull. Amer. Assoc. Petrol. Geol., *51:* 2197-2206.

Grigg, R. W., 1972, Orientation and growth form of sea fans: Limnol. and Oceanog., *17:* 185-192.

Grimsdale, T. R. and F. P. C. M. van Morkhoven, 1955, The ratio between pelagic and benthonic foraminifera as a means of estimating depth of deposition of sedimentary rocks: IV World Petrol. Congr., Proc., sect. I/D Rep. *4:* 473-491.

Gross, M. G., 1964, Variation in the O^{18}/O^{16} ratios of diagenetically altered limestones in the Bermuda Islands: J. Geol. *72:* 170-194.

Hall, C. A., Jr., 1964, Shallow-water marine climates and molluscan provinces: Ecology, *45:* 226-234.

Hall, C. A., Jr., 1975, Latitudinal variation in shell growth patterns of bivalve molluscs: Implications and problems: *In* G. D. Rosenberg and S. K. Runcorn (eds.), Growth rhythms and the history of the earth's rotation: Wiley-Interscience, New York, p. 163-175.

Hall, C. A., Jr., W. A. Dollase, and C. E. Corbató, 1974, Shell growth in *Tivela stultorum* (Mawe, 1823) and *Callista chione* (Linnaeus) (Bivalvia): Annual periodicity, latitudinal differences, and diminution with age: Palaeogeogr., Palaeoclimat., Palaeoecol., *15:* 33-61.

Hallam, A., 1961, Brachiopod life assemblages from the Marlstone rock-bed of Leicestershire: Palaeontology, *4:* 653-657.

Hallam, A., 1967, The interpretation of size-frequency distributions in molluscan death assemblages: Palaeontology, *10:* 25-42.

Hallam, A., 1969, Faunal realms and facies in the Jurassic: Palaeontology, *12:* 1-18.

Hallam, A., 1972a, Models involving population dynamics: *In* T. J. M. Schopf (ed.), Models in paleobiology: Freeman, Cooper, and Co., San Francisco, p. 62-80.

Hallam, A., 1972b, Diversity and density characteristics of Pliensbachian-Toarcian molluscan and brachiopod faunas of the North Atlantic margins: Lethaia, *5:* 389-412.

Hallam, A. (ed.), 1973a, Atlas of paleobiogeography: Elsevier, Amsterdam, 531 p.

Hallam, A., 1973b, Distributional patterns in contemporary terrestrial and marine animals: Palaeontological Assoc. Spec. Pap. *12:* 93-105.

Hallam, A., 1975, Preservation of trace fossils: *In* R. W. Frey (ed.), The study of trace fossils: Springer-Verlag, New York, p. 55-63.

Hallam, A., 1977, Jurassic bivalve biogeography: Paleobiology, *3:* 58-73.

Hallam, A. and M. J. O'Hara, 1962, Aragonitic fossils in the Lower Carboniferous of Scotland: Nature, *195:* 273-274.

Hallam, A. and N. B. Price, 1966, Strontium contents of recent and fossil aragonitic cephalopod shells: Nature, *212:* 25-27.

Hallam, A. and N. B. Price, 1968a, Further notes on the strontium contents of unaltered fossil cephalopod shells: Geol. Mag., *105:* 52-55.

Hallam, A. and N. B. Price, 1968b, Environmental and biochemical control of strontium in shells of *Cardium edule:* Geochim. et Cosmochim. Acta, *32:* 319-328.

Hansen, H. J., 1979, Test structure and evolution in the Foraminifera: Lethaia, *12:* 173-182.

Häntzschel, W., 1962, Trace fossils and problematica: *In* R. C. Moore (ed.), Treatise on invertebrate paleontology. Pt. W, Miscellanea: Geol. Soc. Amer., Boulder, Colo., 245 p.

Häntzschel, W., 1975, Trace fossils and problematica: *In* C. Teichert (ed.), Treatise on invertebrate paleontology, Pt. W, Miscellanea, Suppl. 1: Geol. Soc. Amer., Boulder, Colo., 269 p.

Haq, B, U., 1978, Calcareous nannoplankton: *In* B. U. Haq and A. Boersma (eds.), Introduction to marine micropaleontology: Elsevier, New York, p. 79-107.

Haq, B. U. and A. Boersma (eds.), 1978, Introduction to marine micropaleontology: Elsevier, New York, 376 p.

Haq, B. U., I. Primoli-Silva, and G. P. Lohmann, 1977, Calcareous plankton paleobiogeographic evidence for major climatic fluctuations in the Early Cenozoic Atlantic Ocean: J. Geophys. Res., *82:* 3861-3876.

Hare, P. E., 1963, Amino acids in the protein from aragonite and calcite in the shells of *Mytilus californianus:* Science, *139:* 216-217.

Hare, P. E. and P. H. Abelson, 1965, Amino acid composition of some calcified proteins: Carnegie Inst., Ann. Rep. Direct. Geophys. Lab., *64:* 223-231.

Harman, R. A., 1964, Distribution of foraminifera in the Santa Barbara Basin, California: Micropaleontology, *10:* 81-96.

Harmelin, J. G., 1973, Morphological variations and ecology of the Recent cyclostome bryozoan *"Idmonea" atlantica* from the Mediterranean: *In* G. P. Larwood (ed.), Living and fossil bryozoa: Recent advances in research: Academic, New York, p. 95-106.

Harmelin, J. G., 1975, Relations entre la forme zoariale et l'habitat chez les bryozoaires cyclostomes. Consequences taxonomiques: *In* S. Pouyet (ed.), Bryozoa, 1974, Proc. of the Third Conference International Bryozoology Assoc., Docs. Labs. Geol. Fac. Sci. Lyon, H. S. 3 (fasc. 1), p. 369-384.

Harmon, R. S., H. P. Schwarcz, and D. C. Ford, 1978, Stable isotope geochemistry of speleothems and cave water from the Flint Ridge-Mammoth Cave System, Kentucky: Implication for terrestrial climate change during the period 230,000 to 100,000 years B.P.: J. Geol. *86:* 373-384.

Harper, C. W., Jr., 1977, Groupings by locality in community ecology and paleoecology: Tests of significance: Lethaia, *11:* 251-257.

Harris, R. C. and D. H. Pilkey, 1966, Temperature and salinity control of the concentration of skeletal Na, Mn, and Fe in *Dendraster excentricus:* Pacific Sci., *20:* 235-238.

Hartman, W. D., 1977, Sponges as reef builders and shapers: Amer. Assoc. Petrol. Geol. Stud. Geol., No. 4, p. 127-134.

Hartman, W. T. and T. F. Goreau, 1970, Jamaican coralline sponges: Their morphology, ecology, and fossil relatives: In W. G. Fry (ed.), The biology of Porifera: Academic, New York, p. 205-240.

Hay, W. H., K. M. Towe, and R. C. Wright, 1963, Ultra-microstructure of some selected foraminiferal tests: Micropaleontology, 9: 171-195.

Hazel, J. E., 1971, Ostracode biostratigraphy of the Yorktown Formation (Upper Miocene and Lower Pliocene) of Virginia and North Carolina: U.S. Geol. Surv. Prof. Pap. 704, 13 p.

Hazel, J. E., 1975, Patterns of marine ostracode diversity in the Cape Hatteras, North Carolina, area: J. Paleont., 49: 731-744.

Hecht, A. D., 1976, The oxygen isotopic record of foraminifera in deep-sea sediment: In R. H. Hedley and C. G. Adams (eds.), Foraminifera, v. 2: Academic, London, p. 1-43.

Hecht, A. D. and S. M. Savin, 1972, Phenotypic variation and oxygen isotope ratios in Recent planktonic foraminifera: J. Foram. Res., 2: 55-67.

Hedgpeth, J. W., 1957a, Marine biogeography: In J. W. Hedgpeth (ed.), Treatise on marine ecology and paleoecology, v. 1, Ecology: Geol. Soc. Amer. Mem., 67 (1): 359-382.

Hedgpeth, J. W. (ed.), 1957b, Treatise on marine ecology and paleoecology, v. 1, Ecology: Geol. Soc. Amer. Mem., 67(1): 1296.

Hedley, R. H. and C. G. Adams (eds.), 1976, Foraminifera, v. 2: Academic, London, 265 p.

Herman, Y., 1978, Pteropods: In B. U. Haq and A. Boersma (eds.), Introduction to marine micropaleontology: Elsevier, New York, p. 151-159.

Herman, Y. and P. E. Rosenberg, 1969, Pteropods as bathymetric indicators: Mar. Geol. 7: 169-173.

Hertweck, G., 1972, Georgia coastal region, Sapelo Island, U.S.A.: Sedimentology and biology. V. Distribution and environmental significance of lebensspuren and in-situ skeletal remains: Senckenbergiana Marit., 4: 125-167.

Hertweck, G., 1973, Der Golf von Gaeta (Tyrrhenisches Meer). VI. Lebensspuren einiger Bodenbewohner und Ichnofaziesbereiche: Senckenbergiana Marit., 5: 179-197.

Hesse, R., W. C. Allee, and K. P. Schmidt, 1951, Ecological animal geography: Wiley, New York, 715 p.

Hessler, R. R. and H. L. Sanders, 1967, Faunal diversity in the deep sea: Deep-Sea Res., 14: 65-79.

Heusser, L., 1978, Spores and pollen in the marine realm: In B. U. Haq and A. Boersma (eds.), Introduction to marine micropaleontology: Elsevier, New York, p. 327-339.

Heusser, L. E. and N. J. Shackleton, 1979, Direct marine-continental correlation: 150,000-year oxygen isotope-pollen record for the north Pacific: Science, 204: 837-839.

Hill, M. O., 1973, Diversity and evenness: a unifying notation and its consequences: Ecology, 54: 427-432.

Hoffman, A., 1977, Synecology of macrobenthic assemblages of the Korytnica Clays (Middle Miocene; Holy Cross Mountains, Poland): Acta Geol. Polonica, 27: 227-280.

Hoffman, A., 1979, Community paleoecology as an epiphenomenal science: Paleobiology, 5: 357-379.

Hoffman, A., A. Pisera, and W. Studencki, 1978, Reconstruction of a Miocene kelp-associated macrobenthic ecosystem: Acta Geol. Polonica, 28: 377-387.

Holland, H. D., J. J. Holland, and J. L. Munoz, 1964, The co-precipitation of cations with $CaCO_3$— II: The co-precipitation of Sr^{2+} with calcite between 90° and 100°C: Geochim. et Cosmochim. Acta, 28: 1287-1301.

Hollister, C. D., B. C. Heezen, and K. E. Nafe, 1975, Animal traces on the deep-sea floor: In R. W. Frey (ed.), The study of trace fossils: Springer-Verlag, New York, p. 493-510.

Hollman, R., 1968, Zur Morphologie rezenter Mollusken-Bruchschille: Paläont. Zeit., *42:* 217-235.

Hopson, J. A., 1975, The evolution of cranial display structures in hadrosaurans: Paleobiology, *1:* 21-44.

Horowitz, A. S. and P. E. Potter, 1971, Introductory petrography of fossils: Springer-Verlag, New York, 302 p.

Howard, J. D., 1966, Characteristic trace fossils in Upper Cretaceous sandstones of the Book Cliffs and Wasatch Plateau: Utah Geol. Mineral. Surv. Bull., *80:* 35-53.

Howard, J. D. and R. W. Frey, 1973, Characteristic physical and biogenic sedimentary structures in Georgia estuaries: Bull. Amer. Assoc. Petrol. Geol., *57:* 1169-1184.

Huang, W.-Y. and W. G. Meinschein, 1979, Sterols as ecological indicators: Geochim. et Cosmochim. Acta, *43:* 739-746.

Hubbard, J. A. E. B., 1970, Sedimentological factors affecting the distribution and growth of Visian caninioid corals in northwest Ireland: Palaeontology, *13:* 191-209.

Hubbard, J. A. E. B., 1974, Coral colonies as micro-environmental indicators: Ann. Soc. Geol. Belgique, *97:* 143-152.

Hudson, J. D., 1968, The microstructure and mineralogy of the shell of a Jurassic mytilid (Bivalvia): Palaeontology, *11:* 163-182.

Hudson, J. D., 1977, Oxygen isotope studies on Cenozoic temperatures, oceans, and ice accumulation: Scott. J. Geol., *13:* 313-325.

Hudson, J. H. and E. A. Shinn, 1977, Stress banding in corals: Normal and abnormal periodicity (abs.): J. Paleont., *51* (2, Part III): 15.

Hudson, J. H., E. A. Shinn, R. B. Halley, and B. Lidz, 1976, Sclerochronology: A tool for interpreting past environments: Geology, *4:* 361-364.

Hunt, R. M., Jr., 1978, Depositional setting of a Miocene mammal assemblage, Sioux County, Nebraska (U. S. A.): Palaeogeogr., Palaeoclimat., Palaeoecol., *24:* 1-52.

Hutchinson, G. E., 1957, Concluding remarks: Cold Spring Harbor Symposia on Quantitative Biology, *22:* 415-427.

Hutchinson, G. E., 1959, Homage to Santa Rosalia *or* Why are there so many kinds of animals?: Amer. Naturalist, *93:* 145-159.

Hutchinson, G. E., 1978, An introduction to population ecology: Yale University Press, New Haven, Conn., 260 p.

Imbrie, J. and N. G. Kipp, 1971, A new micropaleontological method for quantitative paleoclimatology: Application to a Late Pleistocene Caribbean core: *In* K. K. Turekian (ed.), The Late Cenozoic glacial ages: Yale University Press, New Haven, Conn., p. 71-181.

Ingle, J. C., Jr., 1967, Foraminiferal biofacies and the Miocene-Pliocene boundary in southern California: Bull. Amer. Paleont., *52:* 217-394.

Ingle, J. C., Jr., 1976, Late Neogene paleobathymetry and paleoenvironments of the Humboldt Basin, northern California: *In* A. E. Fritsche, H. T. Best, Jr., and W. W. Wornardt (eds.), The Neogene symposium: Pacific Section, Soc. Econ. Paleontologists and Mineralogists, San Francisco, p. 53-61.

Israelsky, M. C., 1949, Oscillation chart: Bull. Amer. Assoc. Petrol. Geol., *33:* 92-98.

Jackson, J. B. C., 1968, Bivalves: Spatial and size-frequency distribution of two intertidal species: Science, *161:* 479-480.

Jefferies, R. P. S. and P. Minton, 1965, The mode of life of two Jurassic species of *Posidonia* (Bivalvia): Palaeontology, *8:* 158-185.

Jenkins, G., 1967, Recent distribution, origin, and coiling ratio changes in *Globorotalia pachyderma* (Ehrenberg): Micropaleontology *13:* 195-203.

Johnson, G. A. L. and J. R. Nudds, 1975, Carboniferous coral geochronometers: *In* G. D.

Rosenberg and S. K. Runcorn (eds.), Growth rhythms and the history of the earth's rotation: Wiley-Interscience, New York, p. 27-42.

Johnson, J. G., 1971, A quantitative approach to faunal province analysis: Amer. J. Sci., *270:* 257-280.

Johnson, J. H., 1961, Limestone-building algae and algal limestones: Colorado School of Mines, Golden, 297 p.

Johnson, M. E., 1977, Succession and replacement in the development of Silurian brachiopod populations: Lethaia, *10:* 83-93.

Johnson, R. G., 1960, Models and methods for analysis of the mode of formation of fossil assemblages: Geol. Soc. Amer. Bull., *71:* 1075-1086.

Johnson, R. G., 1965a, Pelecypod death assemblages in Tomales Bay, California: J. Paleont., *39:* 80-85.

Johnson, R. G., 1965b, Temperature variation in the infaunal environment of a sand flat: Limnol. and Oceanog., *10:* 114-120.

Johnson, T. C., 1976, Biogenic opal preservation in pelagic sediments of a small area in the eastern tropical Pacific: Geol. Soc. Amer. Bull., *87:* 1273-1282.

Kaesler, R. L. and J. A. Waters, 1972, Fourier analysis of the ostracod margin: Geol. Soc. Amer. Bull., *83:* 1169-1178.

Kahn, P. G. K. and S. M. Pompea, 1978, Nautiloid growth rhythms and dynamical evolution of the earth-moon system: Nature *275:* 606-611.

Kammer, T. W., 1979, Paleosalinity, paleotemperature, and oxygen isotopic fractionation records of Neogene foraminifera from DSDP site 173 and the Centerville Beach section, California: Mar. Micropaleontology, *4:* 45-60.

Kato, M., 1963, Fine skeletal structure in Rugosa: J. Fac. Sci. Hokkaido University, ser. 4, Geology and Mineralogy, *11:* 571-630.

Kauffman, E. G., 1969a, Form, function, and evolution: *In* R. C. Moore (ed.), Treatise on invertebrate paleontology, Part N, Mollusca 6: Geol. Soc. Amer., Boulder, Colo., p. 129-205.

Kauffman, E. G., 1969b, Cretaceous marine cycles of the Western Interior: Mountain Geol., *6:* 227-245.

Kauffman, E. G., 1977, Geological and biological overview: Western Interior Cretaceous Basin: Mountain Geol., *14:* 75-99.

Kauffman, E. G. and N. F. Sohl, 1974, Structure and evolution of Antillean Cretaceous rudist frameworks: Verhandl. Naturf. Ges. Basel, *84:* 399-467.

Keast, A., 1972, Australian mammals: Zoogeography and evolution: *In* A. Keast, F. C. Erk, and B. Glass (eds.), Evolution, mammals, and southern continents: State University of New York Press, Albany, p. 195-246.

Keen, A. M., 1937, An abridged check list and bibliography of west North American marine mollusca: Stanford University Press, Stanford, Calif., 87 p.

Keen, A. M., 1971, Sea shells of tropical west America: Stanford University Press, Stanford, Calif., 1064 p.

Keen, A. M. and E. Coan, 1974, Marine molluscan genera of western North America, an illustrated key (2nd ed.): Stanford University Press, Stanford, Calif., 208 p.

Kennedy, W. J., J. D. Taylor, and A. Hall, 1969, Environmental and biological controls on bivalve shell mineralogy: Biol. Rev., *44:* 499-530.

Kennett, J. P., 1967, Recognition and correlation of the Kapitean Stage (Upper Miocene, New Zealand): New Zealand Jour. Geol. and Geophys., *10:* 1051-1063.

Kennett, J. P., 1968, Latitudinal variation in *Globigerina pachyderma* (Ehrenberg) in surface sediments of the southwest Pacific Ocean: Micropaleontology, *14:* 305-318.

Kennett, J. P., 1976, Phenotypic variation in some Recent and Late Cenozoic planktonic foraminifera: *In* R. H. Hedley and C. G. Adams (eds.), Foraminifera, v. 2: Academic, New York, p. 111-170.

Kennett, J. P. and N. J. Shackleton, 1975, Laurentide ice sheet melt water recorded in the Gulf of Mexico deep-sea cores: Science, *188:* 147-150.

Kennish, M. J. and R. K. Olsson, 1975, Effects of thermal discharges on the microstructural growth of *Mercenaria mercenaria*: Environmental Geol., *1:* 41-64.

Kern, J. P. and J. E. Warme, 1974, Trace fossils and bathymetry of the Upper Cretaceous Point Loma Formation, San Diego, California: Geol. Soc. Amer. Bull., *85:* 893-900.

Keupp, H., 1977, Ultrafazies und Genese der Solnhofener Plattenkalke (Oberer Malm, Südliche Frankenalb): Abh. der Naturhistorischen Gesellschaft Nürnberg, e.v., v. 37, 128 p.

Kier, P. M., 1972, Upper Miocene echinoids from the Yorktown Formation of Virginia and their environmental significance: Smithsonian Contribs. to Paleobiol., *13,* 41 p.

Kier, P. M., 1974, Evolutionary trends and their functional significance in the post-Paleozoic echinoids: Paleont. Soc. Mem. 5 [J. Paleont., *48*, (3), Suppl.], 96 p.

Kinne, O., 1971, Environmental factors, salinity, animals, invertebrates: *In* O. Kinne (ed.), Marine ecology: A comprehensive, integrated treatise on life in oceans and coastal waters, v. 1, pt. 2: Wiley, New York, p. 821-1083.

Kinsman, D. J. J., 1969, Interpretation of Sr^{2+} concentrations in carbonate minerals and rocks: J. Sedim. Petrol., *39:* 486-508.

Kinsman, D. J. J. and H. D. Holland, 1969, The co-precipitation of cations with $CaCO_3$—IV. The co-precipitation of Sr^{2+} with aragonite between 16° and 96° C: Geochim. et Cosmochim. Acta, *33:* 1-17.

Kitano, Y., 1962, The behavior of various inorganic ions in the separation of calcium carbonate from a bicarbonate solution: Bull. Chem. Soc. Japan, *35:* 1973-1980.

Kitano, Y. and B. W. Hood, 1965, The influence of organic material on the polymorphic crystallization of calcium carbonate: Geochim. et Cosmochim. Acta, *29:* 29-41.

Kitano, Y., N. Kanamori, and S. Yoshoika, 1976, Influence of chemical species on crystal type of calcium carbonate: In N. Watabe and K. M. Wilbur (eds.), The mechanisms of mineralization in the invertebrates and plants: University of South Carolina Press, Columbia, S. C., p. 191-202.

Knutson, D. W., R. W. Buddemeier, and S. V. Smith, 1972, Coral chronometers: Seasonal growth bands in reef corals: Science, *177:* 270-272.

Kolesar, P. T., 1978, Magnesium in calcite from a coralline alga: J. Sedim. Petrol., *48:* 815-820.

Kranz, P. M., 1974, The anastrophic burial of bivalves and its palaeoecological significance: J. Geol., *82:* 237-265.

Krauskopf, K. B., 1967, Introduction to geochemistry: McGraw-Hill, New York, 721 p.

Kriausakul, N. and R. M. Mitterer, 1978, Isoleucene epimerization in peptides and proteins: Kinetic factors and application to fossil protein: Science, *201:* 1011-1014.

Kummel, B. and R. M. Lloyd, 1955, Experiments on relative streamlining of coiled cephalopod shells: J. Paleont., *29:* 159-170.

Kurtén, B., 1954, Population dynamics — A new method in paleontology: J. Paleont., *28:* 286-292.

Kurtén, B., 1964, Population structure in paleoecology: *In* J. Imbrie and N. D. Newell (eds.), Approaches to paleoecology: Wiley, New York, p. 91-106.

LaBarbera, M., 1977, Brachiopod orientation to water movement. 1. Theory, laboratory behavior, and field orientation: Paleobiology, *3:* 270-287.

Labeyrie, L., 1974, New approach to surface sea water paleotemperatures using O^{18}/O^{16} ratios in silica of diatom frustules: Nature, *248:* 40-42.

Labracherie, M. and J. Prud'homme, 1966, Essai d'interprétation de paléomilieux grâce a la méthode de distribution des formes zoariales chez les bryozoaires: Bull. Soc. Geol. de France, *7:* 102-106.

Ladd, H. S. (ed.), 1957, Treatise on marine ecology and paleoecology, v. 2, Paleoecology: Geol. Soc. Amer. Mem., *67*(2): 1077.

Lagaaij, R., 1963, *Cupuladria canariensis* (Busk) — Portrait of a bryozoan: Palaeontology, *6:* 172-217.

Lagaaij, R. and Y. V. Gautier, 1965, Bryozoan assemblages from marine sediments of the Rhone delta, France: Micropaleontology, *11:* 39-58.

Land, L. S., J. C. Lang, and D. J. Barnes, 1975, Extension rate: A primary control on the isotopic composition of West Indian (Jamaican) scleractinian reef coral skeletons: Mar. Biol., *33:* 221-233.

Lane, N. G. and A. Breimer, 1974, Arm types and feeding habits of Paleozoic crinoids: K. Ned. Akad. Wetensch., Proc., ser. B, *77:* 32-39.

Langston, W., Jr., 1974, Nonmammalian Comanchean tetrapods: Geoscience and Man, *8:* 77-102.

Lasker, H., 1976, Effects of differential preservation on the measurement of taxonomic diversity: Paleobiology, *2:* 84-93.

Latimer, W. M., 1952, Oxidation potentials, 2nd ed.: Prentice-Hall, Englewood Cliffs, N. J., 392 p.

Lawrence, D. R., 1968, Taphonomy and information losses in fossil communities: Geol. Soc. Amer. Bull., *79:* 1315-1330.

Lawrence, D. R., 1971a, The nature and structure of paleoecology: J. Paleont., *45:* 593-607.

Lawrence, D. R., 1971b, Shell orientation in recent and fossil oyster communities from the Carolinas: J. Paleont., *45:* 347-349.

Lawton, R., 1977, Taphonomy of the Dinosaur Quarry, Dinosaur National Monument: Contr. Geol., University of Wyoming, *15:* 119-126.

Lecompte, M., 1958, Les recifs Paleozoiques in Belgique: Geol. Rundschau, *47:* 384-401.

Leigh, E. G., Jr., 1971, Adaptation and diversity: Freeman, Cooper, and Co., San Francisco, 288 p.

Lemche, H. and K. G. Wingstrand, 1959, The anatomy of *Neopilina galatheae:* Galathea Rep., *3:* 9-71.

Leutwein, F. and R. Waskowiak, 1962, Geochemische Untersuchungen an rezenten marinen Molluskenschalen: Neues Jahrb. für Mineralogie, *99:* 45-78.

Levinson, S. A., 1961, Identification of fossil ostracodes in thin section: *In* R. C. Moore (ed.), Treatise on invertebrate paleontology, Part Q, Arthropoda 3: Geol. Soc. Amer., Boulder, Colo., p. 70-73.

Levinton, J. S., 1970, The paleoecological significance of opportunistic species: Lethaia, *3:* 69-78.

Levinton, J. S., and R. K. Bambach, 1970, Some ecological aspects of bivalve mortality patterns: Amer. J. Sci., *268:* 97-112.

Li, Y. H., T. Takahashi and W. S. Broecker, 1969, Degree of saturation of $CaCO_3$ in the oceans: J. Geophys. Res., *74:* 5507-5525.

Linsley, R. M., 1978a, Shell form and the evolution of gastropods: Amer. Scientist, *66:* 432-441.

Linsley, R. M., 1978b, Locomotion rates and shell form in the Gastropoda: Malacologia, *17:* 193-206.

Linsley, R. M., E. L. Yochelson, and D. M. Rohr, 1978, A reinterpretation of the mode of life of some Paleozoic frilled gastropods: Lethaia, *11:* 105-112.

Lister, K. R., 1976, Temporal changes in a Pleistocene lacustrine ostracode association; Salt Lake basin, Utah: *In* R. W. Scott and R. R. West (eds.), Structure and classification of paleocommunities: Dowden, Hutchinson, and Ross, Inc., Stroudsburg, Pa., p. 193-211.

Livingstone, D. A., 1963, Chemical composition of rivers and lakes: U. S. Geol. Surv. Prof. Pap. 440-G, 64 p.

Lloyd, M. and R. J. Ghelardi, 1964, A table for calculating the "equitability" component of species diversity: J. Animal Ecol., *33:* 217-225.

Lloyd, R. M., 1969, A paleoecological interpretation of the Caloosahatchee Formation, using stable isotope methods: J. Geol., *77:* 1-25.

Lochman-Balk, C. and J. L. Wilson, 1958, Cambrian biostratigraphy in North America: J. Paleont., *32:* 312-350.

Loeblich, A. R., Jr., and H. Tappan, 1964, Treatise on invertebrate paleontology, Part C, Protista 2: Geol. Soc. Amer., Boulder, Colo., 900 p.

Logan, B. W., P. Hoffman, and C. D. Gebelein, 1974, Algal mats, cryptalgal fabrics, and structures, Hamelin Pool, Western Australia: *In* B. W. Logan et al. (eds.), Evolution and diagenesis of Quaternary carbonate sequences, Shark Bay, Western Australia: Amer. Assoc. Petrol. Geol. Mem., *22:* 140-194.

Longinelli, A., 1966, O^{18}/O^{16} ratios in phosphate and carbonate from living and fossil marine organisms: Nature, *24:* 923-927.

Lorens, R. B. and M. L. Bender, 1976, The physiological exclusion of Mg^{++} from *Mytilus edulis* calcite (abs.): Geol. Soc. Amer. Abs. with Progs., *8:* 986-987.

Lorenz, K. Z., 1974, Analogy as a source of knowledge: Science, *185:* 229-234.

Lowenstam, H. A., 1950, Niagaran reefs of the Great Lakes area: J. Geol., *58:* 430-487.

Lowenstam, H. A., 1954a, Factors affecting the aragonite: calcite ratios in carbonate-secreting marine organisms: J. Geol. *62:* 284-322.

Lowenstam, H. A., 1954b, Environmental relations of modification compositions of certain carbonate-secreting marine invertebrates: Nat. Acad. Sci. Proc., *40:* 39-48.

Lowenstam, H. A., 1957, Niagaran reefs of the Great Lakes area: *In* H. S. Ladd (ed.), Treatise of marine ecology and paleoecology, v. 2: Geol. Soc. Amer. Mem., *67(2):* 215-248.

Lowenstam, H. A., 1961, Mineralogy, O^{18}/O^{16} ratios, and strontium and magnesium contents of Recent and fossil brachiopods and their bearing on the history of the oceans: J. Geol., *69:* 241-260.

Lowenstam, H. A., 1962, Magnetite in denticle capping in Recent chitons (Polyplacophora): Geol. Soc. Amer. Bull., *73:* 435-438.

Lowenstam, H. A., 1963, Biologic problems relating to the composition and diagenesis of sediments: *In* T. W. Donnelly (ed.), The earth sciences; problems and progress in current research: The University of Chicago Press, p. 137-195.

Lowenstam, H. A., 1964a, Coexisting calcites and aragonites from skeletal carbonates of marine organisms and their strontium and magnesium contents: *In* Recent researches in the fields of hydrosphere, atmosphere, and nuclear geochemistry: Maruzen Co., Ltd., Tokyo, p. 373-404.

Lowenstam, H. A., 1964b, Sr/Ca ratio of skeletal aragonites from the recent marine biota at Palau and from fossil gastropods: *In* H. Craig et al. (eds.), Isotopic and cosmic chemistry: North Holland Publ. Co., Amsterdam, p. 114-132.

Lowenstam, H. A., 1964c, Paleotemperatures of the Permian and Cretaceous periods: *In* A. E. M. Nairn, (ed.), Problems in paleoclimatology: Wiley-Interscience, New York, p. 227-248.

Lowenstam, H. A., 1974, Impact of life on chemical and physical processes: *In* E. D. Goldberg (ed.), The sea, v. 5: Wiley-Interscience, New York, p. 715-796.

Lowenstam, H. A. and D. P. Abbott, 1975, Vaterite: A mineralization product of the hard tissues of a marine organism (Ascidiacea): Science, *188:* 363-365.

Lowenstam, H. A. and S. Epstein, 1954, Paleotemperatures of the post-Aptian Cretaceous as determined by the oxygen isotope method: J. Geol., *62:* 207-248.

Lowenstam, H. A. and D. McConnell, 1968, Biologic precipitation of fluorite: Science, *162:* 1496-1498.

Lutz, R. A. and D. Jablonski, 1978, Larval bivalve shell morphology: a new paleoclimatic tool?: Science, *202:* 51-53.

Lutz, R. A. and D. C. Rhoads, 1977, Anaerobiosis and a theory of growth line formation: Science, *198:* 1222-1227.

Ma, T. Y. H., 1958, The relation of growth rate of reef corals to surface temperature of sea water as basis for study of causes of diastrophisms instigating evolution of life: Research on the Past Climate and Continental Drift, *14:* 1-60.

McAlester, A. L. and D. C. Rhoads, 1967, Bivalves as bathymetric indicators: Mar. Geol. *5:* 383-388.

MacArthur, R. H., 1972, Geographical ecology. Patterns in the distribution of species: Harper & Row, New York, 269 p.

MacArthur, R. H. and J. H. Connell, 1966, The biology of populations: Wiley, New York, 200 p.

MacArthur, R. H. and E. O. Wilson, 1963, An equilibrium theory of insular zoogeography: Evolution, *17:* 373-387.

MacArthur, R. H. and E. O. Wilson, 1967, The theory of island biogeography: Princeton University Press, Princeton, N. J., 203 p.

McCammon, H. M., 1969, The food of articulate brachiopods: J. Paleont., *43:* 976-985.

McCarthy, B., 1977, Selective preservation of mollusc shells in a Permian beach environment, Sydney Basin, Australia: Neues Jahrb. Geol. Paläont., Mh., *8:* 466-474.

MacClintock, C., 1967, Shell structure of patelloid and bellerophontoid gastropods (Mollusca): Peabody Mus. Nat. Hist., Yale University Bull., *22,* 140 p.

McCrea, J. M., 1950, On the isotopic chemistry of carbonates and a paleotemperature scale: J. Chem. Phys., *18:* 849-857.

MacDonald, G. J. F., 1956, Experimental determination of calcite-aragonite equilibrium relations at elevated temperatures and pressures: Amer. Mineralogist, *41:* 744-756.

MacGeachy, J. K. and C. W. Stearn, 1976, Boring by macro-organisms in the coral *Montastrea annularis* on Barbados reefs: Int. Revue ges. Hydrobiol., *61:* 715-745.

MacGinitie, G. E. and N. MacGinitie, 1949, Natural history of marine animals: McGraw-Hill, New York, 473 p.

McIntyre, A., W. F. Ruddiman, and R. Jantzen, 1972, Southward penetration of the North Atlantic polar front: Faunal and floral evidence of large-scale surface water mass movements over the last 225,000 years: Deep-Sea Res., *19:* 61-77.

Macintyre, I. G. and O. H. Pilkey, 1969, Tropical reef corals; tolerance of low temperatures on the North Carolina continental shelf: Science, *166:* 374-375.

McKerrow, W. S. (ed.), 1978, The ecology of fossils: G. Duckworth & Co, Ltd., London, 383 p.

McKerrow, W. S., R. T. Johnson, and M. E. Jakobson, 1969, Palaeoecological studies in the Great Oolite at Kirthington, Oxfordshire: Palaeontology, *12:* 56-83.

McKinney, F. K., 1977a, Functional interpretation of lyre-shaped bryozoans: Paleobiology, *3:* 90-97.

McKinney, F. K., 1977b, Paraboloid colony bases in Paleozoic stenolaemate bryozoans: Lethaia, *10:* 209-217.

Macurda, D. B. and D. L. Meyer, 1974, Feeding posture of modern stalked crinoids: Nature, *247:* 394-396.

Majewske, O. P., 1969, Recognition of invertebrate fossil fragments in rocks and thin sections: E. J. Brill, Leiden, 101 p.

Mancini, E. A., 1978a, Origin of micromorph faunas in the geologic record: J. Paleont. *52:* 311-322.

Mancini, E. A., 1978b, Origin of the Grayson micromorph fauna (Upper Cretaceous) of north-central Texas: J. Paleont., *52:* 1294-1314.

Mangerud, J., E. Sonstegaard, and H. P. Sejrup, 1979, Correlation of the Eemian (interglacial) stage and the deep-sea oxygen-isotope stratigraphy: Nature, *277:* 189-192.

Margolis, S. V., P. M. Kroopnick, D. E. Goodney, W. C. Dudley, and M. E. Mahoney, 1975, Oxygen and carbon isotopes from calcareous nannofossils as paleoceanographic indicators: Science, *189:* 555-557.

Marszalek, D. S., 1975, Calcisphere ultrastructure and skeletal aragonite from the alga *Acetabularia antillana:* J. Sedim. Petrol., *45:* 266-271.

Masuda, F. and K. Taira, 1974, Oxygen isotope paleotemperature measurements on Pleistocene molluscan fossils from the Boso Peninsula, central Japan: Geol. Soc. Japan J., *80:* 97-106.

Mayor, A. G., 1924, Causes which produce stable conditions in the depth of the floors of Pacific fringing reef-flats: Papers from the Dept. Mar. Biol., Carnegie Inst., Wash., *19:* 27-36.

Menard, H. W. and A. J. Boucot, 1951, Experiments on the movement of shells by water: Amer. J. Sci., *249:* 131-151.

Menzies, R. J., R. Y. George, and G. T. Rowe, 1973, Abyssal environment and ecology of the world oceans: Wiley-Interscience, New York, 488 p.

Merrill, R. J. and E. S. Hobson, 1970, Field observations of *Dendraster excentricus,* a sand dollar of western North America: Amer. Midl. Nat., *83:* 595-624.

Meyer, D. L., 1973, Feeding behavior and ecology of shallow-water unstalked crinoids (Echinodermata) in the Caribbean Sea: Mar. Biol., *22:* 105-129.

Meyer, D. L. and N. G. Lane, 1976, The feeding behavior of some Paleozoic crinoids and Recent basketstars: J. Paleont., *50:* 472-480.

Middlemiss, F. A., P. F. Rawson, and G. Newall, 1971, Faunal provinces in space and time: Geol. J. Spec. Issue No. 4, Seel House Press, Liverpool, 236 p.

Milliman, J. D., 1974, Marine carbonates: Springer-Verlag, New York, 375 p.

Mitterer, R. M., 1975, Ages and diagenetic temperatures of Pleistocene deposits of Florida based on isoleucine epimerization: Earth and Planet. Sci. Lett., *28:* 275-282.

Moberly, R., Jr., 1968, Composition of magnesian calcites of algae and pelecypods by electron microprobe analysis: Sedimentology, *11:* 61-82.

Möbius, K., 1877, Die Auster und die Austernwirtschaft: Hempel and Parry, Berlin, 126 p.

Monty, C., 1977, Evolving concepts on the nature and the ecological significance of stromatolites: *In* E. Flugel (ed.), Fossil algae: Recent results and developments: Springer-Verlag, Berlin, p. 15-35.

Mook, W. G., 1971, Paleotemperatures and chlorinities from stable carbon and oxygen isotopes in shell carbonate: Palaeogeogr., Palaeoclimat., Palaeoecol., *9:* 245-263.

Morton, J. E., 1967, Molluscs (4th ed.): Hutchinson University Lib., London, 244 p.

Müller, A. H., 1979, Fossilization (Taphonomy): *In* R. A. Robison and C. Teichert (eds.), Treatise on invertebrate paleontology, Part A, Introduction: Geol. Soc. Amer., Boulder, Colo., p. 2-78.

Muller, C. H., 1958, Science and philosophy of the community concept: Amer. Scientist, *46:* 294-322.

Multer, H. G., 1977, Field guide to some carbonate rock environments, Florida Keys and western Bahamas: Kendall/Hunt Publ. Co., Dubuque, Iowa, 415 p.

Murray, J. W., 1973, Distribution and ecology of living benthic foraminiferids: Heinemann Educational Books, Ltd., London, 274 p.

Murray, R. C., 1964, Preservation of primary structures and fabrics in dolomite: *In* J. Imbrie and N. D. Newell (eds.), Approaches to paleoecology: Wiley, New York, p. 388-403.

Mutvei, H. and R. A. Reyment, 1973, Buoyancy control and siphuncle function of ammonoids: Palaeontology, *16:* 623-636.

Nagle, J. S., 1967, Wave and current orientation of shells: J. Sedim. Petrol., *37:* 1124-1138.

Natland, M. L., 1933, The temperature and depth distribution of some recent and fossil foraminifera in the southern California region: Bull. Scripps Inst. Oceanog., Tech. Ser., *3:* 225-230.

Natland, M. L., and P. H. Kuenen, 1951, Sedimentary history of the Ventura Basin, California, and the action of turbidity currents: Soc. Econ. Paleontologists Mineralogists Spec. Publ. 2, p. 76-107.

Naydin, D. P., R. V. Teys, and I. K. Zadorozhnyy, 1966, Isotopic paleotemperatures of the Upper Cretaceous in the Russian platform and other parts of the U.S.S.R.: Geochem. Int., *3:* 1038-1051.

Neale, J. W. (ed.), 1969, The taxonomy, morphology and ecology of recent ostracoda: Oliver and Boyd, Edinburgh, 553 p.

Nelson, D. J., 1963, The strontium and calcium relationships in Clinch and Tennessee river mollusks: *In* V. Schultz and A. W. Klement, Jr. (eds.), Radioecology, Reinhold, New York, p. 204-211.

Nelson, D. J., 1964, Deposition of strontium in relation to morphology of clam (Unionidae) shells: Verh. Internat. Verein. Limnol., *15:* 893-902.

Nelson, P. C., 1975, Community structure and evaluation of trophic analysis in paleoecology--Stone City Formation (Middle Eocene-Texas): Unpublished M.S. thesis, Texas A & M University, 168 p.

Neri, R., G. Schifano, and C. Papanicolaou, 1979, Effects of salinity on mineralogy and chemical composition of *Cerastoderma edule* and *Monodonta articulata* shells: Mar. Geol. *30:* 233-241.

Newell, I. M., 1948, Marine molluscan provinces of western North America: a critique and a new analysis: Proc. Amer. Philos. Soc., *92:* 155-166.

Newell, N. D., J. Imbrie, E. G. Purdy, and D. L. Thurber, 1959, Organism communities and bottom facies, Great Bahama Bank: Bull. Amer. Mus. Nat. Hist., *117:* 183-228.

Newell, R., 1965, The role of detritus in the nutrition of two marine deposit feeders, the prosobranch *Hydrobia ulvae* and the bivalve *Macoma balthica*: Zool. Soc. Lond. Proc., *144:* 25-45.

Nichols, D., 1959, Changes in the chalk heart urchin *Micraster* interpreted in relation to living forms: Royal Soc. London, Philos. Trans., Ser. B., *242:* 347-437.

Nicol, D., 1962, The biotic development of some Niagaran reefs — An example of an ecological succession or sere: J. Paleont., *36:* 172-176.

Nicol, D., 1964, An essay on size of marine pelecypods: J. Paleont., *38:* 968-974.

Nicol, D., 1967, Some characteristics of cold-water marine pelecypods: J. Paleont., *41:* 1330-1340.

Nier, A. O., 1950, A redetermination of the relative abundances of the isotopes of carbon, nitrogen, oxygen, argon, and potassium: Phys. Rev., *77:* 789-793.

Nomland, J. O., 1916, Fauna from the Lower Pliocene at Jacalitos Creek and Waltham Canyon, Fresno County, California: Univ. Calif. Pub., Bull. Dep. Geol., *9:* 199-214.

Odum, E. P., 1971, Fundamentals of ecology, 3rd ed.: W. B. Saunders Co., Philadelphia, 574 p.

Odum, H. T., 1951, Notes on the strontium content of sea water, celestite radiolaria, and strontianite snail shells: Science, *114:* 211-213.

Odum, H. T., 1957a, Strontium in natural waters: Pub. Inst. Mar. Sci., *4:* 22-37.

Odum, H. T., 1957b, Biogeochemical deposition of strontium: Pub. Inst. Mar. Sci., *4:* 38-114.

Oertli, H. J. (ed.), 1971, Colloque sur la paléoécologie des ostracodes: Bull. Centre Rech. Pau—SNPA, v. 5 (Suppl.), Pau, France, 953 p.

Oldfield, S. C., 1976, Surface ornamentation of the echinoid test and its ecological significance: Paleobiology, *2:* 122-130.

Olsen, C. R., 1978, Sedimentation rates: *In* R. W. Fairbridge and J. Bourgeois (eds.), The encyclopedia of sedimentology: Dowden, Hutchinson, and Ross, Stroudsburg, Pa., p. 687-692.

Olson, E. C., 1952, The evolution of a Permian vertebrate chronofauna: Evolution, *6:* 181-196.

Olson, E. C., 1966, Community evolution and the origin of mammals: Ecology *47:* 291-302.

Olson, E. C. and K. Bolles, 1975, Permo-Carboniferous fresh water burrows: Fieldiana Geol., *33:* 271-290.

Olson, E. C. and P. P. Vaughn, 1970, The changes of terrestrial vertebrates and climates during the Permian of North America: Forma et Functio, *3:* 113-138.

Paasche, E., 1968, Biology and physiology of coccolithophorids: Ann. Rev. Microbiol., 22: 71-76.

Palmer, A. R., 1969, Cambrian trilobite distributions in North America and their bearing on Cambrian paleogeography of Newfoundland: In M. Kay (ed.), North Atlantic — Geology and continental drift: Amer. Assoc. Petrol. Geol. Mem., 12: 139-148.

Palmer, A. R., 1973, Cambrian trilobites: In A. Hallam (ed.), Atlas of paleobiogeography: Elsevier, Amsterdam, p. 3-12.

Pannella, G., 1972, Paleontological evidence on the earth's rotational history since Early Precambrian: Astrophys. Space Sci., 16: 212-237.

Pannella, G. and C. MacClintock, 1968, Biological and environmental rhythms reflected in molluscan shell growth: J. Paleont., 42(5, Part II): 64-80.

Pannella, G., C. MacClintock, and M. N. Thompson, 1968, Paleontologic evidence of variations in length of synodic month since Late Cambrian: Science, 162: 792-796.

Papp, A., H. Zapfe, F. Bachmayer, and A. F. Tauber, 1947, Lebensspuren mariner Krebse: K. Akad. der Wissen., Vienna, Mathem. Naturwiss. Klasse. Sitzber., 155: 281-317.

Park, R., 1976, A note on the significance of lamination in stromatolites: Sedimentology, 23: 379-393.

Parker, R. H., 1956, Macro-invertebrate assemblages as indicators of sedimentary environments in east Mississippi Delta region: Bull. Amer. Assoc. Petrol. Geol., 40: 295-376.

Patterson, B. and R. Pascual, 1972, The fossil mammal fauna of South America: In A. Keast, F. C. Erk, and B. Glass (eds.), Evolution, mammals, and southern continent: State University of New York Press, Albany, p. 247-309.

Pawson, D. L., 1966, Ecology of holothurians: In R. A. Boolootian (ed.), Physiology of Echinodermata: Wiley-Interscience, New York, p. 63-71.

Pearse, A. S. and G. Gunter, 1957, Salinity: In J. W. Hedgpeth (ed.), Treatise on marine ecology and paleoecology, v. 1 Ecology: Geol. Soc. Amer., Mem., 67: 129-158.

Pearse, J. S. and V. B. Pearse, 1975, Growth zones in the echinoid skeleton: Ann. Zool., 15: 731-753.

Pedley, H. M., 1976, A palaeoecological study of the upper coralline limestone, Terebratula-Aphelesia bed (Miocene, Malta) based on bryozoan growth-form studies and brachiopod distribution: Palaeogeogr., Palaeoclimat., Palaeoecol., 20: 209-234.

Peel, J. S., 1975, A new Silurian gastropod from Wisconsin and the ecology of uncoiling in Palaeozoic gastropods: Bull. Geol. Soc. Denmark, 24: 211-221.

Peel, J. S., 1978, Faunal succession and mode of life of Silurian gastropods in the Arisaig Group, Nova Scotia: Palaeontology, 21: 285-306.

Peet, R. K., 1974, The measurement of species diversity: Ann. Rev. Ecol. Systematics, 5: 285-307.

Perkins, R. D. and C. I. Tsentas, 1976, Microbial infestation of carbonate substrates planted on the St. Croix shelf, West Indies: Geol. Soc. Amer. Bull., 87: 1615-1628.

Peryt, T. M. and T. S. Piatkowski, 1977, Stomatalites from the Zechstein Limestone (Upper Permian) of Poland: In E. Flügel (ed.), Fossil algae: Recent results and developments: Springer-Verlag, Berlin, p. 124-135.

Petersen, C. G. J., 1915, On the animal communities of the sea bottom in the Skagerrak, the Christiana Fjord, and the Danish waters: Rep. Danish Biol. Sta., 23: 3-28.

Peterson, C. H., 1976, Relative abundance of living and dead molluscs in two California lagoons: Lethaia, 9: 137-148.

Peterson, C. H., 1977, The paleoecological significance of undetected short-term temporal variability: J. Paleont., 51: 976-981.

Philcox, M. E., 1971, Growth form and role of colonial coelenterates in reefs of the Gower Formation (Silurian), Iowa: J. Paleont., 45: 338-346.

Phleger, F. B., 1960, Ecology and distribution of Recent foraminifera: Johns Hopkins Press, Baltimore, 297 p.

Pianka, E. R., 1978, Evolutionary ecology, 2nd ed.: Harper & Row, New York, 397 p.

Pielou, E. C., 1969, An introduction to mathematical ecology: Wiley-Interscience, New York, 286 p.

Pilkey, O. H. and J. Hower, 1960, The effect of environment on the concentration of skeletal magnesium and strontium in *Dendraster*: J. Geol., *68:* 203-216.

Piper, D. J. W., W. R. Normark, and J. C. Ingle, 1976, The Rio Dell Formation: A Plio-Pleistocene basin slope deposit in northern California: Sedimentology, *23:* 304-328.

Playford, P. E. and A. E. Cockbain, 1969, Algal stromatolites: deepwater forms in the Devonian of Western Australia: Science, *165:* 1008-1010.

Playford, P. E., A. E. Cockbain, E. C. Druce, and J. L. Wray, 1976, Devonian stromatolites from the Canning Basin, Western Australia: *In* M. R. Walter (ed.), Stromatolites: Elsevier, Amsterdam, p. 543-564.

Pohowsky, R. A., 1978, The boring ctenostomate bryozoa: taxonomy and paleobiology based on cavities in calcareous substrata: Bull. Amer. Paleont., *73:* 1-192.

Pokorný, V., 1978, Ostracodes: *In* B. V. Haq, and A. Boersma (eds.), Introduction to marine micropaleontology: Elsevier, New York, p. 109-149.

Poluzzi, A. and R. Sartori, 1974, Report on the carbonate mineralogy of bryozoa: Docum. Lab. Géol. Fac. Sci. Lyon, H. S. *3:* 193-210.

Pryor, W. A., 1975, Biogenic sedimentation and alteration of argillaceous sediments in shallow marine environments: Geol. Soc. Amer. Bull., *86:* 1244-1254.

Purchon, R. D., 1968, The biology of the Mollusca: Pergamon Press, New York, 560 p.

Purdy, E. G., 1964, Sediments as substrates: *In* J. Imbrie and N. D. Newell (eds.), Approaches to paleoecology: Wiley, New York, p. 238-271.

Ragland, P. C., O. H. Pilkey, and B. W. Blackwilder, 1969, Comparison of the Sr/Ca ratio of fossil and Recent mollusc shells: Nature, *244:* 1223-1224.

Raiswell, R. and P. Brimblecombe, 1977, The partition of manganese into aragonite between 30 and 60°C: Chem. Geol., *19:* 145-151.

Raup, D. M., 1956, *Dendraster*: A problem in taxonomy: J. Paleont., *30:* 685-694.

Raup, D. M., 1966a, The exoskeleton: *In* R. A. Boolootian (ed.), Physiology of Echinodermata: Wiley-Interscience, New York, p. 379-395.

Raup, D. M., 1966b, Geometric analysis of shell coiling: General problems: J. Paleont., *40:* 1178-1190.

Raup, D. M., 1968, Theoretical morphology of echinoid growth: Paleont. Soc. Mem. 2 [J. Paleont., *42*(5) Supp.], p. 50-63.

Raup, D. M., 1972, Approaches to morphologic analysis: *In* T. J. M. Schopf (ed.), Models in paleobiology: Freeman, Cooper, and Co., San Francisco, p. 28-44.

Raup, D. M., 1977, Probabilistic models in evolutionary paleobiology: Amer. Scientist, *65:* 50-57.

Raup, D. M. and S. M. Stanley, 1978, Principles of paleontology (2nd ed.): W. H. Freeman and Co., San Francisco, 481 p.

Reyment, R. A., 1971, Introduction to quantitative paleoecology: Elsevier, New York, 226 p.

Rhoads, D. C., 1966, Depth of burrowing by benthic organisms: A key to nearshore-offshore relations: Geol. Soc. Amer. Ann. Mtng. Prog., p. 176.

Rhoads, D. C., 1970, Mass properties, stability, and ecology of marine muds related to burrowing activity: *In* T. P. Crimes and J. C. Harper (eds.), Trace fossils: Geol. J., Spec. Issue 3, p. 391-406.

Rhoads, D., 1975, The paleoecological and environmental significance of trace fossils: *In* R. W. Frey (ed.), The study of trace fossils: Springer-Verlag, New York, p. 147-160.

Rhoads, D. C. and G. Pannella, 1970, The use of molluscan shell growth patterns in ecology and paleoecology: Lethaia, *3:* 143-161.

Rhoads, D. C., I. G. Speden, and K. M. Waage, 1972, Trophic group analysis of Upper Cretaceous

(Maestrichtian) bivalve assemblages from South Dakota: Bull. Amer. Assoc. Petrol. Geol., *56:* 1100-1113.

Rhoads, D. C. and D. K. Young, 1970, The influence of deposit-feeding organisms on bottom-sediment stability and community trophic structure: J. Mar. Res., *28:* 150-178.

Richards, R. P., and R. K. Bambach, 1975, Population dynamics of some Paleozoic brachiopods and their paleoecological significance: J. Paleont., *49:* 775-798.

Richter, R., 1929, Gründung and Aufgaben der Forschungsstelle für Meeres geologie "Senckenberg" in Wilhelmschaven: Natur u. Museum, *59:* 1-30.

Ricketts, E. F., J. Calvin, and J. W. Hedgpeth, 1968, Between Pacific tides: Stanford University Press, Stanford, Calif., 614 p.

Rider, J. and R. Cowan, 1977, Adaptive architectural trends in incrusting Ectoprocta: Lethaia, *10:* 29-41.

Riley, J. P. and M. Tongudai, 1967, The major cation/chlorinity ratios in sea water: Chem. Geol., *2:* 263-269.

Robison, R. A. and C. Teichert (eds.), 1979, Treatise on invertebrate paleontology, Part A, Introduction: Geol. Soc. Amer. and University of Kansas, Boulder, Colo., 569 p.

Rodriguez, J. and R. C. Gutschick, 1970, Late Devonian-Early Mississippian ichnofossils from western Montana and northern Utah: *In* T. P. Crimes and J. C. Harper (eds.), Trace fossils: Geol. J. Spec. Issue 3, p. 407-438.

Rollins, H. B., M. Carothers, and J. Donahue, 1979, Transgression, regression and fossil community succession: Lethaia, *12:* 89-104.

Romer, A. S., 1961, Palaeozoological evidence of climate. Vertebrates: *In* A. E. M. Nairn (ed.), Descriptive paleoclimatology: Wiley-Interscience, New York, p. 183-206.

Rosa, D., 1931, L'ologénèse. Nouvelle théorie de l'évolution et de la distribution géographique: Felix Alcon, Paris, 368 p.

Rosenberg, G. D. and S. K. Runcorn (eds.), 1975, Growth rhythms and the history of the earth's rotation: Wiley-Interscience, New York, 559 p.

Ross, C. A. (ed.), 1974, Paleogeographic provinces and provinciality: Soc. Econ. Paleontologists and Mineralogists, Spec. Pub. 21, Tulsa, Okla., 233 p.

Ross, R. J., Jr., 1975, Early Paleozoic trilobites, sedimentary facies, lithospheric plates, and ocean currents: Fossils and Strata, *4:* 307-329.

Ross, J. R. P., 1976, Body wall ultrastructure of living cyclostrome ectoprocts: J. Paleont., *50:* 350-353.

Rudwick, M. J. S., 1961, The feeding mechanism of the Permian brachiopod *Prorichthofenia:* Palaeontology, *3:* 450-471.

Rudwick, M. J. S., 1964a, The function of zigzag deflections in the commissure of fossil brachiopods: Palaeontology, *7:* 135-171.

Rudwick, M. J. S., 1964b, The inference of function from structure in fossils: British J. Phil. Sci., *15:* 27-40.

Rudwick, M. J. S., 1968, Some analytic methods in the study of ontogeny in fossils with accretionary skeletons: Paleont. Soc. Mem. 2 [J. Paleont., *42* (5), Suppl.], p. 35-49.

Rudwick, M. J. S., 1970, Living and fossil brachiopods: Hutchinson and Co. Ltd., London, 199 p.

Runcorn, S. K., 1975, Paleontological and astronomical observations on the rotational history of the earth and moon: *In* G. D. Rosenberg and S. K. Runcorn (eds.), Growth rhythms and the history of the earth's rotation: Wiley-Interscience, New York, p. 285-291.

Ryland, J. S., 1970, Bryozoans: Hutchinson, London, 175 p.

Sackett, W. M. and R. P. Thompson, 1963, Isotopic organic carbon composition of recent continental derived clastic sediments of eastern Gulf Coast, Gulf of Mexico: Bull. Amer. Assoc. Petrol. Geol., *47:* 525-528.

Saidova, K. M., 1961, Ehkologia foraminfer i paleogeografia dal'nevostochnykh morej SSSR i Severozapadnoj chasti Tikhogo Okeana: Akad. Nauk SSSR, Inst. Okeanol., 182 p.

Saito, T. and J. van Donk, 1974, Oxygen and carbon isotope measurements of Late Cretaceous and Early Tertiary foraminifera: Micropaleontology, 20: 152-177.

Sandburg, P. A., 1964, The ostracod genus *Cyprideis* in the Americas: Stockholm Contribs. in Geol., 12: 178 p.

Sandburg, P. A., 1977, Ultrastructure, mineralogy, and development of bryozoan skeletons: In R. M. Woollacott and R. L. Zimmer (eds.), Biology of bryozoans: Academic, New York, p. 143-181.

Sanders, H. L., 1956, Oceanography of Long Island Sound, 1952-1954. X. The biology of marine bottom communities: Bingham Oceanogr. Coll. Bull., 15: 345-414.

Sanders, H. L., 1968, Marine benthic diversity: a comparative study: Amer. Naturalist, 102: 243-282.

Sanders, H. L., P. C. Mangelsdorf, Jr., and G. R. Hampson, 1965, Salinity and faunal distribution in the Pocasset River, Massachusetts: Limnol. and Oceanog., 10 (suppl.): R216-R228.

Sanders, W. B. and P. D. Ward, 1979, Nautiloid growth and lunar dynamics: Lethaia, 12: 172.

Savilov, H. L., 1957, Biological aspect of the bottom fauna groupings of the North Okhotsk Sea: In B. N. Nikitin (ed.), Trans. Inst. Oceanol. Marine Biol., U.S.S.R., Acad. Sci. Press, 20: 67-136. (Published in U.S. by Amer. Inst. Biol. Sci., Washington, D.C.).

Savin, S. M., R. G. Douglas, and F. G. Stehli, 1975, Tertiary marine paleotemperatures: Geol. Soc. Amer. Bull., 86: 1499-1510.

Schäfer, W., 1962, Actuo-Paläontologie nach Studien in der Nordsee: Kramer, Frankfort, 666 p.

Schäfer, W., 1972, Ecology and paleoecology of marine environments: University of Chicago Press, Chicago, 568 p.

Schenck, H. G. and A. M. Keen, 1936, Marine molluscan provinces of western North America: Proc. Amer. Phil. Soc., 76: 921-938.

Scherer, M., 1975, Fe-Anreicherung der Rogensteinzone des Norddeutschen unteren Buntsandsteins (Trias): Ein Hinweis auf die diagenetische Geschichte: Neues Jahrb. Geol. Paläont. Monatsh., (9): 568-576.

Schink, D. R. and N. L. Guinasso, Jr., 1977, Effects of bioturbation on sediment-seawater interaction: Mar. Geol., 23: 133-154.

Schlager, W. and N. P. James, 1978, Low-magnesium calcite limestones forming at the deep-sea floor, Tongue of the Ocean, Bahamas: Sedimentology, 25: 675-702.

Schmidt, K. P., 1954, Faunal realms, regions, and provinces: Quat. Rev. Biol., 29: 322-331.

Schmidt, W. J., 1924, Die Bausteine des Tierkörpers in polarisiertem Licht: Bonn, 528 p.

Scholle, P. A., 1977, Chalk diagenesis and its relation to petroleum exploration: oil from chalks, a modern miracle: Amer. Assoc. Petrol. Geol. Bull., 61: 982-1009.

Schopf, T. J. M., 1969, Paleoecology of ectoprocts (bryozoans): J. Paleont., 43: 234-244.

Schroeder, J. H., E. J. Dwonik, and J. J. Papike, 1969, Primary protodolomite in echinoid skeletons: Geol. Soc. Amer. Bull., 80: 1613-1616.

Scotese, C., R. K. Bambach, C. Barton, R. Van der Voo, and A. Ziegler, 1979, Paleozoic base maps: J. Geol., 87: 217-277.

Scott, G. H., 1963, Uniformitarianism, the uniformity of nature, and paleoecology: New Zealand J. Geol. Geophys., 6: 510-527.

Scott, R. W., 1974, Bay and shoreface benthic communities in the Lower Cretaceous: Lethaia, 7: 315-330.

Scott, R. W., 1978, Approaches to trophic analysis of paleocommunities: Lethaia, 11: 1-14.

Scott, R. W. and R. R. West (eds.), 1976, Structure and classification of paleocommunities: Dowden, Hutchinson, and Ross, Stroudsburg, Pa., 291 p.

Scrutton, C. T., 1964, Periodicity in Devonian coral growth: Palaeontology, *7:* 552-558.

Segerstråle, S. G., 1957, Baltic Sea: *In* J. W. Hedgpeth (ed.), Treatise on marine ecology and paleoecology, v. 1, Ecology: Geol. Soc. Amer. Mem., *67*(1): 751-800.

Seilacher, A., 1953, Studien zur Palichnologie. I. Über die Methoden der Palichnologie: Neues Jahrb. Geol. Paläont. Abh., *96:* 421-452.

Seilacher, A., 1960, Strömungsanzeichen im Hunsrückschiefer: Hesse. Landesamt f. Bodenforschung Notizblatt, *88:* 88-106.

Seilacher, A., 1963, Lebensspuren und Salinitätsfazies: Fortschr. Geol. Rheinld, u. Westf., *10:* 81-94.

Seilacher, A., 1967, Bathymetry of trace fossils: Mar. Geol., *5:* 413-428.

Seilacher, A., 1970, Arbeitskonzept zur Konstruktions-morphologie: Lethaia, *3:* 393-396.

Seilacher, A., 1973, Biostratinomy: The sedimentology of biologically standardized particles: *In* R. N. Ginsburg (ed.), Evolving concepts in sedimentology: Johns Hopkins University Press, Baltimore, p. 159-177.

Seilacher, A., 1974, Flysch trace fossils: Evolution of behavioral diversity in the deep-sea: Neues Jahrb. Geol. Paläont., Monatsh., (4): 233-245.

Seilacher, A., 1977, Evolution of trace fossil communities: *In* A. Hallam (ed.), Patterns of evolution as illustrated by the fossil record: Elsevier, Amsterdam, p. 359-376.

Seilacher, A., 1979, Constructional morphology of sand dollars: Paleobiology, *5:* 191-221.

Selley, R. C., 1976, An introduction to sedimentology: Academic, New York, 408 p.

Shaak, G. D., 1975, Diversity and community structure of the Brush Creek marine interval (Conemaugh Group, Upper Pennsylvanian), in the Appalachian Basin of western Pennsylvania: Bull. Florida State Mus., Biol. Sci., *19:* 69-133.

Shackleton, N. J., 1967, Oxygen isotope analyses and Pleistocene temperatures reassessed: Nature, *215:* 15-17.

Shackleton, N. J., 1973, Attainment of isotopic equilibrium between ocean water and the benthonic foraminifera genus *Uvigerina:* Isotopic changes in the ocean during the last glacial age: Colloq. Inter. du Cent. Natl. Res. Sci., *219:* 203-209.

Shackleton, N. J. and J. P. Kennett, 1975a, Late Cenozoic oxygen and carbon isotopic changes at DSDP site 284. Implications for glacial history of the northern hemisphere and Antarctica: *In* J. P. Kennett, R. E. Houtz, et al. (eds.), Initial reports of the Deep Sea Drilling Project: Washington (U.S. Government Printing Office), *29:* 801-807.

Shackleton, N. J. and J. P. Kennett, 1975b, Paleotemperature history of the Cenozoic and the initiation of Antarctic glaciation: Oxygen and carbon isotope analysis in DSDP sites 277, 279, and 281: *In* J. P. Kennett, R. E. Houtz, et al. (eds.), Initial reports of the Deep Sea Drilling Project: Washington (U.S. Government Printing Office), *29:* 743-755.

Shackleton, N. J. and R. K. Matthews, 1977, Oxygen isotope stratigraphy of Late Pleistocene coral terraces in Barbados: Nature, *268:* 618-620.

Shackleton, N. J. and N. D. Opdyke, 1973, Oxygen isotope and paleomagnetic stratigraphy of equatorial Pacific core V28-238: Oxygen isotope temperatures and ice volumes on a 10^5 and 10^6 year scale: Quat. Res., *3:* 39-55.

Shackleton, N. J. and N. D. Opdyke, 1976, Oxygen isotope and paleomagnetic stratigraphy of Pacific core V28-239 Late Pliocene to latest Pleistocene: *In* R. M. Cline and J. D. Hayes (eds.), Investigation of Late Quaternary paleoceanography and paleoclimatology: Geol. Soc. Amer. Mem., *145:* 449-464.

Shackleton, N. J. and E. Vincent, 1978, Oxygen and carbon studies in Recent foraminifera from the southwest Indian Ocean: Mar. Micropaleont., *3:* 1-13.

Shaver, R. H., 1974, Silurian reefs of northern Indiana: Reef and interreef macrofaunas: Amer. Assoc. Petrol. Geol. Bull., *58:* 934-956.

Sheldon, A. L., 1969, Equitability indices: dependence on the species count: Ecology, *50:* 466-467.

Shelford, V. E., 1963, The ecology of North America: University of Illinois Press, Urbana, 610 p.

Shinn, E. A., 1963, Spur and groove formation on the Florida reef tract: J. Sedim. Petrol., *33:* 291-303.

Shinn, E. A., 1968, Burrowing in Recent lime sediments of Florida and the Bahamas: J. Paleont., *42:* 879-894.

Shinn, E. A., R. B. Halley, J. H. Hudson, and B. H. Lidz, 1977, Limestone compaction: An enigma: Geology, *5:* 21-24.

Shotwell, J. A., 1964, Community succession in mammals of the Late Tertiary: *In* J. Imbrie and N. D. Newell (eds.), Approaches to paleoecology: Wiley, New York, p. 135-150.

Siever, R. and R. A. Scott, 1963, Organic geochemistry of silica: *In* I. Berger (ed.), Organic geochemistry of silica: Pergamon, Oxford, p. 579-595.

Simberloff, D., 1972, Properties of the rarefaction diversity measurement: Amer. Naturalist, *106:* 414-418.

Simkiss, K., 1964, Phosphates as crystal poisons of calcification: Biol. Rev. *39:* 487-505.

Simonsen, R. (ed.), 1972, First symposium on Recent and fossil marine diatoms: Nova Hedwigia, *39:* 1-294.

Simpson, G. G., 1940, Mammals and land bridges: J. Wash. Acad. Sci., *30:* 137-163.

Simpson, G. G., 1960, Notes on the measurement of faunal resemblance: Amer. J. Sci., *258:* 300-311.

Skelton, P. W., 1976, Functional morphology of the Hippuritidae: Lethaia, *9:* 83-100.

Skelton, P. W., 1979, Preserved ligament in a radiolitid rudist bivalve and its implication of mantle marginal feeding in the group: Paleobiology, *5:* 90-106.

Skougstad, M. W. and C. A. Horr, 1963, Occurrence and distribution of strontium in natural water: U.S. Geol. Surv. Water Supply Paper 1496-D, 97 p.

Slobodkin, C. B., 1962, Growth and regulation of animal populations: Holt, Rinehart, and Winston, New York, 184 p.

Smith, A. B., 1978, A comparative study of the lifestyle of two Jurassic irregular echinoids: Lethaia, *11:* 57-66.

Smith, A. G., J. C. Briden, and G. E. Drewry, 1973, Phanerozoic world maps: Palaeontological Assoc. Spec. Pap., *12:* 1-42.

Smith, S. V., R. W. Buddemeier, R. C. Redalje, and J. G. Houck, 1979, Strontium-calcium thermometry in coral skeletons: Science, *204:* 404-407.

Sokal, R. R. and F. J. Rohlf, 1969, Biometry: Freeman, San Francisco, 778 p.

Solbrig, O. T. and G. H. Orians, 1977, The adaptive characteristics of desert plants: Am. Sci., *65:* 412-421.

Sorauf, J. E., 1971, Microstructure in the exoskeleton of some Rugosa (Coelenterata): J. Paleont., *45:* 23-32.

Sorauf, J. E., 1972, Skeletal microstructure and microarchitecture in Scleractinia (Coelenterata): Palaeontology, *15:* 88-107.

Sorauf, J. E., 1974a, Observations on microstructure and biomineralization in coelenterates: Biomineralization, *7:* 37-55.

Sorauf, J. E., 1974b, Growth lines on tabulae in *Favosites* (Silurian, Iowa): J. Paleont., *48:* 553-555.

Spaeth, C., J. Hoefs, and V. Vetter, 1971, Some aspects of isotopic composition of belemnites and related paleotemperatures: Geol. Soc. Amer. Bull., *82:* 3139-3150.

Stanley, K. O. and J. A. Fagerstrom, 1974, Miocene invertebrate trace fossils from a braided river environment, western Nebraska, U.S.A.: Paleogeogr., Paleoclim., Paleoecol., *15:* 63-82.

Stanley, S. M., 1968, Post-Paleozoic adaptive radiation of infaunal bivalve mollusks—A consequence of mantle fusion and siphon formation: J. Paleont., *42:* 214–229.

Stanley, S. M., 1970, Relation of shell form to life habits of the Bivalvia (Mollusca): Geol. Soc.
Amer. Mem., *125:* 296 p.

Stanley, S. M., 1972, Functional morphology and evolution of byssally attached bivalve mollusks:
J. Paleont., *46:* 165–212.

Stanley, S. M., 1977, Coadaptation in the Trigoniidae, a remarkable family of burrowing
bivalves: Palaeontology, *20:* 869–899.

Stanton, R. J., Jr., 1967, The effects of provenance and basin-edge topography on sedimentation
in the basal Castaic Formation (Upper Miocene, marine), Los Angeles County, California: Calif.
Div. Mines and Geol., Spec. Rep. *92:* 21–31.

Stanton, R. J., Jr., 1976, The relationship of fossil communities to the original communities of
living organisms: *In* R. W. Scott and R. R. West (eds.), Structure and classification of paleocommunities: Dowden, Hutchinson, and Ross, Stroudsburg, Pa., p. 107–142.

Stanton, R. J., Jr., and J. R. Dodd, 1970, Paleoecologic techniques—Comparison of faunal and
geochemical analyses of Pliocene paleoenvironments, Kettleman Hills, California: J. Paleont.,
44: 1092–1121.

Stanton, R. J., Jr., and J. R. Dodd, 1972, Pliocene cyclic sedimentation in the Kettleman Hills,
California: *In* E. W. Rennie, Jr. (ed.), Geology and oil fields, west side central San Joaquin
Valley: Pacific Section Amer. Assoc. Petrol. Geol. Guidebook 1972, p. 50–58.

Stanton, R. J., Jr., and J. R. Dodd, 1976a, Pliocene biostratigraphy and depositional environment
of the Jacalitos Canyon area, California: *In* A. E. Fritsche, H. Terbest, Jr., and W. W. Wornardt
(eds.), The Neogene Symposium: Soc. Econ. Paleontologists and Mineralogists, Pacific section,
p. 84–85.

Stanton, R. J., Jr., and J. R. Dodd, 1976b, The application of trophic structure of fossil communities in paleoenvironmental reconstruction: Lethaia, *9:* 327–342.

Stanton, R. J., Jr., J. R. Dodd, and R. R. Alexander, 1979, Eccentricity in the clypeasteroid
echinoid *Dendraster:* environmental significance and application in Pliocene paleoecology:
Lethaia, *12:* 75–87.

Stanton, R. J., Jr., and I. Evans, 1971, Environmental controls of benthic macrofaunal patterns
in the Gulf of Mexico adjacent to the Mississippi Delta: Trans. Gulf Coast Assoc. Geol. Soc.,
21: 371–378.

Stanton, R. J., Jr., and I. Evans, 1972a, Recognition and interpretation of modern molluscan
biofacies: *In* R. Rezak and V. J. Henry (eds.), Contributions on the geological and geophysical
oceanography of the Gulf of Mexico: Texas A & M University Studies, *3:* 203–222.

Stanton, R. J., Jr., and I. Evans, 1972b, Community structure and sampling requirements in
paleoecology: J. Paleont. *46:* 845–858.

Stanton, R. J., Jr., and P. C. Nelson, 1980, Reconstruction of the trophic web in paleontology:
community structure in the Stone City Formation (Middle Eocene, Texas): J. Paleont., *54:*
118–135.

Stanton, R. J., Jr., and J. E. Warme, 1971, Stop 1: Stone City Bluff: *In* B. F. Perkins (ed.),
Trace fossils, a field guide: Louisiana State University, School of Geosci. Misc. Publ. 71-1,
p. 3–10.

Stearn, C. W. and R. Riding, 1973, Forms of hydrozoan—*Millepora* on a Recent coral reef:
Lethaia, *6:* 187–199.

Stehli, F. G., 1956, Shell mineralogy in Paleozoic invertebrates: Science, *123:* 1031–1032.

Stehli, F. G., 1968, Taxonomic diversity gradients in pole location: the Recent model: *In* E. T.
Drake (ed.), Evolution and environment: Yale University Press, New Haven, Conn., p. 163–277.

Stehli, F. G., 1970, A test of the earth's magnetic field during Permian time: J. Geophy. Res.,
75: 3325–3342.

Stehli, F. G. and W. B. Creath, 1964, Foraminiferal ratios and regional environments: Amer. Assoc. Petrol. Geol. Bull., *48:* 1810-1827.

Stehli, F. G. and C. E. Helsley, 1963, Paleontologic technique for defining ancient pole positions: Science, *142:* 1057-1059.

Stein, R. S., 1975, Dynamic analysis of *Pteranodon ingens:* A reptilian adaptation to flight: J. Paleont., *49:* 534-548.

Stenzel, H. B., 1935, A new formation in the Claiborne Group: Univ. Texas Bull., *3501:* 267-279.

Stenzel, H. B., 1963, Aragonite and calcite as constituents of adult oyster shells: Science, *142:* 232-233.

Stenzel, H. B., 1964, Living *Nautilus: In* R. C. Moore (ed.), Treatise on invertebrate paleontology, Part K, Mollusca 3: Geol. Soc. Amer., Boulder, Colo., p. 59-93.

Stephenson, W., W. T. Williams, and S. D. Cook, 1972, Computer analyses of Petersen's original data on bottom communities: Ecol. Mon., *42:* 387-415.

Stevens, G. R. and R. N. Clayton, 1971, Oxygen isotope studies on Jurassic and Cretaceous belemnites from New Zealand and their biogeographic significance: New Zealand J. Geol. Geophys., *14:* 829-897.

Stewart, R., 1946, Geology of Reef Ridge, Coalinga District, California: U.S. Geol. Surv. Prof. Pap., 205C: 81-115.

Stitt, J. H., 1976, Functional morphology and life habits of the Late Cambrian trilobite *Stenopilus pronus* Raymond: J. Paleont., *50:* 561-576.

Stratton, J. F., 1975, Studies of *Polypora* from the Speed Member, North Vernon Limestone (Eifelian, Middle Devonian) in southern Indiana: Unpublished Ph.D. thesis, Indiana University, 237 p.

Stratton, J. F. and A. S. Horowitz, 1975, Studies of the flow of water through models of *Polypora: In* S. Pouyet (ed.), Bryozoa, 1974, Proceedings of the Third Conference, International Bryozoology Association: Docs. Lab. Géol. Fac. Sci. Lyon, H.S.3 (fasc. 2), p. 425-438.

Strauch, F., 1968, Determination of Cenozoic sea-temperatures using *Hiatella arctica* (Linné): Palaeogeogr., Palaeoclimat., Palaeoecol., *5:* 213-233.

Surlyk, F., 1972, Morphological adaptations and population structures of the Danish chalk brachiopods (Maastrichtian, Upper Cretaceous): Kongelige Danske Vidensk., Selsk. Biol. Skrifter, *19:* 1-57.

Surlyk, F., 1974, Life habit, feeding mechanism and population structure of the Cretaceous brachiopod genus *Aemula:* Palaeogeogr., Palaeoclimat., Palaeoecol., *15:* 185-203.

Sverdrup, H. V., M. W. Johnson, and R. H. Fleming, 1942, The oceans: Their physics, chemistry, and general biology: Prentice-Hall, Englewood Cliffs, N.J., 1087 p.

Sweet, W. C., 1964, Cephalopoda—general features: *In* R. C. Moore (ed.), Treatise on invertebrate paleontology, Part K, Mollusca 3: Geol. Soc. Amer., Boulder, Colo., p. 4-13.

Sylvester-Bradley, P. C., 1971, Dynamic factors in animal palaeogeography: *In* F. A. Middlemiss, P. F. Rawson, and G. Newall (eds.), Faunal provinces in space and time: Geol. J. Spec. Issue No. 4, Seel House Press, Liverpool, p. 1-18.

Tan, F. C. and J. D. Hudson, 1974, Isotopic studies on the paleoecology and diagenesis of the Great Estuarine Series (Jurassic) of Scotland: Scot. J. Geology, *10:* 91-128.

Tappan, H., 1968, Primary production, isotopes, extinctions, and the atmosphere: Palaeogeogr., Palaeoclimat., Palaeoecol., *4:* 187-210.

Tappan, H. and A. R. Loeblich, Jr., 1973, Evolution of the oceanic plankton: Earth Sci. Rev., *9:* 207-240.

Tasch, P., 1973, Paleobiology of the invertebrates: Wiley & Sons, Inc., New York, 946 p.

Tavener-Smith, R. and A. Williams, 1972, The secretion and structure of the skeleton of living and fossil bryozoa: Philos. Trans. Roy. Soc. London, B. Biol. Sci., *264:* 97-159.

Taylor, J. D., W. J. Kennedy, and A. Hall, 1969, The shell structure and mineralogy of the Bivalvia, introduction. Nuculacea-Trigonacea: Bull. British Mus. Nat. Hist. (Zool.) Suppl. 3, 125 p.

Taylor, J. D. and M. Layman, 1972, The mechanical properties of bivalve (Mollusca) shell structures: Palaeontology, *15:* 73–87.

Termier, H. and G. Termier, 1975, Rôle des ésponges hypercalcifiées en paléoécologie et en paléobiogéographie: Bull. Soc. Geol. France, (7), *17:* 803–819.

Textoris, D. A. and A. V. Carozzi, 1964, Petrography and evolution of Niagaran (Silurian) reefs, Indiana: Amer. Assoc. Petrol. Geol. Bull., *48:* 397–426.

Thayer, C. W., 1975a, Morphologic adaptations of benthic invertebrates to soft substrata: J. Mar. Res., *33:* 177–189.

Thayer, C. W., 1975b, Strength of pedicle attachment in articulate brachiopods: Ecological and paleoecological significance: Paleobiology, *1:* 388–399.

Thayer, C. W., 1975c, Size-frequency and population structure of brachiopods: Palaeogeogr., Palaeoclimat., Palaeoecol., *17:* 139–148.

Thayer, C. W., 1977, Recruitment, growth, and mortality of a living articulate brachiopod, with implications for the interpretation of survivorship curves: Paleobiology, *3:* 98–109.

Thomas, R. D. K., 1975, Functional morphology, ecology, and evolutionary conservatism in the Glycymerididae (Bivalvia): Palaeontology, *18:* 217–254.

Thomas, R. D. K., 1978, Shell form and the ecologic range of living and extinct Arcoida: Paleobiology, *4:* 181–194.

Thompson, T. G. and J. Chow, 1955, The strontium-calcium atom ratio in the carbonate-secreting marine organisms: Deep Sea Res., Suppl. to v. 3, p. 20–39.

Thomson, K. S., 1976, On the heterocercal tail in sharks: Paleobiology, *2:* 19–38.

Thorson, G., 1957, Bottom communities (sublittoral or shallow shelf): *In* J. W. Hedgpeth (ed.), Treatise on marine ecology and paleoecology, v. 1, Ecology: Geol. Soc. Amer. Mem., *67*(1): 461–534.

Tipper, J. C., 1975, Lower Silurian animal communities—Three case histories: Lethaia, *8:* 287–299.

Toomey, D. F. and J. M. Cys, 1979, Community succession in small bioherms of algae and sponges in the Lower Permian of New Mexico: Lethaia, *12:* 65–74.

Toots, H., 1965, Random orientation of fossils and its significance: Wyoming University Contr. Geol., *4:* 59–62.

Tourtelot, H. A. and R. O. Rye, 1969, Distribution of oxygen and carbon isotopes in fossils of Late Cretaceous age, western interior region of North America: Geol. Soc. Amer. Bull., *80:* 1903–1922.

Towe, K. M., 1972, Invertebrate shell structure and the organic matrix concept: Biomineralization Res. Reports, *4:* 1–14.

Towe, K. M. and R. Cifelli, 1967, Wall ultrastructure in the calcareous foraminifera: crystallographic aspects and a model for calcification: J. Paleont., *41:* 742–762.

Travis, D. F., 1960, Matrix and mineral deposition in skeletal structures of decapod crustacea: *In* R. F. Sognnaes (ed.), Calcification in biological systems: Amer. Assoc. Adv. Sci., Publ. 64, p. 57–116.

Trewin, N. H. and W. Welsh, 1976, Formation and composition of a graded estuarine shell bed: Palaeogeogr., Palaeoclimat., Palaeoecol., *19:* 219–230.

Trueman, A. E., 1940, The ammonite body-chamber, with special reference to the buoyancy and mode of life of the living ammonite: Quat. J. Geol. Soc. London, *96:* 339–383.

Trueman, E. R., 1964, Adaptive morphology in paleoecological interpretation: *In* J. Imbrie and N. D. Newell (eds.), Approaches to paleoecology: Wiley, New York, p. 45–74.

Turekian, K. K., 1964, The marine geochemistry of strontium: Geochim. et Cosmochim. Acta, *28:* 1479-1496.

Turekian, K. K., 1969, The oceans, streams, and atmosphere: *In* W. H. Wedepohl (ed.), Handbook of geochemistry, v. 1: Springer-Verlag, Berlin, p. 297-323.

Turekian, K. K. and R. L. Armstrong, 1960, Magnesium, strontium, and barium concentrations and calcite-aragonite ratios of some Recent molluscan shells: J. Mar. Res., *18:* 133-151.

Turpaeva, E. P., 1957, Food interrelationships of dominant species in marine benthic biocoenoses: *In* B. N. Nikitin (ed.), Trans. Inst. Oceanology, v. 20, Marine Biology. U.S.S.R. Acad. Sciences Press, translated and published in U.S. by Amer. Inst. Biological Sciences, Washington, 1959, p. 137-148.

Urey, H. C., 1947, The thermodynamic properties of isotopic substances: J. Chem. Soc., *1947:* 562-581.

Urey, H. C., H. A. Lowenstam, S. Epstein, and C. R. McKinney, 1951, Measurement of paleo-temperatures and temperatures of the Upper Cretaceous of England, Denmark, and southeastern United States: Geol. Soc. Amer. Bull., *62:* 399-416.

Valentine, J. W., 1961, Paleoecological molluscan geography of the Californian Pleistocene: University of California Publ. Geol. Sci., *34:* 309-442.

Valentine, J. W., 1963, Biogeographic units as biostratigraphic units: Amer. Assoc. Petrol. Geol. Bull., *47:* 457-466.

Valentine, J. W., 1966, Numerical analysis of marine molluscan ranges on the extratropical northeastern Pacific shelf: Limnol. and Oceanog., *11:* 198-211.

Valentine, J. W., 1969, Patterns of taxonomic and ecological structure of the shelf benthos during Phanerozoic time: Palaeontology, *12:* 684-709.

Valentine, J. W., 1971a, Plate tectonics and shallow marine diversity and endemism, an actualistic model: Syst. Zool., *20:* 253-264.

Valentine, J. W., 1971b, Resource supply and species diversity patterns: Lethaia, *4:* 51-61.

Valentine, J. W., 1973a, Evolutionary paleoecology of the marine biosphere: Prentice-Hall, Englewood Cliffs, N.J., 511 p.

Valentine, J. W. 1973b, Phanerozoic taxonomic diversity: a test of alternate models: Science, *180:* 1078-1079.

Valentine, J. W. and R. F. Meade, 1961, California Pleistocene paleotemperatures: University California Publ. Geol. Sci., *40:* 1-46.

Van Straaten, L. M. J. U., 1952, Biogene textures and the formation of shell beds in the Dutch Wadden Sea: Koninkl. Nederl. Akad. Wettensch. Proc. Ser. B., *55:* 500-516.

Vaughan, T. W., 1919, Corals and the formation of coral reefs: Smithsonian Inst. Ann. Rep. 1917, p. 189-238.

Veizer, J., 1977, Diagenesis of pre-Quaternary carbonates as indicated by tracer studies: J. Sedim. Petrol., *47:* 565-581.

Vermeij, G. J., 1978, Biogeography and adaptation patterns of marine life: Harvard University, Cambridge, Mass., 332 p.

Vinogradov, A. P., 1953, The elementary chemical composition of marine organisms: Sears Found. for Mar. Res., Mem. 2, 647 p.

Virnstein, R. W., 1977, The importance of predation by crabs and fishes on benthic infauna in Chesapeake Bay: Ecology, *58:* 1199-1217.

Voorhies, M. R., 1969, Taphonomy and population dynamics of an Early Pliocene vertebrate fauna, Knox County, Nebraska: Wyoming Contr. Geol., Spec. Pap. No. 1, 69 p.

Waage, K. M., 1964, Origin of repeated fossiliferous concretion layers in the Fox Hills Formation: Kansas Geol. Survey Bull. 169, *2:* 541-563.

Waage, K. M., 1968, The type Fox Hills Formation, Cretaceous (Maestrichtian), South Dakota. Part 1. Stratigraphy and paleoenvironments: Peabody Mus. Bull., Yale University, *27:* 1-171.

Wagner, C. W., 1957, Sur les ostracodes du Quaternaire Récent des Pays-bas et leur utilisation dans l'etude geologique des depots Holocènes: Mouton and Co., the Hague, Netherlands, 259 p.

Walker, K. R., 1972, Trophic analysis: A method for studying the function of ancient communities: J. Paleont., *46:* 82-93.

Walker, K. R. and L. P. Alberstadt, 1975, Ecological succession as an aspect of structure in fossil communities: Paleobiology, *1:* 238-257.

Walker, K. R. and R. K. Bambach, 1971, The significance of fossil assemblages from fine-grained sediments: Time-averaged communities: Geol. Soc. Amer. Abs. with Progs., *3:* 783-784.

Walker, K. R. and L. F. Laporte, 1970, Congruent fossil communities from Ordovician and Devonian carbonates of New York: J. Paleont., *44:* 928-944.

Walker, K. R. and W. C. Parker, 1976, Population structure of a pioneer and a later stage species in an Ordovician ecological succession: Paleobiology, *2:* 191-201.

Wallace, A. R., 1876, The geographical distribution of animals: Macmillan, London, 2 v., 503 and 607 p.

Walter, M. R. (ed.), 1976, Developments in sedimentology 20, Stromatolites: Elsevier Scientific, Amsterdam, 790 p.

Walter, M. R., 1977, Interpreting stromatolites: Amer. Scientist, *65:* 563-571.

Walton, W. R., 1964, Recent foraminiferal ecology and paleoecology: *In* J. Imbrie and N. D. Newell (eds.), Approaches to paleoecology: Wiley, New York, p. 151-237.

Warme, J. E., 1967, Graded bedding in the Recent sediments of Mugu Lagoon, California: J. Sedim. Petrol., *37:* 540-547.

Warme, J. E., 1969, Live and dead molluscs in a coastal lagoon: J. Paleont., *43:* 141-150.

Warme, J. E., 1975, Borings as trace fossils, and the processes of marine bioerosion: *In* R. W. Frey (ed.), The study of trace fossils: Springer-Verlag, New York, p. 181-227.

Warme, J. E. and R. J. Stanton, Jr., 1971, Stop 2: Rockdale railroad cut: *In* B. F. Perkins (ed.), Trace fossils, a field guide: Louisiana State University, School, of Geosci. Misc. Publ. 71-1, Baton Rouge, La., p. 11-15.

Warner, G. F., 1977, On the shapes of passive suspension feeders: *In* B. F. Keegan, P. O. Ceidigh, and P. J. S. Boaden (eds.), Biology of benthic organisms: Pergamon, New York, p. 567-579.

Waterhouse, J. B. and G. F. Bonham-Carter, 1975, Global distribution and character of Permian biomes based on brachiopod assemblages: Canadian J. Earth Sci., *12:* 1085-1146.

Watkins, R., W. B. N. Berry, and A. J. Boucot, 1973, Why "communities"?: Geology, *1:* 55-58.

Watkins, R. and A. J. Boucot, 1975, Evolution of Silurian brachiopod communities along the southeastern coast of Acadia: Geol. Soc. Amer. Bull., *86:* 243-254.

Weber, J. N., P. Deines, E. W. White, and P. H. Weber, 1975, Seasonal high and low density bands in reef coral skeletons: Nature, *255:* 697-698.

Weber, J. N. and D. M. Raup, 1966, Fractionation of the stable isotopes of carbon and oxygen in marine calcareous organisms—The Echinoidea. Part I. Variation of ^{13}C and ^{18}O content within individuals: Geochim. et Cosmochim. Acta, *30:* 681-703.

Weber, J. N. and E. W. White, 1977, Caribbean reef corals *Montastrea annularis* and *Montastrea cavernosa*—Long-term growth data as determined by skeletal x-radiography: Amer. Assoc. Petrol. Geol. Stud. in Geol., *4:* 171-179.

Weber, J. N., E. W. White, and P. H. Weber, 1975, Correlation of density banding in reef coral skeletons with environmental parameters: The basis for interpretation of chronological records preserved in the coralla of corals: Paleobiology, *1:* 137-149.

Weber, J. N. and P. M. Woodhead, 1970, Carbon and oxygen isotope fractionation in the skeletal carbonate of reef-building corals: Chem. Geol., *6:* 93-117.

Weimer, R. J. and J. H. Hoyt, 1964, Burrows of *Callianassa major* Say, geologic indicators of littoral and shallow neritic environments: J. Paleont., *38:* 761-767.

Weiner, S., 1975, The carbon isotopic composition of the eastern Mediterranean planktonic foraminifera *Orbulina universa* and the phenotypes of *Globigerinoides ruber:* Palaeogeogr., Palaeoclimat., Palaeoecol., *17:* 149-156.

Weiner, S. and L. Hood, 1975, Soluble protein of the organic matrix of mollusk shells: A potential template for shell formation: Science, *190:* 987-989.

Welch, J. R., 1978, Flume study of simulated feeding and hydrodynamics of a Paleozoic stalked crinoid: Paleobiology, *4:* 89-95.

Wells, J. W., 1957, Corals: *In* J. W. Hedgpeth (ed.), Treatise on marine ecology and paleoecology, v. 1, Ecology: Geol. Soc. Amer. Mem., *67*(1): 1087-1104.

Wells, J. W., 1963, Coral growth and geochronometry: Nature, *197:* 948-950.

Wells, J. W., 1967, Corals as bathometers: Mar. Geol., *5:* 349-365.

West, R. M. (ed.), 1977, Paleontology and plate tectonics with special reference to the history of the Atlantic Ocean: Milwaukee Pub. Mus. Spl. Publ. in Biol. and Geol., No. 2, 109 p.

Westermann, G. E. G., 1973, Strength of concave septa and depth limits of fossil cephalopods: Lethaia, *6:* 383-403.

Weyl, P. K., 1978, Micropaleontology and ocean surface climate: Science, *202:* 475-481.

Weymouth, F. W., 1923, The life-history and growth of the Pismo Clam (*Tivela stultorum* Mawe): State of Calif. Fish and Game Comm. Fish Bull., v. 7, 120 p.

Whitehead, D. R., 1973, Late Wisconsin vegetational changes in unglaciated eastern North America: Quat. Res., *3:* 621-631.

Whittaker, R. H., 1972, Evolution and measurement of species diversity: Taxon, *21:* 213-251.

Wiedenmayer, F., 1978, Modern sponge bioherms of the Great Bahama Bank: Eclogae Geol. Helv., *71:* 699-744.

Wilbur, K. M., 1964, Shell formation and regeneration: *In* K. M. Wilbur and C. M. Yonge (eds.), Physiology of Mollusca: Academic, New York, p. 243-282.

Wilbur, K. M., 1976, Recent studies in invertebrate mineralization: *In* N. Watabe and K. M. Wilbur (eds.), The mechanisms of mineralization in the invertebrates and plants: University of South Carolina Press, Columbia, p. 79-108.

Wilbur, K. M. and N. Watabe, 1963, Experimental studies on calcification in molluscs and the alga *Coccolithus huxleyi*: Ann. New York Acad. Sci., *109:* 82-112.

Williams, A., 1968, Evolution of the shell structure of articulate brachiopods: Spec. Paps. in Palaeont., Palaeont. Assoc. London, *2:* 1-55.

Williams, A. and A. D. Wright, 1970, Shell structure of the Craniacea and other calcareous inarticulate Brachiopoda: Spec. Paps. in Palaeont., Palaeont. Assoc. London, *7:* 1-51.

Willis, J. C., 1922, Age and area. A study of geographical distribution and origin of species: Cambridge University Press, Cambridge, 259 p.

Wilson, E. O., 1975, Sociobiology: Belknap Press, Cambridge, Mass., 697 p.

Wilson, J. L., 1975, Carbonate facies in geologic history: Springer-Verlag, New York, 471 p.

Wilson, J. T., 1966, Did the Atlantic close and re-open?: Nature, *211:* 676-681.

Winland, H. D., 1969, Stability of calcium carbonate polymorphs in warm, shallow seawater: J. Sedim. Petrol., *39:* 1579-1587.

Wolfe, J. A., 1978, A paleobotanical interpretation of Tertiary climates in the northern hemisphere: Amer. Scientist, *66:* 694-703.

Wolff, T., 1970, The concept of the hadal or ultra-abyssal fauna: Deep-Sea Res., *17:* 983-1003.

Wolman, M. G. and J. P. Miller, 1960, Magnitude and frequency of forces in geomorphic processes: J. Geol., *68:* 54-74.

Woodring, W. P., R. Stewart, and R. W. Richards, 1940, Geology of the Kettleman Hills oilfield, California; stratigraphy, paleontology, and structure: U.S. Geol. Surv. Prof. Pap. *195:* 1-170.

Woodward, S. P., 1856, A manual of the Mollusca: A treatise on Recent and fossil shells: Virtue Brothers, London, 542 p.

Woollacott, R. M. and R. L. Zimmer, 1977, Biology of bryozoans: Academic, New York, 566 p.

Wornardt, W. W., Jr., 1969, Diatoms, past, present, future: *In* H. H. Renz and P. Brönnimann (eds.), Proceedings of the first international conference on planktonic microfossils, v. 2: E. J. Brill, Leiden, Netherlands, p. 690-714.

Wray, J. L., 1971, Ecology and geologic distribution: *In* R. Ginsburg, R. Rezak, and J. L. Wray (eds.), Geology of calcareous algae (notes for a short course): Comparative Sedim. Lab., University of Miami, p. 5.1-5.6.

Wray, J. L., 1977, Calcareous algae: Elsevier, Amsterdam, 185 p.

Wright, R. P., 1974, Jurassic bivalves from Wyoming and South Dakota: A Study of feeding relationships: J. Paleont., *48:* 425-433.

Wyckoff, R. W. G., 1972, The biochemistry of animal fossils: Scientichnica, Bristol, 127 p.

Yancey, T. and P. Wilde, 1970, Faunal communities on the central California shelf near San Francisco—A sedimentary environmental study: Calif. University Hydraulic Engrg. Lab. Tech. Report HEL-2-29, 65 p.

Yonge, C. M., 1957, Symbiosis: *In* J. W. Hedgpeth (ed.), Treatise on marine ecology and paleoecology, v. 1, Ecology: Geol. Soc. Amer. Mem., *67* (1): 429-442.

Ziegler, A. M., L. R. M. Cocks, and R. K. Bambach, 1968, The composition and structure of Lower Silurian marine communities: Lethaia, *1:* 1-27.

Ziegler, A. M., C. R. Scotese, W. S. McKerrow, M. E. Johnson, and R. K. Bambach, 1977, Paleozoic biogeography and continents bordering the Iapetus (pre-Caledonian) and Rheic (pre-Hercynian) oceans: *In* R. M. West (ed.), Paleontology and plate tectonics with special reference to the history of the Atlantic Ocean: Milwaukee Pub. Mus. Spec. Publ. in Biol. and Geol., No. 2, p. 1-22.

Zolotarev, V. N., 1974, Magnesium and strontium in the shell calcite of some modern pelecypods: Geochem. Int., *11:* 347-353.

Index

Page numbers referring to illustrations are in **boldface** type.